MECHANICAL ENGINEERING THEORY AND APPLICATIONS

HEAT PIPES

DESIGN, APPLICATIONS AND TECHNOLOGY

MECHANICAL ENGINEERING THEORY AND APPLICATIONS

Additional books and e-books in this series can be found on Nova's website under the Series tab.

MECHANICAL ENGINEERING THEORY AND APPLICATIONS

HEAT PIPES

DESIGN, APPLICATIONS AND TECHNOLOGY

YUWEN ZHANG
EDITOR

NOTICE TO THE READER

The Publisher has taken reasonable care in the preparation of this book, but makes no expressed or implied warranty of any kind and assumes no responsibility for any errors or omissions. No liability is assumed for incidental or consequential damages in connection with or arising out of information contained in this book. The Publisher shall not be liable for any special, consequential, or exemplary damages resulting, in whole or in part, from the readers' use of, or reliance upon, this material. Any parts of this book based on government reports are so indicated and copyright is claimed for those parts to the extent applicable to compilations of such works.

Independent verification should be sought for any data, advice or recommendations contained in this book. In addition, no responsibility is assumed by the publisher for any injury and/or damage to persons or property arising from any methods, products, instructions, ideas or otherwise contained in this publication.

This publication is designed to provide accurate and authoritative information with regard to the subject matter covered herein. It is sold with the clear understanding that the Publisher is not engaged in rendering legal or any other professional services. If legal or any other expert assistance is required, the services of a competent person should be sought. FROM A DECLARATION OF PARTICIPANTS JOINTLY ADOPTED BY A COMMITTEE OF THE AMERICAN BAR ASSOCIATION AND A COMMITTEE OF PUBLISHERS.

Additional color graphics may be available in the e-book version of this book.

Library of Congress Cataloging-in-Publication Data
Names: Zhang, Yuwen, 1965- editor.
Title: Heat pipes : design, applications and technology / Yuwen Zhang, Ph.D., Department of Mechanical & Aerospace Engineering, University of Missouri, Columbia, MO, US, editor.
Other titles: Heat pipes (Nova Science Publishers)
Description: Hauppauge, New York : Nova Science Publishers, Inc., [2018] | Series: Mechanical engineering theory and applications | Includes bibliographical references and index.
Identifiers: LCCN 2018029866 (print) | LCCN 2018033123 (ebook) | ISBN 9781536139099 (ebook) | ISBN 9781536139082 (hardcover) | ISBN 9781536139099 (ebook)
Subjects: LCSH: Heat pipes.
Classification: LCC TJ264 (ebook) | LCC TJ264 .H38 2018 (print) | DDC 621.402/5--dc23
LC record available at https://lccn.loc.gov/2018029866

Published by Nova Science Publishers, Inc. † New York

CONTENTS

PREFACE

Heat pipes are efficient passive devices that can transfer large amounts of heat over long distances with small temperature differences between the heat sources and sinks by evaporation and condensation of the working fluid. Heat can be transferred without the use of any mechanically moving parts such as pumps and active controls in heat pipes. For convention heat pipes, including thermosyphons, the vapor and liquid circulate in the heat pipe via evaporation/condensation and capillary or gravitational forces. For pulsating heat pipes, liquid slug and vapor plugs in capillary tube oscillate due to evaporation and condensation. The effective thermal conductivity of a heat pipe can be three orders of magnitude higher than that of a copper rod with the same size. Heat pipe can find its applications in many sectors of industries including electronics cooling, energy systems, spacecraft thermal control, permafrost cooling, and manufacturing. This book presents current research and development related to the design, applications and technology of various heat pipes.

Chapter 1 – A number of passive natural circulation single and two-phased closed and closed-loop thermosyphon-type heat pipe applications are highlighted. The complexity of natural circulation behavior and its departure from pool boiling and film condensation theory are illustrated by way splashing, geysering and smacking. Experimentally determined heat transfer coefficients and maximum heat transfer rate correlations are given in terms of their dependence on their orientation relative to gravity, working fluid properties, liquid charge fill ratio, temperature and wall heat flux. A number of applications are presented relating to the cooling of nuclear reactors and electronic components and equipment, waste heat recovery and energy saving, open, closed and looped oscillatory or pulsating heat pipes, water pumping and night-sky radiation cooling and day time solar heating for buildings.

Chapter 2 – High-powered 3D microelectronics devices, such as RF power amplifier and laser diodes, have brought emerging demands for removing highly concentrated heat from small device areas. As a passive cooling method, heat pipe heat spreaders capable of high density phase change are highly desired. In the past twenty years, this technical demand has constantly driven development of advanced wick structures to provide more effective heat and mass transfer. As a result of these efforts, cooling heat flux phase change > 1kW/cm^2 has been demonstrated in lab studies. Heat pipe platforms capable of a few hundred W/cm^2 cooling have also been developed for industry uses. In these advances, more tolerance is given to phase change superheat and operating temperature, with less emphasis on isothermal characteristics stressed in the conventional heat pipes. This mission shift renders designers to design wick structures with large particles to allow occurrence of nucleate boiling and fast vapor ventilation from the wick structures. Accordingly, some traditional heat pipe limits (e.g., boiling limit) are either inapplicable or not considered significant in high heat flux thermal management. Heat and mass transfer in the heated wick structures becomes the major concern. In addition, high heat flux phase change requires high density heat conduction from the wick subtract to phase change interfaces, dramatically increasing superheat and complexing evaluation of the heat transfer coefficient.

Chapter 3 – The chapter is devoted to the solution of an important scientific and applied problem of creating highly efficient means for ensuring the thermal regime (METR) of promising electronic devices based on scientifically grounded ways of constructive and technological implementation of two-phase technologies in the designs of the main types of electronic devices built on a new elemental base taking into account the features of their construction. The reliability of their work is increased by reducing the temperature of the elemental base, as well as reducing power consumption of with METR based on heat pipes (HP) of complex electronic systems by reducing the hydraulic resistance of the water path. Scientific ground is provided for improvement of the existing and creation of new design solutions of HPs, heat transfer devices and METR based on two-phase technologies for the main types of advanced electronic equipment: complex modular computing systems in basic load-bearing structures with water cooling with a thermal power of up to 30 kW in one instrument cabinet, means of electronic computing equipment and computer control, infrared equipment, promising microlaser devices of information systems and controls, transmitting modules, LED lighting devices. New technological solutions are developed and validated which allow the production and commercialization of the proposed HP designs for METR of modern and advanced electronic devices.

Chapter 4 – Pulsating heat pipe (PHP) is a relatively new and promising addition to the family of passive two-phase energy transport devices. By charging with water, methanol, ethanol, acetone and binary mixtures of them at various volume mixing ratios, a vertical closed-loop PHP of ten turns with inner/outer diameter of 2.0/4.0mm has been experimentally investigated in bottom-heating mode. Three levels of filling ratios have been categorized, small (35%, 45%), medium (55%) and large (62%, 70%) with heat input ranged from 10W to 100W. It was found that due to the zeotropic properties and the complex molecular interactions between the components, the PHPs charged with the mixtures were quite more complex than those with pure fluids. At small or medium filling ratios, most of the binary mixtures have better anti-dry-out performance than at least one of the pure fluids (even both). At relatively high heat input and large filling ratio, the thermal performances of the PHPs charged with the binary mixtures at various mixing ratios had no superiority to the pure fluids especially the pure water.

Chapter 5 – Numerical study of the chaotic behavior in pulsating heat pipes has been implemented using quantitative approaches such as spectral analysis of time series, correlation dimension, autocorrelation function (ACF), Lyapunov exponent, and phase space reconstruction. In addition, volume fraction of liquid and vapor has been simulated and studied under different operating conditions. Constant temperature and heat flux have been employed to investigate thermal performance such as thermal resistance and axial mean temperature for several boundary conditions. Volume of Fluid (VoF) method was used for liquid-vapor two-phase flow simulation. Water and ethanol have been used as working fluids in pulsating heat pipe. This chapter is divided into three major parts depending on the PHP's geometry and dimension. First part studies a simple two dimensional geometry. Second part studies a multi-turn two dimensional geometry and third part investigates a three dimensional geometry for the pulsating heat pipe. An experimental study from other work has been used for comparison and validation in three dimensional simulation.

Chapter 6 – The aims of this chapter are to depict the various shaped heat pipes which work based on two-phase heat transfer devices and the HSHPTM (Heat Sink-Heat Pipe Thermal Module) software. The programs and methods for some thermal module designs are introduced in the chapter. The developed complex method provides for high quality and short time of development owing to the lesser number of iterations of the design procedures. Based on the theoretical models with empirical formula, the author had developed a computer program coded by Virtual Basic version 6.0 to develop window programs and convenience for industrials named HSHPTM software. These results show that the prediction by the HSHPTM software agrees with the experimental data.

Chapter 7 – There are a number of applications that could use heat pipes or loop heat pipes (LHPs) in the intermediate temperature range of 450 to 750 K, including space nuclear power system radiators, fuel cells, geothermal power, waste heat recovery systems, and high temperature electronics cooling. Previous life tests conducted with 30 different intermediate temperature working fluids, including elements, organic working fluids, and halides and over 60 different working fluid/envelope combinations are reviewed. During more recent investigation, life tests have been run with three elemental working fluids: sulfur, sulfur-iodine mixtures, and mercury. Other fluids offer benefits over these three liquids in this temperature range. Three sets of organic fluids stand out as good intermediate temperature fluids: (1) diphenyl, diphenyl oxide, and eutectic diphenyl/diphenyl oxide, (2) naphthalene, and (3) toluene. In addition, life tests have been conducted at temperatures up to 550 K with water and titanium and Monel alloys and at temperatures up to 673 K with titanium and superalloy envelopes paired with halides. The long-term life tests also established that Titanium/TiBr4 at 653 K, and Hastelloy B-3, C-22 and C-2000/$AlBr_3$ at 673 K were compatible. The results indicate that the tested envelope materials and working fluids can form viable material/working fluid combinations.

Chapter 8 – Variable Conductance Heat Pipes (VCHPs) are used to passively control the evaporator (and associated electronics temperature) while the power and heat sink conditions vary widely. Essentially all VCHPs use an electrically heated, cold Non-Condensable Gas (NCG) reservoir, and typically control the evaporator temperature to within ±2°C. The first warm-reservoir VCHPs actually fabricated and tested are described here. In these devices, the NCG reservoir is located next to the evaporator, so the need for electrical power is eliminated, while maintaining the same ±2°C control. Eliminating power is a requirement for missions that rely on batteries during long periods of darkness, such as Lunar landers and rovers, as well as research balloons in the polar regions during the long periods of winter darkness. Pressure Controlled Heat Pipes (PCHPs) achieve tighter temperature control by changing the amount of NCG in the system, either by changing the reservoir volume, or adding/subtracting non-condensable gas. A PCHP suitable for operation in microgravity was developed using a grooved aluminum heat pipe, and controlled temperatures within ±50mK using a flexible bellows to change the reservoir volume. A PCHP was developed that can control the temperature within ±1mK.

Chapter 9 – Thermal performances of heat pipes are dependent upon type of fill liquid, fill ratio, power input, pipe inclination and pipe dimensions. The boiling and condensation processes that occur inside heat pipes are quite complex. A visual study of the internal flow patterns would be most helpful to understand the internal heat transfer phenomena. Visualisation by various investigators to observe the flow patterns inside thermosyphons, pulsating, loop heat pipes and forced circulation are presented.

Yuwen Zhang

ACKNOWLEDGMENTS

The editor would like to express his gratitude to all contributing authors, who are leading experts in the area of heat pipe design, applications and technology, for their dedications to this book besides their heavy demanding everyday works. Their countless efforts, unusual cooperation, and patience made this book possible. The editor would also like to thank the staffs at Nova Science Publishers for their efforts to ensure high-quality presentation of the book.

In: Heat Pipes: Design, Applications and Technology ISBN: 978-1-53613-908-2
Editor: Yuwen Zhang

Chapter 1

SINGLE- AND TWO-PHASE CLOSED- AND OPEN-LOOP THERMOSYPHON-TYPE HEAT PIPES: THEORETICAL AND EXPERIMENTAL SIMULATION AND APPLICATIONS

Robert Thomas Dobson*

Department of Mechanical and Mechatronic Engineering,
University of Stellenbosch, Stellenbosch, South Africa

ABSTRACT

This chapter highlights a number of passive natural circulation single and two-phased closed and closed-loop thermosyphon-type heat pipe applications conducted by the author over the past 25 years. In these devices heat is transferred without the use of any mechanically moving parts such as pumps and active controls. The basic one-dimensional theory needed to simulate their behaviour is given. The complexity of natural circulation behaviour and its departure from pool boiling and film condensation theory are illustrated by way splashing, geysering and smacking. Experimentally determined heat transfer coefficients and maximum heat transfer rate correlations are given in terms of their dependence on their orientation relative to gravity, working fluid properties, liquid charge fill ratio, temperature and wall heat flux. A number of applications are presented relating to the cooling of nuclear reactors and electronic components and equipment, waste heat recovery and energy saving, open, closed and looped oscillatory or pulsating heat pipes, water pumping, night-sky radiation cooling and day time solar heating for buildings. A comprehensive case study for an entirely passive, inherently safe nuclear reactor cooling and heat removal system is given.

* Corresponding Author Email: rtd@sun.ac.za.

Keywords: natural circulation, two-phase flow, heat transfer, cooling, heating, water pumping, energy saving, geysering, heat transfer coefficients, solar heating, night-sky radiation cooling

NOMENCLATURE

A Area, m^2
B breadth
B buoyancy force, N
Bo Bond number, $Bo = \sqrt{E\ddot{o}}$
C coefficient
C_f coefficient of friction
c specific heat, J/kg°C
c_p specific heat at constant pressure, J/kg°C
c_v specific heat at constant volume, J/kg°C
d diameter, m
$E\ddot{o}$ Eötvös number, $E\ddot{o} = d^2 g(\rho_\ell - \rho_v)/\sigma$
F force, N
f friction factor, $f = 4C_f$
g force due to gravity, N
G volumetric flow rate, m^3/s
Gr Grashof number $Gr = \beta \Delta T g \rho^2 L_{fin}{}^3/\mu$
FC Film condensation number
FR liquid charge fill ratio defined as: V_ℓ/V_e
FC Film condensation number, $FC = g\rho_\ell(\rho_\ell - \rho_v)d^3 h_{fg}/(\mu_\ell k_\ell(T_{sat} - T_{wall}))$
g acceleration due to gravity, m/s^2
H height, m
h specific enthalpy J/kg
h height, head, m
h heat transfer coefficient, W/m^2°C
h_{fg} latent heat of vaporization, J/kg
I Solar *Insolation* or radiation, W/m^2
i specific enthalpy, J/kg
Ja Jacob number, $Ja = c_{p\ell}(T_{sat} - T_{wall})/h_{fg}$
Ku Kutateladze number, $Ku = [\dot{Q}'']/(\rho_v h_{fg} \sigma g[(\rho_\ell - \rho_v)/\rho_v{}^2])^{\frac{1}{4}}$
k thermal conductivity, W/m°C

L length, m
L_c characteristic length, m
M Mouton Number, $M = g\mu^4(\rho_l - \rho_v)/(\rho_l^2\sigma^3)$
m mass
$\dot{m}$ mass flow rate, kg/s
P pressure, Pa or N/m^2; Perimeter, m
Pr Prandtl number, $Pr = c_p\mu/k$
$\dot{Q}$ heat flow rate, W
$\dot{Q}'$ power input per unit length, W/m
$\dot{Q}''$ power input per unit area, W/m^2
$\dot{Q}'''$ heat transfer rate per unit volume, W/m^3
R radius
R r-direction
Re Reynolds number, $Re = \rho vd/\mu = \dot{m}d/\mu A_x$
R specific gas constant, J/kgK
R thrust, N
R electrical resistance, Ω
R thermal resistance, °C/W
R^2 correlation coefficient
S splashing factor
t time, s
T temperature, K, °C; period, s
$\overline{T}$ average temperature, °C
U internal energy, J/kg
U overall heat transfer coefficient, W/m^2°C
u specific internal energy, J/kg
V volume m^3
$\dot{V}$ volumetric flow rate, m^3/s
V voltage, V
v velocity, m/s
x mass fraction, $m_{vapour}/(m_{vapour} + m_{liquid})$
W width, m
$\dot{W}$ rate of work, W
x distance, m
z control volume length
z axial-direction (in cylindrical coordinates)

Greek Symbols

α Void fraction, $V_{\text{vapour}}/(V_{\text{vapour}} + V_{\text{liquid}})$
α absorptivity
α solar azimuth, degrees
β Volumetric ecpansion coefficient, °C^{-1}
X Martinelli parameter
ϕ^2 two-phase frictional multiplier
ρ density, kg/m^3
α void fraction
τ shear stress, N/m^2
θ inclination angle, rad.
μ dynamic viscosity, kg/ms or Pas
σ surface tension N/m
δ thickness, m
ρ density, kg/m^3
ρ reflectivity
μ dynamic viscosity, kg/ms
τ shear stress, N/m^2
σ surface tension, N/m
η efficiency
ϕ inclination to the horizontal, °
Δ difference
ϕ relative humidity, %
μ viscosity, kg/ms
ρ density, kg/m^3
η fin efficiency
σ surface tension N/m
σ surface tension N/m
σ Stefan- Boltzmann constant, $\sigma = 5.67\text{x}10^{-8}$ W/m^{2}K^4

Subscripts

a air adiabatic, ambient
avg average
b bottom
c condenser, condensation, cold e evaporator, exit, evaporation

CON	convector
cond	conduction
conv	convection
cold	cold
crit	critical
end	end
d	discharge
dep	deposit
e	exit, environment, evaporator, electrical
exp	experimental
evap	evaporation
f	saturated liquid
f	friction, film, fin
g	saturated vapour
H	horizontal
h	hot, heated
h	heating, hot, heater
hp	heat pipe
in	in
i	inlet, inside
L	leading
l	liquid
ℓ	liquid
m	momentum length
M	manifold
minor	minor losses
o	outlet, only, outside, out
P	pressure
p	plate, plug, pipe, constant pressure
P	pipe
R	radiator
r	recovery
rad	radiation
sky	sky
sol	solar azimuth
surr	surroundings
s	surface
s	solids
sat	saturated
T	top, trailing, tank

t	thermal
v	velocity, vapour
w	wall, water
v	vapour, constant volume
V	vertical
w	water, wall
x	cross section
z	z-direction
z	zenith

Abbreviations

ACU	air-conditioning unit
BTS	bent thermosyphon
CLTS	closed loop thermosyphon
G	gravity
GHG	greenhouse gas
F	friction
P	pressure
PCD	pitch circle diameter
M	mass
MF	momentum flux
MC	moisture content

1. Introduction and Glossary

In this introduction we will introduce a few defining words and concepts in as much as they relate to single and two-phase closed and closed-loop natural-circulation thermosyphon-type heat pipes.

Consider a fluid that is contained in a pipe-like closed loop, much like a so-called hula-hoop and which is inclined at some angle, relative to gravity, as shown in Figure 1. If a lower portion of one side of the loop is exposed to a heat source and the other side, but at a higher elevation relative to gravity, to a lower temperature, or heat sink, the fluid will tend to flow around the loop; thereby emulating the age-old adage that "hot air rises and cold air sinks." This flow is termed *natural circulation* and the pipe-like loop, a thermosyphon. This process is further termed *passive* as no mechanical devices such as a pump are needed to circulate the fluid. The process is also *self-controlling* in as much that as the temperature difference between that heat source and sink increases, so too does the flow rate. For these reasons thermosyphon loops can be applied in many technological applications, notably solar heaters, air conditioning and ventilation, nuclear power plant and thermal management of electrical and electronic devices. In the nuclear industry, in particular, the thermosyphonic flow process is considered as being a passive system and hence also *inherently safe*.

This natural circulation or thermosyphonic flow process may be explained in terms of simple fundamental principles considering a small package, or so-called *control volume* of fluid as shown in Figure 2. A pressure force acts on the control volume due to the weight of the fluid on top of it; the lower down in the fluid, the greater the pressure force p acting on the control volume. The result of all these pressure forces (P) acting over the entire surface of the control volume is a net upwards force, and is given by $P = \int_S pds = (p_2 - p_1)A = \rho_f g(h_2 - h_1)A$. Counteracting this force is the *weight* G of the control volume; this is the force exerted on the control volume by virtue of its mass m and the acceleration due to the earth's gravity g, and always acts downwards, and is given by $G = mg = \rho_f A(h_2 - h_1)\text{g}$. The often called *buoyancy* force B may now be defined as $B = P - G$. If $B = 0$, the control volume will remain stationary, if $B > 0$, the control volume will tend to rise (we say that the control volume is *buoyant*), and vice versa, if $B < 0$, the control volume will *sink*. If we now sum up all the control volumes in the loop and if $\sum B > 0$, the fluid in the loop will flow in the positive z-direction as shown in Figure 1, and if $\sum B < 0$, it will flow counter-clockwise.

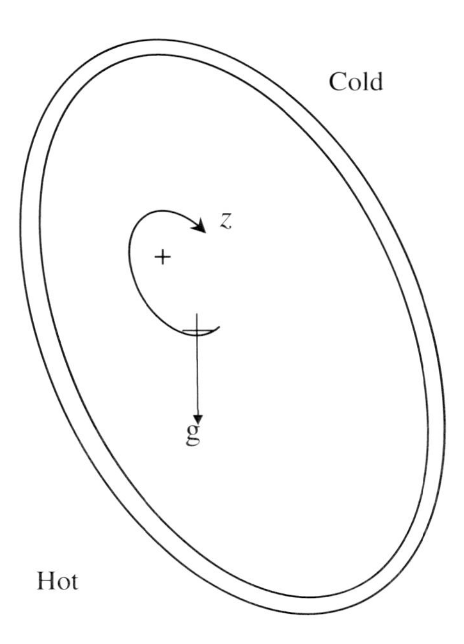

Figure 1. A looped pipe.

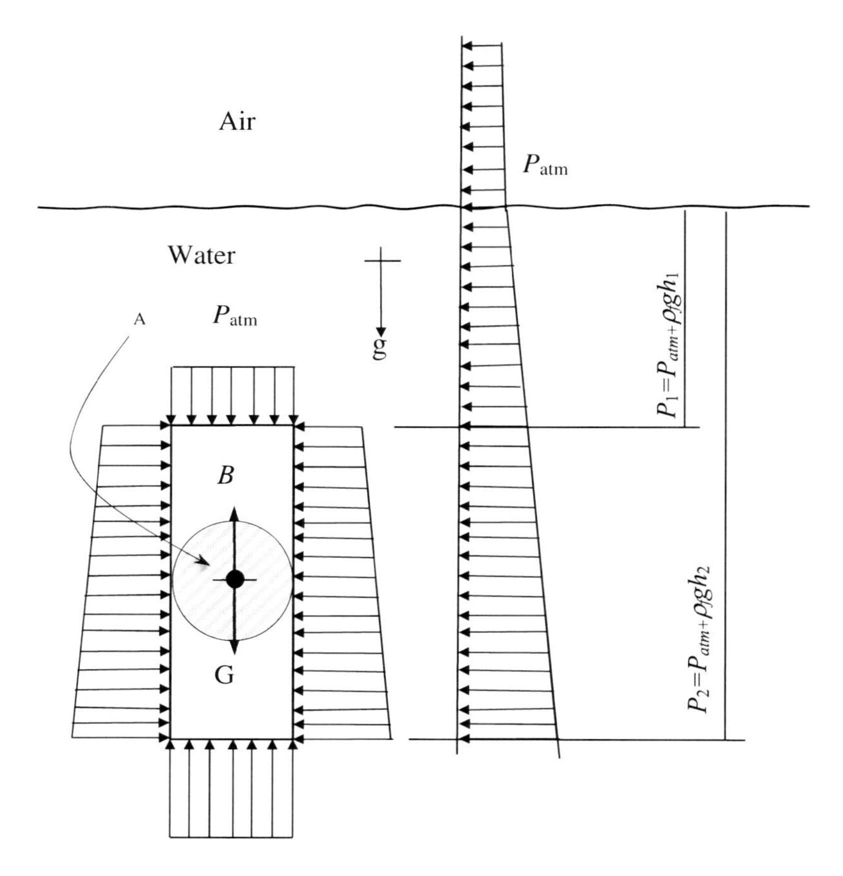

Figure 2. Simple example of buoyancy and gravity forces acting on a one-dimensional axially-symmetrical fluid control volume of density ρ_{cv} and cross-sectional diameter A in a surroundings of fluid of density ρ_f.

Liquid may start to boil in the heated portion, or *evaporator* portion, of the loop and consequently condense in the cooled portion or condenser; boiling occurring when the saturated pressure, corresponding to the liquid temperature, is higher than the local pressure in the liquid. If boiling has not yet occurred in the heated portion, the fluid is termed *sub-cooled*; it can however commence to boil as it rises as soon as the saturated pressure of the liquid is higher than the local pressure, which is essentially determined by the local *hydrostatic* pressure of the liquid. Boiling that occurs in this way is often called *flashing*. In both cases a *two-phase* flow is established and depending on the visually observed shape of the bubbles, may be characterised as sub-cooled, bubble, plug, churn, or annular flow. In horizontal and inclined pipes, if the flow rate is low enough, the liquid tends to flow in the bottom of the pipe and the less dense vapour in the top of the pipe. This is termed *stratified* flow. When single phase liquid circulation takes place the flow is often termed *density-driven* to distinguish it from the more ebullient case of two-phase *flashing-driven* flow.

1.1. Closed Two-Phase Thermosyphon-Type Heat Pipe

Consider a small diameter, relatively long pipe which is sealed at both ends as shown in Figure 3. The pipe is evacuated and then charged with a small amount of liquid called the *working fluid,* for example water; the liquid itself flows to the bottom due to gravity, and its vapour fills the top portion of the pipe. Upon being heated at the bottom and cooled at the top, the liquid boils in the bottom of the pipe and vapour flows upwards and condenses in the top; the condensate then flows, naturally, under the influence of gravity back to the bottom of the pipe. The heat required to boil and vaporize the liquid is then transferred by the flowing vapour to the cooled side. On reaching the cooled section, the vapour condenses and the heat of vaporization is in turn now released during the condensation process and transferred, through the pipe wall, to the coolant on the outside. This device is called a *closed* two-phase thermosyphon-type heat pipe.

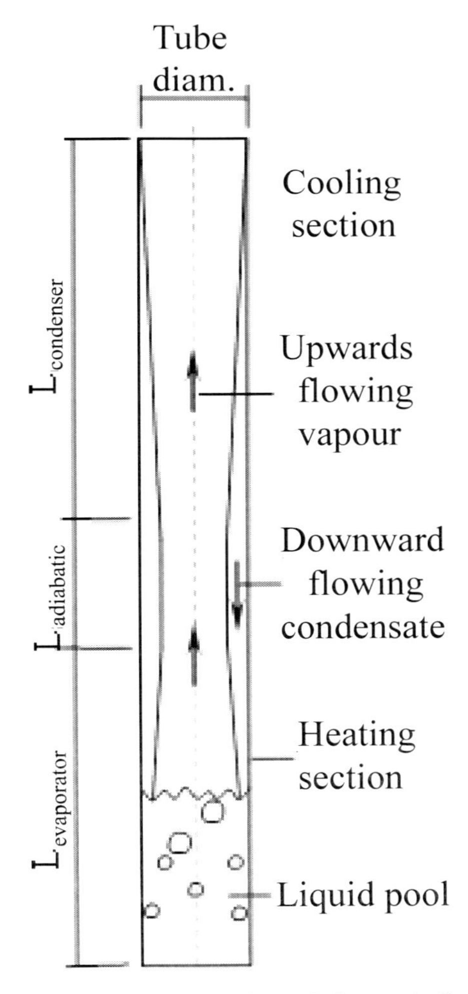

Figure 3. Simple representation of the working fluid in a tubular vertically orientated closed two-phase thermosyphon with heated evaporator and cooled condenser.

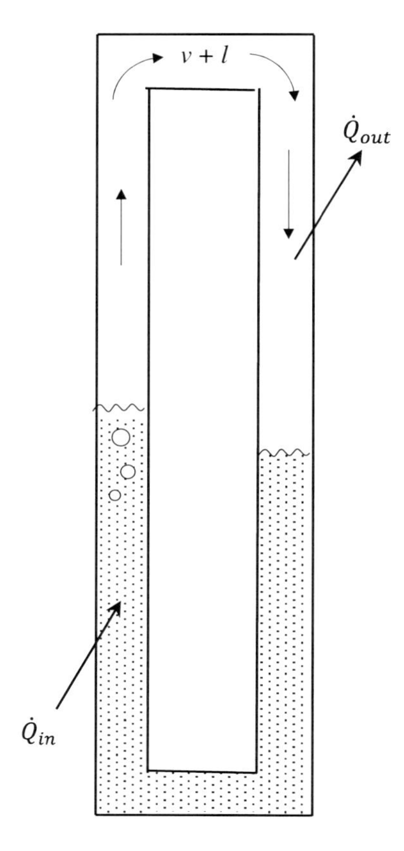

Figure 4. Closed-loop two-phase thermosyphon-type heat pipe.

1.2. Closed-Loop Two-Phase Thermosyphon-Type Heat Pipe

If the closed two-phase thermosyphon-type heat pipe is bent into a single loop (analogous to a "hula hoop," but vertically orientated) and heated on the one side and cooled on the other, as shown in Figure 4, the vapour that is then formed in the boiling process in the lower portion flows upwards in the one side of the looped pipe and the condensate formed at the top then flows back down to the bottom and the process then repeats itself. This type of heat pipe is called *a closed-loop* two-phase thermosyphon-type heat pipe.

1.3. Heat Pipe

If the closed two-phase thermosyphon-type heat pipe together with its working fluid also contains a capillary surface tension driven structure or a wick, exactly like a candle's wick, then it is possible to transfer the heat from the top downwards, so to speak naturally against the force of gravity as shown in Figure 5. [The flow in the capillary structure is also known as capillary-driven flow and is as a direct result of the physical phenomenon of *surface energy/tension*. This is a naturally driven process and was first formulated as being "free" energy by Gibbs in 1876.] In this case the vapour condenses at the bottom and the condensate is transported back upwards against gravity to the top by a so called surface tension force as a result of the liquid "pumping" effect of the wick. This sort of device is simply called a heat pipe and in this form, patented by Grover in 1963 and motivated by him as a *natural* transfer heat device in aerospace applications. Heat pipe technology may thus not only take into account the driving mechanism (gravity) that makes apples fall from an apple tree but also take into account surface tension driven forces that got the apple up there in the first place.

1.4. Passive (or Naturally Driven) Flow and Heat Transfer Devices

A passive or naturally driven flow and heat transfer device has no mechanically moving parts such as pumps and neither does it have any mechatronically operated devices such as sensors, servo-mechanisms, switches, and electronicly activated controls. As such they are deemed very reliable. Dramatic examples of naturally driven flow include cloud formation in which hot air rises and cold air sinks, and deep-sea thermocline flow.

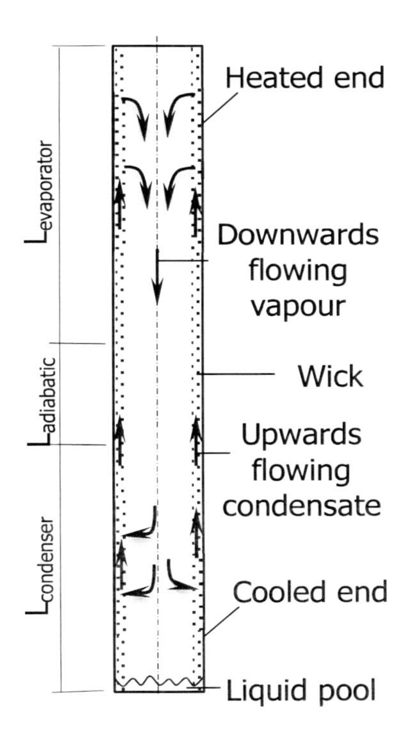

Figure 5. A simple representation of the wick and working fluid flow in a heat pipe.

1.5. Pulsating Heat Pipes

A pulsating heat pipe is similar to the closed-loop two-phase thermosyphon-type heat pipe except that its diameter is smaller between 1 and 5 mm, depending on the surface tension of the working fluid, and it is fashioned so as to meander between 10 to 30 times, back and forth from a heated to a colder region, as shown in Figure 6. Depending on a number of operating assumptions it is also able to transfer heat from a high level, against gravity, to a lower elevation and without a capillary structure or wick.

1.6. Heat Pipe Heat Exchanger

A heat pipe heat exchanger is essentially a closed two-phase thermosyphon-type heat pipe that has been "stretched out" as shown in Figure 7. The need for such a configuration in the high-temperature (1000°C) nuclear industry is to ensure that tritium does not diffuse from the primary reactor helium coolant, through the steel and, into the process water stream.

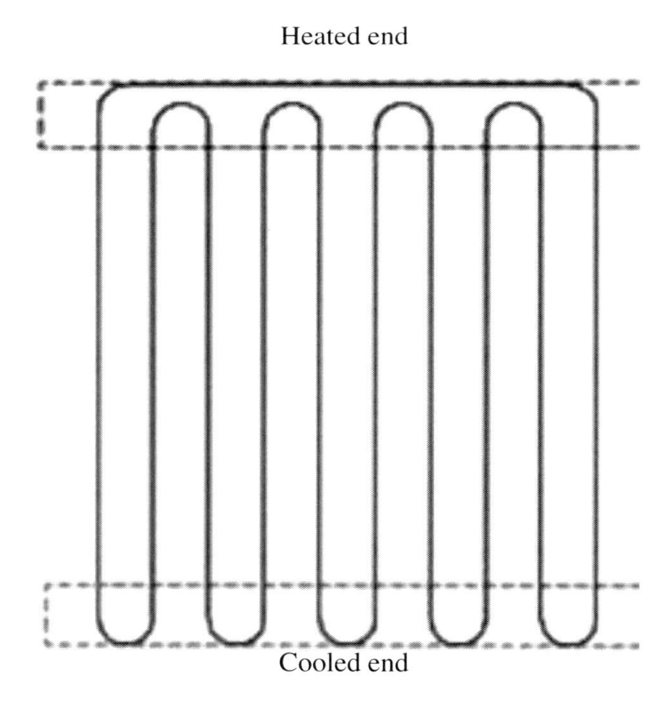

Figure 6. Pulsating heat pipe.

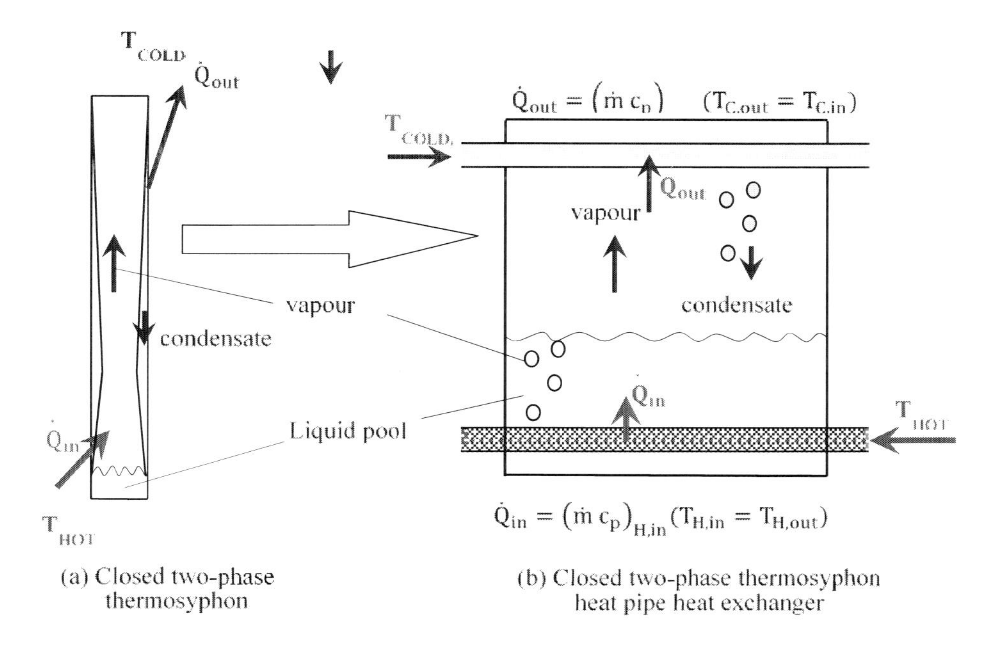

Figure 7. Difference between a closed two-phase thermosyphon heat pipe (a) and a closed two-phase thermosyphon heat pipe heat exchanger (b).

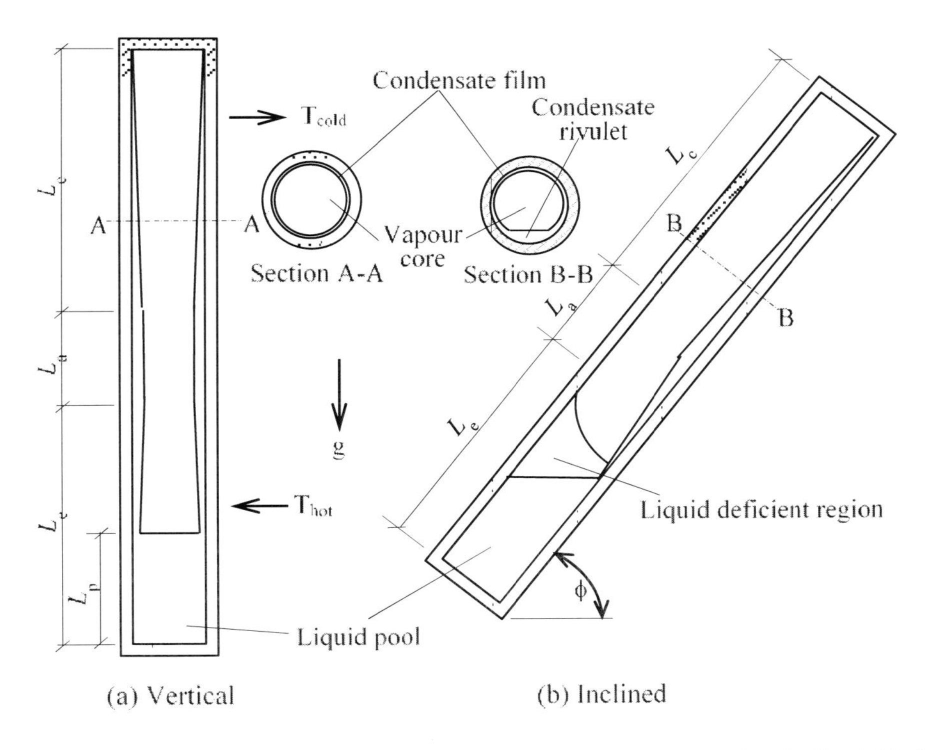

Figure 8. Simple two-phase flow and heat transfer depiction of a vertical (a) and an inclined (b) closed tubular thermosyphon.

1.7. Splashing

Earlier two-phase thermosyphon heat transfer theory analyses tended to be based on an extension of two-phase flow and heat transfer concepts as described in Whalley [1], Lock [2], Carey [3] and Faghri [4]. The thermal-hydraulic behaviour is characterised making use of the *pool boiling* and *film-wise condensation* theory in which the thermosyphonic flow and heat transfer may be idealized as consisting of a liquid pool in the lower evaporator portion and in which nucleate boiling may occur. The condensate is assumed to form as a relatively thin film on the inside wall with a central vapour core. In the case of a vertically orientated thermosyphon, the film is formed as a thin symmetrical annular layer film. A vapour core is separated from, in the case of a vertically orientated pipe, a thin annular film of condensate formed on the wall of increasing thickness in the condenser section, constant thickness in the adiabatic section and then of decreasing thickness on the wall in the evaporator section above the liquid pool as depicted in Figure 8(a). In the case of an inclined pipe, the liquid-film of condensate on the wall empties into a rivulet flowing in the bottom of the pipe as shown in Figure 8(b). A *maximum heat transfer rate* was considered to occur when the shear stress due to the upwards-flowing vapour at the vapour-condensate interface resulted in the downwards-flowing liquid being dragged upwards and as they say, flooding the pipe with condensate and thereby preventing liquid from flowing back into the evaporator section.

In the power and process industries this approach was used to simulate so-called *reflux* condensers in which the condensate formed in the pipe, returns back to the inlet against the flow of the upwards flowing vapour. Care has also to be taken to ensure that no non-condensable gases are collected in the top of the pipe. If non-condensable gas was present it would tend to act as a thermally insulating blanket and the rate of condensation would be seriously compromised.

By assuming a liquid-pool and thin *film theory,* researchers tended to try and find explicit analytical mathematical expressions for the heat transfer rate and the maximum heat transfer rate for thermosyphons. The liquid film (and rivulet) and vapour would be subdivide into control volumes. Then, by *applying* the equations of change (conservation of mass, momentum and energy and appropriate property functions) to each control volume the thermal transport behaviour could be analytically or numerically characterised. The effects of so called suction and blowing of the vapour and liquid-entrainment and -detrainment could also be included in the analysis. [Blowing refers to evaporation where the vapour is idealized as being formed as if it was blown into the vapour core in the pipe through small holes in the surface. Suction on the other hand is the reverse process and is as if vapour is sucked into the liquid. Entrainment referred to liquid (droplets) being drawn into the vapour core by viscous-shear friction forces, and detrainment is when the entrained liquid deposits back into the liquid film.]

This simplified theoretical approach however does necessarily give good results!

The liquid pool, vapour core, and liquid-film and -rivulet flow and heat transfer theoretical modelling approach does not necessarily give good heat transfer results and neither does it represent the actual flow pattern and thermal behaviour of a two-phase thermosyphon. For instance visualization-experiments using transparent-walled pipes show the process somewhat differently. Using water as working fluid, any small bubbles emanating from initiation site(s) on the wall (so-called nucleate boiling) and moving into the fluid, quickly coalesce to form a relatively large and ever expanding bullet-like shaped vapour bubble which propels and drags liquid out of the liquid-pool and high up into the condenser section. The now liquid deficient evaporator section tends to dry out and overheat. When the somewhat subcooled condensate returns back from the condenser section and replenishes the liquid pool, there is a quiescent period before "erupting" again, reminiscent of a geyser such as the famous Old Faithful in Yellowstone National Park. It is further seen that the heat input boiling episode, is out of phase with the heat removal phase. The net result of this visually observed and seemingly random but chaotic behaviour within the thermosyphon is an increase in the heat transfer rate from the hot to the cold side. This departure from the liquid pool and condensate film theory may however be taken into account by introducing an experimentally correlated *splashing enhancement factor S* [5] and which may be defined as the ratio of the experimentally determined heat transfer rate to the theoretically determined heat transfer rate using pool boiling and film theory as $S = \dot{Q}_{experimental}/\dot{Q}_{theoretical}$

Groenewald [5] then proceeded to evaluate a splashing factor *S* based on approximately 100 experimental data points for an ammonia-charged two-phase closed thermosyphon of total length L_{total} = 6.2 m and inside diameter d_i = 32 mm. The evaporator of length L_e ranging from 0.4 to 2 m, was heated with hot water ranging between 30 and 80°C and cooled with cold water at 10°C and with an adiabatic length of essentially zero; hence with corresponding condenser section length of $L_c = L_{total} - L_e$. Care was taken to ensure that the thermosyphon was free of any non-condensable gas; care also had to be taken to ensure that no air-bubbles accumulated on the outer surface of the condenser pipe cooling the water. Unless both of these conditions are pertinently and explicitly addressed it is most unlikely that the experimentally determined heat transfer coefficient results are correct.

The net result of Groenewald's [5] finding, which is supported by visual observations using transparent pipe sections, is that flow in a closed two-phase thermosyphon is *complex* and not really amenable to conventional and standard mathematical analysis. Rather, experimentally correlated and validated heat transfer coefficients using regression analysis is necessary to appropriately characterize the flow and heat transfer behaviour. It is found that in the evaporator section the evaporator heat transfer coefficient may be reasonably, accurately correlated in terms of the evaporator wall heat flux or Kutateladze number, a variable capturing the working fluid properties such as the temperature or pressure or a Prandtl number and the ratio of the total length to the evaporator section length and the inclination angle. For the condenser heat transfer coefficient, the length of the condenser section needs also to be taken into account; the Kutateladze number is not applicable but, rather, must be replaced by a Reynolds number given as $Re = 4(\dot{Q}/h_{fg})/\pi d\mu$.

1.8. Flooding

It is found that as the temperature difference between the hot and cold sides of a thermosyphon increases, there is a *maximum heat transfer rate* at which the heat transfer rate does not further increase due to a further increase in the temperature difference. This is due to so-called *flooding* which occurs when the returning condensate film is hindered from entering into the evaporator by the upwards flowing vapour. As the heat transfer rate increases so too does the vapour velocity and so too does the coefficient of friction between the vapour and the liquid film. As the liquid film thickness increases, waves tend to form which then quickly bridge across the pipe diameter blocking the vapour flow. The pressure builds up in the evaporator and then blasts the liquid upwards in bursts and spurts, so to say, at more-or-less equal intervals. In a vertical thermosiphon, liquid bridging occurs at a lower vapour flow rate than for an inclined thermosyphon. In this way the inclined thermosyphon can transfer more heat, provided that the temperature difference is increased yet further. For inclination angles of between 10 and 75 ° to the horizontal, an increase of about 40 per cent compared to the vertical thermosyphon is achieved.

Note that the heat transfer rate before flooding for both the vertical and inclined thermosyphons is more-or-less the same and that both increase linearly as the temperature difference increases thereby indicating a more or less constant and equal overall heat transfer coefficient (slope of the graph). However, once flooding has started a relatively large scatter in the heat transfer rate is noticed.

1.9. Geysering

After a flooding episode in which the somewhat super-heated liquid and vapour is blasted up into the condenser, the liquid and condensate is cooled and the now somewhat subcooled liquid flows back into the evaporator and boiling in the now replenished evaporator liquid pool ceases momentarily before the process repeats itself in cycle. This cycle might be termed "geysering."

1.10. Smacking

Especially in the case of relatively long vertical thermosyphon when geysering occurs, liquid slugs/plugs of say a few diameters in length may be formed in the condenser section.

If the conditions are "right" both a downwards flowing liquid slug may collide with a newly formed upwards flowing slug of liquid. When this occurs a smacking sound is heard, the sound of which may be accurately duplicated by slapping the fingers of one hand against the palm of the other hand. This must be an undesirable occurrence, as when visualizing this happening in a glass evaporator tube, it has been seen to break the glass tube.

1.11. Complexity

From our discussion so far it is clear that the internal fluid and heat transport with a thermosyphon is a complex phenomenon. The ability to capture complexity in a simple water-charged loop thermosyphon as depicted in Figure 9 is possible [6]. Assume one-dimensional control volumes, quasi-static equilibrium, single phase flow and the Boussinesq approximation. The Boussinesq approximation assumes that the density ρ is constant in the equation of motion except in the buoyancy term where the density is given as a function of temperature T, a reference density ρ_{ref} at a reference temperature T_{ref} and temperature dependent coefficient of volumetric expansion β_{ref} as $\rho = \rho_{ref}\left[1 + \beta_{ref}\left(T - T_{ref}\right)\right]$. A simple coefficient of friction for laminar flow $C_f = 16 / Re$ where Reynolds number $Re = \rho v d/\mu$ and v is the velocity and μ is the dynamic viscosity at T_{ref}. Note that as the flow tends to stand still, Re tends to zero and the coefficient of friction tends to infinity. This division by a zero may be circumvented by assuming that if Re becomes less than 0.01, the Re must then be equal to 0.01 in the simulation model.

The thermosyphon is heated on the left hand side by an electrical resistance heating element and is cooled on the right hand side by a tank containing cold water. The volumetric flow rate $\dot{V}$ of the fluid in such a loop may be given in terms of a driving buoyancy term B and a restraining frictional term $F\dot{V}$ as $\frac{\Delta \dot{V}}{\Delta t} = -B - F\dot{V}$ where

$B = \frac{\frac{\pi g}{4\rho}\sum_{i=1}^{N} \rho_i L_i \mathrm{Sin}\theta}{\sum_{i=1}^{N} \frac{L_i}{d_i^2}}$, $F = \frac{\frac{128\mu}{4\rho}\sum_{i=1}^{N} \frac{L_{eq,i} L_i}{d_i^3}}{\sum_{i=1}^{N} \frac{L_i}{d_i^2}}$, L_i is the length, d_i the diameter and equivalent length L_e (to take into account the so-called minor losses) and θ_i is the inclination to the horizontal of each ith control volume. Care must be taken to always ensure that $(-F\dot{V})$ always act opposite to the flow direction and that the gravity term $(-B)$ depends on whether the acceleration due to gravity g tends to drive fluid clockwise (the assumed positive flow direction) or resist the flow if it tends to retard the flow.

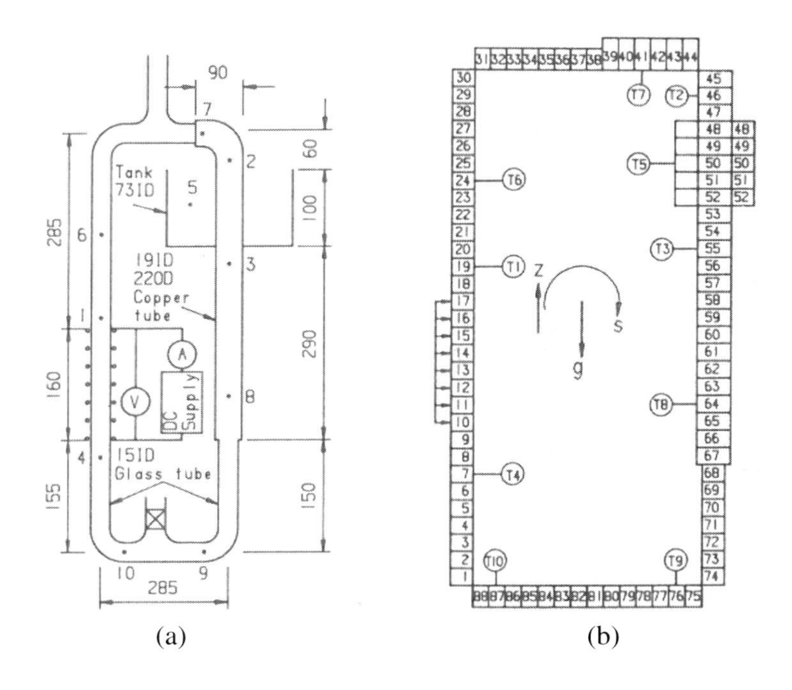

Figure 9. Thermosyphon loop dimensions (in mm) (a) and then subdivided into a number of control volumes (b) [6].

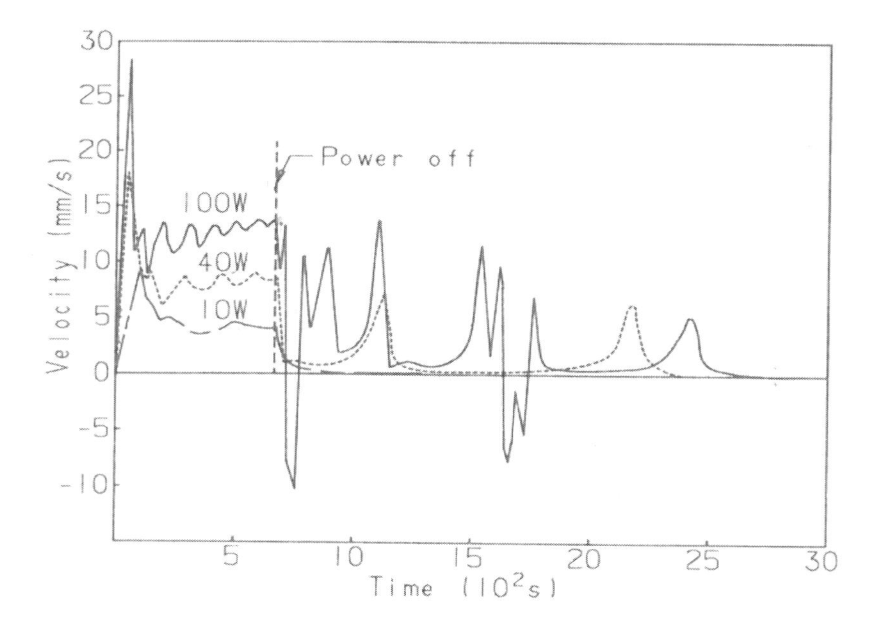

Figure 10. Loop working fluid velocity as a function of time for different power inputs [6].

The theoretical simulation based on the above assumptions indeed captures a reasonable amount of complexity as shown in Figure 10, in which the (average) fluid velocity, initially at rest, is plotted as a function of time for different heating rates varying from 10 to 100 W.

2. HEAT TRANSFER COEFFICIENTS AND MAXIMUM HEAT TRANSFER RATE

In this section we will concentrate on the inside evaporator and condenser section, heat transfer coefficients and the maximum heat transfer rate formulations for closed thermosyphons. At the outset it should however be noted that there are almost just as many formulations as there have been researchers. This is not surprising due to the relatively complex nature of two-phase flow and heat transfer. However, we will consider only three cases in detail here:

2.1. Ammonia charged water-heated and -cooled closed thermosyphon.
2.2. Ammonia-charged steam heated and air-cooled closed thermosyphon.
2.3. Butane and R123 charged heat transfer coefficients.

These natural circulation concepts are all particularly suitable for heating and/or cooling applications in energy-consumption intensive industries such as the electrical power and chemical processing industries in which energy may be recovered from a waste stream and transferred to gainfully preheat another process, heat steam.

Many thermal performance characterizing formulations give a non- dimensionless form of the heat transfer characteristic in terms of a number of dimensionless independent variables. A number of popularly used, dimensionless variables are given in Table 1. A particularly common form of such a correlating equation, or more simply put, *correlation* used in capturing the heat transfer coefficient in terms of a number of independent variables takes the form of a *power series* as given by Eq. (1). The heat transfer coefficient h being given in terms of a Nusselt number $Nu = hL_c/k$ as the dependent variable y, and, wherein to make it dimensionless, it includes a characteristic length L_c, which for a pipe would normally be the diameter, and the thermal conductivity of the working fluid k.

$$y = c_0 x_1^{c_1} x_2^{c_2} x_3^{c_3} \cdot \quad (1)$$

The correlating coefficient constants c, are obtained by multi-linear regression after having *linearized* Eq. (1) by taking logs on both sides and thereby getting Eq. (2)

$$\log y = \log c_0 + c_1 \log x_1 + c_2 \log x_2 + c_3 \log x_3 \cdot \quad (2)$$

The independent variables x_i should then be arranged from left to right depending on their relative importance. Note that such correlations are statistically determined using a specific experimentally determined data-set. As such, it is thus important that the operating conditions and geometrical dimensions, for which this statistically correlated equation is valid, be specified.

There are many experimentally correlated equations given in the literature for the evaporator and condenser heat transfer coefficients and for the maximum heat transfer rate. There are a number of relatively definitive textbooks giving such correlations. For example Pioro and Pioro [7] give some 53 equations; Reay and Kew [8] also give a number of relevant heat transfer coefficient correlations. There are also a number of review articles, one by Bhattacharyya et al. [9] and another by Jafari et al. [10], to name but only two; Jafari, for instance, gives some 34 correlations. The problem in using an already published correlation is that it is unlikely that one will find a correlation that relates precisely to one's own geometric and operating conditions. Each published correlation relates to a specific set of operating parameters and limits and geometries, working fluids and orientations relative to gravity. The correlating coefficients have been invariably obtained using data regression techniques as applied to a unique set of experimental data points. It is thus also not advisable to use extrapolated values of these correlations as implying a theoretical basis when they are, in reality, simply an interpolating function of a set of experimentally-obtained tabulated data.

Table 1. Popularly-used dimensionalless variables used in heat transfer performance correlating equations

Name	Formula	Representing
Kutateladze *Ku*	$\dot{Q}''/\left(\rho_l h_{fg}\left(\sigma g(\rho_l-\rho_v)/\rho_v^2\right)^{0.25}\right)$	Heat flux
Reynolds *Re*	$\rho v d/\mu$	Forced convection
Prandtl *Pr*	$c_p\mu/k$	Properties or temperature
Jacob *Ja*	$c_{pl}\lvert T_w-T_f\rvert/h_{fg}$	Temperature diff.
Coef. of friction C_f	$\tau_w/(\rho v^2/2)$	Wall shear stress
Grashoff *Gr*	$\beta\lvert\Delta T\rvert g L_c^3/(\mu/\rho)^2$	Natural convection
Rayleigh *Ra*	$GrPr$	Natural convection
Weber *We*	$\rho v L_c/\sigma$	Surface tension
Bond *Bo*	$(\rho_l-\rho_v)gLL_c^2/\sigma$	Bubble formation
Nusselt	hL_c/k	Heat transfer rate

[Note that C_f must not be confused with the friction factor f which for round tubes is $f = 4C_f$.].

It is probably wise and especially in the long-run lifespan of a commercially attractive thermosyphon product, if at all possible, to determine its own specific heat transfer correlating coefficients. By doing so, one adds credibility to one's own product and its operating performance and it is less likely that a competitor can discredit the product based on its thermal performance predictions. Note that in most air-cooled or heated thermosyphons, the inside heat transfer coefficients are relatively unimportant as it is the air side heat transfer coefficients that are the so-called controlling heat transfer coefficient. For instance, typical evaporator and condenser heat transfer coefficients are some 3000 and 2000 W/m^{2}K, respectively; whereas the natural and forced convective air-cooled heat transfer coefficients are only about 3 and 15 W/m^{2}K. Under these conditions, varying the inside heat transfer coefficients by 50% to 100% has little effect on the overall thermal performance of the thermosyphon, and that is even if the air-side is finned.

2.1. Ammonia-Charged Water Heated and Cooled Closed Thermosyphon

The inside heat transfer coefficients and maximum heat transfer rate of a relatively long ammonia charged grade 304 stainless steel thermosyphon of length 6.2 m and internal diameter 39.1 mm were experimentally determined [11]. The thermosyphon was supplied with five 400 mm long and one 4200 mm long water heating/cooling jackets as shown in Figure 11 and could be inclined at angles of 30, 45, 60, 75 and 90 degrees to the horizontal. Heating water could be supplied up to a temperature of 80°C and the cooling water could be varied between 10 and 20°C.

The first step in experimentally determining the steady state thermal performance parameters of a thermosyphon operation is to check that the heat input, in accordance with the conservation of energy, is indeed equal to the heat output. The heat transfer rate into the evaporator is calculated by $\dot{Q}_{hw} = \dot{m}_{hw} c_{p,hw}(T_{hwi} - T_{hwo})$ and out of the condenser by $\dot{Q}_{cw} = \dot{m}_{cw} c_{p,cw}(T_{cwo} - T_{cwi})$. It was found that the experimentally supplied heat into the thermosyphon corresponds to within 10% of the heat removed and thereby instilling a degree of confidence in the validity of the experimental results. A scatter of about ± 10% is to be expected as the evaporator and condenser heat transfer rates at any instant in time will be somewhat out of time-phase with each other due to splashing and periodic geyser-like eruptions out of liquid out of the evaporator section and up into the condensing section.

The evaporator heat transfer coefficient was calculated as $h_e = \dot{Q}_{hw}/A_e(\bar{T}_{hw} - T_i)$, where $A_e = \pi d_i L_e$, $\bar{T}_{hw} = (T_{hwi} + T_{hwo})/2$ and similarly for the condenser heat transfer coefficient as $h_c = \dot{Q}_{cw}/A_c(\bar{T}_{cw} - T_i)$, $A_c = \pi d_i L_c$ and $\bar{T}_{cw} = (T_{cwi} + T_{cwo})/2$. The heat transfer coefficients were correlated using a power series correlation and the coefficients determined using multilinear regression (to the as-indicated coefficients of determination R^2) as:

For the vertical thermosyphon:

$$h_{ei} = 0.0404\left(\dot{Q}_e''\right)^{0.931} P_i^{0.615} (L_l/L_e)^{-0.290} \qquad [R^2 = 98.2\%\,] \tag{3}$$

$$h_{ci} = 0.3706(\dot{Q}_c'')^{0.0861} P_i^{-0.2429} (L_c)^{1.3523} (L_l/L_e)^{-0.3474} \qquad [R^2 = 72.2\%\ ] \tag{4}$$

And for the inclined thermosyphon:

$$h_{ei} = 0.4.3(\dot{Q}_e'')^{0.980} P_i^{-0.455} (L_l/L_e)^{0.305} (\sin\phi)^{0.351} \qquad [R^2 = 94.4\%] \tag{5}$$

$$h_{ci} = 29.78(\dot{Q}_c'')^{0.03941} P_i^{-0.2717} (L_c)^{-0.6661} (L_l/L_e)^{-0.01131} (\sin\phi)^{-0.3821} \qquad [R^2 = 46.9\%\ ] \tag{6}$$

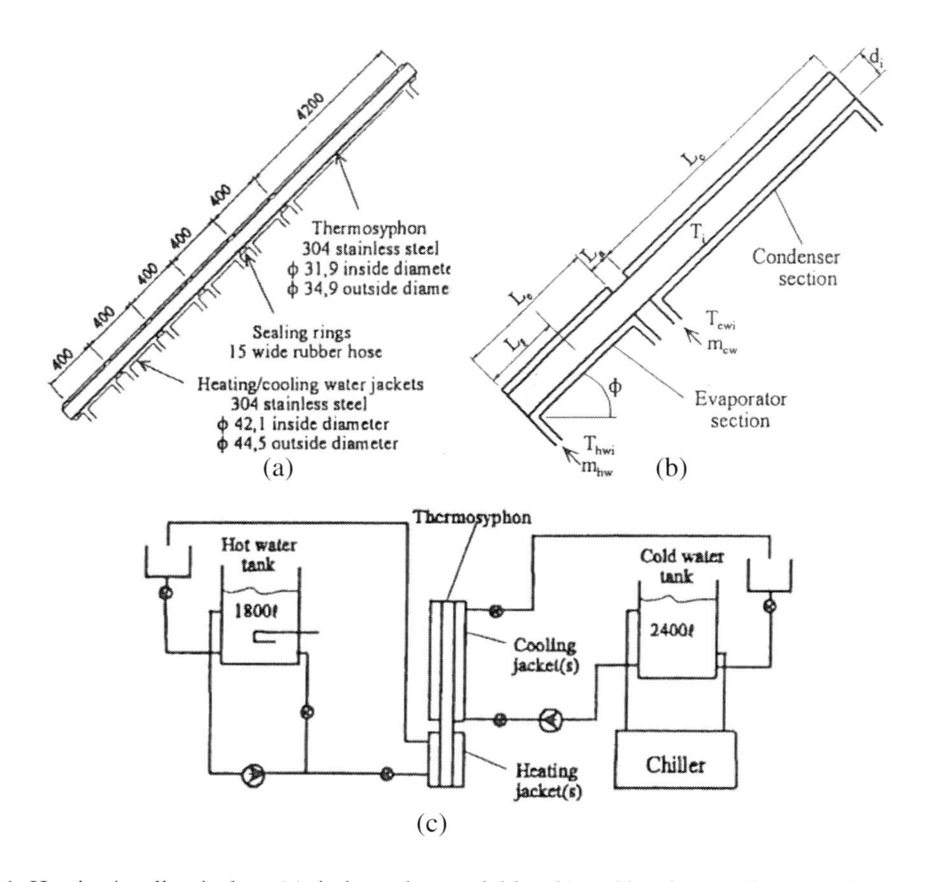

Figure 11. Heating/cooling jackets (a), independent variables (b) and heating cooling experimental test system (c) for the ammonia charged 6.2 m long thermosyphon [11].

In Eqs. (3) to (6) $\dot{Q}_{e,c}'' = \dot{Q}_{e,c}/A_{e,c}$, $A_{e,c} = \pi d_i L_{e,c}$ and the heat transfer coefficients h are in W/m^{2}K if the heat flux $\dot{Q}_{e,c}'' = \dot{Q}_{e,c}/A_{e,c}$ is in W/m^2, the internal pressure P_i is in bar (absolute), the inclination to the horizontal ϕ is in degrees, $A_{e,c}$ is in m^2, and the total length $L_l = L_e + L_a + L_c$ and with lengths in m.

The maximum heat transfer $\dot{Q}_{max}$ may be determined in terms of a maximum so-called superficial velocity $v_{sv,max}$, cross-sectional area $Ax = \pi d_i^2/4$, the working fluid vapour density ρ_v and the latent heat of evaporation h_{fg} as:

$$\dot{Q}_{max} = Ax\rho_v h_{fg} v_{sv,max} \tag{7}$$

where the maximum superficial velocity is given as an experimentally correlated function given as:

$$v_{sv,max} = 136.2T_i^{-0.729}\phi^{-0.429}(L_l/L_l)^{-0.108} \tag{8}$$

2.2. Ammonia-Charged Steam Heated and Air-Cooled Closed Thermosyphon

In the power and process industries there is a need for relatively large steam dephlegmators and reflux condensers. The temperature of the steam being condensed is typically about 60°C and the heat transfer rates may vary up to many thousands of kW. In regions where water could be a potentially scarce resource (such is the case in South Africa), the use of air-cooled condensers is prevalent. The air-cooled condenser tubes are typically up to 10 m in length and with diameters of about 30 mm. In this section (section 2.2) the use of a two-phase closed thermosyphon suitable for meeting these condensing requirements will be considered. All the applicable correlations to determine the heat transferred from the steam to the air, the maximum heat transfer rate as well as the airside frictional flow resistance, will be given.

The experimental set-up used in arriving at the air-side heat transfer correlations is given by way of Fig 12. The thermosyphon is made up of a 6.08 m mild steel seamless pipe of inside diameter 34.7 mm and outside diameter 38.2. The fin pitch is 10 fins per 25.4 mm, the fin thickness is 0.2 mm and they are manufactured by extruding an aluminium pipe to give an effective structure as shown in Fig 13. The fins were removed from the lower 1.5 m length of the pipe to form the evaporator. The evaporator is heated by condensing steam on it from a steam generator. The steam generator is supplied with hot water, up to a maximum of about 84°C, from a separate water heating system. The condensing section was cooled using ambient temperature air at about 26 ± 2°C and is drawn across the fins through four separate sections.

The set-up shown in Figure 12 was used to determine the inside heat transfer coefficient correlations; the inside diameter was 32 mm, the liquid charge length L_l was 400 mm and heating temperatures were between 40 and 80°C. The total length was held constant at $L = 6.20$ m and as the evaporator length was varied from 400 mm to 2000 mm, the condenser length varied as $L_c = L_i - L_e$. In this way, liquid charge fill ratios L_l/L_e varying from 20 to 100% were obtained.

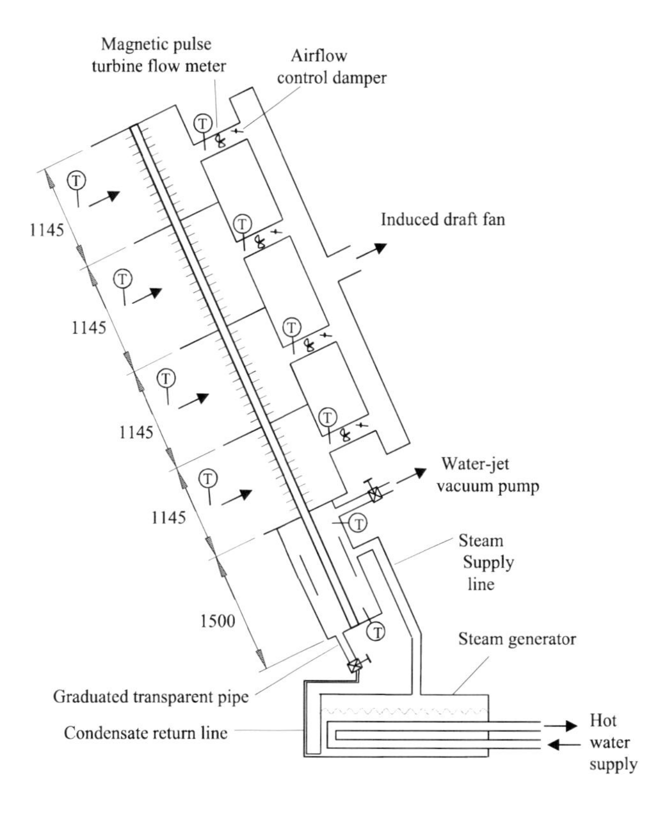

Figure 12. Important features of the thermosyphon experimental set-up (Dimensions in mm) [11].

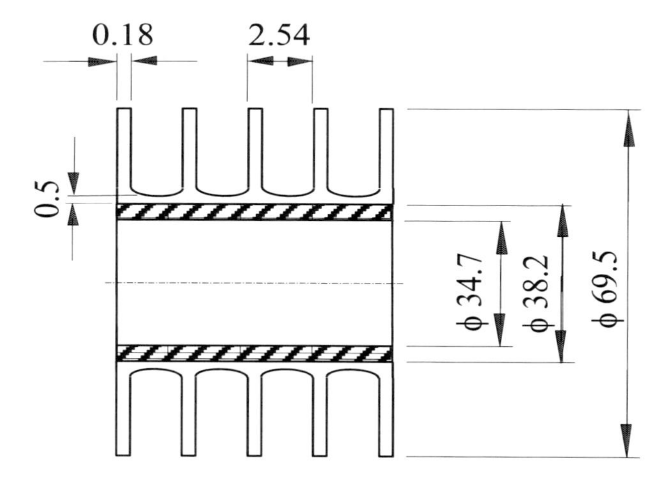

Figure 13. Tube and fin dimensions (in mm).

The performance of the thermosyphon may be simulated by considering the thermal resitance drawing as shown in Figure 14 for the basic thermosyphon dimensions given in Figure 15. The heat transfer rate $\dot{Q}$ is then given by:

$$\dot{Q} = \frac{T_{eo} - T_{co}}{\sum R} \tag{9}$$

where the individual resistances are given by:

$$R_{eo} = 1/h_{eo}A_{eo}, R_{ew} = ln(d_{eo}/d_i)/2\pi k_w L_e$$
$$R_{ei} = 1/h_{ei}A_{ei}, R_{ci} = 1/h_{ci}A_{ci}, R_{cw} = ln(d_o/h_i)/2\pi k_w L_c \text{ and } R_{co} = 1/h_{co}A_{co}$$
$$A_{eo} = \pi d_o L_e, A_{co} = \pi d_o L_c, A_{ei} = \pi d_i L_e \text{ and } A_{ci} = \pi d_i L_c$$

The areas in Eq. (9) may be readily determined from the thermosyphon dimensions and the thermal conductivity of the wall material k_w and evaporator and condenser outside hear transfer coefficients h_{eo} and h_{co} may be obtained from most heat transfer texts. The internal heat transfer coefficients h_{ei} and h_{ci} are however not readily available, and thus have to be experimentally determined and given in terms of appropriate correlations as follows:

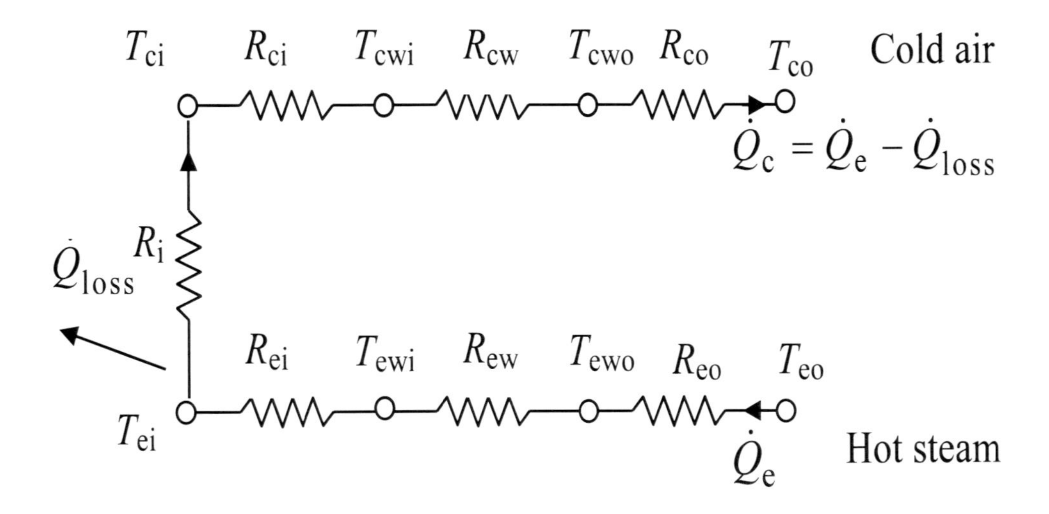

Figure 14. Thermal resistance diagram for a thermosyphon.

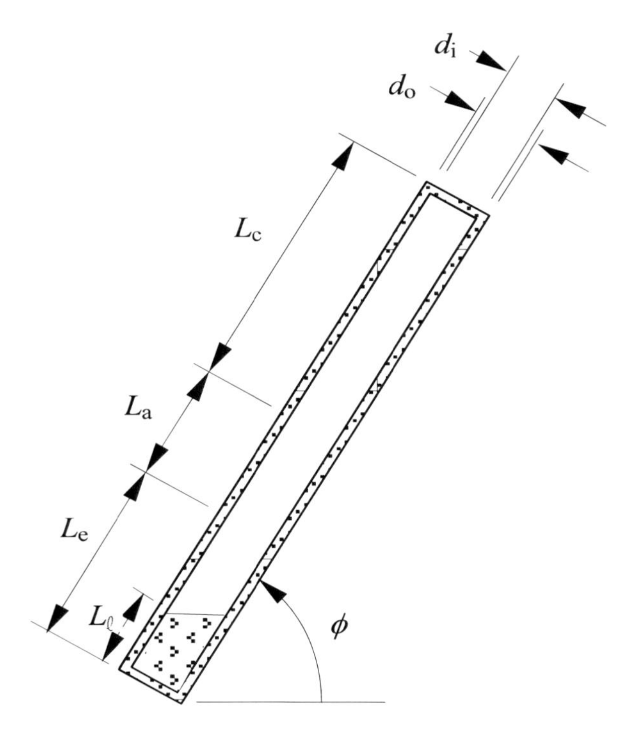

Figure 15. Basic dimensions for a closed two-phase thermosyphon.

The inside heat transfer coefficient in the evaporator for inclination angles of 30 to 75 ° may be given as:

$$h_{\mathrm{ei}} = 0.413\left(\dot{Q}''_{\mathrm{ei}}\right)^{0.980}(P_{\mathrm{i}})^{-0.455}(L_l - L_e)^{0.305}(\sin\phi)^{0.351} \tag{10}$$

where h is in kW/m^{2}K if $\dot{q}''$ is in kW/m^2, P_i in bar and L_ℓ/L_e in m/m, and for a vertically orientated geometry as:

$$h_{\mathrm{ei}} = 0.0404\left(\dot{Q}''_{\mathrm{ei}}\right)^{0.931}(P_{\mathrm{i}})^{0.615}(L_l/L_{\mathrm{e}})^{0.290} \tag{11}$$

Predictions using Eqs (10) and (11) were shown to be within ± 30% of experimentally determined values.

The inside heat transfer coefficient in the condenser for inclination angles of 30 to 75° may be given as:

$$h_{\mathrm{ei}} = 29.78\left(\dot{Q}''_{\mathrm{ci}}\right)^{-0.0394}(P_{\mathrm{i}})^{-0.2717}(L_l - L_e)^{-0.0113}(\sin\phi)^{-0.3821} \tag{12}$$

where h is in kW/m²K if $\dot{q}''_{ci}$ is in kW/m², P_i in bar and $L_{\square}/L_e$ in m/m, and for a vertically orientated geometry:

$$h_{ci} = 0.3706\left(\dot{Q}''_{ei}\right)^{0.08611}(P_i)^{-0.2429}(L_c)^{-1.3523}(L_l/L_c)^{-0.3474} \tag{13}$$

The variation of the predicted condenser heat transfer coefficients using Eqs. (12) and (13) were within ± 15% of the experimentally determined values.

As the steam condenses on the outside wall of the thermosyphon, a relatively thin layer of condensate forms, the liquid runs around and down the wall under the influence of gravity in a *film-wise* fashion and drips off at the lowest end. This film of liquid can be regarded as additional resistance to the heat transfer required to sustain the condensation process; the thicker the liquid film, the greater the insulating effect and the less the rate of condensation of steam. For a laminarly flowing liquid film, a number of analytical condensation heat transfer coefficients may be derived, depending on whether a constant temperature difference across the liquid film or a constant heat flux is assumed. If a constant uniform temperature difference between the steam and the wall is assumed, the following equation for an average heat transfer coefficient for a relatively long inclined pipe can then be derived using the so-called *Nusselt film theory:*

$$\bar{h}_{\Delta T} = \frac{1}{A}\int_A h\mathrm{d}A = 0.728\left[\frac{k^3\rho^2 h_{fg}\,g\cos\phi}{\mu d_o\left(T_v - T_w\right)}\right]^{1/4} \tag{14}$$

Similarly, but for a vertical pipe (or wall) of length (or height) L:

$$\bar{h}_{\Delta T} = \frac{1}{A}\int_A h\mathrm{d}A = 0.943\left[\frac{k^3\rho^2 h_{fg}\,g}{\mu L\left(T_v - T_w\right)}\right]^{1/4} \tag{15}$$

If a constant uniform heat flux across the wall is assumed, the average heat transfer coefficient for a relatively long inclined pipe can be derived as:

$$\bar{h}_q = \frac{1}{A}\int_A h\mathrm{d}A = 0.693\left[\frac{k^3\rho^2 h_{fg}\,g\cos\phi}{\mu d_o \dot{q}''}\right]^{1/3} \tag{16}$$

Similarly, but for a vertical pipe (or wall) of length (or height) L:

$$\bar{h}_{\mathrm{q}} = \frac{1}{A}\int_A h\mathrm{d}A = 1.040\left[\frac{k^3\rho^2 h_{\mathrm{fg}}\mathrm{g}}{\mu\dot{q}''L}\right]^{1/3} \tag{17}$$

The problem in using Eqs. (14) to (17) is that in practice both the temperature difference and the heat flux are not constant but will vary along the pipe length. It is suggested that a more reasonable approach would be to define the heat transfer coefficient in terms of an average heat transfer flux and an average temperature difference as:

$$\bar{h}_{\mathrm{mq}} = \frac{\frac{1}{A}\int_A \dot{q}''\mathrm{d}A}{\frac{1}{A}\int_A (T_{\mathrm{v}} - T_{\mathrm{w}})\mathrm{d}A} \tag{18}$$

In this case the heat transfer coefficient for a relatively long inclined pipe can be derived as:

$$\bar{h}_{\mathrm{mq}} = 0.605\left[\frac{k^3\rho^2 h_{\mathrm{fg}}\mathrm{g}\cos\phi}{\mu d_{\mathrm{o}}\dot{q}''}\right]^{1/3} \tag{19}$$

Similarly, but for a vertical pipe (or wall) of length (or height) *L:*

$$\bar{h}_{\mathrm{mq}} = 0.924\left[\frac{k^3\rho^2 h_{\mathrm{fg}}\mathrm{g}}{\mu\dot{q}''L}\right]^{1/3} \tag{20}$$

The airside of the thermosyphon has fins (see figures 12 and 13) and the heat transfer coefficient for these fins was determined experimentally (using a specially manufactured fin-bundle made from identical finned tube in a wind tunnel) for face velocities V_{af} of between 2 and 10 m/s and air temperatures of about 31°C as:

$$h_{\mathrm{af}} = 507.2 + 80.42V_{\mathrm{af}} \tag{21}$$

where h_{af} is in W/m^2°C if V_{af} is in m/s.

The airside pressure loss across the finned portion of the pipe was experimentally determined as:

$$\Delta P = 30.3\rho_{\mathrm{a}}\frac{V_{\mathrm{af}}^{1.89}}{2} \tag{22}$$

where ΔP is in Pa if ρ_a is in kg/m^3 and the face velocity is in m/s.

A correlation was determined for the superficial vapour velocity at the maximum heat transfer rate as:

$$V_{sv,max} = 1362 (T_i)^{-0.718} (\phi)^{-0.429} (L_\ell / L_e)^{-0.108} \quad (23)$$

where $V_{sv,max}$ is in m/s if T_i is in °C and ϕ in °. $V_{sv,max}$ increases significantly as ϕ decreases, for instance at T_i = 30°C, $V_{sv,max}$ = 1.6 m/s for the vertically orientated thermosyphon, whereas for ϕ = 30° $V_{sv,max}$ is some 72% greater at 2.75 m/s. The maximum heat transfer rate is then readily calculated using:

$$\dot{Q}_{max} = \dot{m}_{v,max} h_{fg} \quad (24)$$

and

$$\dot{m}_{v,max} = \rho_v V_{sv,max} \pi d_i^2 / 4 \quad (25)$$

The variation of the predicted maximum heat transfer rate using Eq. (24) was shown to be within ± 15% of the experimentally determined values.

The steam side evaporator heat transfer rate is determined by:

$$\dot{Q}_e = \dot{m}_{condensate} h_{fg} - \dot{Q}_{loss} \quad (26)$$

where the heat loss was experimentally determined as a function of the temperature difference of the steam in the heating jacket and the surrounding air (on average for the different inclinations) as:

$$\dot{Q}_{loss} = 5{,}01 (T_s - T_a) \quad (27)$$

The airside condenser heat transfer rate was determined as:

$$\dot{Q}_c = \dot{m}_a c_{pa} (T_{a,outlet} - T_{a,inlet}) \quad (28)$$

The evaporator heat input rate corresponded to within ±10% of the condenser heat rate out. The experimentally determined heat transfer rate is well correlated (R^2 = 0.9848) for the vertical thermosyphon in terms of the independent experimental variables as:

$$\dot{Q} = 0.137(T_s - T_a)^{1.05}(V_{af})^{0.257}(L_l/L_c)^{0.005} \tag{29}$$

For the inclined thermosyphon tests the experimental data is correlated ($R^2 = 0.8618$) as:

$$\dot{Q} = 0.185(T_s - T_a)^{1.01}(V_{af})^{0.257}(L_l/L_c)^{0.324}(\phi)^{0.025} \tag{30}$$

A sensitivity analysis was conducted to identify the most important variables on the heat transfer rate of the thermosyphon. This was done by varying the more important variables by ± 25% and then determining the percentage variation of the heat transfer rate from the base case values for a set of typical (base case) operating conditions. It was seen that with a variation in heat transfer rate from the base case of between 40 and 45% that the steam temperature is the most sensitive variable. The second most sensitive variable is the airside heat transfer coefficient with variations of between 10 and 18%. None of the other operating parameters influenced the heat transfer by more that 7%.

2.3. Heat Transfer Coefficients and Maximum Heat Transfer Rate for R123 and Butane

In this section experimentally determined correlations for the evaporator and condenser heat transfer coefficients and the maximum heat transfer rate of two-phase closed thermosyphons using R134a and Butane as working fluids are given. Inside diameters of 14.9, 17.3, 20.2 and 31.9 mm and lengths of 2.0 and 6.2 m were tested for vertical and 45° inclined orientations and an evaporator liquid charge fill ratio of 50% [12].

Refrigerant R134a and butane were chosen as the working fluids as both these fluids have low ozone depletion and global warming potentials and hence make them attractive working fluids from an ecological point of view. R134a is increasing in popularity in the refrigeration industry and legislation in certain countries prescribes its use. Butane is relatively cheap and is commercially available. It is even available in many supermarkets where it is sold as gas-lighter fuel (a mass percentage mixture of related hydrocarbons of 50% n-butane, 25% iso-butane and 25% propane). The saturated vapour pressure of both butane and R134a at room temperatures is significantly higher than atmospheric pressure and relatively small vapour specific volumes (in the order of respectively 220 kPa and 0.18 m^3/kg for n-butane, for instance). A saturated vapour pressure at room temperature has an important practical implication as it obviates the need of a high quality vacuum pump to charge the system. After charging, the non-condensable gases may be removed by boiling-off a little of the liquid charge. The successful removal of the non-condensable gasses using this method can be verified by checking that the working fluid temperature at the top of the condenser is only slightly less than the temperature in the evaporator. Water, although arguably thermally superior, but with a sub-atmospheric saturated pressure at room temperature of only 2.4 kPa would require high quality vacuum equipment to charge the system at room temperatures. Also, because of water's large specific volume of about 58 m^3/kg at low saturation temperature, it is considerably more sensitive to the thermally insulating effect of non-condensable gas collecting on the surface on which the condensation is taking place. Limited research has been undertaken using R134a whilst no formal database on the heat transfer of butane as a thermosyphon working fluid was available.

The important thermal characteristics, unique to two-phase closed thermosyphons, needed to theoretically or analytically characterise their performance include the inside evaporator and condenser heat transfer coefficients and the maximum heat transfer rate at different inclination angles and liquid charge fill ratios. It is thus the express objective of section 2.3 to experimentally determine these thermal characteristics for a number of closed two-phase thermosyphons of varying diameters for cooling temperatures in the region of 5 to 20°C and heating temperatures of up to 80°C for both R134a and butane as working fluids. These test conditions and temperatures are particularly suitable for recovering heat from low-grade temperature waste streams to preheat an incoming stream at ambient conditions. This is typical of thermosyphons that might be included in heat recovery heat exchangers for heat recovery applications in many food processing and industrial processes. Further, these experimentally determined heat transfer coefficients will be presented as correlations suitable for use in theoretical performance calculations and will also be compared with other published thermal performance correlations.

Knowing the heating and cooling water inlet and outlet temperatures and the mass flow rates of the heating and cooling streams $\dot{m}_e$ and $\dot{m}_e$ he evaporator and condenser section (see Figure 16), the heat transfer rates can be calculated in accordance with the conservation of energy as:

$$\dot{Q}_e = \dot{m}_e c_p (T_{hi} - T_{ho}) \pm \dot{Q}_{loss/gain} \tag{31}$$

and

$$\dot{Q}_c = \dot{m}_c c_p (T_{ci} - T_{co}) \pm \dot{Q}_{loss/gain} \tag{32}$$

The right hand side terms $\dot{Q}_{loss/gain}$ in Eqs. (31) and (32) account for the heat that is not transferred to the working fluid in the evaporator and from the working fluid in the condenser but that which is lost or gained from the environment through the heating/cooling jacket walls as well as through the structure supporting the thermosyphon. These losses are normally small compared with the actual heat transferred.

Eq. (9) may be rearranged to give the inside heat transfer coefficients for the evaporator and condenser sections of the thermosyphon as:

$$h_{ei} = \left[A_{ei} \left(\frac{\overline{T}_h - T_i}{\dot{Q}_e} - \frac{1}{h_{eo} A_{eo}} - \frac{\ln(d_o/d_i)}{2\pi k L_e} \right) \right]^{-1} \tag{33}$$

$$h_{ci} = \left[A_{ci} \left(\frac{T_i - \overline{T}_c}{\dot{Q}_c} - \frac{1}{h_{co} A_{co}} - \frac{\ln(d_o/d_i)}{2\pi k L_c} \right) \right]^{-1} \tag{34}$$

Knowing the inside working fluid temperatures and the heating and cooling water inlet and outlet temperatures, the average temperatures $\overline{T}_h$ and $\overline{T}_c$ may be calculated. Also, knowing the wall thermal conductivity, the wall thermal resistances may also be calculated. The outside heat transfer coefficients h_{eo} and h_{co} may also be relatively accurately determined using well known correlations as given in any standard heat transfer textbook. The inside heat transfer coefficients h_{ei} and h_{ci} are more problematic and would normally have to be experimentally determined (using Eqs. (33) and (34) for instance).

Four different diameter thermosyphons (three copper and one 304 stainless steel) were tested. Table 2 gives their basic dimensions. The thermosyphons were fitted with tubular jackets (see Figure 16) through which heating and cooling water could be circulated using a hot and cold water supply system shown in Figure 11. The thermosyphon support structures could be inclined at different angles. In all cases, a liquid charge fill ratio of 50% (liquid volume to evaporator volume) was used at inclinations of 45° and 90° to the horizontal and cooling water temperatures of between 5 to 20°C and heating water temperatures of up to 80°C. Circulating water mass flow rates were determined by measuring the time it took for a known volume of water to flow.

To demonstrate compliance with the conservation of energy, the condenser heat transfer rate and evaporator heat transfer rate are compared with each other. As shown in Figure 17 a reasonable heat balance was achieved for the experimentally determined heat transfer rates of about ± 10 to ± 20% but with the the inclined thermosyphon yielding less scatter. Scatter of this order is typical due to the complexity (due to partial evaporator dryout, geysering and splashing) of the heat transfer process as well as that the heat input is necessarily of phase with heat output.

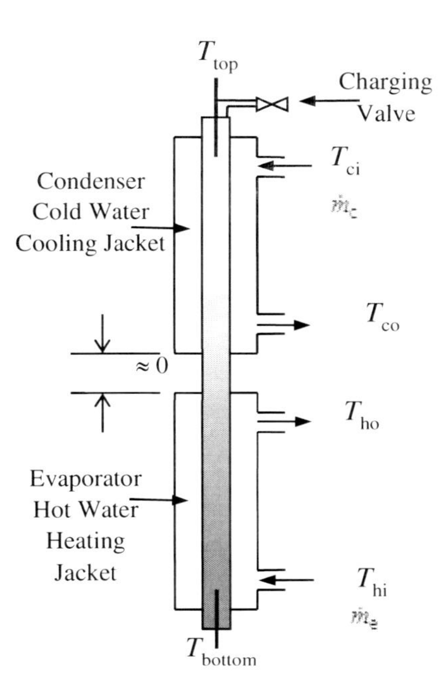

Figure 16. Temperature measurement locations.

The condenser heat transfer coefficients were taken as a Reynolds number and a property grouping as shown in Eqs. (35) and (36).

$$\phi = 90^\circ \quad h_{ci} = 4.61561 \times 10^{9}\, \mathrm{Re}_l^{0.364} \left[k_l \Big/ \left(\frac{\mu_l^2}{g\rho_l(\rho_l - \rho_v)} \right)^{1/3} \right]^{-2.05}$$

$$[R^2 = 0.425] \tag{35}$$

$$\phi = 45^\circ \quad h_{ci} = 3.7233 \times 10^{-5}\, \mathrm{Re}_l^{0.136} \left[k_l \Big/ \left(\frac{\mu_l^2}{g\rho_l(\rho_l - \rho_v)} \right)^{1/3} \right]^{1.916}$$

$$[R^2 = 0.121] \tag{36}$$

On average, the condenser section inside heat transfer coefficients for the inclined orientation is some 50% higher than for the vertical orientation. This is to be expected as the average condensate film thickness for a relatively long pipe decreases as the pipe angle to the horizontal decreases. Because the heat transfer coefficient for a laminar film may be approximated by the thermal conductivity divided by the film thickness, and hence the smaller the film thickness, the larger the heat transfer coefficient. The predicted condenser inside heat transfer coefficients using Eqs. (35) and (36) are compared with published correlations in Figure 18. All the given correlations give widely differing heat transfer coefficients. This figure is displayed to emphasise that care must be taken in selecting a particular correlation of the many available and that it might even be advisable to experimentally determine ones own correlation for ones own thermosyphon.

Table 2. Sizes and dimensions of the experimental thermosyphons

Material	d_i mm	d_o mm	L_e m	L_c m
Copper	14.90	15.88	1.00	1.00
Copper	17.27	19.05	1.03	1.03
Copper	20.19	22.22	1.03	1.03
Stainless Steel	31.90	34.90	1.20	5.00

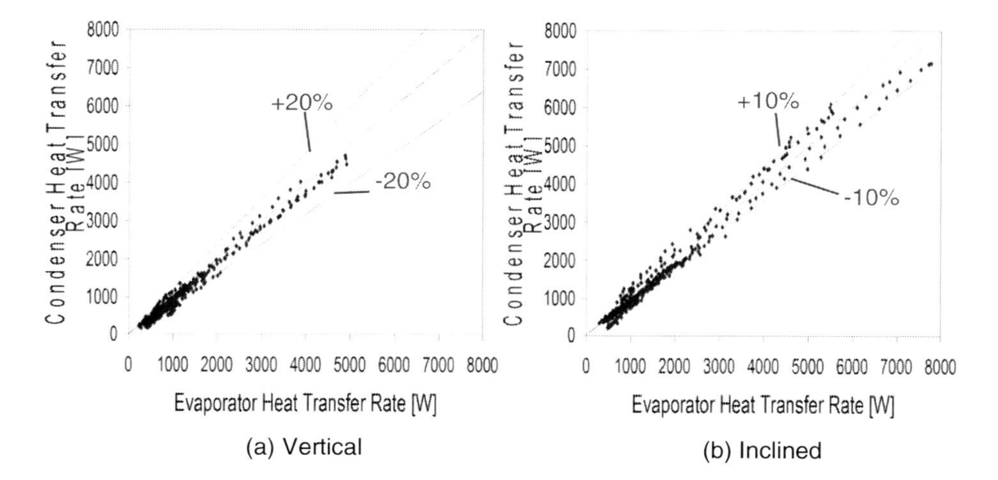

Figure 17. Heat output as a function of the heat input for a vertical (a) and inclined at 45 ° closed thermosyphon.

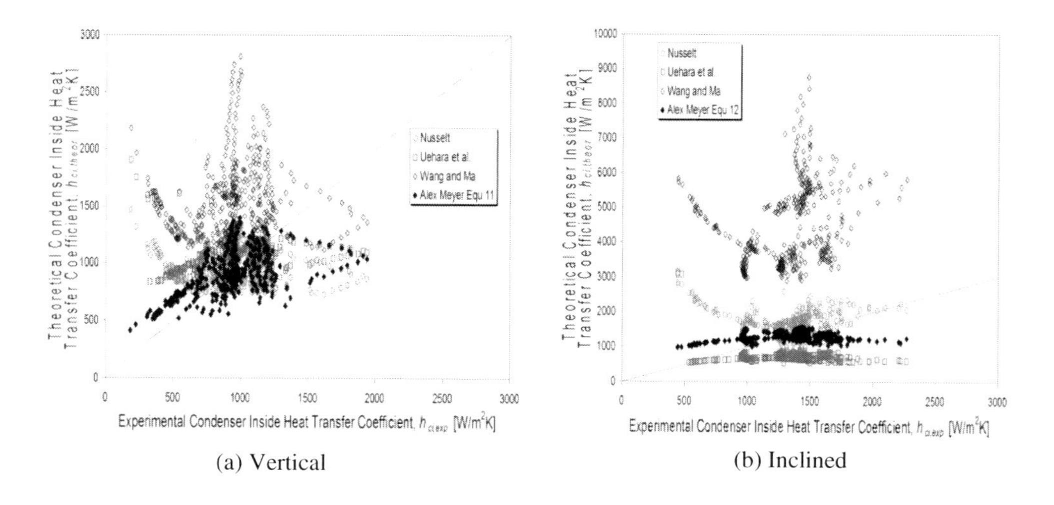

(a) Vertical (b) Inclined

Figure 18. Comparison of various correlations for vertically and an inclined thermosyphon.

The maximum heat transfer rates for the vertical and inclined orientations of the two-phase closed thermosyphons were found to be highly dependent on the diameter and the heat transfer rate and hence are given in terms of a Bond and a Kutaladze number as shown in Eqs. (37) and (38).

$$\phi=90^\circ \; \dot{Q}_{max} = 1.66 \times 10^6 Bo^{0.316} Ku^{1.60} \; [R^2 = 0.928] \tag{37}$$

$$\phi=45^\circ \; \dot{Q}_{max} = 7.47 \times 10^6 Bo^{0.210} Ku^{1.92} \; [R^2 = 0.962] \tag{38}$$

By comparing the predicted values using these equations with the experimentally determined values, relativley high correlation coefficients were obtained as indicated above. It was also found that most of the predicted and experimental values were all within ± 15% of each other. The maximum heat transfer rate for the inclined thermosyphon is some 40% higher than for the vertical thermosyphon. It is however interesting to note that if the maximum heat transfer rate has not been attained for the same temperature difference between the heat source and the heat sink, then the heat transferred by both the vertical the inclined thermosyphons is essentially the same. The reason for this is that the increased condenser heat transfer coefficient for the inclined case is offset by the decreased evaporator heat transfer coefficient; and visa versa for the vertical thermosyphon.

3. Cooling of Nuclear Reactors

In section 3 we will consider a number of natural circulation loops or thermosyphons of potential importance for cooling (and heating) applications for nuclear reactors and associated technologies.

3.1. Water-Cooled Nuclear Reactor Cooling and Heat Removal System

In section 3.1 we will consider the case of an entirely passive, inherently safe nuclear reactor heat removal system. We will present it as a relatively comprehensive case study. It will focus on the applicable background information, the basic underlying theory relating to natural circulation loops, two-phase flow, and flow-pattern and flow-behavior characterization. A theoretical simulation model will be formulated. A computer program solution algorithm capable of capturing the complex transient and dynamic behavior of a natural circulation loop will be given, and finally, theoretical and experimental results will be given.

3.1.1. Background Information

As of July 2015, 30 countries worldwide were operating 438 nuclear reactors for electricity generation and 67 new nuclear plants were under construction in 15 countries. Nuclear power plants provided 10.9 percent of the world's electricity production in 2012 [13]. As such, nuclear power must be regarded not only as a proven technology and commercially viable power industry, but, in an age when global warming and pollution is becoming of universal concern, it must also be regarded as a potentially environmentally friendly source of energy. However, as a result of the widely publicized Three Mile Island, Chernobyl and Fukushima accidents, its public acceptance has been somewhat tarnished. In an attempt to make nuclear energy a more acceptable alternative to fossil fuel sources for the generation of electrical power and process heat, it thus appears justified to investigate nuclear power and heat solutions with the potential of a more convincing safety philosophy. GEYSER represents one such an approach. Vécseyy and Doroszlai [14] intends on fulfilling a number of basic requirements in favour of public acceptance. These design requirements relating to safety issues and analysis may include:

- Credible, convincing, passive safety based on simple natural laws (that is *inherently safe*),
- Safe engineering instead of engineering safety, involving simple deterministic processes instead of probabilistic considerations,
- Simple approved components and systems, wide-spread and well known technology,
- Simple unmanned operation with walkaway safety,
- Complete exclusion of human influence on safety related processes,
- Nuclear maintenance at site minimized to simple and safe replacement of the core, and
- Maximum available core life.

While none of these ideas are new, Vécsey and Doroszlai (1988) claim that their unique combination, as represented by the GEYSER concept, leads for the first time to an entirely passive light water reactor (LWR) system of high inherent safety.

The concepts advocated by Vécseyy and Doroszlai [14] have also led to the popularization of the concept of a small modular reactor (SMR) that is a reactor of less than 300 MW_e, and there is at present than more than 45 SMR designs under development in the world [15]. Of these, the Argentine developed system, CAREM, appears to be the most advanced from a construction point of view; concrete foundations having already been poured [16]. Other reactors in a relatively advanced stage of construction are the Korean 100 MW_e System Integrated Modular Advanced Reactor (SMART) which is the first integrated PWR having received certification. The China National Nuclear Corporation is developing the ACP100 design and will soon be submitting its preliminary safety analysis report to the National Nuclear Safety Commission. The Russian Federation is building two units of the KLT-40S series to be mounted on a barge and used for cogeneration of process heat and electricity. The construction is to be completed by the end of 2016 and expected electricity production in 2017. Most of these SMR designs make use of water as this allows for a natural circulation loop for the primary coolant heat removal system. Gas-cooled reactors on the other hand make use of blowers to circulate the primary coolant. To achieve high temperature without a high-pressure reactor containment barrier, the natural circulation of a liquid metal coolant may also be considered. One such concept is the Small Secure Transportable Autonomous Reactor (SSTAR) and is a lead-cooled natural circulation reactor with an 18 MW_e (45 MW_t) supercritical carbon dioxide (S-CO_2) Brayton cycle power conversion unit [17]. Another lead-cooled inherently safe reactor concept is DYONISOS (DYnamic Nuclear Inherently-Safe Reactor Operating with Spheres). In order to be inherently safe a nuclear reactor must have the following properties [18]:

i. The core must be critical only under normal heat transfer conditions; in all other situations it must be subcritical.
ii. Outside the envelope of design operation, even in the case of a loss-of-coolant accident (LOCA), the reactor system must be capable of removing decay heat from the fuel without the aid of active devices, solely by means of the final heat sink, the atmosphere and under natural forces, such as gravitation and thermal expansion [and surface tension].
iii. The reactor core should remain in a stable state subsequent to a severe accident, controlled only by natural forces, without the intervention of active devices and human operators. All possible configurations of the fuel outside the moderator structure must be sub-critical.
iv. Highly pressurized coolants and those which can generate high pressures following interactions during an accident must be avoided. This is because of natural hazards and pressure gradients which may result in the removal or movement of liquids and gases in undesirable and even unpredictable ways. They further intimate that design-dependent and desirable, but not essential, features of such a reactor are:
    a. The fuel should have a large surface-to-volume ratio, to facilitate decay heat removal.
    b. The thermal conductivity of the fuel ought to be high to achieve heat removal, even in the case of a loss of coolant accident (LOCA).
    c. The moderator should not generate gases or vapours, and should be chemically inert when in contact with the coolant, fuel, atmosphere and structural components.

d. The coolant should also be chemically inert, with a boiling point which is well above the operating design temperature.
e. In the case of a LOCA, the design of the core should enable decay heat to be removed by natural convection of the cover gas or incoming air over the fuel elements.
f. The vessel and the internal construction should have a high thermal conductivity to facilitate removal of heat following a LOCA.
g. The containment should allow unimpeded circulation of air within it, and have a thin wall of high thermal conductivity to maximize heat transfer to its outer surface.

Yadigaroglu and Zeller [19] considered a gravity- and flashing-driven natural circulation loop as shown in Figure 19. The height of the diffuser is 12 m and the diameter at the top 0.8 m. The system is filled with working fluid to a level just above the diffuser outlet which also protrudes a little into the steam drum. The fluid is heated in the reactor but to below its boiling temperature at the local hydrostatic pressure. As the water rises in the diffuser, the hydrostatic pressure decreases and the fluid starts boiling as soon as the local pressure corresponds to the saturated pressure corresponding to the fluid temperature; this vaporization process is commonly called *flashing*. The fluid in the steam drum is saturated and hence the fluid entering the down-comer is saturated; however, as it flows downward, the working fluid's degree of sub-cooling progressively increases.

Based on all the foregoing discussion, nuclear safety design criteria and the gravity- and flashing-driven natural circulation loop as depicted in Figure 19 is a concept for an inherently safe nuclear reactor thermal-hydraulic system and is shown schematically [19, 20]. Such a system might be termed entirely passive as no mechanical pumps and active electronic/electrical control devices are needed. Use is made however of a passive flow diverting valve. Should the pressure in the steam supply drum decrease below a certain level or its pressure or/and temperature increase beyond a certain predetermined high-level, the valve will automatically open and allow steam to flow into the air-cooled condenser loop. This valve and air-cooled loop would thus also serve as an automatic reactor decay heat removal initiation system in case of a normal or unplanned reactor shutdown. The secondary and tertiary fission product containments are also shown in Figure 20. If PBMR-type TRISO® coated nuclear fuel is used [21], the primary containment is inherent in the fuel itself by virtue of the high temperature resistant triple coated (a silicon carbide layer between two layers of pyrolytic carbon layers) UO_2 spherical fuel kernels.

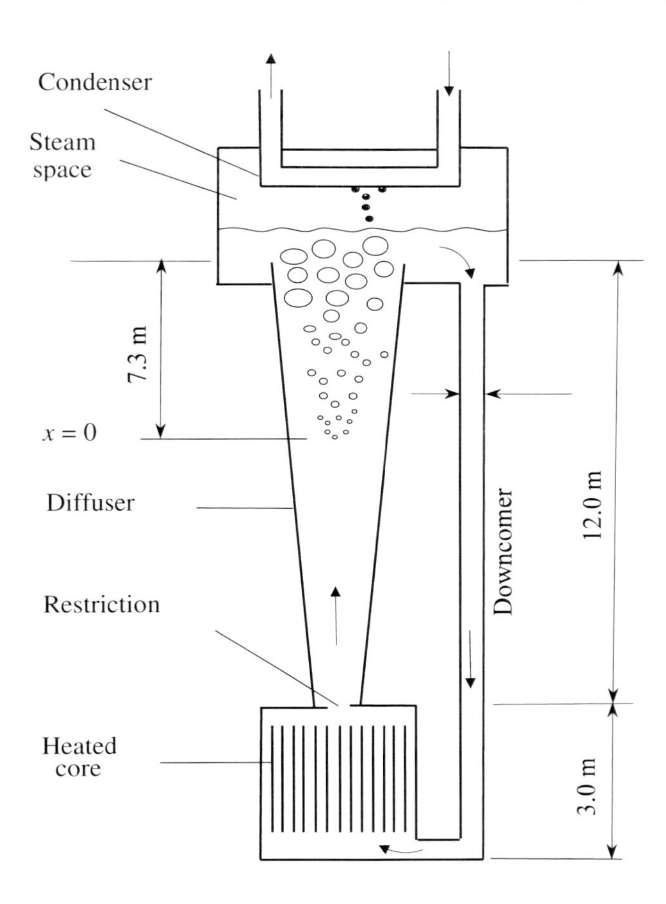

Figure 19. 10 MW heated gravity- and flashing-driven natural circulation loop [20].

In order to better understand the efficacy of the thermal-hydraulic behaviour of the tertiary looped and tertiary fission product contained nuclear reactor cooling system depicted in Figure 20, an experimental simulation model has been constructed. The basic details of this experimental model are shown by way of Figure 21. The height of the experimental model is limited by the laboratory height of 8.8 m. Also, sufficient height clearance between the air outlet cowling and the laboratory ceiling is needed so as not to hinder the natural convective flow though the natural draught air-cooled condenser and then out through an open skylight window. The reactor, flashtube riser and both steam drums have transparent polycarbonate observation ports. This has the advantage of allowing the two-phase flow to be visually observed, but this advantage limited the maximum temperature to 120°C and a maximum internal pressure to 200 $kPa_{absolute}$. For safety sake, both steam drums are fitted with pressure relief valves (PRV1 and PVR3). Air relief valves (AVRs) are fitted to the highest points of the loops to remove non-condensable gas (essentially air) from working fluid. Methanol is used as working fluid in the separated thermosyphon-type heat pipe secondary loop. Liquid working-fluid flow rates are measured using orifice plate differential pressure flow meters; thermocouples (T-type) temperature measuring points are indicated by "T"s.

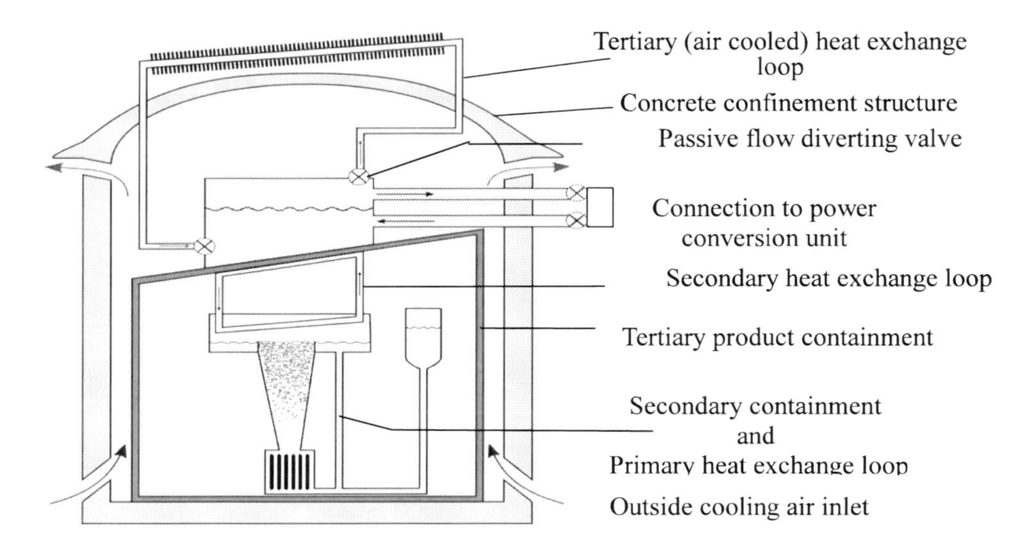

Figure 20. Schematic representation of an inherently safe passive Natural-circulation thermal-hydraulic heat transfer system for a nuclear reactor [20].

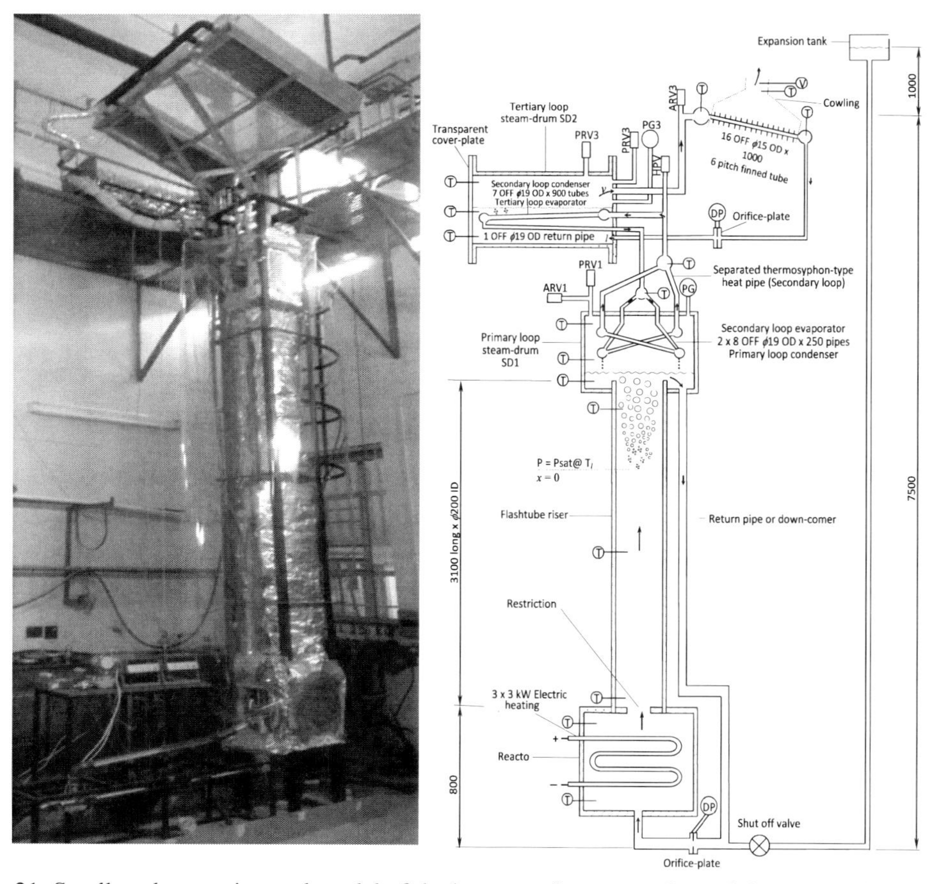

Figure 21. Small-scale experimental model of the heat transfer system for an inherently safe nuclear reactor steam supply system and consisting of a primary, secondary and tertiary natural circulation thermosyphon heat pipe systems. Dimensions in mm.

The expansion tank can be isolated using a shut off valve at the bottom of the system. This allows for three operating modes. Firstly, a single phase mode with valve open and allowing no boiling to occur. Secondly, a two-phase mode but leaving the valve open and operating the system at a maximum pressure of $P = P_{atm} + \rho gh$ where h in this case is 8.5 m. Thirdly, while operating in a two-phase mode and after having displaced a sufficient amount of liquid out of the steam drum and into the expansion tank, closing the shut off valve. In this way the expansion of the working fluid is taken up by the vapour space in the steam drum. In this operating mode the maximum pressure, at the bottom of the reactor, is the pressure of the vapour in the steam space plus ρgh, where h in this case is now 3.9 m. This mode of operation will be called the *heat pipe operating mode*. A more-detailed discussion and specification is given later on in the experimental-work section.

Having presented a background context for this section relating to water-cooled nuclear reactor cooling and heat removal systems, we will now consider in detail a single and two-phase flow natural circulation thermosyphon-type heat transfer loop with a large-diameter riser. Applicable literature relating to natural circulation loops and two-phase flow characterization will be presented.

3.1.2. Literature Survey/Study/Evaluation

There is a plethora of published literature relating to nuclear reactor heat transfer theory and not an inconsiderable amount of research has been conducted on natural circulation loops as applied in/to nuclear technology. This review will concentrate on published literature relating to single and two-phase flow in the primary, secondary and tertiary loops of the natural cooling in the reactor heat removal system pre-proposed by way of Figure 21. The following sub-sections will be considered: natural circulation loops, the so-called *instabilities* that occur in such loops and the expected two-phase flow patterns that may occur.

3.1.3. Natural Circulation Loops

Many geometric configurations of thermosyphons have been studied. Some of the shapes considered have been circular toroidal [22], arbitrary toroidal [23], double thermosyphon [24, 25] and rectangular [26, 27, 28]. Natural circulation loops in so far as they relate to reactor core heat removal systems however may be termed *integrated*, *separated*, or *heat pipe*. A natural circulation system may be said to be *integrated* when the circulation takes place entirely within a single containment vessel, for example the hot reactor core coolant flows upward inside the riser and then downward on the outside where it cooled by the secondary flow through the heat exchanger positioned between the riser and the reactor vessel; as shown schematically by way of Figure 22(a).

A great deal of published literature relates to the case of a rectangular loop as shown in Figure 22(d) but with the lower horizontal portion heated and the upper horizontal portion cooled. This geometry is of significance in the emergency cooling of many pressurized water reactors but the flow is prone to bifurcation and stability problems as the flow in a thermosyphon heated and cooled in this way has an equal chance of flowing either clockwise or counter clockwise. This case will not however be considered here as this configuration should be avoided by making sure that the heating and subsequent cooling portions are positioned in such a way to ensure that the desired flow direction is guaranteed; for instance by heating the lower portion of one vertical side and cooling the top of the other vertical side.

Natural circulation loops are finding applications in a wide variety of fields starting with large scale nuclear reactors [29, 30] to medium-scale chemical reactors [31] to the mini-scale of closed-loop pulsating heat pipes [32, 33].

The basic approach in theoretically simulating the two-phase flow is to discretize the loop into a series of parallel one-dimensional axially symmetrical parallel control volumes. The liquid and vapour in each control volume are both treated as two separate fluids in which the two phases are treated as two separate fluids, and the equations of change (conservation of mass, momentum and energy, and appropriate propertiy functions) are then applied to the vapour and liquid in cooling. This results in six conservation equations for each separate control volume. Computer simulation codes that emulate this approach are called state-of-the-art or *best-estimate* codes, and include a number of well-known codes such as RELAP, TRAC, MONO, etc, and they consider thermal non-equilibrium and allow for "slip" between the phases.

The disadvantage of this approach lies, beside the complexity of the numerical schemes and the computational efforts required in the need of several closure relations to couple the balance equations for the two phases. Closure relations are mostly empirical and flow-pattern dependent. The complexity of this type of models makes it difficult if not impossible to determine cause-effect relationships in the results of the simulations. On the other hand, state-of-the-art codes have the advantage of being quite general so that they can be used to simulate a wide range of geometrical configurations and a large variety of thermal-hydraulic problems [34].

Simpler approaches are however also possible; the simplest being the homogeneous equilibrium model (HEM). In this model the two-phase flow is imagined as a single fluid in which both phases are well-mixed and cannot be distinguished from each other and travel at the same velocity and has a density ρ given in terms of the mass fraction x as $\rho = \left(x/\rho_v + (1-x)/\rho_l\right)^{-1}$ (1, 35]. A homogeneous model is able to predict the occurrence of flashing. However once boiling commences, a void fraction α defined as the ratio of the vapour to liquid cross-sectional flow area A_v/A_l, is needed to more accurately predict the buoyancy term. In this case the vapour and liquid are considered to flow as two separately flowing fluids, and is called a *separated* two-phase flow model, and assumes that both phases are in thermal equilibrium with each other at any cross-sectional position in the loop. In this case the density is given as $\rho = \alpha\rho_v + (1-\alpha)\rho_l$ and the void fraction as $\alpha =$

$\left(1+\frac{v_v}{v_l}\frac{1-x}{x}\frac{\rho_v}{\rho_l}\right)^{-1}$ where $v_v/\,v_l$ may be given by an experimentally determined correlation [3].

Another one-dimensional approach is to use a so-called drift flux model in the conservation of momentum equation together with a void fraction and two-phase frictional multiplier to simulate density wave oscillations [37]. This model is relatively difficult to understand but the separated model appears to be more popular and according to Manera [34], the separated flow model is a good compromise taking into account complexity and its relative simplicity.

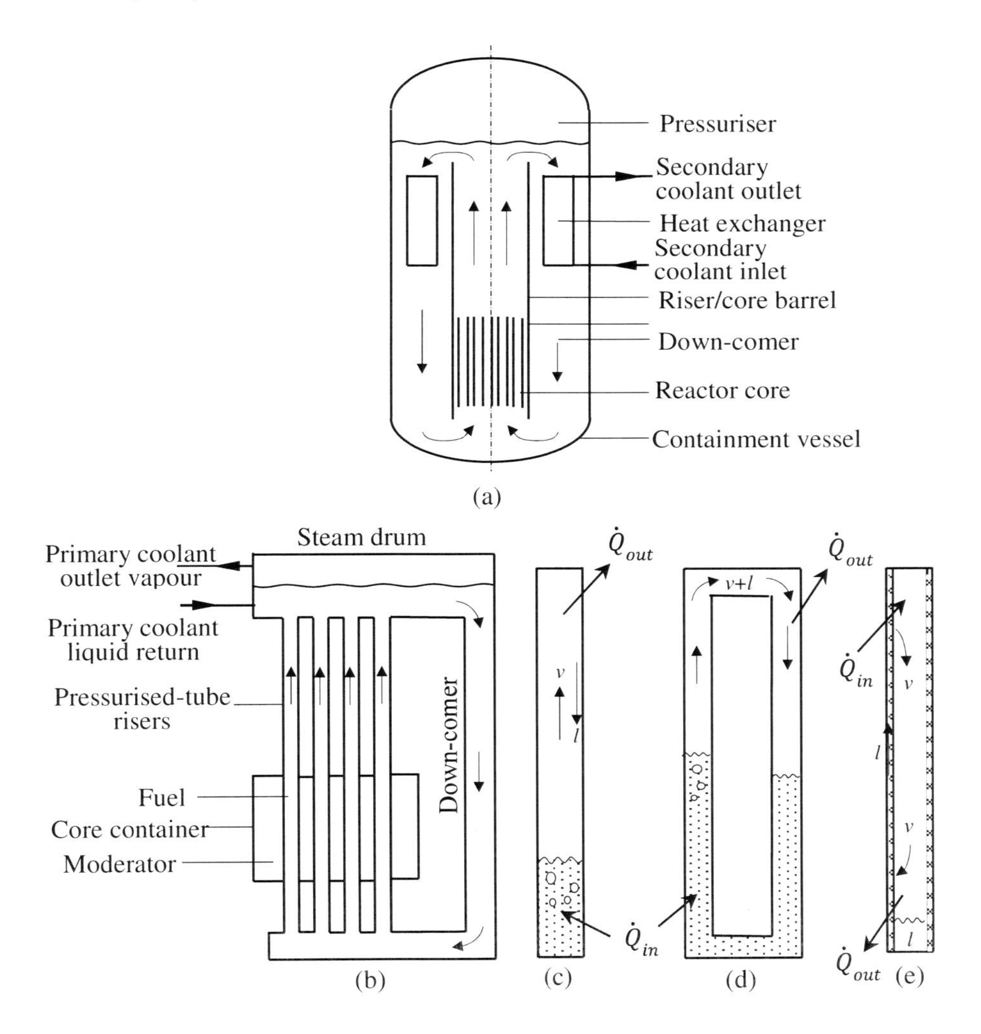

Figure 22. Natural circulation loops, integrated (a), separated (b), closed (c) and looped thermosyphon-type heat pipe (d) respectively and a surface tension capillary structure or wicked heat pipe (e).

3.1.4. Flow Instabilities

A vast amount of two-phase flow and heat transfer work has been done using various forms of the mass, momentum and energy conservation equations. Yadigaroglu and Lahey [38] intimate that studies have shown that, although it is possible to write exact conservation equations, the degree of complexity of such forms, and the large amount of detailed local and statistical information required for their modelling, prohibit their use in practical applications. Thus it is necessary to use simplified forms of the conservation equations for problems of practical significance. Further, a review of the presently published literature will reveal that it is unlikely that any two author's solution procedures of any two ostensible similar problems are precisely the same. Rather than establish a systematic review and appraisal of all the versions of the published mathematical formulations and solutions procedures and their mathematical formulations, the simulation model is given later in this section under "theoretical simulation model." It is essentially a separated flow model and according to Manera [34] includes sufficient complexity to fully take into account the essential physical phenomena encountered during both single and/or flashing conditions. Moreover, the number of empirical correlations (the two-phase frictional multiplier and the void fraction correlations) and cause-effect relationships can be more readily assessed.

A comprehensive review of recent advancements over the last two decades in two-phase natural circulation loops has been undertaken by Bhattacharyya, et al. (2012). It is clear from this review that the single most important research activity revolves around what is termed "instabilities"; a concept not of unimportance in so far as the perception of what constitutes the safe and reliable operation of nuclear reactors is concerned. In this regard instabilities are universally classified as either a *static* or a *dynamic instability*.

[Note that in the published literature, detailed flow behavior is more often than not summarily termed "flow instabilities." This is due (in the author's opinion) to the difficulty in correctly interpreting and simulating the complex transient and dynamic nature of two-phase flow. One might argue that in accordance with the complex or chaos theory that all natural phenomena are completely deterministic, whatever the many different outcomes may be provided, that the initial or initiating conditions are known.]

A static instability is defined by some as being when the operating conditions are changed by a small step from the original, it is not possible to regain a steady state close to the previous one. Vijayan and Nayak (2010) in their course notes on an introduction to instabilities in natural circulation systems further classify instabilities as depending on:

- Analysis method (or the governing equations of change used).
- Propagation method; here two instabilities are considered, density waves instability, and acoustic instabilities. Density wave instabilities occur when there is a time varying density around the loop due to the change in density (due to change in temperature and the consequent change in density due to thermal

expansion of the liquid in single phase flows and/or in two-phase flow the change in void fraction) around a loop. The frequency of these instabilities is proportional to the time it takes for a particle or small package (so to say) of fluid to flow around the loop and wherein in its transit around the loop its density also changes with time and at different axial positions around the loop. Acoustic instability on the other hand has been suggested as being a visually observable movement of a small package of fluid that is due to perturbations moving at the speed of sound in the fluid. For example, consider a 10 m circumferential loop; a density wave travelling at 5 m/s will have a frequency of about 0.5 cycles per second. On the other hand, changes in flow rate as manifested by the time rate of change of pressure difference across a flow meter are said to be acoustic instabilities if their time period is in the order of 10 m divided bu 1000 m/s being 0.01s or 100 cycles per second (assuming that the average speed of sound in the fluid as being somewhere near that of water and air of about 1500 m/s and 350 m/s, respectively or $(1500+350)/2 \approx 1000$ m/s).

- Nature and or number of unstable zones; depending on operating conditions and natural frequency of the fluid supporting pipes and structures the oscillations may be periodic or chaotic.
- Loop geometry; for example, symmetrically heated and cooled loops tend to be unstable.
- Disturbances; such as boiling inception, flashing, flow pattern transition, laminar to turbulent transitions, slip-stick that is the friction factor tends to infinity as the velocity tends to zero similar to the the slip-stick concept associated with the static and dynamic friction between solids).

Vijayan et al. [39] give an experimentally determinedi s*tability map* as shown in Figure 23 for a ϕ9.1 diameter pipe. Note that the pressure difference is across a flowmeter and hence is indicative of the volumetric flow rate in the loop. Note also that the heating method is by means of an electrical resistance heating element which, at a given power, is essentially a so-called *constant heat flux* heating load.

To get a better physical appreciation of a two-phase *buoyancy* driven flow, consider a simple small-diameter air-lift water pump operating at different submergence *h*/*L* ratios as shown in Figure 23(b). As the air flow rate increases so too does the water flow rate, and an air flow rate of about 5 m^3/ hour, in this case, the water flow reaches a maximum at the different submergence ratios. At this air flow rate, the flow pattern is essentially a plug flow as depicted in Figure 24(a) and consists of relatively large bullet-shaped bubbles flowing at a relatively low frequency. As the bubbles rise, water is displaced upwards intermittently in pulses corresponding more-or-less to the flow rate of the bubbles. Up until this point the flow is said to be buoyancy driven or gravity dominated. If the flow rate is measured using an electronic flow metering device and the plotted graphically as a function of time, it is seen to flow in an oscillatory wave-like manner at a frequency of about 1 Hz;

this region is thus also called the *density wave* (or type-I instability) region. As the air flow rate increases, the water flow rate further decreases; in this operating region of the air-lift pump, the flow is said to be friction dominated.

Referring back to Figure 23 for what is, shall we say, a wall heat flux driven flow; as the heat flux is slowly increased, the flow rate exhibits a single phase start-up transient in which the flow rate increases slowly up to a maximum before falling down and after a number smaller oscillations settles down and increases proportional to the slowly increasing heat flux. As the temperature slowly increases in the heating section, vapour bubbles start to form and the flow rate enters a single to two-phase transition (shown in Figure 23 as the density wave region), but on a further increase in power, the amplitude of the oscillations tend to reduce and up to a certain power the flow is relatively stable with only small oscillations. However at a certain higher power level, the flow is again oscillatory, the flow tends to decrease but the amplitude of the oscillations increase dramatically; the frequency of the oscillations in this region are in the order of about 100 Hz which also tends to correspond with the natural frequency of sound waves of the fluid flow, hence they may be termed acoustic waves. On increasing the heat flux further, the amplitude of the oscillations tend to decrease but the flow rate now increases again; this region is termed the stable two-phase flow region. [Actually the flow at these high heat fluxes tends to become separated with a thin liquid film around the inside periphery of the pipe and a relatively high speed vapour flowing in the core of the pipe.]

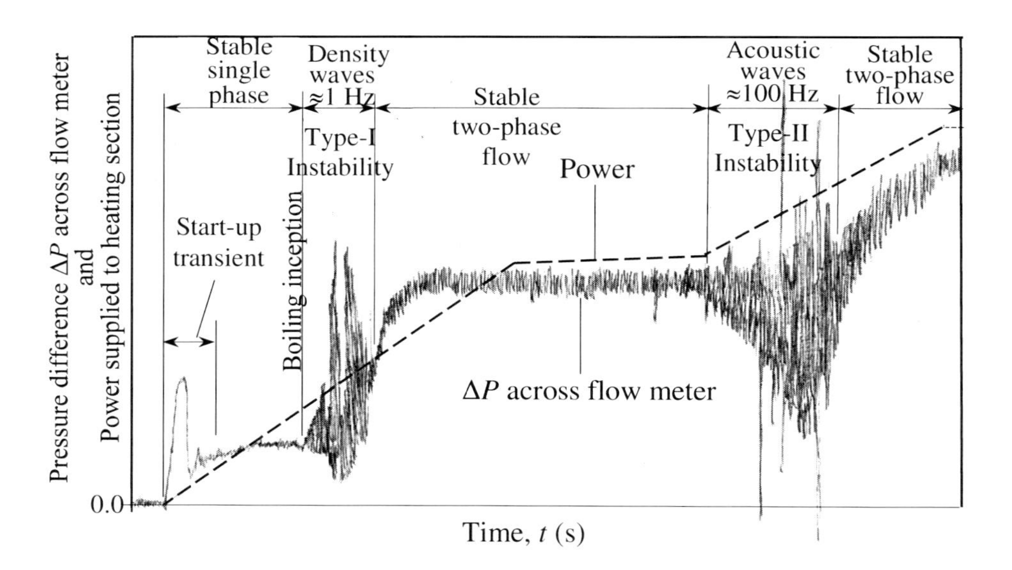

Figure 23. Qualitative representation of the pressure difference ΔP across the flow meter and the power supplied to heating section as a function of time. (Adapted from [39]).

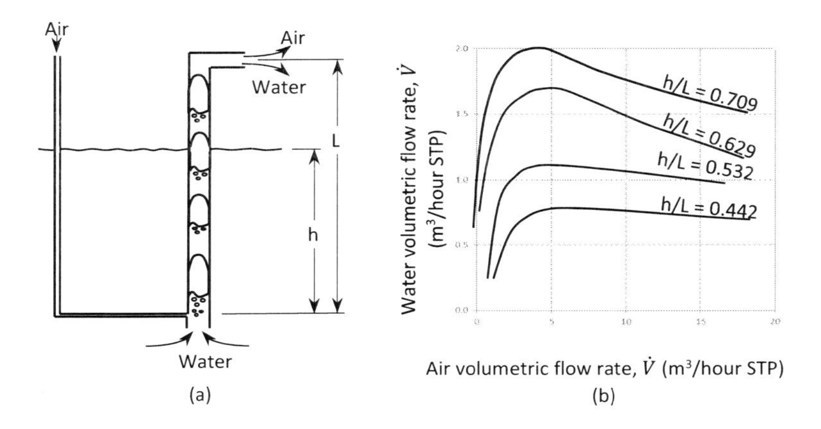

Figure 24. Experimentally-determined water flow rate dependence on air flow rate at different submergence ratios *h/L* (b) for a simple water air-lift pump (a) [42].

The pipe diameters of the loops that Vijayan et al. [39] tested varied from 6 to 40 mm loops. For larger diameter pipes, say from 200 mm upwards, two-phase flow and instability behaviour cannot be extrapolated. Very little published literature is available for larger diameter (> 70 mm) tubes, especially with water and steam. Yoneda et al. [40, 41] considered a 155 mm diameter pipe of about three meters in length as depicted schematically in Figure 24.

3.1.5. Flow Pattern Characterisation

Flow patterns in small diameter pipes (that is if the Bond number $Bo = \left(g(\rho_l - \rho_v)L_c^2\right)/\sigma$ is less than 40 or about 70 mm, for water) are typically characterised in terms of so-called bubbly, plug (or slug), churn and annular flow patterns as shown in Figure 25 by way of sample photographs and stylised sketches. For spherical bubbles, Whalley [1] derives an equation (by equating the buoyancy and drag forces acting on the bubble) for the terminal rising velocity, for low Reynolds numbers of less than one, as $u_b = \frac{d_b^2\, g(\rho_l - \rho_v)}{12\mu_l}$.

Larger bubbles are however not spherical, nor do they obey Stokes' law, but Whalley [1] gives a useful *map* for the characterization of bubble-type flow in what he calls "intermediate-sized" diameter pipes in terms of the Bond (or Eötvös) *Bo*, Reynolds *Re* and Mouton *M* numbers, respectively as:

$$Bo = \left(g(\rho_l - \rho_v)L_c^2\right)/\sigma,\ Re = \left(\rho_l u_b d_e\right)/\mu_l \text{ and } M = g\mu_l^4(\rho_l - \rho_v)/\left(\rho_l^2\sigma^3\right)$$

where in this case d_e is the equivalent diameter in m and is the diameter the bubble would have if it were spherical, and the surface tension σ is in N/m. The rising velocity of the bubble can therefore be found by calculationg the Mourton and Bond numbers, looking up the Reynolds number in Figure 26 and hence the calculating rising velocity u_b.

3.1.6. Large Diameter Pipes

Very little published literature is available for larger diameter (> 70 mm) tubes, especially with water and steam. Yoneda et al. [40, 41], however, considered a 155 mm diameter pipe of about three meters in length. He found, based on a statistical analysis of experimental data, that the terminal velocity could be well-correlated in terms of the water viscosity and the water-steam surface tension, and that the flow reached a quasi- developed state within a relatively short height to diameter aspect ratio of about four.

3.1.7. Theoretical Simulation Model

In this section we consider, by way of a specific example, the theoretical simulation of the natural circulation loop as schematically depicted in Figure 27. This figure is essentially a schematic drawing showing the discretization into discrete control volumes of the primary circulation loop that is shown in Figures 19 and 20 of an as conceptually envisaged natural circulation nuclear reactor cooling system.

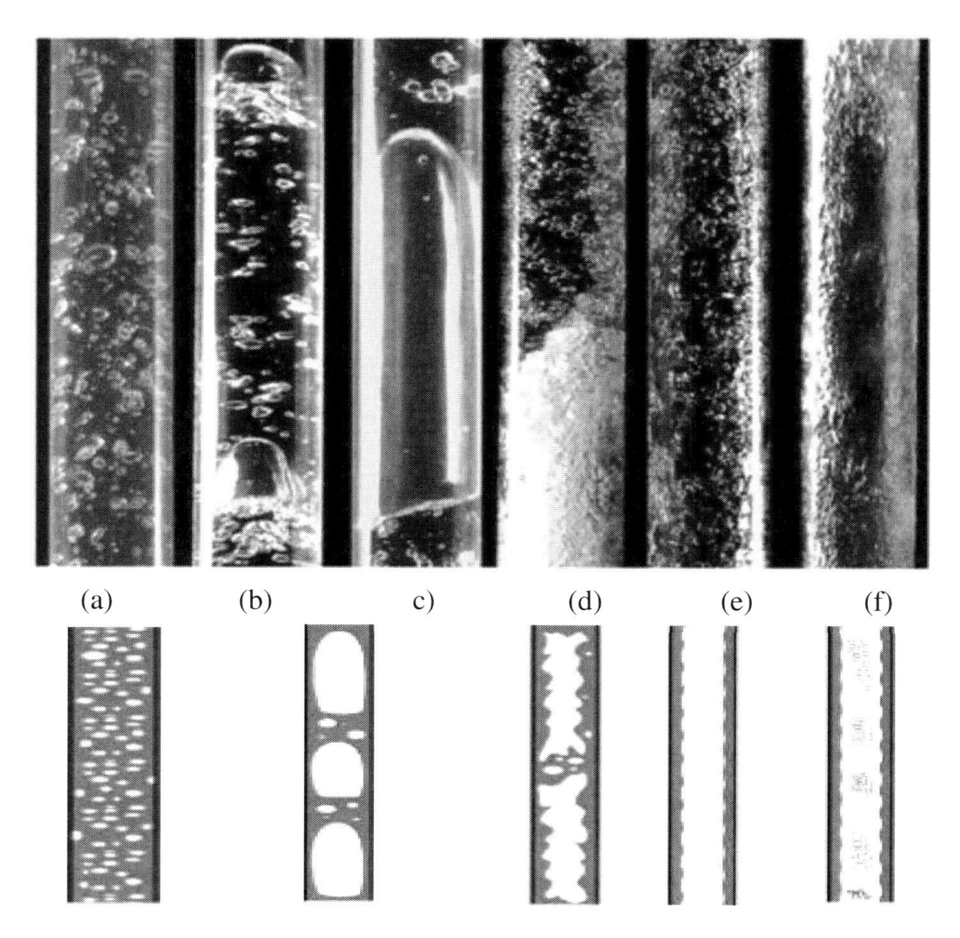

Figure 25. Two-phase flow patterns, (a) bubbly flow, (b) and (c) bubbly, small plugs and plug flow, (d) churn flow, (e) annular, wispy-annular (f).

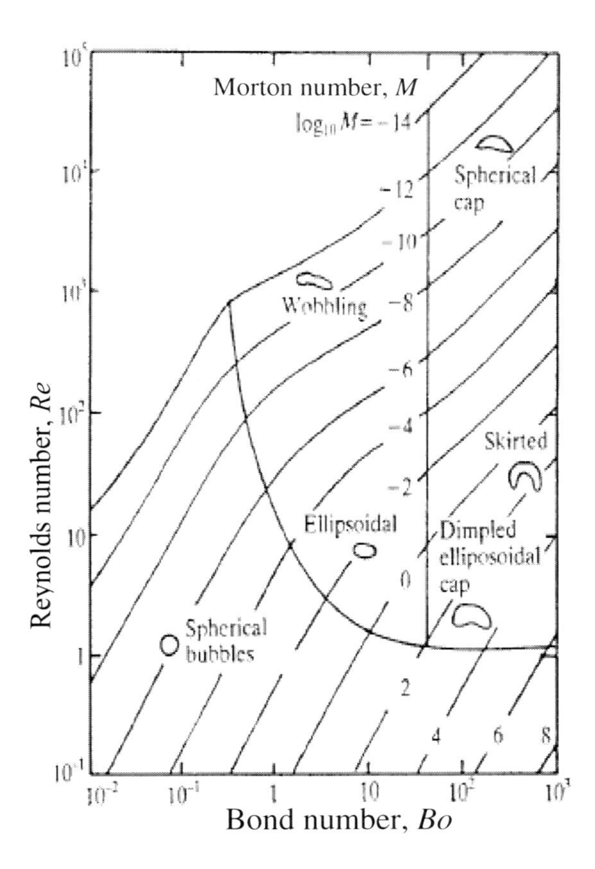

Figure 26. Reynolds, Bond and Mouton numbers for single rising bubbles [1].

3.1.7.1. Thermal-hydraulic Theoretical Simulation Assumptions

The system as schematically depicted in Figure 27(a) may be discretized into a number of cylindrical-shaped control volumes as shown in Figure 27(b). Figure 28 shows schematically, the application of the conservation of mass, momentum and energy to the ith control volume shown Figure 27(b). The assumptions made in modelling the behavior of this natural circulation loop are:

i. The cylindrically-shaped control volumes are one-dimensional; that is, at any cross-section area along the axis of flow $\dot{m} = \int_0^R \rho v 2\pi r dr = \rho v A$, where $A = \pi R^2$.

ii. At any instant in time the mass flow rate in the loop is independent of its position in the loop; this implies that at any instant in time that $\frac{\partial \dot{m}}{\partial z} = 0$ but $\frac{\partial \dot{m}}{\partial t} \neq 0$; this condition is often called *quasi-equilibrium* or *quasi-static* equilibrium. For this assumption to be valid the fluid particle velocity needs to be very much slower than the speed of sound in the fluid; this implies that at any instant in time the fluid, both the liquid and vapour, is essentially incompressible and that as steam forms in the control volume, liquid is displaced instantaneously out of the control volume; this displaced liquid, and its internal energy is then assumed to find itself instantaneous and directly into the steam drum without its internal energy having changed. In so doing, the liquid finds its way into the steam drum without affecting the energy of the control volumes between the control volume in which it was formed and the steam drum; and visa verse if steam condenses.

iii. At any cross-section, both the liquid and vapour phases are in thermodynamic equilibrium with each other.

iv. The amount of liquid in the steam drum is "large" and as such the liquid displaced into it as outlined in iii) above does not affect its energy. Neither does it affect the height of the liquid in the steam drum; that is, ΔL_{sdl} in Figure 27(a) is essentially zero at all times.

v. A so-called *separated two-phase flow model* [3] may be used to represent the properties and behaviour of the control volumes containing both liquid and steam. Such a two-phase model requires for a control volume that $m = m_v + m_l, \dot{m} = \dot{m}_v + \dot{m}_l, \dot{m}_v = \rho_v v_v A_v$ and $\dot{m}_v = \rho_v v_v A_v$, a mass fraction $x = \frac{m_v}{m_v+m_l} = \frac{\dot{m}_v}{\dot{m}_v+\dot{m}_l}$, and a density $\rho = \alpha\rho_v + (1-\alpha)\rho_l$ where the so-called void fraction α is defined by $\alpha = \frac{V_v}{V_v+V_l} = \frac{\dot{V}_v}{\dot{V}_v+\dot{V}_l} = \frac{A_v}{A_v+A_l}$. The void fraction may also be expressed as $\alpha = \left(1 + \frac{v_v}{v_l}\frac{1-x}{x}\frac{\rho_v}{\rho_l}\right)^{-1}$. The ratio v_v/v_l is termed the *slip factor* and needs be given as an experimentally correlated expression. One such correlation (Chisholm, 1983) for the void fraction for a two-phase water system is given by the Lockart-Martinelli correlation as $\alpha = (1 + 0.28X)^{-1}$ where X, the so-called Martinelli parameter, is given as $X = 1 + \left(\frac{\rho_v}{\rho_l}\right)^{0.5}\left(\frac{\mu_l}{\mu_v}\right)^{0.125}\left(\frac{1-x}{x}\right)^{0.875}$. Further, an experimentally determined correlation for a so-called frictional multiplier ϕ is needed. One such correlation, *the Lockart-Martinelli liquid-only two-phase frictional multiplier* is given as $\phi_{lo}^2 = \left(1 + \frac{20}{X} + \frac{1}{X^2}\right)(1-x)^{1.75}$ and in this case it implies that there is a *liquid only* velocity defined as $v_{lo} = \frac{\dot{m}_v+\dot{m}_l}{\rho_l(A_v+A_l)}$ and with a *liquid-only* Reynolds number given as $Re_{lo} = \rho_l\left(\frac{\dot{m}_v+\dot{m}_l}{\rho_l(A_v+A_l)}\right)d_h/\mu_l = \dot{m}'' d_h/\mu_l$ where the $d_h = 4A\,/\,\wp\Delta z$, $\wp$ being the *wetted* perimeter and $\dot{m}'' = \dot{m}/A$ is the mass flux. These definitions allow the shear stress to be expressed in terms of a liquid-only shear stress τ_{lo} and a two-phase liquid-only frictional multiplier ϕ_{lo}^2 as $\tau = \tau_{lo}\phi_{lo}^2$ where $\tau_{lo} = C_{flo}\frac{\dot{m}^2}{2\rho_l A^2}$. A Blasius-type liquid-only coefficient of friction (for a smooth tube) may be given as $C_{flo} = 0.079Re_{lo}^{-0.25}$ for turbulent flow and $C_{flo} = 16/Re_{lo}$ for laminar flow, and to ensure non-division, by zero if

$Re_{lo} < 1$ as, say $C_{flo} = 16$. [Note that by taking the Reynolds number as 1181 for the transition from turbulent to laminar flow, and visa verse, ensures that coefficient of friction C_{flo} is a continuous function of Re_{lo}, if deemed appropriate. Maybe one would what to build in some "extra complexity" into the simulation model by introducing some sort of hysteresis, assuming, say, that laminar flow becomes turbulent at $Re = 2300$ and that turbulent flow changes to laminar flow at $Re = 1800$].

vi. Axial heat conduction in the fluid and in the pipe walls is neglected. It is assumed that radiation and conduction heat loss to the environment is taken into account by including these losses in an appropriately adjusted convective heat coefficient. It is further assumed that the volume of the control volume boundaries are indeed rigid and hence their volumes are independent of pressure and thermal expansion.

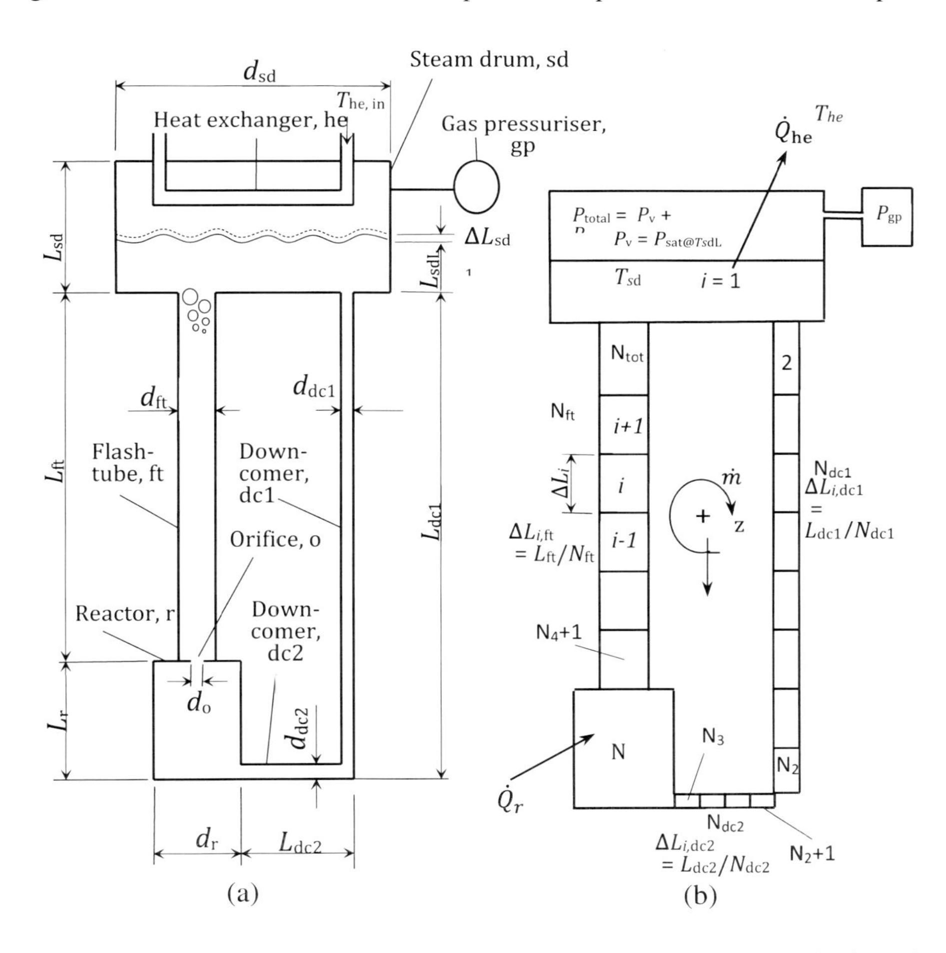

Figure 27. Flash-tube-type nuclear reactor cooling system basic geometry (a) and discretisation scheme (b).

3.1.7.2. Equations of Change

The equations of change are obtained by applying the general statements of the conservation of mass, momentum and energy to the applicable finite sized control volumes depicted in Figure 28. In solving for the set of time dependent difference equations, an explicit numerical solution method will be used with so-called *upwind* differences.

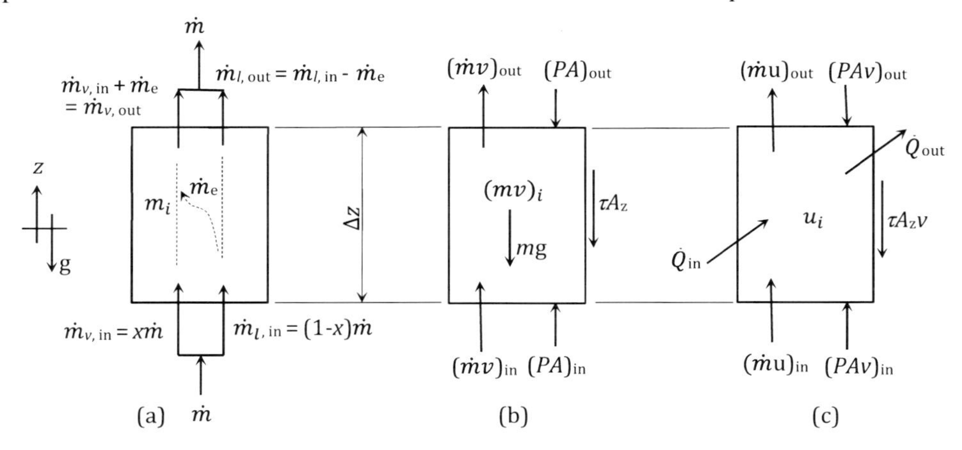

Figure 28. Conservation of mass (a), momentum (b), energy (c) as applied to a representative control volume.

3.1.7.3. Conservation of Mass

Apply the conservation of mass to the control volume given in Figure 28(a) to get:

$$\frac{\Delta m}{\Delta t} = \dot{m}_{in} - \dot{m}_{out} \qquad \text{[kg/s]} \tag{39}$$

or, explicitly as:

$$m_i^{t+\Delta t} = m_i^t + \Delta t(\dot{m}_{in} - \dot{m}_{out})_i^t \tag{40}$$

The density of the control volume is thus readily determined knowing the volume of the control volume V_i as:

$$\rho_i^{t+\Delta t} = m_i^{t+\Delta t}/V_i \tag{41}$$

For a two-phase control volume and in terms of the mass fraction x and void fraction α, by:

$$\dot{m} = \dot{m}_v + \dot{m}_l = x\dot{m} + (1-x)\dot{m} \tag{42}$$

$$V = V_v + V_l = \alpha V + (1-\alpha)V \tag{43}$$

$$\rho = \alpha\rho_v + (1-\alpha)\rho_l \tag{44}$$

3.1.7.4. Conservation of Energy

Apply the conservation of energy, in terms of the so-called *thermal energy* equation, as formally derived from the more commonly used *total energy* equation as given in Appendix A, to the control volume in Figure 28(c), to get:

$$\frac{\Delta}{\Delta t}(mu) = (\dot{m}u)_{in} - (\dot{m}u)_{out} + \dot{Q}_{in} - \dot{Q}_{out} + (PA\,v)_{in} - (PA\,v)_{out} - \tau_w A_z v \quad \text{[W]} \tag{45}$$

or, explicitly as:

$$u_i^{t+\Delta t} = \left[(mu)_i^t + \Delta t \Sigma \dot{E}_i^t\right] / m_i^{t+\Delta t} \tag{46}$$

where

$$\Sigma \dot{E}_i^t = \left[(\dot{m}u)_{in} - (\dot{m}u)_{out} + \dot{Q}_{in} - \dot{Q}_{out} + (PA\,v)_{in} - (PA\,v)_{out} - \tau_w A_z v\right]_i^t \quad \text{[W]} \tag{47}$$

Having now $u_i^{t+\Delta t}$ and $\rho_i^{t+\Delta t}$ the remaining thermodynamic properties may then readily be determined, given the appropriate property equations as functions of internal energy and density $f\left(u_i^{t+\Delta t}, \rho_i^{t+\Delta t}\right)$.

In the numerical solution algorithm computer program (see later), we have elected to use the following procedure. In this procedure:

$$m_i^{t+\Delta t} = m_i^t + \Delta t \left(\frac{\Delta m}{\Delta t}\right)_i^{t-\Delta t/2} \tag{48}$$

where

$$m_i^{t+\Delta t} = \left(\frac{\Delta m}{\Delta t}\right)_i^{t-\Delta t/2} = \frac{m_i^t - m_i^{t-1}}{\Delta t}$$

and

$$P_i^{t+\Delta t} = P_i^t + \Delta t \left(\frac{\Delta P}{\Delta t}\right)_i^{t-\Delta t/2} \tag{49}$$

where

$$\left(\frac{\Delta P}{\Delta t}\right)_i^{t-\Delta t/2} = \frac{P_i^t - P_i^{t-1}}{\Delta t}$$

Having now determined $u_i^{t+\Delta t}$ using Eq. (46) with $m_i^{t+\Delta t}$ as given by Eq. (40) then the remaining thermodynamic properties may be determined as functions of $u_i^{t+\Delta t}$and $P_i^{t+\Delta t}$, that is $T_i^{t+\Delta t} = f\left(u_i^{t+\Delta t}, P_i^{t+\Delta t}\right)$ and $\rho_i^{t+\Delta t} = f\left(u_i^{t+\Delta t}, P_i^{t+\Delta t}\right)$, for a so-called sub-cooled or superheated single phase control volume. For a two-phase liquid-vapour control volume the mass fraction $x_i^{t+\Delta t}$ is given as:

$$x_i^{t+\Delta t} = \left(u_i^{t+\Delta t} - u_{f,i}^{t+\Delta t}\right) / \left(u_{g,i}^{t+\Delta t} - u_{f,i}^{t+\Delta t}\right)$$

where $u_{f,i}^{t+\Delta t} = f\left(P_{sat,i}^{t+\Delta t}\right)$ and $u_{g,i}^{t+\Delta t} = f\left(P_{sat,i}^{t+\Delta t}\right)$,

and the volume or void fraction $\alpha = V_v/V_l$ is given by:

$$\alpha_i^{t+\Delta t} = \left(1 + SF\frac{\rho_{g,i}^{t+\Delta t}}{\rho_{f,i}^{t+\Delta t}}\frac{1 - x_i^{t+\Delta t}}{x_i^{t+\Delta t}}\right)^{-1} \tag{50}$$

where SF is the so-called slip factor, being the ratio of the vapour to liquid velocities v_v / v_l and given by an experimentally determined correlation in terms the vapour and liquid densities and dynamic viscosities and the mass fraction and surface tension [3]. Another somewhat simpler correlation (Martinelli, 19xx) is given as a function, in terms of the so-called Martinelli parameter X, as:

$$\alpha = \left(1 + 0.28X^{0.71}\right)^{-1} \tag{51}$$

where $X = \left(\frac{\rho_v}{\rho_l}\right)^{0.5}\left(\frac{\mu_l}{\mu_v}\right)^{0.5}\left(\frac{1-x}{x}\right)^{0.875}$

Having the mass and void fractions x and α, the density and temperature at $t + \Delta t$ (and $u_i^{t+\Delta t}$ and $P_i^{t+\Delta t}$) the remaining thermodynamic properties are then necessarily given by:

$$\rho_i^{t+\Delta t} = \alpha_i^{t+\Delta t}\rho_{g,i}^{t+\Delta t} + \left(1 - \alpha_i^{t+\Delta t}\right)\rho_{f,i}^{t+\Delta t} \tag{52}$$

where $\rho_{f,i}^{t+\Delta t} = f\left(P_{sat,i}^{t+\Delta t}\right)$ and $\rho_{g,i}^{t+\Delta t} = f\left(P_{sat,i}^{t+\Delta t}\right)$

and

$$T_i^{t+\Delta t} = f\left(P_{sat,i}^{\,t+\Delta t}\right) \tag{53}$$

Note that although neglected, the axial conduction in the fluid could readily have been included as additional heat transfer rate terms $\dot{Q}_{\text{conduction,in}} = \frac{T_{i-1}-T_i}{\frac{\Delta z_{i-1}+\Delta z_i}{2kA}}$ or $\dot{Q}_{\text{conduction,out}} = \frac{T_i-T_i+1}{\frac{\Delta z_i+\Delta z_{i+1}}{2kA}}$ where A is the smaller of A_{i-1} and A_i or A_i and A_{i+1}, respectively.

The term $[(PA\,v)_{in} - (PA\,v)_{out}]$ is the reversible rate of work done on the fluid and is positive if the fluid in the control volume is being compressed; and negative if the fluid in the control volume is expanding. The term $[-\tau_w A_z v]$ is the work done against friction. It is irreversible and thus necessarily always positive and thus always manifests itself as an increase in the temperature of the fluid.

For a separated flow model the equations are somewhat more elaborate. The left hand term in Eq. (47) becomes:

$$\begin{aligned}\frac{\Delta}{\Delta t}(mu) = \frac{\Delta}{\Delta t}(m_v u_v + m_l u_l) &= \frac{\Delta}{\Delta t}(xmu_v + (1-x)mu_l) \\ &= \frac{\Delta}{\Delta t}\big(m(xu_v + (1-x)u_l)\big) \\ &= \frac{\Delta}{\Delta t}(mu)\end{aligned} \tag{54}$$

where $u = xu_v + (1-x)u_l$

The convective energy flow terms on the right hand side becomes:

$$\begin{aligned}(\dot{m}u)_{in} - (\dot{m}u)_{out} = x\dot{m}u_{v,in} + (1-x)\dot{m}u_{l,in} - x\dot{m}u_{v,out} \\ -(1-x)\dot{m}u_{l,out} = \dot{m}(u_{in} - u_{out})\end{aligned} \tag{55}$$

where $u = xu_v + (1-x)u_l$

The heat transfer rate terms on the right hand side of Eq. (47) include the thermal conduction along the flow path, heat loss/gain by a temperature difference driven convection through the solid boundary, by a wall heat flux through the boundary, as by way of an example an electrical resistance heating element and a heat transfer rate by a heat pipe in the control volume, for example the heat pipe in the steam drum.

The reversible work terms on the right hand side of Eq. (45) become:

$$\begin{aligned}(PA\,v)_{in} - (PA\,v)_{out} &= \big(P(A_v v_v + A_l v_l)\big)_{in} - \big(P(A_v v_v + A_l v_l)\big)_{out} \\ &= \left[P\left(\alpha A\frac{x\dot{m}}{\rho_v \alpha A} + (1-\alpha)A\frac{(1-x)\dot{m}}{\rho_l(1-\alpha)A}\right)\right]_{in}\end{aligned}$$

$$-\left[P\left(A_v\frac{x\dot{m}}{\rho_v\alpha A}+A_l\frac{(1-x)\dot{m}}{\rho_l(1-\alpha)A}\right)\right]_{out}$$

$$=\dot{m}\left[\left(P\left(\frac{x}{\rho_v}+\frac{1-x}{\rho_l}\right)\right)_{in}-\left(P(x/\rho_v+(1-x)/\rho_l)\right)_{out}\right]$$

$$=\dot{m}\left[(P/\rho_h)_{in}-(P/\rho_h)_{out}\right] \quad (56)$$

where $\rho_h = x/\rho_v + (1-x)/\rho_l{}^{-1}$

The irreversible work done on the fluid as a result of the friction term on the right hand side may be given as $-\tau_w A_z v = -\tau_{lo}\phi_{lo}^2 A_z v$ where $\tau_{lo} = C_{flo}\frac{1}{2}\rho_l\left(\frac{\dot{m}}{\rho_l A}\right)^2$, for a cylindrical control volume $A_z = \wp(\Delta z + z_{\mathrm{minor}})$ where z_{minor} is the equivalent length of pipe of the same diameter and represents the so-called *minor losses* or the irreversible work done on the fluid as a result of there being inlets or contractions, bends, outlets or expansions, etc. Typical values for these losses, for an inlet, outlet and elbow may be taken as 18*d*, 30*d* and 52*d*, respectively. The average velocity, to keep it simple, may be taken as $v = \frac{\dot{m}}{((A\rho)_{in}+(A\rho)_{out})/2}$ where, $\rho = \alpha\rho_v + (1-\alpha)\rho_l$. The term ϕ_{lo}^2 is experimentally-established and is termed the *liquid-only two-phase frictional multiplier*. The friction coefficient C_{flo} is, conveniently, given by $C_{flo} = 0.079Re_{lo}^{-0.25}$ for turbulent flow and $C_{flo} = 16/Re$ for laminar flow. Note that if the simulation of the fluid movement is started from standstill, the Reynolds number is infinitely large and hence so too is the coefficient and there will be no flow. There will of course be flow and there will also be a start-up transient as indicated and seen in Figure 23. The so-called slip-stick theory describing this phenomenon is complex but for the sake of simplicity may be arbitrarily determined by assuming for instance that if *Re* < 1 then, say, $C_{flo} = 16$. Further, the transition between laminar and turbulent flow may be taken as *Re* = 1181; at this Reynolds number the laminar and turbulent coefficients of friction as given by the two equal preceding equations for the coefficient of friction are equal to each other. Another facet of this transition is that there is a definite degree of hysteresis associated with transition. Text normally quoted shows that the transition takes place when the Reynolds number is between 1800 and 2300. However if the flow is laminar, then as the velocity is increases there is a rather sudden transition to turbulent flow at a Reynolds number closer to the upper limit. However if the flow is turbulent and the velocity is decreased, a similarly-sudden transition to laminar flow occurs, but at a Reynolds number closer to the lower of the two limits.

For a homogeneous two-phase flow model, the frictional multiplier is given as $\phi_{lo}^2 = \frac{C_{fh}}{C_{flo}}\frac{\rho_l}{\rho_v}$, but if it is assumed that $C_{fh} = C_{flo}$ then (Whalley, 1990):

$$\phi_{lo}^2 \approx \rho_l/\rho_v \tag{57}$$

For two-phase flow, a frictional multiplier may be given (Chisholm, 19xx) as:

$$\phi_{lo}^2 = \left(1+\frac{20}{X}+\frac{1}{X^2}\right)(1-x)^{1.75} \tag{58}$$

where, once again, the Martinelli parameter is given as $X = \left(\frac{\rho_v}{\rho_l}\right)^{0.5}\left(\frac{\mu_l}{\mu_v}\right)^{0.125}\left(\frac{1-x}{x}\right)^{0.875}$

3.1.7.5. Conservation of Momentum

Apply the conservation of momentum to the control volume depicted Figure 28 (b) to get:

$$\frac{\Delta}{\Delta t}(mv) = (\dot{m}v)_{in,i} - (\dot{m}v)_{out,i} + (PA)_{in,i} - (PA)_{out,i} - m_i g\sin\theta_i - \tau A_z \tag{59}$$

for a single phase flow; or using $v = \dot{m}/\rho A$ and dividing throughout by the cross sectional area A, as:

$$\frac{\Delta}{\Delta t}\left(\frac{m\dot{m}}{A\rho}\right) = \left(\frac{\dot{m}^2}{A\rho}\right)_{in,i} - (\dot{m}v)_{out,i} + (PA)_{in,i} - (PA)_{out,i} - m_i g\sin\theta_i - \tau A_z$$

For a separated two-phase flow model, after dividing throughout by the cross sectional area A of the control volume and making use of the identities $\rho = \alpha\rho_v + (1-\alpha)\rho_l$, $v = \dot{m}/\rho A$ the momentum equation may be expressed in terms of the mass, mass flow rate, void fraction and mass fraction as:

$$\frac{\Delta}{\Delta t}\left[\frac{m\dot{m}}{A^2}\left(\frac{x^2}{\alpha\rho_v}+\frac{(1-x)^2}{(1-\alpha)\rho_l}\right)\right]_i = \frac{\dot{m}^2}{A_i}\left[\left(\frac{x^2}{\alpha A\rho_v}+\frac{(1-x)^2}{(1-\alpha)A\rho_l}\right)_{in} - \left(\frac{x^2}{\alpha A\rho_v}+\frac{(1-x)^2}{(1-\alpha)A\rho_l}\right)_{out}\right]$$
$$+P_{in,i} - P_{out,i_i} - \left[\left(\alpha\rho_v + (1-\alpha)\rho_l\right)g\sin\theta\Delta z\right]_i - \left[\frac{\tau_{lo}\phi_{lo}^2 A_z}{A}\right]_i \tag{60}$$

where $m = \left(\alpha\rho_v + (1-\alpha)\rho_l\right)A\Delta z$ and in accordance with the concept of upwind differencing for the momentum flux term (the first term on the right hand side of Eq. (59), "in" $= i-1$ if $\dot{m} \geq 0$ and $in = i+1$ if $\dot{m} < 0$, and "out" always equals "i"; Note that, for convenience, the mass m in the left hand term has not been expressed in terms of x and α; had it been however, the time dependent term on the left hand side of Eq. (59) could have be written as:

$$\frac{\Delta}{\Delta t}\left[\frac{m\dot{m}}{A^2}\left(\frac{x^2}{\alpha\rho_v}+\frac{(1-x)^2}{(1-\alpha)\rho_l}\right)\right]_i=\frac{\Delta}{\Delta t}\left(\dot{m}(\Delta z)_i\left[\left(1+\frac{1-\alpha}{\alpha}\frac{\rho_l}{\rho_v}\right)x^2+\left(1+\frac{1-\alpha}{\alpha}\frac{\rho_l}{\rho_v}\right)(1-x)^2\right]_i\right) \quad (61)$$

Note further that the sign of the gravity term in the above equations is accounted for by multiplying g by $\sin\theta$, where θ is the inclination of the control volume axis to the horizontal, for example, $\theta = -\pi/2$ for the vertical portion of the down-comer and $\theta = \pi/2$ for the riser and reactor but $\theta = 0$ for the horizontally-orientated sections of the loop.

The mass flow rate of the fluid in and around the loop, at any instance in time, is then given by summing for all control volumes "around the loop" from 1 to N_{tot}+1 and noting that the pressure terms all cancel out:

$$\frac{\Delta}{\Delta t}(\dot{m}M)=\dot{m}^2MF-G-F \quad (62)$$

or explicitly as

$$\dot{m}^{t+\Delta t}=[\dot{m}^tM^t+\Delta t(\dot{m}^2MF-G-F)^t]/M^{t+\Delta t} \quad (63)$$

where

$$M_i^t=\sum_{i=1}^{N_{tot}+1}[M1_i^t(M2_i^t+M3_i^t)]$$

$$M_i^{t+\Delta t}=\sum_{i=1}^{N_{tot}+1}[M1_i^{t+\Delta t}(M2_i^{t+\Delta t}+M3_i^{t+\Delta t})]$$

$$M1_i^t=m_i^t/A_i^2 \text{ and } M1_i^{t+\Delta t}=m_i^{t+\Delta t}/A_i^2,\ M2_i^t=\left(\frac{x^2}{\alpha\rho_v}\right)_i^t \text{ and } M2_i^{t+\Delta t}=\left(\frac{x^2}{\alpha\rho_v}\right)_i^{t+\Delta t}$$

$$M3_i^t=\left(\frac{(1-x)^2}{(1-\alpha)\rho_l}\right)_i^t \text{ and } M3_i^{t+\Delta t}=\left(\frac{(1-x)^2}{(1-\alpha)\rho_l}\right)_i^{t+\Delta t}$$

$$MF=\sum_{i=1}^{N_{tot}+1}[MF1((MF2+MF3)_{in}-(MF4+MF5)_{out})]_i$$

$$MF1=\frac{1}{A_i},\ MF2=\frac{x^2}{\alpha A\rho_v},\ MF3=\frac{(1-x)^2}{(1-\alpha)A\rho_l},\ MF4=\frac{x^2}{\alpha A\rho_v} \text{ and } MF5=\frac{(1-x)^2}{(1-\alpha)A\rho_l},$$

$$G=\sum_{i=1}^{N_{tot}+1}\left[(\alpha\rho_v+(1-\alpha)\rho_l)\mathrm{g}\sin\theta\Delta z\right]_i$$

$$F=\sum_{i=1}^{N_{tot}+1}\left[\tau_{lo}\phi_{lo}^2\ \wp(\Delta z+\Delta z_{minor})/A\right]_i$$

where $\Delta z_{\mathrm{minor}}$ is the equivalent length of pipe of the same diameter and represents the so-called *minor losses* or the irreversible work done on the fluid as a result of there being inlets or contractions, bends, outlets or expansions, etc. Typical values for these losses, for an inlet, outlet and elbow may be taken as 18*d*, 30*d* and 52*d*, respectively.

Having now determined the mass flow rate at any instant in time $t + \Delta t$, the control volume pressure may now be determined at the so-to-say new-new time $t + \Delta t$ using Eq. (58), but re-arranged as:

$$P_{out,i} = P_{in,i} + \frac{\dot{m}^2}{A}\left[\left(\frac{x^2}{\alpha A\rho_v} + \frac{(1-x)^2}{(1-\alpha)A\rho_l}\right)_{in,i} - \left(\frac{x^2}{\alpha A\rho_v} + \frac{(1-x)^2}{(1-\alpha)A\rho_l}\right)_{out,i}\right]$$
$$-\left[\left(\alpha\rho_v + (1-\alpha)\rho_l\right)g\sin\theta\Delta z\right]_i$$
$$-\left[\frac{\tau_{lo}\phi_{lo}^2\ \wp(\Delta z+\Delta z_{minor})}{A}\right]_i$$
$$-\frac{\Delta}{\Delta t}\left[\frac{m\dot{m}}{A}\left(\frac{x^2}{\alpha A\rho_v} + \frac{(1-x)^2}{(1-\alpha)A\rho_l}\right)\right]_i^{t-\Delta t/2} \quad (64)$$

The last term in Eq. (58) is calculated as:

$$\frac{\left[\frac{m}{A^2}\left(\frac{x^2}{\alpha\rho_v} + \frac{(1-x)^2}{(1-\alpha)\rho_l}\right)\right]_i^{t+\Delta t} - \left[\frac{m}{A^2}\left(\frac{x^2}{\alpha\rho_v} + \frac{(1-x)^2}{(1-\alpha)\rho_l}\right)\right]_i^{t}}{\Delta t}$$

The inlet pressure to the loop, that is for control volume 2 in Figure 28(b), is given by $P_{in,1} = P_{gp} + P_{sdV} + \rho_l g(L_{sdL} + \Delta L_{sdL})$, where $P_{sdV} = P_{sat@TsdL}$ and ΔL_{sdL} depends on the amount of liquid that has been displaced by the volume of the vapour that has formed in the loop. Knowing now the inlet pressure, the outlet pressure for the control volume is determined using Eq. (64), and so on, around the loop until the last control volume $N_{tot}+1$ in Figure 27. Having now the inlet and outlet pressures of each of the control volume its pressure may be given as $P_i = (P_{in} + P_{out})/2$.

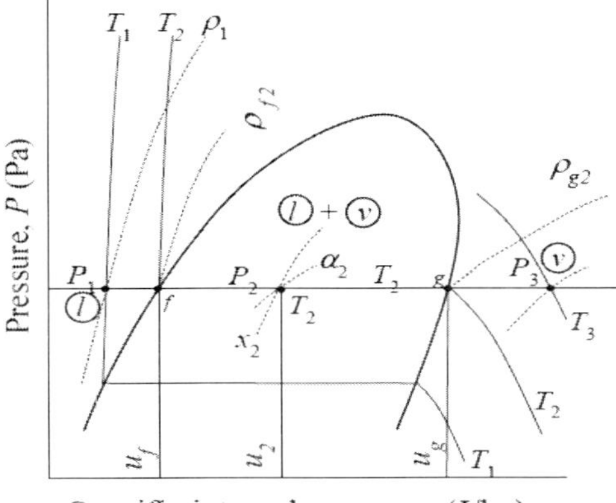

Figure 29. Pressure-specific internal energy diagram showing a sub-cooled liquid 1, a two-phase vapour plus liquid 2 and a vapour 3 state point.

[Once again, the sign of the gravity term in the above equation is accounted for by multiplying g by $\sin\theta$, where θ is the inclination the control volume axis to the horizontal, for example if $\dot{m} > 0$, $\theta = \pi/2$ for the vertical portion of the down-comer, $\theta = -\pi/2$ for the riser and reactor but $\theta = 0$ for the horizontally-orientated sections of the loop.]

3.1.7.6. Property Functions

Subcooled liquid region, that is $x \leq 0$ and $\alpha \leq 0$

$u = f(T), T = f(u, P)$ and $\rho = f(u, P)$

Two-phase region, that is $0 < x < 1$ and $0 < \alpha < 1$

$u_f = f(P), u_g = f(P), T_{sat} = f(P), \rho_f = f(P)$ and $\rho_g = f(P)$

Superheated region, that is $x \geq 1$ and $\alpha \geq 1$

$T = f(u, P)$ and $\rho = f(u, P)$

Sundry property functions

$\sigma = f(T)$ and $\mu = f(T)$

3.1.8. Numerical Simulation Model Computer-Solution Program Algorithm

An explicit numerical solution method is used to solve for the set of time dependent temperature coupled partial differential equations constituting the theoretical simulation model that was described in Section 3.1.7. Figure 27 (a) gives a drawing of the system, Figure 27(b) shows how the system is discretized into a series of control volumes. Figure 28 gives the applicable i^{th} control volume after applying the conservation of mass, momentum and energy and the assumptions stipulated in section 3.1.7. For closure, the property functions as given in appendix B are applied.

The solution algorithm is shown, step-wise, as follows:

Step 1. Referring to Figure 27 define the geometry of the loop (which is assumed to consist of a series of cylindrical control volumes).

Defining dimensions

$L_{sdL} = 0.3$: $L_{sdV} = 0.3$: $L_{sd} = L_{sdL} + L_{sdV}$: $L_{ft} = 3$: $L_r = 0.6$: $L_{dc1} = L_{ft} + L_r$: $L_{dc2} = 0.25$

$d_{sd} = 0.4$: $d_{dc1} = 0.030$: $d_{dc2} = 0.030$: $d_r = 0.2$: $d_{ft} = 0.2\ d_{orifice}$

Control volume numbers

$N_{sd} = 1$: $N_{dc1} = 7$: $N_{dc2} = 3$: $N_r = 1$: $N_{ft} = 7$: $N_1 = Nsd$: $N_2 = N_{sd} + N_{dc1}$: $N_3 = N_{sd} + N_{dc1} + N_{dc2}$: $N_4 = N_{sd} + Ndc1 + Ndc2 + N_r$: $N_5 = N_{sd} + N_{dc1} + N_{dc2} + Nr + N_{ft}$: $N_{tot} = N_5$

Control volume lengths

$\Delta L_{sdL} = L_{sdL}/N_{sd}$: $\Delta L_{dc1} = L_{dc1}/N_{dc1}$: $\Delta L_{dc2} = L_{dc2}/N_{dc2}$: $\Delta L_r = L_r/N_r$: $\Delta L_{ft} = L_{ft}/N_{ft}$

Step 2. Define the initial values and heat pipe parameters.

$T_{amb} = 20$: $T_{beginRHS} = 19.9$: $T_{beginLHS} = 20.1$

$x_{begin} = 0$: $ALPHA_{begin} = 0$: $\dot{m} = 0$: $QDOT_{reactor}$ $\dot{Q} = 6.5$

Heat pipe heat exchanger

$U_{hp} = 250$: $A_{hp} = PIE * 0.015 * 7$: $T_{hpC} = 20$

Gas pressure in steam drum

$P_{gp} = 25$ [in kPa]

Step 3. Define the problem-specific constants and stability criteria.

ΔTime = 0.1 : SF = 400 : GIPPOF1 = 1 : GIPPOF2 = 1 : GIPPOF3 = 1 : GippoHOMOff = 2.5

1 Start the time-step loop

$t = t + \Delta t$

$u_{i=0} = u_{i=N_{tot}}$: $u_{i=N_{tot+1}} = u_{i=1}$

If $\dot{m}^t \geq 0$ then

Step 4. Estimate the new mass and pressure

For $i = 1$ to N_{tot} : $m_i^{t+\Delta t} = m_i^t + \Delta t \left(\frac{\Delta m}{\Delta t}\right)_i^{t-\Delta t/2}$: $P_i^{t+\Delta t} = P_i^t + \Delta t \left(\frac{\Delta P}{\Delta t}\right)_i^{t-\Delta t/2}$: next i

[Alternatively, the new mass could have been determined using $m_i^{t+\Delta t} = m_i^t + \Delta t(\dot{m}_{in} - \dot{m}_{out})_i^t$]

Step 5. Calculate the heat losses to the environment

Calculate the heat transfer coefficients for control volumes 1 to N_{tot}

For $i = 1$ to N_{tot}

$h_i = 5$: $R_i = \frac{1}{h_i A z_i}$

$k_w = 18 : t_w = R_{wo} - R_{wo} : Az_w = (Az_{wi} + Az_{wo})/2 : R_w = \frac{t_w}{k_w Az_w}$

$k_{ins} = 0.5 : t_{ins} = R_{inso} - R_{inso} : Az_{ins} = (Az_{insi} + Az_{inso})/2 : R_{ins} = \frac{t_{ins}}{k_{ins} Az_{ins}}$

$h_o = 5 : R_i = \frac{1}{h_o Az_o}$

Calculate the thermal resistances

$R_i = R_i + R_w + R_{ins} + R_o$

Calculate the heat losses

$\dot{Q}_{loss,i} = (T_i - T_a)/R_i$
Next i

Step 6. Calculate the heat removed from the steam drum by the heat pipe heat exchanger

Assume that the heat pipe heat transfer coefficients are known and are not dependent on the temperature, Reynolds, Prandtl and Kutadeladze numbers, for now:

$d_{hpeo} = 0.014 : \wp_{hpeo} = \pi d_{hpeo} : L_{hpeo} = (2 \text{ x } 7 \text{ x } 0.45) : Az_{hpeo} = \wp_{hpeo} L_{hpeo}$
$d_{hpei} = 0.015 : \wp_{hpei} = \pi d_{hpei} : L_{hpei} = L_{hpeo} : Az_{hpei} = \wp_{hpei} L_{hpei}$
$d_{hpco} = 0.020 : \wp_{hpco} = \pi d_{hpco} : L_{hpco} = (7 \text{ x } 0.9) : Az_{hpco} = \wp_{hpco} L_{hpco}$
$d_{hpci} = 0.019 : \wp_{hpci} = \pi d_{hpci} : L_{hpci} = L_{hpco} : Az_{hpci} = \wp_{hpci} L_{hpci}$

$t_{hpew} = (d_{hpeo} - d_{hpei})/2 : d_{hpew} = (d_{hpei} + d_{hpeo})/2 : \wp_{hpew} = \pi d_{hpew} : L_{hpew} = L_{hpeo} : Az_{hpew} = \wp_{hpew} L_{hpew}$
$t_{hpcw} = (d_{hpco} - d_{hpci})/2 : d_{hpcw} = (d_{hpci} + d_{hpco})/2 : \wp_{hpcw} = \pi d_{hpcw} : L_{hpcw} = L_{hpco} : Az_{hpcw} = \wp_{hpcw} L_{hpcw}$

$h_{hpeo} = 5\ 000 : R_{hpeo} = 1/(h_{hpeo} Az_{hpeo})$
$h_{hpei} = 2\ 000 : R_{hpei} = 1/(h_{hpei} Az_{hpei})$
$h_{hpci} = 2\ 000 : R_{hpci} = 1/(h_{hpci} Az_{hpci})$
$h_{hpco} = 3\ 000 : R_{hpco} = 1/(h_{hpco} Az_{hpco})$
$k_{hpew} = 400 : h_{hpew} = t_{hpew} / (k_{hpew} Az_{hpew}) : R_{hpew} = t_{hpew}/(k_{hpew} Az_{hpew})$
$k_{hpcw} = 20 : h_{hpcw} = t_{hpcw} / (k_{hpcw} Az_{hpcw}) : R_{hpcw} = t_{hpcw}/(k_{hpcw} Az_{hpcw})$

$R_{hpe} = R_{hpeo} + R_{hpew} + R_{hpei}$
$R_{hpc} = R_{hpeo} + R_{hpew} + R_{hpei}$

$R_{hp} = R_{hpe} + R_{hpc}$

$$\dot{Q}_{hp} = (T_{sd1} - T_{hp})/R_{hp} : \dot{Q}_{loss,1} = \dot{Q}_{loss,1} + \dot{Q}_{hp}$$

Step 7. Calculate the heat supplied by the reactor

$$\dot{Q}_{reactor} = f(t) : \dot{Q}_{loss,i=N_{reactor}} = \dot{Q}_{loss,i=N_{reactor}} - \dot{Q}_{reactor}$$

Step 8. Calculate the heat transfer rates for each control volume

For i =1 : $\dot{Q}_i = -\dot{Q}_{loss,i}$

Step 9. Calculate the change in internal energy due to convection

$$(-\Delta\dot{m}u)_i^t = -\dot{m}^t(u_i - u_{i-1})_i^t$$

Step 10. Calculate the reversible work rate terms

$$\rho_{h,i}^t = \left((x/\rho_v + (1-x)/\rho_l)_i^t\right)^{-1}$$

$$\rho_{h,i-1}^t = \left((x/\rho_v + (1-x)/\rho_l)_{i-1}^t\right)^{-1}$$

$$\left(\frac{P}{\rho_h}\right)_{i,out}^t = \frac{P_{i+1/2}^t}{\rho_{h,i}^t}$$

$$\left(\frac{P}{\rho_h}\right)_{i,in}^t = \frac{P_{i-1/2}^t}{\rho_{h,i-1}^t}$$

$$(-\Delta PA\,v)_i^t = -\dot{m}\left(\left(\frac{P}{\rho_h}\right)_{i,out}^t - \left(\frac{P}{\rho_h}\right)_{i,in}^t\right)$$

Step 11. Calculate the irreversible work rate terms $\tau A_z v$

$$-\tau_{w,i}^t = -\left(\tau_{lo}\,\phi_{lo}^2\right)_i^t$$

$$\tau_{lo,i}^t = \left(C_{flo}\rho_l\left(\frac{\dot{m}}{\rho_l A}\right)^2/2\right)_i^t$$

$$\left(\phi_{lo}^2\right)_i^t = f(X)$$

$$\rho_i^t = \left(\alpha\rho_v + (1-\alpha)\rho_l\right)_i^t$$

$$v_i^t = \frac{\dot{m}^t}{\rho_i^t A_i}\ v_i^t = \frac{\dot{m}^t}{\rho_i^t A_i}$$

$$F_i^t = \left(\tau_{lo,i}\phi_{lo,i}^2 \wp_i\left(\Delta z_i + z_{i,minor}\right)\right)^t,$$
$$(Fv)_i^t = F_i^t v_i^t$$

Step 12. Calculate the sum of the thermal energy terms

$$\Sigma\dot{E}_i^t = (-\Delta\dot{m}u)_i^t + \Sigma\dot{Q}_i^t + (-\Delta PA\,v)_i^t + (Fv)_i^t$$

Step 13. Calculate the internal energy

$$u_i^{t+\Delta t} = \left[(mu)_i^t + \Delta t\Sigma\dot{E}_i^t\right]/m_i^{t+\Delta t}$$

Step 14. Calculate the saturated liquid and vapour internal energies at the estimated new pressure.

$$u_{f,i}^{t+\Delta t} = f\left(P_i^{t+\Delta t}\right)$$
$$u_{g,i}^{t+\Delta t} = f\left(P_i^{t+\Delta t}\right)$$

If $u_i^{t+\Delta t} \le u_{f,i}^{t+\Delta t}$ then $x_i^{t+\Delta t} = 0$: $x_i^{t+\Delta t} = 0$: $T_i^{t+\Delta t} = f\left(u_i^{t+\Delta t}, P_i^{t+\Delta t}\right)$: $\rho_i^{t+\Delta t} = f\left(u_i^{t+\Delta t}, P_i^{t+\Delta t}\right)$

If $u_i^{t+\Delta t} > u_{f,i}^{t+\Delta t}$ and $u_i^{t+\Delta t} \le u_{g,i}^{t+\Delta t}$ then

$$x_i^{t+\Delta t} = \left(u_i^{t+\Delta t} - u_{f,i}^{t+\Delta t}\right)/\left(u_{g,i}^{t+\Delta t} - u_{f,i}^{t+\Delta t}\right)$$
$$T_i^{t+\Delta t} = f\left(P_{sat,i}^{t+\Delta t}\right)$$
$$\rho_{f,i}^{t+\Delta t} = f\left(P_i^{t+\Delta t}\right) \text{ and } \rho_{g,i}^{t+\Delta t} = f\left(P_i^{t+\Delta t}\right)$$
$$SF = f(Re, Fr, We, d)$$
$$\alpha_i^{t+\Delta t} = \left(1 + SF\frac{\rho_{g,i}^{t+\Delta t}}{\rho_{f,i}^{t+\Delta t}}\frac{1-x_i^{t+\Delta t}}{x_i^{t+\Delta t}}\right)^{-1}$$
$$\rho_i^{t+\Delta t} = \alpha_i^{t+\Delta t}\rho_{g,i}^{t+\Delta t} + \left(1 - \alpha_i^{t+\Delta t}\right)\rho_{f,i}^{t+\Delta t}$$
$$m_i^{t+\Delta t} = \rho_i^{t+\Delta t} V_i$$
$$\left(\frac{\Delta m}{\Delta t}\right)_i^{t+\Delta t/2} = \frac{m_i^{t+\Delta t} - m_i^t}{\Delta t}$$

end if

If $u_i^{t+\Delta t} > u_{g,i}$: $T_i^{t+\Delta t} = f\left(u_i^{t+\Delta t}, P_i^{t+\Delta t}\right)$, $\rho_i^{t+\Delta t} = f\left(u_i^{t+\Delta t}, P_i^{t+\Delta t}\right)$, $x_i^{t+\Delta t} = 1$ and $\alpha_i^{t+\Delta t} = 1$

Step 15. Calculate the new mass flow rate $\dot{m}^{t+\Delta t}$

Consider the momentum equation for the ith control volume expressed in terms of the mass and void fractions as (Eq. (60) repeated for convenience).

$$\frac{\Delta}{\Delta t}\left[\frac{m\dot{m}}{A^2}\left(\frac{x^2}{\alpha\rho_v}+\frac{(1-x)^2}{(1-\alpha)\rho_l}\right)\right]_i = \frac{\dot{m}^2}{A_i}\left[\left(\frac{x^2}{\alpha A\rho_v}+\frac{(1-x)^2}{(1-\alpha)A\rho_l}\right)_{in}-\left(\frac{x^2}{\alpha A\rho_v}+\frac{(1-x)^2}{(1-\alpha)A\rho_l}\right)_{out}\right]_i$$
$$+P_{in,i}-P_{out,i_i}-\left[\left(\alpha\rho_v+(1-\alpha)\rho_l\right)g\sin\theta\Delta z\right]_i-\left[\tau_{lo}\phi_{lo}^2\, P(\Delta z+z_{minor})/A\right]_i \quad (60)$$

The mass flow rate at $t+\Delta t$ is now obtained by summing all the control volumes around the loop from the top of the liquid in the steam drum (where the velocity is essentially zero) and the control volumes in the down comer, reactor flashtube and back to the top of the liquid in the steam drum. Using an explicit finite difference numerical solution method and so-called *upwind differencing,* and noticing that the pressure terms all cancel out, required mass flow rate is given by:

$$\dot{m}^{t+\Delta t}=\left[\dot{m}^t\sum M_i^t+\Delta t(\dot{m}^2\sum MF+\sum G-\sum F)_i^t\right]/\sum M_i^{t+\Delta t}$$

where, if the control volume is subcooled or superheated:

$$M_i^{t+\Delta t}=\frac{(m/\rho)_i^{t+\Delta t}}{A_i^2},\ M_i^t=\frac{(m/\rho)_i^t}{A_i^2}\text{ and } MF=\frac{1}{A_i}\left(\frac{1}{(A\rho)_j}-\frac{1}{(A\rho)_i}\right)^t$$

And for a two-phase liquid and vapour control volume:

$$M_i^{t+\Delta t}=\frac{m_i^{t+\Delta t}}{A_i^2}\left(\frac{x^2}{\alpha\rho_g}+\frac{(1-x)^2}{(1-\alpha)\rho_f}\right)_i^{t+\Delta t},\ M_i^t=\frac{m_i^t}{A_i^2}\left(\frac{x^2}{\alpha\rho_g}+\frac{(1-x)^2}{(1-\alpha)\rho_f}\right)_i^t,\text{ and}$$

$$MF=\frac{1}{A_i}\left[\left(\frac{x^2}{\alpha A\rho_v}+\frac{(1-x)^2}{(1-\alpha)A\rho_l}\right)_j-\left(\frac{x^2}{\alpha A\rho_v}+\frac{(1-x)^2}{(1-\alpha)A\rho_l}\right)_i\right]^t,$$

where $j=i-1$ if $\dot{m}\geq 0$ or $j=i+1$ if $\dot{m}<0$

$$G=\rho_i^t\Delta z_i g\sin\theta_i$$

where $\rho = \alpha\rho_v + (1-\alpha)\rho_l$, $\theta = \pi/2$ for the vertical LHS of the loop, that is the reactor and the flashtube, $\theta = -\pi/2$ for the vertical RHS of the loop, that is the vertical portions of the down-comer and $\theta = 0$ for the horizontally orientated portions of the circulation loop.

$$F = \wp\, \tau_i^t (\Delta z + z_{\text{minor}})_i / A_i$$

where $\tau_i^t = \tau_{lo}\phi_{lo}^2$, $\tau_{lo} = \frac{1}{2} C_{flo}\rho_l \left(\frac{\dot{m}}{\rho_l A}\right)^2 = \frac{C_{flo}}{2\rho_l A^2}\dot{m}^2$, and to ensure that that the friction always acts against the flow $\dot{m}^2$ is replaced by $\dot{m}|\dot{m}|$, $\phi_{lo}^2 = f(X)$, $\wp = \pi d$, $X = \left(\frac{\rho_v}{\rho_l}\right)^{0.5}\left(\frac{\mu_l}{\mu_v}\right)^{0.5}\left(\frac{1-x}{x}\right)^{0.875}$ and

if $Re < SS_1$ then $c_{flo} = 16/SS_1$
if $Re \geq SS_1$ and $Re \leq SS_2$ then $Cf_{lo} = 16/Re$
if $Re > SS_2$ then $Cf_{lo} = 0.079Re^{-0.25}$

where $Re = \frac{|\dot{m}^t| d_h}{\mu A}$, $\mu = f(T^t)$ and SS_1 is a correlating coefficient that attempts to identify the transition point between static and dynamic friction (viscosity) or the so-called concept of *slip-stick*, and SS2 is a correlating coefficient that attempts to define the transition between laminar and turbulent flow conditions. They may be taken arbitrarily as 1 and 1181, respectively. [1181 is the Reynolds number corresponding to the intersection of the smooth wall Blassius type coefficient of friction $Cf_{lo} = 0.079Re^{-0.25}$ and the laminar coefficient of friction given as $Cf_{lo} = 16/Re$.]

Step 16. Calculate the new pressures $P_i^{t+\Delta t}$ for each control volume around the loop

The pressure at the top of the liquid in the steam drum is given by the sum of the partial pressure of the vapour and the pressurizer pressures as:

$$P_{sdL,top} = P_{g@TsdL} + P_{\text{gp}}$$

The pressure at the bottom of the steam drum depends essentially on the saturated steam vapour pressure at the steam drum liquid temperature, the gas pressure and the height of the liquid in the steam drum and is given by:

$$P_{sd,bottom} = P_{sdL,top} + \rho_{sdL} g(L_{sd} + \Delta L_{sd})$$

where $\Delta L_{sd}^{t+\Delta t} \approx 0$ if the volume of the steam drum is relatively large compared to the volume of the liquid that is displaced by the formation of the vapour in the flashtube.

Having now a new mass flow rate (and the density and temperatures for each control volume) and the pressure at one end of the control volume, the pressure at the other end can be calculated using Eq. 3.25, and which, for convenience, is repeated below:

$$\begin{aligned}P_{out,i} &= P_{in,i} + \left[\frac{\dot{m}^2}{A}\left(\frac{x^2}{\alpha A\rho_v} + \frac{(1-x)^2}{(1-\alpha)A\rho_l}\right)\right]_{in,i} \\ &- \left[\frac{\dot{m}^2}{A}\left(\frac{x^2}{\alpha A\rho_v} + \frac{(1-x)^2}{(1-\alpha)A\rho_l}\right)\right]_{out,i} \\ &+ \left[(\alpha\rho_v + (1-\alpha)\rho_l) g\sin\theta\Delta z\right]_i \\ &- \left[\frac{\tau_{lo}\phi_{lo}^2 \wp(\Delta z + \Delta z_{minor})}{A}\right]_i \\ &- \frac{\Delta}{\Delta t}\left[\frac{m\dot{m}}{A}\left(\frac{x^2}{\alpha A\rho_v} + \frac{(1-x)^2}{(1-\alpha)A\rho_l}\right)\right]_i\end{aligned}$$

or simply as:

$$P_{out,i}^{t+\Delta t} = P_{in,i}^{t+\Delta t} + (MF)_i^{t+\Delta t} - G_i^{t+\Delta t} - F_i^{t+\Delta t} - \left(\frac{\Delta M}{\Delta t}\right)_i^{t-\Delta t/2}$$

where

$$P_{in,i}^{t+\Delta t} = P_{out,i-1}^{t+\Delta t}$$

And if control volume is subcooled or superheated:

$$\begin{aligned}M_i^t &= \frac{(m\dot{m}/\rho)_i^t}{A_i^2} \\ M_i^{t+\Delta t} &= \frac{(m\dot{m}/\rho)_i^{t+\Delta t}}{A_i^2} \\ \left(\frac{\Delta M}{\Delta t}\right)_i^{t-\Delta t/2} &= \left(M_i^{t+\Delta t} - M_i^t\right)/\Delta t \\ MF_i^{t+\Delta t} &= \frac{\left(\dot{m}^{t+\Delta t}\right)^2}{A_i}\left(\frac{1}{(A\rho)_j} - \frac{1}{(A\rho)_i}\right)^{t+\Delta t}\end{aligned}$$

where $j = i - 1$ if $\dot{m} \geq 0$ (or $j = i + 1$ if $\dot{m} < 0$)

If the control volume is a two-phase (liquid and vapour) control volume:

$$M_i^t = \frac{(m)_i^t \dot{m}^t}{A_i^2}\left(\frac{x^2}{\alpha\rho_g} + \frac{(1-x)^2}{(1-\alpha)\rho_f}\right)_i^t$$

$$M_i^{t+\Delta t} = \frac{(m)_i^{t+\Delta t}\dot{m}^{t+\Delta t}}{A_i^2}\left(\frac{x^2}{\alpha\rho_g} + \frac{(1-x)^2}{(1-\alpha)\rho_f}\right)_i^{t+\Delta t}$$

$$MF_i^{t+\Delta t} = \frac{\left(\dot{m}^{t+\Delta t}\right)^2}{A_i}\sum\left[\left(\frac{x^2}{\alpha A\rho_g} + \frac{(1-x)^2}{(1-\alpha)A\rho_f}\right)_j - \left(\frac{x^2}{\alpha A\rho_g} + \frac{(1-x)^2}{(1-\alpha)A\rho_f}\right)_i\right]^{t+\Delta t}$$

where $j = i - 1$ if $\dot{m} \geq 0$ or $j = i + 1$ if $\dot{m} < 0$

$$G_i^{t+\Delta t} = \rho_i^{t+\Delta t}\Delta z_i g\sin\theta_i$$

where $\rho = \alpha\rho_g + (1-\alpha)\rho_f$, $\theta = \pi/2$ for the vertical LHS of the loop, that is the reactor and flashtube, $\theta = -\pi/2$ for the vertical RHS of the loop, that is the vertical portions of the down-comer and $\theta = 0$ for the horizontal portions of the circulation loop (and the steam drum).

$$F_i^{t+\Delta t} = \wp_i \tau_i^{t+\Delta t}(\Delta z + z_{\text{minor}})_i / A_i$$

where $\tau_i^{t+\Delta t} = \left(\tau_{lo}\phi_{lo}^2\right)_i^{t+\Delta t}$, $\tau_{lo} = \frac{1}{2}C_{flo}\rho_l\left(\frac{\dot{m}}{\rho_l A}\right)^2 / 2 = \frac{C_{flo}}{2\rho_l A^2}\dot{m}|\dot{m}|$, where to ensure that that the friction always acts against the flow, $\dot{m}^2$ is replaced by $\dot{m}|\dot{m}|$, $\phi_{lo}^2 = f(\chi)$ and $\wp$ = perimeter of the control volume,

and

if $Re < SS_1$ then $C_{flo} = 16/SS_1$

if $Re \geq SS_1$ and $Re \leq SS_2$ then $Cf_{lo} = 16/Re$

if $Re > SS_2$ then $Cf_{lo} = Re$

where $Re = \frac{|\dot{m}^t| d_h}{\mu A}$ and $\mu = f(T^t)$

The pressure of the control volume is then given as:

$$P_i^{t+\Delta t} = \left(P_{in,i}^{t+\Delta t} + P_{out,i}^{t+\Delta t}\right)/2$$

Step 17. Replace the old values by the new values and go back to "1"

3.1.8. Theoretical and Experimental Results

Using the theory as described in the previous sections, the simulation of the thermal-hydraulic behaviour of the system as shown in Figure 21 was attempted. This system simulated is a smallscale experimental model of a possible inherently safe natural circulation cooling system for the nuclear reactor as schematically conceptualised in Figure 20. To theoretically simulate it, it was simplified as shown in Figure 42. The primary loop is considered as a number of control volumes but both the secondary (thermosyphon-type heat pipe) and the tertiary (natural draught air-cooled condenser) were taken as single control volumes. The essential objective of the simulation was to emulate the experimentally obtained mass rate in the primary loop and the power input as a function of time shown in Figure 30. A degree of similarity was indeed obtained (but only after judicially adjusting the void fraction and friction multiplier) as shown in Figure 31. It is seen that a close agreement in primary loop mass flow rate between the experimental and theoretical has been achieved for the specified input power in both cases of about 3 800 W. (A reasonable agreement for the temperatures was also obtained, but not shown.) Refinement to the theoretical model is still an ongoing work at this time.

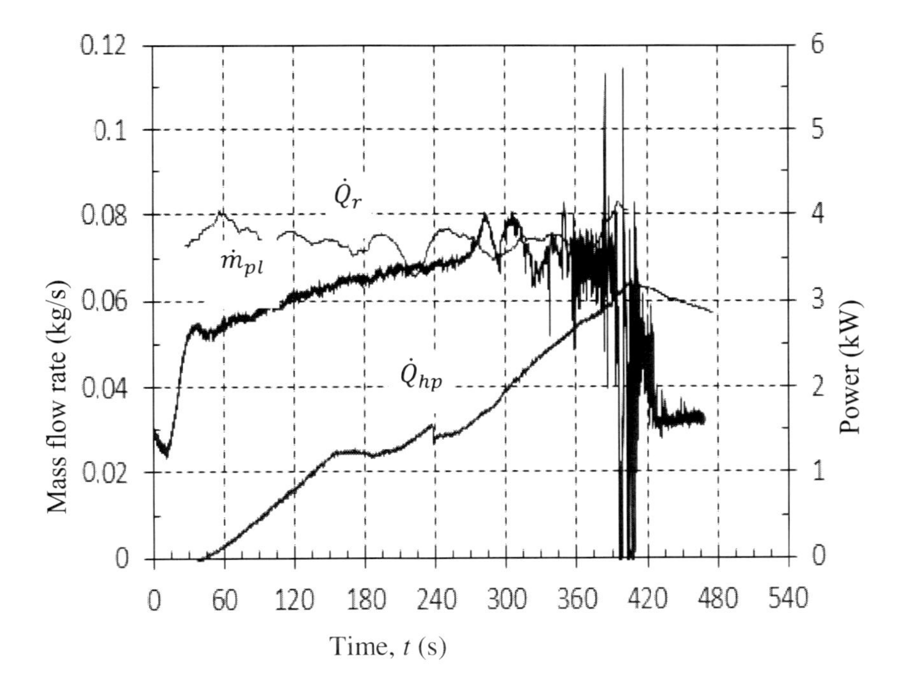

Figure 30. Experimentally determined primary loop mass flow rate and power input for the small-scale experimental setup given in Figure 21.

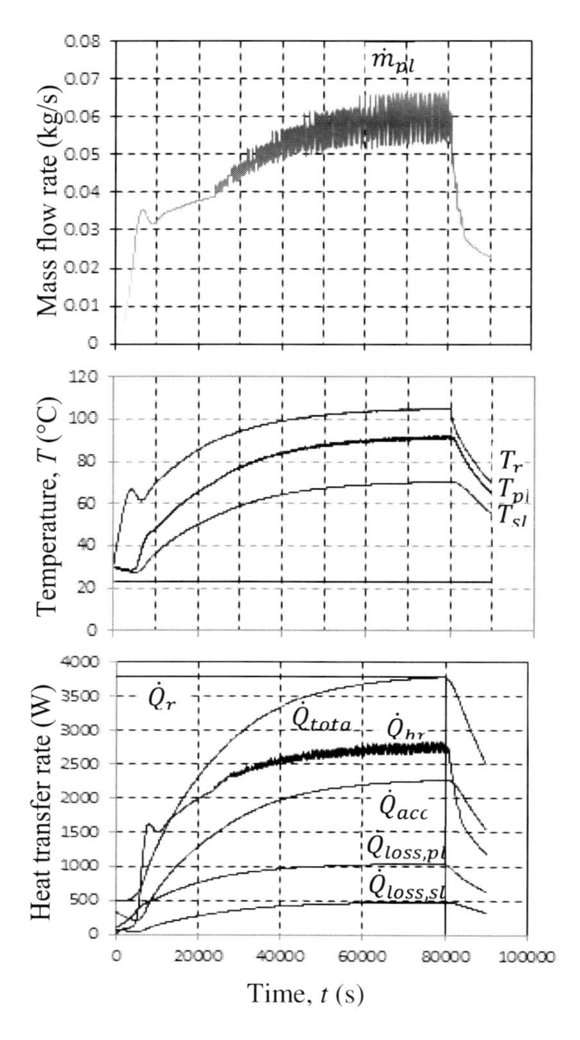

Figure 31. Theoretically determined mass flow rate, temperatures, and heat transfer rates for the simplified system depicted in Figure 42.

3.2. Entirely-Passive Reactor Cavity Cooling System (RCCS)

A reactor cavity cooling system (RCCS) can be likened to a series of rectangular vertically orientated closed loop thermosyphon-type heat pipes with the one vertical leg located between the reactor and the concrete confinement building and the other vertical leg being cooled by either tube-in-tube heat exchangers or by immersion in a an external heat sink as shown in Figure 32. The prime object of the RCCS is to ensure that the temperature of the concrete structures surrounding the reactor at no time exceeds 65°C under normal operating conditions and then also in the event of a loss of primary reactor coolant, and that the concrete does not exceed 125°C for more than an hour. In this section we will consider a number of small scale loops as well as a full scale prototype loops suitable for including in the RCCS cooling system of a high temperature pebble bed modular reactor (PBMR).

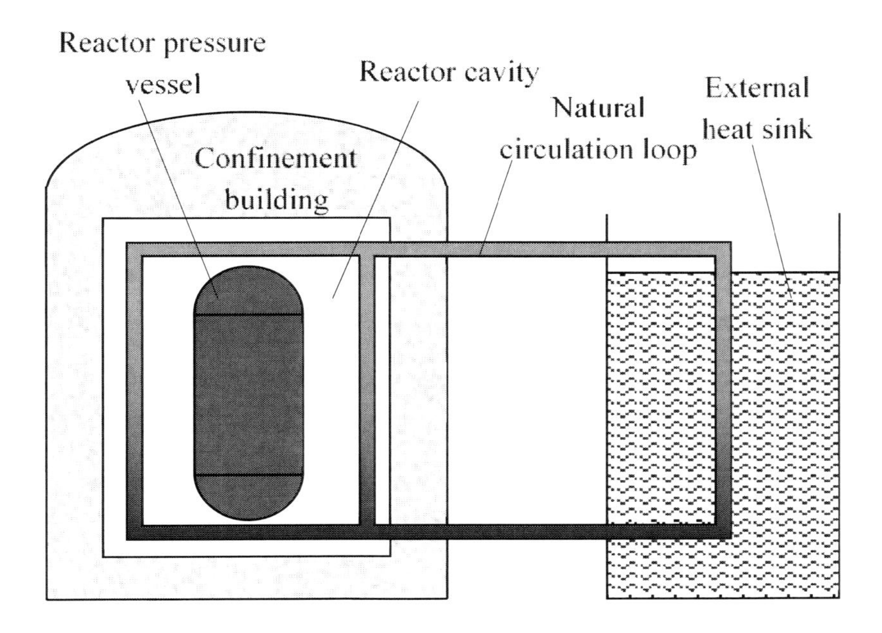

Figure 32. Concept natural circulation reactor cavity cooling system.

In addressing natural circulation loops for reactor cooling, three loops were built and tested, two small scale single loops varying in height from 2.4 to 8.7 and one fullsize 27 m high. Three sized single loops were considered: a small-scale 2.4 m high loop, a 7.5 m high system and a full-size 27 m high loop, and two multichannel loops were built and tested. These loops are depicted by way of Figure 32 to Figure 37. Each of these loops had transparent pipe inserts through which flow patterns could be observed. An idea of the average flow rate could also be obtained by directly determining the speed at which small bits of dirt moved along with the flow. The outcome of the testing of these loops was to validate their theoretical transient simulation programs experimentally and to establish any dimensional size-scaling implications. Further, it was deemed important (from a system vibration point of view) that the single and or two-phase flow pattern in the flow be known. This criteria was met by measuring the static and dynamic pressure pulses in the single flow region of the loop where, if it were a real reactor, the measurements predicting the flow behaviour in a radioactive region of the cooling loop could be safely taken and analyzed in a non-radiation region of the reactor complex.

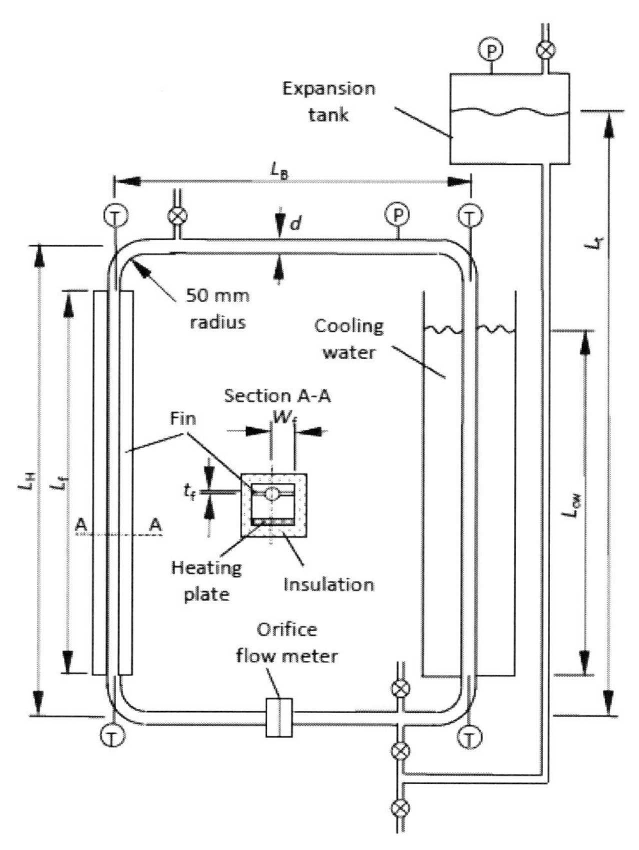

Figure 33. Schematic representation of the 2.4 m high loop [43].

All five experimental loops depicted were furnished with temperature sensors, a 10 Hz sampling rate differential pressure orifice flow meter, an absolute pressure meter and high frequency up to 20 kHz sensors. The basic research approach was to develop a theoretical simulation program by representing the system by a series of control volumes. Then, by applying the equations of change (conservation of mass, momentum and energy) to each of the control volumes, generate a set of time dependent finite difference equations. Then, with property functions for closure, write a computer program solving the equations using an explicit numerical solution procedure. The transient theoretical thermal-hydraulic behaviour so captured is then experimentally verified, making sure that the energy is accurately accounted for. In developing the theoretical simulation program, a number of assumptions have been made of which some of the major assumptions include:

- One-dimensional control volumes, as the pipe diameters are not expected to exceed about 50 mm in diameter.
- Quasi-equilibrium; this implies that the speed of sound in the heat pipe working fluid is much greater than the actual velocity of the working fluid and that thermal expansion of both liquid and vapour occurs instantaneously.

- Liquid and vapour phase at any elevation in the loop are in thermal equilibrium with each other.

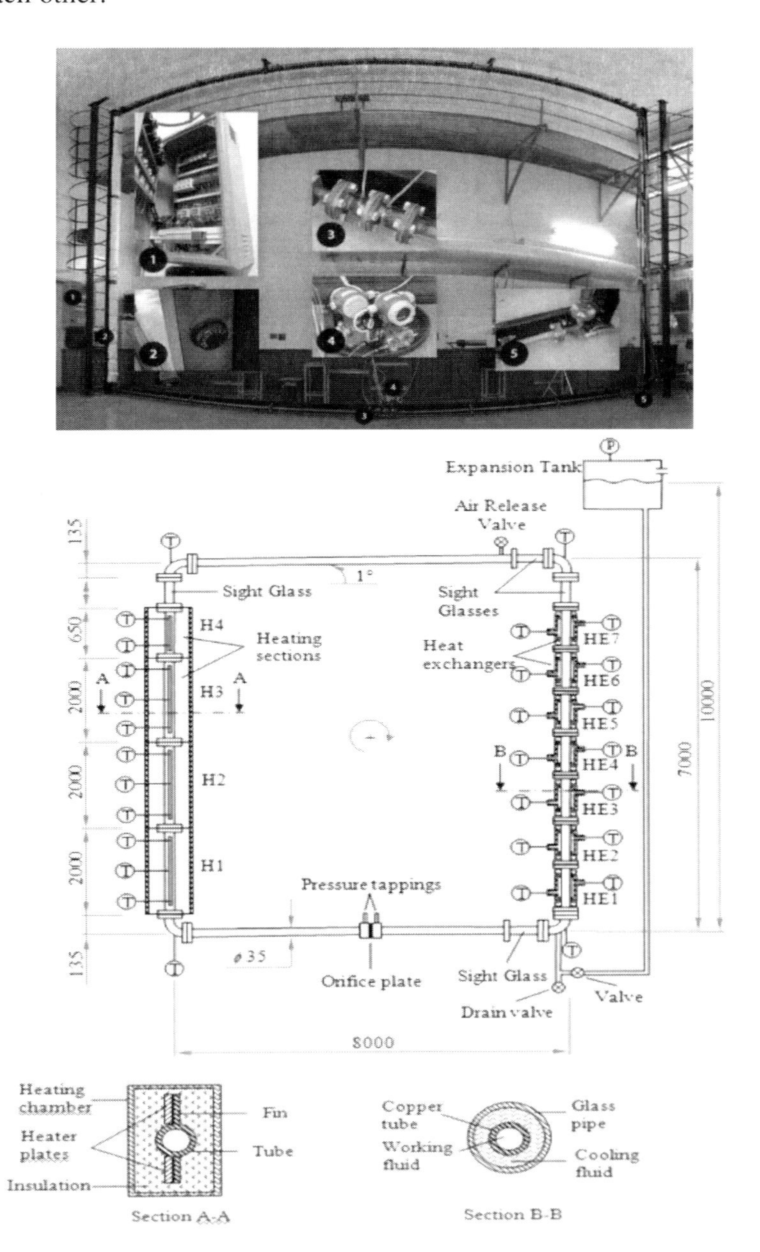

Figure 34. Wide angle photograph of the 7.5 m high loop (a), and construction details (b) [44].

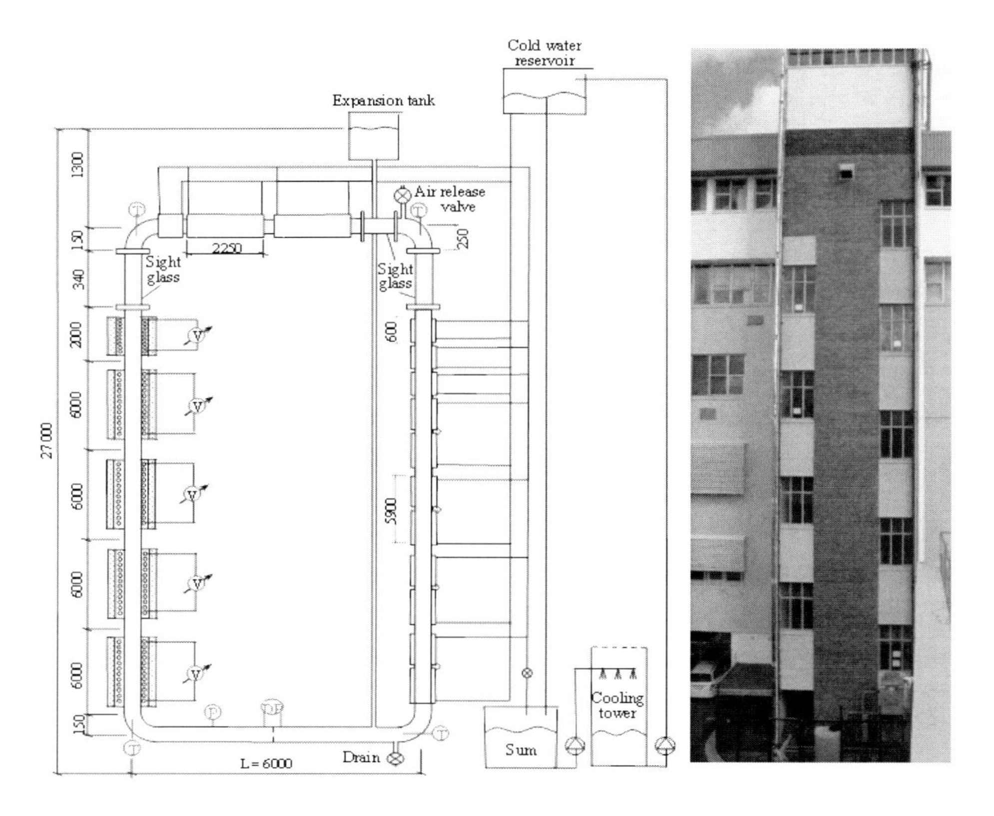

Figure 35. The 27 m high loop with 47.8 mm inside diameter [43].

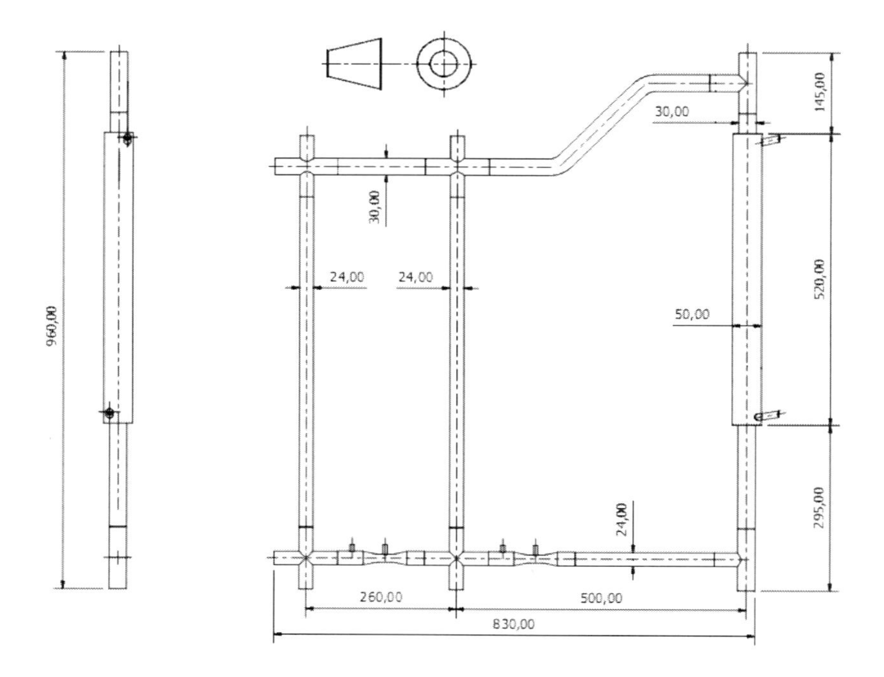

Figure 36. Two-phase two heated channel transparent (White, 2010).

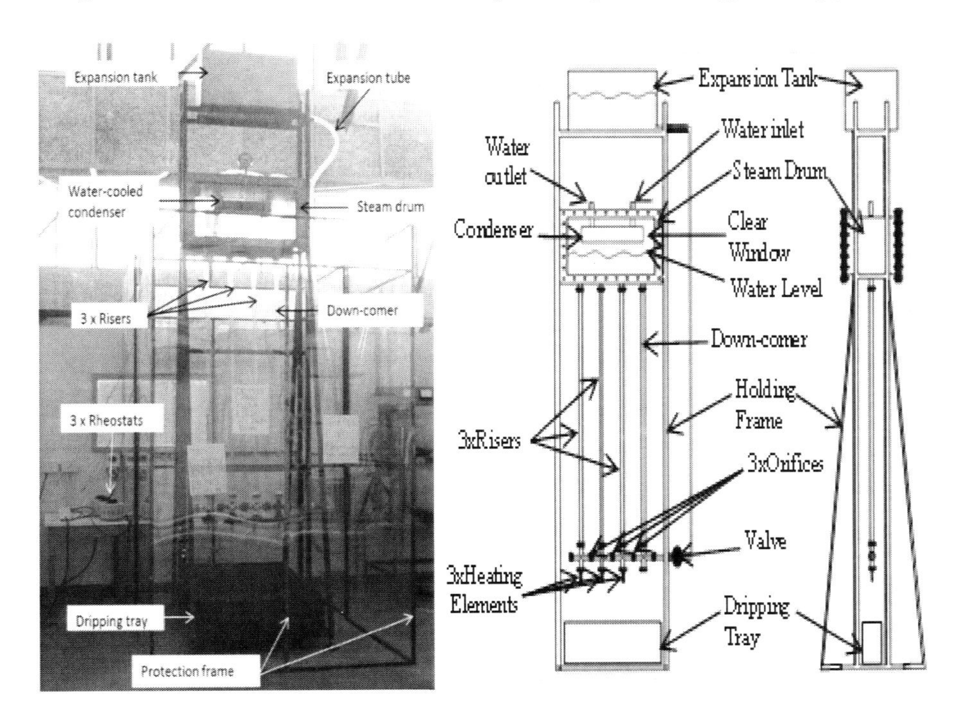

Figure 37. Two-phase multi-channel natural circulation loop with transparent risers [46].

3.3. Entirely-Passive Spent and Used Fuel Tank Cooling System

Figure 38 gives an idea of a possible layout of an entirely-passive spent and used fuel tank cooling system. Figure 39 gives a schematic drawing of the small-scale experimental loop test setup shown photographically in Figure 40. This work is captured by way of a thesis by Senda [47] and by two conference proceedings [48, 49].

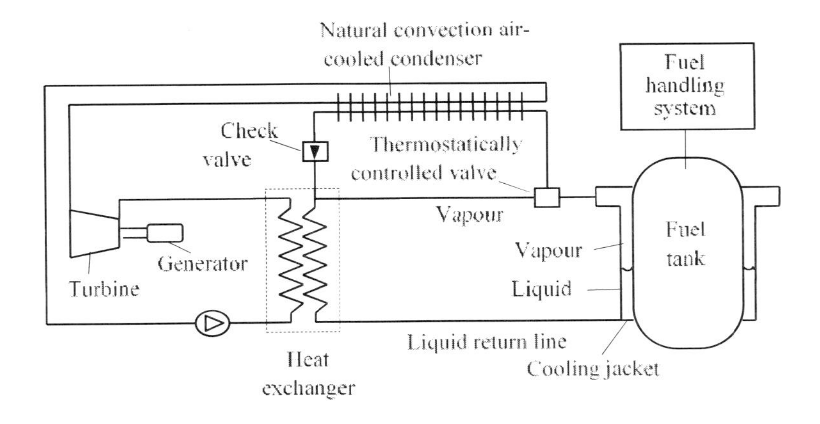

Figure 38. Schematic drawing of a conceptual spent fuel tank cooling and Rankine cycle power generation system [47].

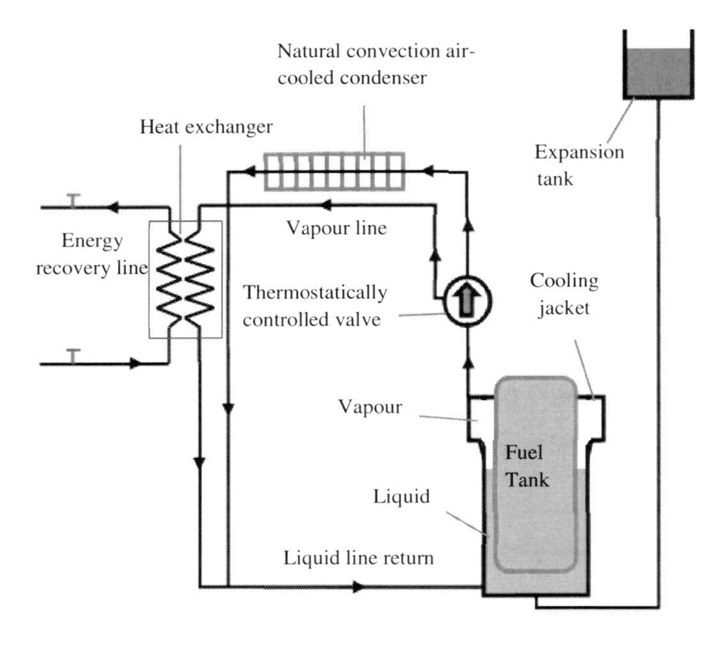

Figure 39. Experimental test loop of a spent fuel cooling system [48].

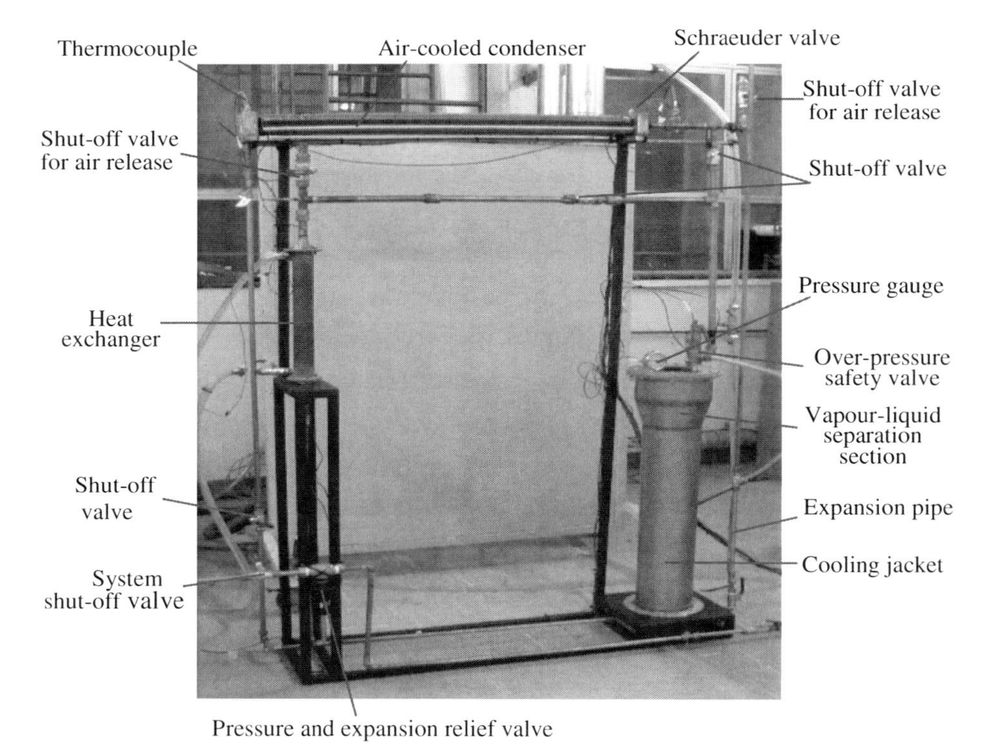

Figure 40. Waste heat recovery and utilization system experimental set-up [49].

3.4. High-Temperature Tritium-Diffusion-Resistant Helium-to-Water Heat Pipe Heat Exchanger

This is a proof-of-concept experimental facility that uses combustion gases at about 600°C to simulate the helium and Dowtherm-A, instead of liquid sodium, as the working fluid. Figure 41 shows the helium-to-water heat pipe heat exchanger experimental test set-up and facility (a), the heat pipe heat exchanger positioned in the laboratory (b), with and without its insulation (c) and (d), respectively. Both steady-state and transient time dependent theoretical simulation procedures were developed. The theoretical simulations on average captured the experiment very well but with a relatively large scatter. Visual observations (though the windows provided for) clearly show the complex behaviour of the boiling and condensation process in the heat exchanger, under actual operating conditions. These results are fully documented by way of Dobson and Laubscher [50, 51] and Laubscher and Dobson [52, 53, 54] and a thesis by Laubscher [55].

3.5. Steady State Natural Circulation Nuclear Reactor Cooling System

Consider the relatively simple thermal resistance diagram representation for a natural circulation reactor cooling system as given by way of figure 42. The system has been specifically so-configured for a natural circulation water cooled nuclear reactor such that heat released in the reactor core $\dot{Q}_r$ will be transferred to the outside air at temperature T_{ao} without any mechanically moving parts and without the possible release of any unwanted radioactive nuclide into the atmosphere. Radioactive fission products that might have been released into the primary coolant and then through the reactor pressure retaining barrier of the *primary* (circulation) loop *pl* are then prevented from finding their way into the power conversion unit (PCU) by the inclusion of a second closed loop thermosyphon-type heat pipe, the *secondary* loop *sl* and second steam drum *sd*2. Afterheat may then be safe without the use of fans but by a naturally air-cooled condenser; the *tertiary* loop *tl* into the atmosphere. Note that a non-condensable gas removal system, secondary steam drum pressurizer and the power conversion unit have not been included in this analysis.

[In high temperature nuclear reactors, radioactive tritium is formed and at these high temperatures may diffuse though the hot coolant steel container and find its way into the power conversion working fluid or even into a process heat steam and conceivable even find its way into a consumer product. To circumvent the possibility of such contamination, the use of natural circulation heat pipe technology may be considered; such a nuclear reactor cooling system may be then termed *entirely passive and inherently safe*; as cooling in this way is achieved without the use of any mechanically moving equipment or active control systems.]

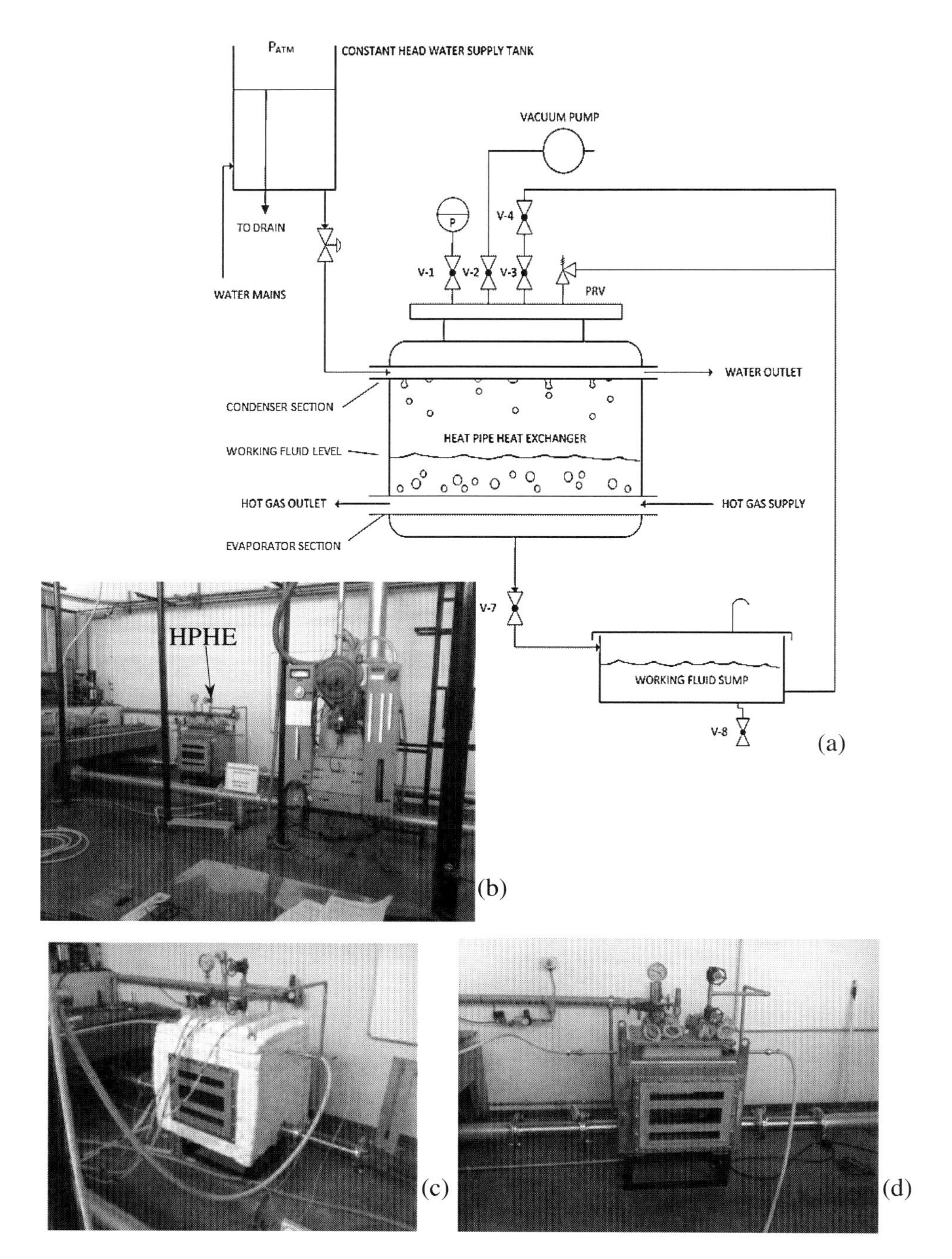

Figure 41. Helium-to-water heat pipe heat exchanger experimental test set-up and facility (a), the heat pipe heat exchanger positioned in the laboratory (b), with and without its insulation (c) and (d) [52].

Given a reactor heat release $\dot{Q}_r$ and an air temperature $\overline{T}_a$ and typical values for the various heat transfer coefficients there are thus essentially only three unknown temperature-dependent equations that need to be solved simultaneously. These three temperatures are the reactor primary loop temperature T_r and the secondary and the tertiary loop steam drum temperatures T_{sd1} and T_{sd2}. We will further assume that the various heat transfer coefficients are all more-or-less constant. This is so because the natural convection heat transfer coefficient h_a from the system boundary to the surrounding T_a is about 5 to 15 W/m^2°C is at least two orders of magnitude less than the other applicable heat transfer coefficients which are all at least greater than about 1000 W/m^2°C). The air-side thus influences the overall heat transfer coefficient the most; or, as often termed is the *controlling* heat transfer coefficient.

The three unknown temperatures T_{sd1}, T_{sl} and T_{sd2} may be expressed in terms of the various thermal resistances $R = 1/hA$ as given by Eqs. (65), (66) and 67), and hence, although they are highly non-linear (due to the radiation heat transfer coefficients being dependent on temperature to the fourth power) they are cumbersome, but are readily solved with a Gauss-Seidel numerical iteration method.

$$T_{sd1} = \frac{\dot{Q}_r + T_{sl}/R_{sl1} + T_a/R_{loss,pl}}{1/R_{sl1} + 1/R_{loss,pl}} \tag{65}$$

$$T_{sl} = \frac{T_{sd1}/R_{sl1} + T_{sl2}/R_{sl2} + T_a/R_{loss,sl}}{1/R_{sl1} + 1/R_{sl2} + 1/R_{loss,sl}} \tag{66}$$

$$T_{sd2} = \frac{T_{sl}/R_{sl2} + \bar{T}_a/R_{tl} + T_a/R_{loss,tl}}{1/R_{sl2} + 1/R_{tl} + 1/R_{loss,tl}} \tag{67}$$

where

$$R_{sl1} = R_{sleo} + R_{slew} + R_{slei},$$
$$R_{sl2} = R_{slco} + R_{slcw} + R_{slci},$$
$$R_{sl2} = R_{tlco} + R_{tlcw} + R_{tlci},$$
$$R_{sleo} = \frac{1}{h_{sleo}A_{sleo}},\ R_{slew} = \frac{1}{h_{slew}A_{slew}},\ R_{slei} = \frac{1}{h_{slei}A_{slei}}$$
$$R_{slco} = \frac{1}{h_{slco}A_{slco}},\ R_{slcw} = \frac{1}{h_{slcw}A_{slcw}},\ R_{slci} = \frac{1}{h_{slci}A_{slci}}$$
$$R_{tlco} = \frac{1}{h_{tlco}A_{tlco}},\ R_{tlew} = \frac{1}{h_{tlew}A_{tlew}},\ R_{tlci} = \frac{1}{h_{tlci}A_{tlci}}$$
$$R_{loss,pl} = \frac{1}{h_{loss,pl}A_{loss,pl}},\ R_{loss,sl} = \frac{1}{h_{loss,sl}A_{loss,sl}},\ R_{loss,tl} = \frac{1}{h_{loss,tl}A_{loss,tl}}$$

$$h_{sleo} = 5000\ , A_{sleo} = ,\ h_{slew} = , A_{slew} = ,\ h_{slei} = 3000\ , A_{slei} = 0.25$$

$$h_{slco} = 2000\ ,\ A_{slco} = 0.25\ ,\ h_{slcw} = ,\ A_{slcw} = ,\ h_{slci} = ,\ A_{slci} =$$

$$h_{tlco} = 5\ , A_{tlco} = ,\ h_{tlew} = \ , A_{tlew} = , h_{tlci} = , A_{tlci} =$$

$$h_{loss,pl} = 1\ , A_{loss,pl} = 15\ ,\ h_{loss,sl} = 1\ , A_{loss,sl} = 0.1\ , h_{loss,tl} = 1\ , A_{loss,tl} = 3$$

Having a heat transfer coefficient for heat transfer from the fuel cladding to the coolant R_c, the cladding temperature T_c is given by:

$$T_c = T_{sl1} + R_c\dot{Q}_r \tag{68}$$

Note that in this simple analysis of a natural circulation nuclear reactor cooling system, that the power conversion unit (PCU) has not been considered. If however it is included, the heat supplied from the steam drum to the power conversion unit could conveniently and simply be included as a heat loss term from the steam drum as a change in enthalpy h and given by:

$$\dot{Q}_{sd2,PCU} = \dot{m}_{PCU}\left(h_{v,sd2,out} - h_{l,sd2,in}\right) \tag{69}$$

Values used in program:

$h_{sleo} = 5000\frac{W}{m^2K}$ Steam condensation on copper pipes

$A_{sleo} = 0.238761\ m^2$

$R_{slew} = \frac{\ln\frac{r_{sleo}}{r_{slei}}}{2\pi L_{sle}k_{cu}}$, $r_{sleo} = 0.0095\ m^2$, $r_{slei} = 0.0075\ m^2$,

$L_{sle} = 4\ m, k_{cu} = 400\frac{W}{mK}$

$h_{slei} = 3000\frac{W}{m^2K}$ Methanol boiling in copper pipes

$A_{slei} = 0.1884956\ m^2$

$h_{slci} = 2800\frac{W}{m^2K}$ Methanol condensation in copper pipes

$A_{slci} = 0.1884956\ m^2$

$R_{slcw} = \frac{\ln\frac{r_{slco}}{r_{slci}}}{2\pi L_{slc}k_{cu}}$, $r_{slco} = 0.0095\ m^2$, $r_{slci} = 0.0075\ m^2$,

$L_{slc} = 6.3\ m, k_{cu} = 400\frac{W}{mK}$

$h_{slco} = 2000\frac{W}{m^2K}$ Water boiling on copper pipes

$A_{slco} = 0.238761\ m^2$

$h_{tlci} = 4000\frac{W}{m^2K}$ Steam condensation in copper pipes

$A_{tlci} = 0.3518584\ m^2$

$R_{tlcw} = \frac{\ln\frac{r_{tlco}}{r_{tlci}}}{2\pi L_{slc}k_{cu}}$, $r_{tlco} = 0.0075\ m^2$, $r_{tlci} = 0.007\ m^2$, $L_{tlc} = 8\ m, k_{cu} = 400\frac{W}{mK}$

$R_{tlcfin} = \frac{1}{h_{fin}A_{fin}\eta_{fin}}$

$h_{fin} = 20\frac{W}{m^2K}$ Natural convection between fins and air

$$A_{fin} = \frac{L}{p} \times L_{fin} \times w_{fin} \times 2 = 6.67\ m^2$$

$$\eta_{fin} = \frac{\tanh(mL)}{mL} \text{ where } m = \sqrt{\frac{2h_{fin}}{t_{fin}}}$$

Results for 5kW input: $T_{sd1} = 106.7$ °C; $T_{sl} = 92.4$ °C; $T_{sd2} = 74.9$ °C

Results for 9kW input: $T_{sd1} = 172.1$ °C; $T_{sl} = 146.4$ °C; $T_{sd2} = 114.9$ °C

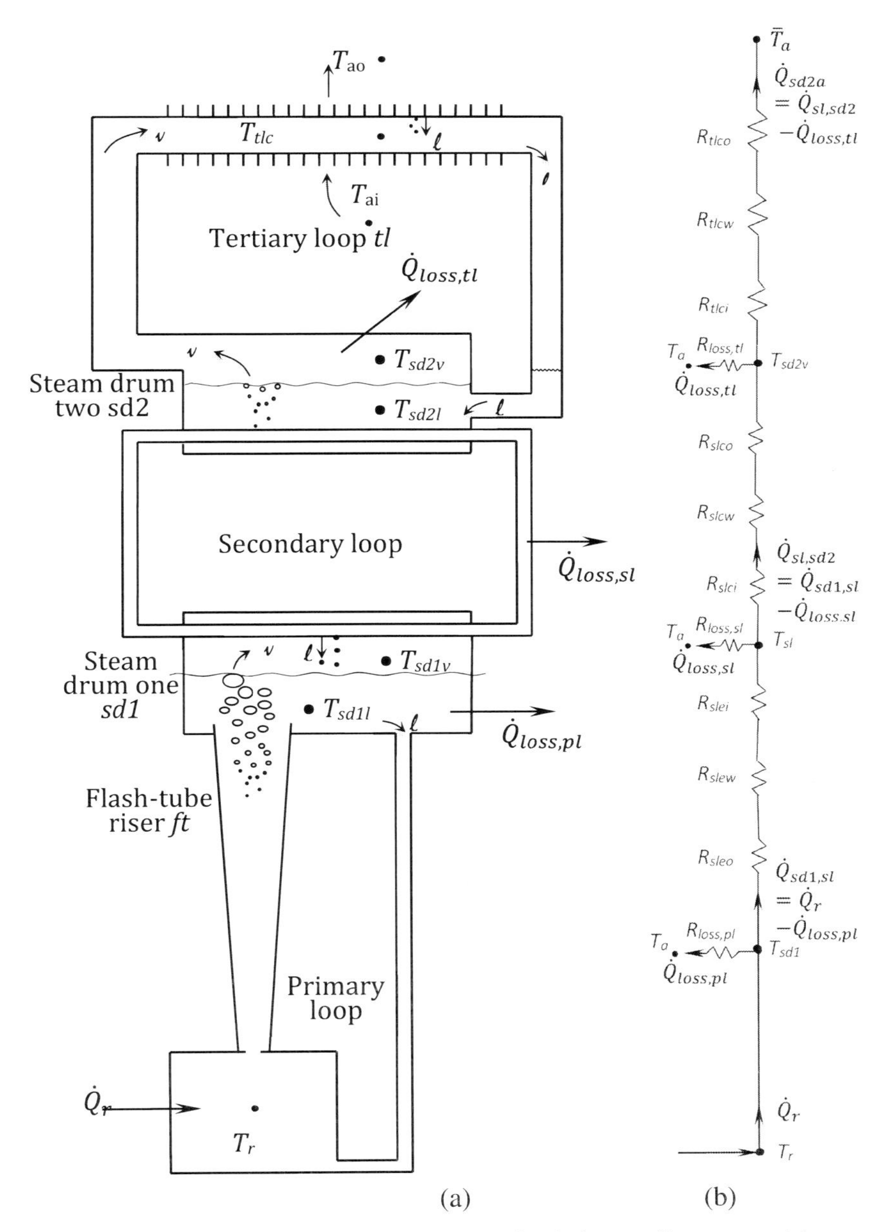

Figure 42. Thermal resistance diagram (b) of a natural circulation cooling system (a).

4. Energy Saving Using Heat Pipe Heat Recovery Heat Exchangers

In the light of an ever increasing demand for energy the need for energy savings has become an important economic consideration. One means of saving energy is to recover a portion of the energy in a warm waste stream and then to use the energy so recovered to preheat another colder stream. A heat pipe heat exchanger (HPHE) is an effective device whereby heat may be recovered from a hot fluid stream and used to preheat another colder stream without the use of any mechanically moving parts such as a pump. There are many such potential applications in the hydrocarbon, process, chemical, food, power and manufacturing and air-conditioning industries where a HPHE can be effectively used to conserve energy, use less electricity, oil or coal and also by so doing reduce pollution. For example, a characteristic of all drying operations is the need for large quantities of energy for the evaporation of water from the product and the subsequent release of large quantities of moist hot air back into the atmosphere.

The HPHE considered in this section is made up of a series of two-phase closed thermosyphon-type heat pipes. This type of heat pipe is simple to construct and can be manufactured and maintained using standard refrigeration technology. Commercial production of HPHEs' began in the mid-1970s and has since found many applications, particularly in process and agricultural air-drying and the heating ventilation and air conditioning industries [8, 56, 57].

A number of heat recovery situations are presented in this section where significant energy savings can be made. It may be concluded that the use of a HPHE offers a practical method with a realisable and significant potential for energy saving in South Africa, or for that matter anywhere in the world. The theory of operation of a single two-phase closed thermosyphon type heat pipe will be formulated, and it will be shown how they may be grouped together to form a typical HPHE. There are many different instances where waste heat can be recovered. A particularly attractive application is to recover heat from the exhaust air in drying of food products in order to deactivate bacterial growth and thereby extend product shelf life. The situations considered in this section include a milk spray drier, a mini food drier, and a heat pump drier. A heat pipe heat exchanger for heating hydrofluoric acid contaminated fumes from the pickling process at a stainless steel factory will also be also considered.

A heat pipe heat exchanger consists of a series of individual thermosyphon heat pipes in a staggered arrangement as shown in Figure 43 to form the single heat exchanger shown in Figures 44 and 45. The heat pipes may be inclined but then the angle of inclination between the heat pipe and the horizontal should be greater than about 30°. At this angle the heat transfer rate increases the most as the temperature difference between the hot and cold stream increase to a maximum before leveling off. Any acceptable refrigerant (for example water, ammonia, methanol, R134a, butane, toluene, naphthalene, etc) may be used to charge a thermosyphon heat pipe. The final choice would depend on a number of factors such as the operating temperatures and pressures, the phase (gas or liquid) of the heating and cooling streams, material compatibility and corrosion properties, environmental conditions and legal requirements. There will thus be many possible liquid charge, pipe material, pipe diameter and length and fin material and fin and pipe spacing combinations. Many different combinations of heating and cooling streams are also possible; in this case air is considered for both the hot and cold streams. Furthermore, to improve the airside heat transfer, the heat pipes are usually finned.

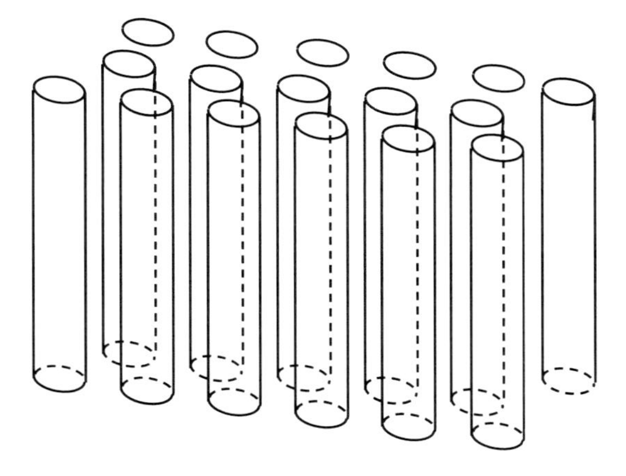

Figure 43. Thermosyphon heat exchanger as a series of individual pipes.

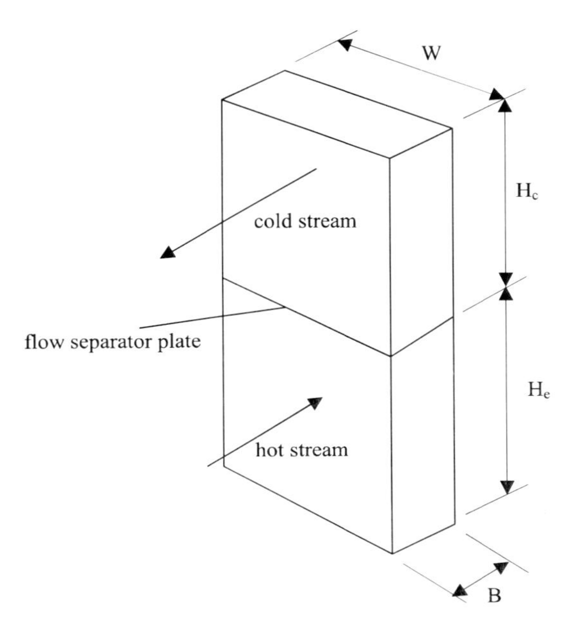

Figure 44. Overall dimensionsof an air-to-air thermosyphon heat exchanger [12].

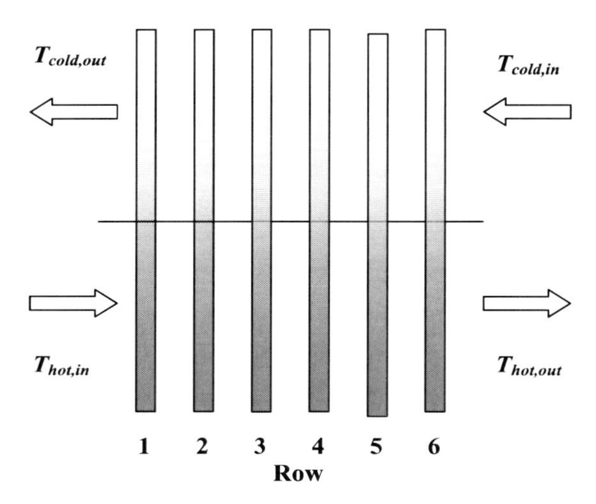

Figure 45. Basic design configuration of the HPHE.

The inside heat transfer coefficients for the evaporator and condensing sections would normally have to be experimentally determined [11, 12, 58, 59, 60, 61]. Although many correlations are available they give largely varying heat transfer coefficients for ostensibly the same geometry and operating conditions. Typically they would vary between about 900 $W/m^{2}°C$ for naphthalene to about 2000 to 3000 $W/m^{2}°C$ for water as working fluids.

Many different combinations of heating and cooling streams are also possible. In this paper however, only air streams have been considered. The outside heat transfer coefficients can be calculated using standard heat transfer textbooks [62], (Incropera and Dewitt, 2002). If the heat transfer rate increases beyond a certain level, the condensate is hindered from returning back to the evaporator by the counterflowing vapour resulting in a *maximum heat transfer rate* (Dobson, 2000). The maximum heat transfer rate for a thermosyphon inclined at 45° is about 40% greater that for a vertical thermosyphon. When the airside pressure drop cannot be calculated using standard heat transfer textbooks it has to be determined experimentally [65].

For any specific set of operating conditions and a given overall heat pipe length there will be an optimum evaporator length to condenser length ratio. The heat transfer rate $\dot{Q}_{\mathrm{HPHE}} = (\overline{T}_h - \overline{T}_c)/\Sigma R$ is given by the overall temperature difference $\overline{T}_h - \overline{T}_c$ divided by the overall thermal resistance ΣR. The overall thermal resistance can in turn be given by the sum of two separate thermal resistances, a resistance between the hot stream and the working fluid temperature and a thermal resistance between the working fluid and the cold stream. These resistances also depend on the evaporator and condenser lengths; if either the evaporator or condenser lengths tend to zero so to does the heat transfer rate. If for instance the evaporator was heated with a liquid stream (with its associated low thermal resistance) whilst the condenser was cooled with an air stream (with an associated relatively high thermal resistance), then for optimum heat transfer the evaporator length would be much shorter than the condenser length. In the case considered in this paper both streams are air and hence both the resistances will be more or less the same. The optimum heat transfer condition would thus also require that both the evaporator and the condenser lengths be more or less equal. Thus for an air heated and cooled situation and to limit the number of possible heat exchanger size options, both evaporator and condenser lengths may be taken as being equal.

The thermal design specifications need to be specified for the HPHE and include the atmospheric pressure, hot stream inlet temperature, cold stream inlet temperature, a desired cold stream outlet temperature and the hot and cold stream mass flow rates. The next step requires an assumed heat exchanger geometry including the pipe diameter length and evaporator to condenser length ratio, the number of rows and the number of tubes per row and whether staggered or not or whether finned or not. The assumed counter flow arrangement/solution shown in Figure 46 now requires a series of nested trial and error iterations such that the evaporator heat transfer rate equals the condenser heat transfer rate and that the desired cold steam outlet condition has been met.

Given or assuming the physical definition of the HPHE (for example the dimensions given in Table 5), the heat transfer coefficients and thermal conductivities (h_{eo}, η_{eo}, A_{eo}, k_w, h_{ei}, h_{ci}, k_c, h_{co}, and also see section 2 of this chapter), and the temperatures (T_{hi}, T_{ho}, η_{eo}, A_{eo}, T_{ci} and T_{co}), the internal temperature T_i can then be found by trial and error by guessing values for T_i such that $\dot{Q}_c = \dot{Q}_e$. In a similar way, temperature and heat flux dependent variables may be taken into account as well. The basic solution procedure requires that the hot and cold stream inlet temperatures be specified. Starting from the 1st row (of the counter flow heat exchanger configuration shown in Figure 46), cold stream outlet temperatures are guesses, and then by "marching" from one row to the next, the inlet cold stream temperature is calculated. This iteration procedure is repeated for different values of cold stream outlet until the calculated cold steam inlet temperature corresponds to the initially specified value.

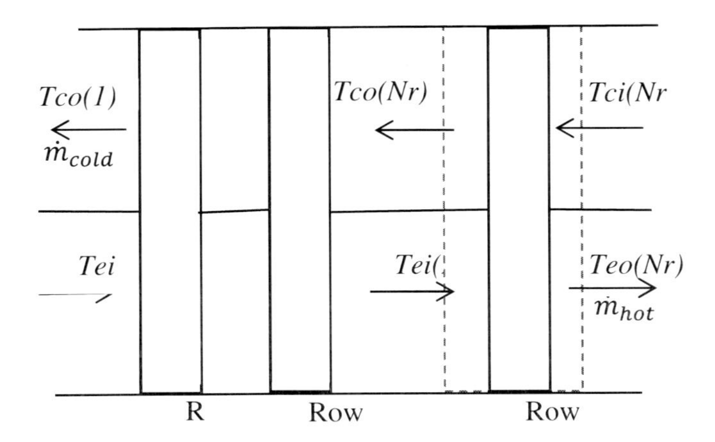

Figure 46. Side view of a counter flow heat pipe heat recovery heat exchanger showing a row of evaporator e sections and their corresponding condenser c sections for N_r rows [66].

A simple iterative numerical solution method may be used to solve the heat transfer equations for each control volume shown in Figure 47.

The energy balance equation is solved at each row of pipes and the obtained values used as the input values for the next row. The procedure is as follows:

1. Input geometric properties, known and desired temperatures

$T_{hotin}, T_{coldin}, T_{desired}, \dot{m}_{hot}, \dot{m}_{cold}, \; L_{evap}, L_{cond}, d_o, d_i, S_t, S_h, N_p, N_r$

2. Create empty arrays of size N_r for thermodynamic values and dimensionless parameters

3. Assign $T_{ei}(1) = T_{hotin}$, $T_{ci}(N_r) = T_{coldin}$, guess $T_{eo}(Nr)$ and from this guess calculate corresponding $T_{co}(1)$ using $\dot{Q}_{hot} = \dot{m}_{cold} c_{p,coldin} (T_{co} - T_{coldin})$

4. Assuming a linear temperature profile through the heat exchanger,

$$grad_{hot} = (T_{ei}(1) - T_{eo}(N_r))/N_r$$

$$grad_{cold} = (T_{co}(1) - T_{ci}(N_r))/N_r$$

$$T_{co}(N_r) = T_{ci}(N_r) + grad_{cold}$$

5. Choose tube bank configuration and add extra geometric variables

$$d_f, t_f, P_f$$

6. Error between calculated outlet temperature and initial inlet temperature must be minimised, thus assign *err = 1*

7. While *err* > 0.1

Start calculation procedure

For j = N_r to j = 1

i. Calculate thermodynamic properties

$$P_{atm} = f(altit_{searef}), cp_i = f(T_i), \rho_i = f(T_i, P_{atm}), k_i = f(T_i), \mu_i = f(T_i), Pr_i = f(cp_i, \mu_i, k_i)$$

Calculate Reynolds number

$$Re_i = f(T_i, \dot{m}_i, d_o)$$

Calculate outside heat transfer resistances: $R_{io} = {}^1/_{h_{io}A_{io}}$

Calculate inside heat transfer properties and resistances: $R_{ii} = {}^1/_{h_{ii}A_{ii}}$

Calculate heat transfer for each evaporator/condensor and thermosyphon

$$\dot{Q}_{evap/cond} = \frac{abs(\bar{T}_{h/c} - T_{inside})}{R_{evap/cond}}$$

$$\dot{Q}_{hp} = \frac{\bar{T}_h - \bar{T}_c}{R_{tot}}$$

$$\dot{Q}_{hot/cold} = \dot{m}_{hot/cold} cp_{hot/cold} (T_{hotin/coldout} - T_{hotout/coldin})$$

ii. While $\left(\dot{Q}_{hot/cold} - \dot{Q}_{hp}\right) > 0.001$

$$T_{conew} = T_{co} - 0.001 \times (\dot{Q}_{hot/cold} - \dot{Q}_{hp})$$

$$T_{einew} = T_{eo} + \frac{\dot{m}_{cold} cp_{cold} (T_{conew} - T_{ci})}{\dot{m}_{hot} cp_{hot}}$$

Repeat section i.i with adjusted temperatures, until the heat transfer rates are within the loop limit

Move on to the next row of pipes

$T_{eo}(i\text{-}1) = T_{ei}(i)$, $T_{ci}(i\text{-}1) = T_{co}(i)$

Repeat until Row 1 is reached

Compare calculated inlet evaporator temperature to actual evaporator inlet temperature

$err = T_{hotin} - T_{ei\_calculated}$

if $T_{hotin} > T_{ei\_calculated}$

increase $T_{eo}(Nr)$, go to Step 7

if $T_{hotin} < T_{ei\_calculated}$

decrease $T_{eo}(Nr)$, go to Step 7

if $err < 0.1$

$$\dot{Q}_{des} = \dot{m}_{cold} cp_{cold} (T_{des} - T_{coldin})$$

If calculated temperatures are correct, go to Step 8

8. Evaluate whether desired heat transfer is attained

if $\dot{Q}_{hp\_total} \geq \dot{Q}_{desired}$ then display values and end program

if $\dot{Q}_{hp\_total} < \dot{Q}_{desired}$ then tell user desired heat transfer is not attainable for selected geometry, reset program

4.1. Milk Spray Drying

A mathematical model to quantitatively determine the energy recovered from the waste hot stream has been developed [59]. The recovered energy is then used to preheat the incoming cold air stream. For a milk spray drier as shown in Figure 48 and operating under the (typical) operating conditions as given in table 1, 798 kW (26.5% of the total energy of 3013 kW needed) can be recovered to preheat the incoming air. A suitable heat pipe heat exchanger for this duty is specified in Table 5 for the operating conditions specified in Table 4. The amount of energy that can be recovered is even more significant in milk spray drying as they have to be operated every day throughout the year.

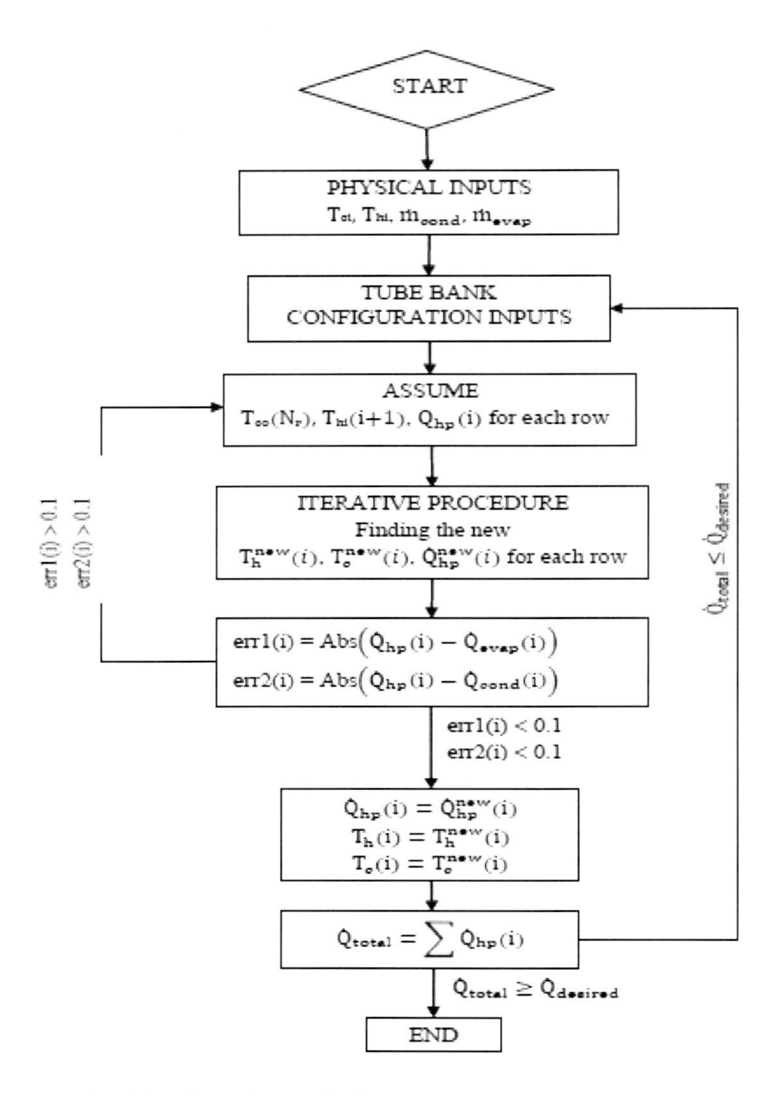

Figure 47. Computer algorithm flow diagram [66].

For the spray drier operating for 18 hours per day and an oil fuel energy cost of R2.50/kg /11.25 kWh/kg = R0.23/kWh, the annual savings would be 798kW·18 h/day·365days/year·R0.23/kWh = R1 206 000 per year. It is estimated that the heat exchangers would cost in the order of R234 000, and assuming additional installation costs of about R650 000 and maintenance costs R200 000/year, a payback period in the order of slightly less than one year is possible.

The model also shows that a significant amount of energy at a rate of 254 kW is lost from the drying chamber to the surroundings. This indicates that further energy savings may be possible by appropriately insulating the drying chamber. Appropriate insulation in this case would be to use one or more radiation shields, spaced close enough to each other in order to suppress convection heat transfer.

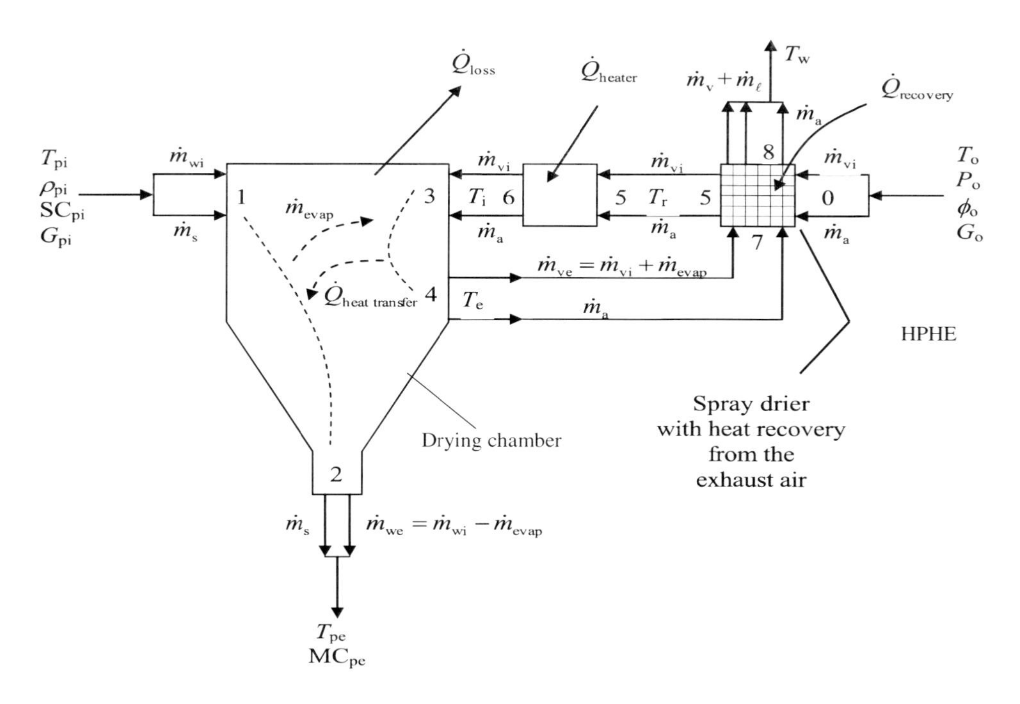

Figure 48. Schematic drawing of a milk spray drier [59].

4.2 Mini Food Drier

In this case a HPHE may be used to recover the waste heat from a relatively small commercially available air drier and then used to preheat the incoming cold air (Dobson and Meyer, 2006). This mini drier is typically used for small scale drying of fruits, vegetables, herbs, meat and other products and its wet loading capacity is between 50 and 250 kg. The energy saving potential was evaluated for the drier user's required specification for the HPHE as given in Table 7.

The experimental set-up consisted of the HPHE (shown in Figure 49) and it was retrofitted to a standard drier using flexible ducting as shown in Figures 50 and 51. The drier's overall dimensions are 2.8 m long, 1.4 m wide and 1.9 m high. The overall dimensions of the HPHE are shown in Figure 49 and its detailed specifications are given in Table 5 for the operating conditions given in Table 4. It was manufactured in accordance with standards and technology for copper pipe and aluminium plate finned heat exchangers as normally applied in the HVAC industry. The air-drier typically evaporates water from the product being dried and exhausts this moist warm air into the atmosphere. With the HPHE installed, this warm moist air is then fed through the evaporator section of the HPHE. Fresh ambient air is then drawn through the condenser section of the HPHE where it is preheated.

Temperature measurements were taken at the inlets and outlet of the respective hot and cold streams and an anemometer was used to measure the flow velocities from which the air mass flow rates could be calculated. A kWh-meter was used to measure the electrical energy consumption. To ensure accuracy and repeatable drying operations, the product to be dried was simulated using wet towels laid out on the drying racks.

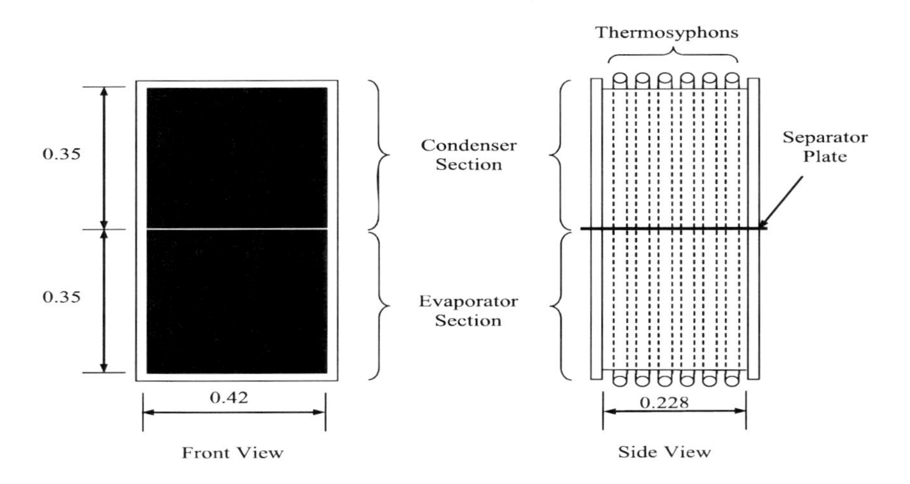

Figure 49. The as-designed and manufactured HPHE (also, see Table 4).

The results of an economic evaluation are given in Table 8 for the drier operating parameters given in Table 6 and the HPHE specifications given in Table 7. The actual material and labour costs including a nominal mark-up incurred to retrofit the HPHE are reflected in Table 8, which amounted to R7 469, and with the energy saving minus additional running costs of R2 321 yielded a simple payback period of 3.2 years. The anti-corrosion epoxy protective coating cost of R2231 constituted a significant additional cost. The manufacturer of the HPHE claimed that their experience showed that it would not be necessary for typical agricultural produce drying. The manufacturer of the drier on the other hand required the protective coating. Had this cost not been incurred, the payback period would have been only 2.3 years. Had the HPHE been included in the design as a standard production feature, an even lower payback period of about 1.8 years is deemed possible?

Table 4. Spray drier operation conditions

T_{pi} = 60°C, ρ_{pi} = 1171 kg/m^3, G_{pi} = 4000 ℓ/hour
SC_{pi} = 55% $kg_{solids}/kg_{solids+water}$,
T_o = 20°C, P_o =100 kPa, ϕ_o =50%, G_o = 17.5 m^3/s, T_i = 162°C
Dependent variables:
$T_{pe} \approx T_e \approx$ 87°C, $MC_{pe} \approx$ 6%

Table 5. Specification (for one of the four in total needed) of the heat pipe heat recovery heat exchanger for the milk spray drier case study

Working Fluid	Water
Fins material and type	Al Plate
Pipe mate`l (staggered)	Cu tube
Evaporator length	1.00 m
Condenser length	1.00 m
Number of tube rows	9
Number of tubes per row	50
Longitudinal pitch	0.039 m
Transverse pitch	0.039 m
Fin pitch	445 fins/m
Fin thickness	0.0004 m
Outside diameter of tubes	0.0159 m
Inside diameter of tubes	0.0149 m
Pressure drop	165 Pa

Table 6. Mini food drier parameters

Inlet hot temperature	40 – 60°C
Inlet cold temperature	Ambient air
Desired outlet temperature	Whatever is attainable
Mass flow of the air into the condenser section	0.72 kg/s
Mass flow of the air into the evaporator section	0.72 kg/s

Table 7. Design specification for the HPHE retrofitted to the mini food drier

Working Fluid	R134a
Fin	Al Plate
Pipe material (staggered grid)	copper
Evaporator length	0.35 m
Condenser length	0.35 m
Number of tube rows	6
Number of tubes per row	11
Longitudinal pitch	0.0381 m
Transverse pitch	0.0381 m
Fin pitch	10/25.4 mm
Fin thickness	0.0002 m
Outside diameter of tubes	0.01588 m
Inside diameter of tubes	0.01490 m

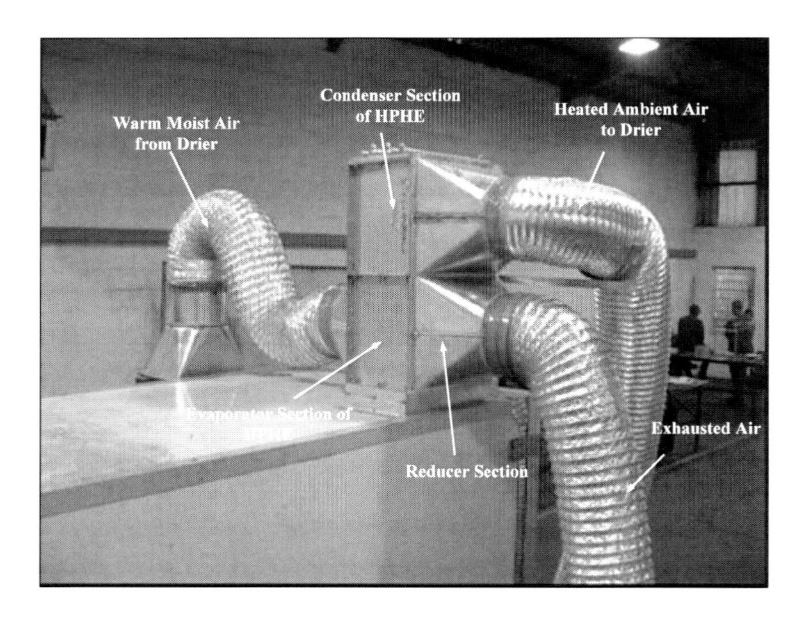

Figure 50. Experimental HPHE retrofitted to a drier [66].

Table 8. Economic evaluation for the mini-drier (in 2004-Rands)

Additional costs with HPHE:		
Standard HPHE	R	2660
Anti-corrosion epoxy coating	R	2231
Variable speed fan	R	1000
Ducting	R	78
Reducer sections	R	500
Labour	R	1000
Total HPHE installation cost	R	7469
Additional HPHE running cost	R/year	400
Energy consumption without HPHE:		
Electrical energy consumption per hour	kWh/h	10.4
Number of operating shifts per annum	shifts/year	250
Number of hours per shift	h/shift	18
Cost of electricity	R/kWh	0.18
Annual cost of energy without HPHE	R/year	8456
Energy consumption with HPHE:		
Electrical energy consumption per hour	kWh/h	7.08
Number of operating shifts per annum	shifts/year	250
Number of hours per shift	h/shift	18
Cost of electricity	R/kWh	0.18
Annual cost of energy with HPHE	R/year	5735
Saving per year (3 – 4 – 2)	R/year	2321
Simple payback period (1/5)	years	3.2

4.3. Heat Pump Drying

There are a number of advantages in using heat pump driers. High energy efficiencies are achievable because both the sensible and the latent heat of vaporization are recovered, drying can be carried out at relatively low temperatures and drying can be conducted independently of the ambient weather conditions. Figure 52 shows a schematic drawing of a drier using an electrical resistance heater. Fresh air is heated to state 6 and passes over the product where its temperature drops and the humidity increases to a state 1. The moisture evaporated from the product leaves the drier system in the relatively humid outlet air. In this system all the latent heat required to evaporate the water from the product and which was supplied by the electrical resistance heater is lost to the surroundings in the vented outlet air.

A portion of the sensible heat can be recovered by re-circulating some of the outlet air as shown in Figure 53. Although the energy recovered in this way is significant, it is nevertheless only a small fraction of the total electrical energy supplied to the system. By making use of a heat pump all the latent heat of vaporization can be recovered as well as all the electrical energy supplied to the compressor. Such a system is shown in Figure 54. If the drier is well insulated it is possible that the energy supplied to the system will be more than enough to keep it in equilibrium and provision must be made to vent some of this excess energy in order to not exceed the specified drying temperature condition. There are basically two ways of doing this: by including an external condenser in the system or by taking in fresh air and venting some the heated air. If an external condenser is used and no air is vented, it is called a closed system; and this would be the indicated method if a controlled atmosphere was being used. On the other hand, when inlet air to the system is vented to the atmosphere it is called an open system.

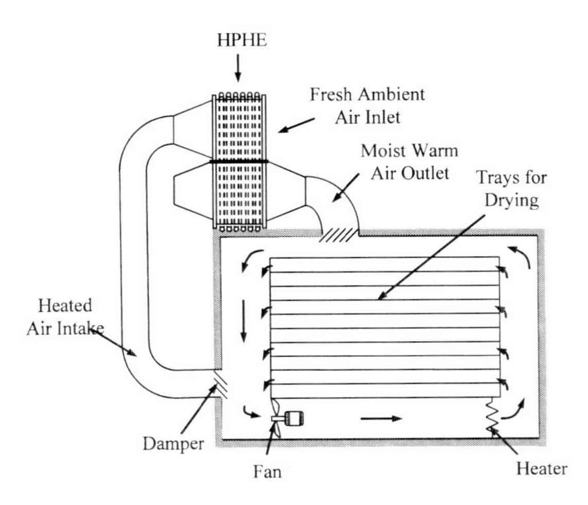

Figure 51. Schematic diagram of the HPHE retrofitted to a mini-drier [61].

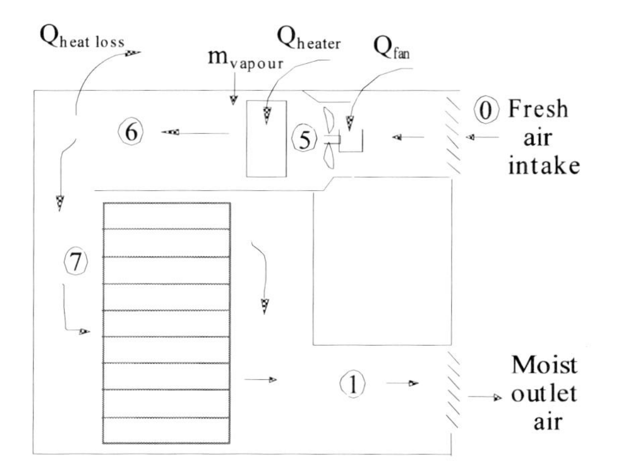

Figure 52. Once through drier.

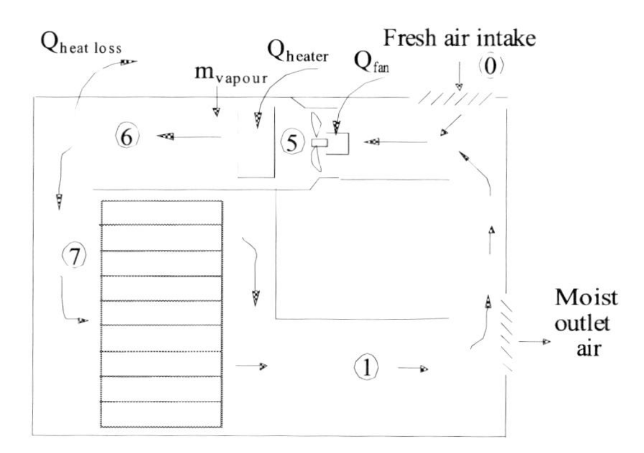

Figure 53. Drier with re-circulation.

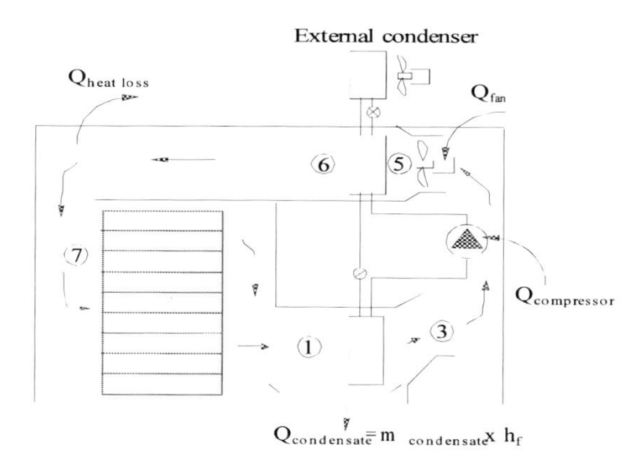

Figure 54. Heat pump Drier [61].

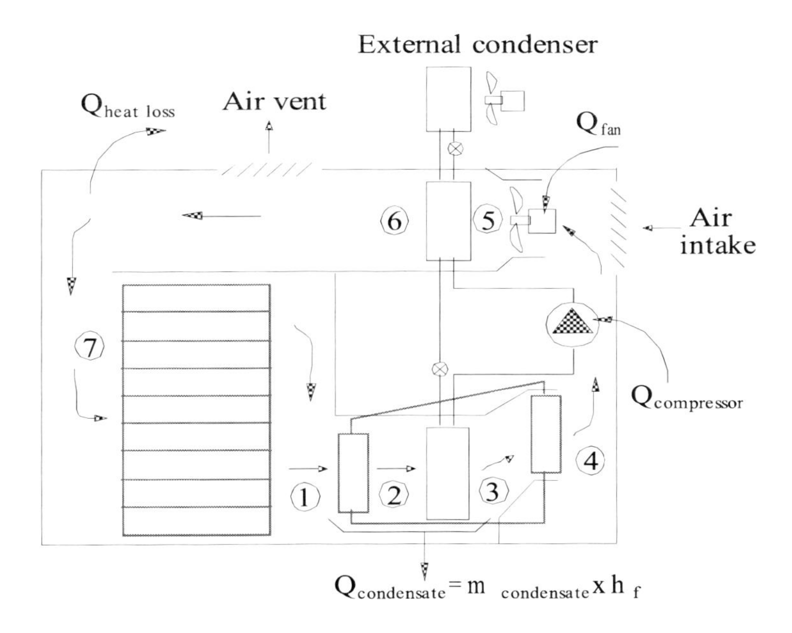

Figure 55. Heat pump Drier with a *wrap around* heat pipe heat exchangers (and air recirculation option).

In Figure 55 the use of a loop heat pipe heat exchanger is shown as a means of recovering energy. In the evaporator (between states 1 and 2) of the loop, the air is cooled and in the upper condensing portion (between state 3 and 4) the air is heated. In this loop HPHE refrigerant is boiled in the lower evaporator and the vapour flows up in the riser to the condenser, the condensate so formed then flows back under the influence of gravity in the return line back to the evaporator. In this way the latent heat absorbed in the evaporator is recovered in the condenser.

For the same drying conditions (at sea level) and with ambient air 24°C, 50% RH, heated air entering the product trays 45°C, 20% RH (state 7), and air leaving the product 30°C, 70% (state 1) then for a 1 kW heater power consumption for the once through drier the heater load can be reduced to 0.71 kW using recirculation, to 0.35 kW using a heat pump and 0.29 using a *wrap around* loop heat pipe heat exchanger. In Figure 92 an often-called *wrap around* (or a *separated* heat pipe heat exchangers as shown in Figure 56) extracts heat from the incoming moisture laden air prior to it being passed over the colder evaporator coil of a heat pump refrigeration cycle. In this way a smaller evaporator coil is required as well as reducing the amount of cooling needed to extract the moisture from the air. The now relatively dry but cold air is then pre-heated using the energy extracted from the incoming air thus reducing the heating load required to heat the air up to the hot drying temperature.

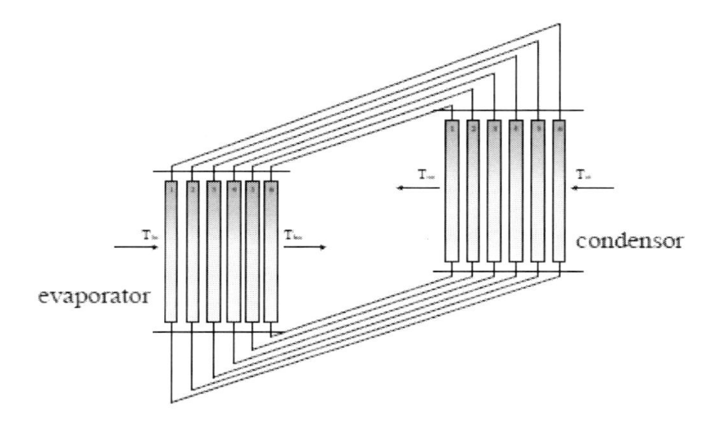

Figure 56. A separated heat pipe heat recovery heat exchanger [66].

Figure 57. A 500kW heat pipe heat recovery heat exchanger for a stainless steel pickling plant.

4.4. Acid Pickling Process Plant

Figure 57 shows an image of a 500 kW heat pipe exchanger during its installation at a stainless steel mill. Highly corrosive, hot hydrofluoric acid-contaminated exhaust fumes from the pickling process need to be heated to 450°C prior to a further cleaning process before being released back into the environment. To do the pre heating, a waste hot air stream of 480°C is used. To meet this corrosive high temperature requirement, the HPHE was manufactured in accordance with strict quality assurance procedures using only stainless steel 316L. To meet the temperature requirements, Toluene was used as the working fluid in the low temperature heat pipes whilst Naphthalene had to be used in the high temperature heat pipes. A high temperature experimental rig was built to verify the inside heat transfer coefficients and a special working fluid charging apparatus was developed.

5. Pulsating Heat Pipes

A pulsating (or oscillatory or oscillating) heat pipe (PHP), a heat transfer device invented some twenty five years ago by Akarchi [67, 68] had already been intensively researched and many application areas were considered. In particular, its potential in the cooling of electronic equipment had been emphasized by Akarchi and Polášek [69, 70).

A PHP consists of a relatively long sealed tube that has been wound back and forth in one plane; reminiscent of a river as it meanders along a relatively flat plain. It is evacuated and then charged with a working fluid (for example water, ammonia, methanol or Freon) at a liquid fill ratio (liquid to total volume) of about 50%. The liquid in the pipe then tends to naturally distribute itself into many ostensibly randomly spaced and sized liquid-slugs and vapour-bubbles as shown in Figure 58 and which depicts a number of typical PHP configurations. If the diameter is too large, the liquid will tend to stratify into a single liquid layer and a vapour layer, and if the diameter is too small then the intensity of the relative movement between the liquid and vapour may be inhibited due to capillarity. The ideal diameter is in the order of 0.5 to 1.8 of the so-called capillary length L, as given by $L = \sqrt{\sigma/[(\rho_l - \rho_v)g]}$ where σ is the surface tension in N/m, ρ_l and ρ_v are the liquid and vapour densities in kg/m^3 and g is the gravitational acceleration in m/s^2. Typical tube diameters meeting this requirement for water at 20°C are from about 2 to 5 mm.

If the one end of the PHP is heated and the other end cooled (see Figure 58), heat is transferred from the heat source to the cooler heat sink. This is seen to always spontaneously occur in the so-called *bottom heat mode* (that is when the heated end is below the cooled end) even at low (< 40°C) temperature differences. The liquid plug vapour bubble behaviour in this heat mode is characterized (as observed in glass PHPs) by a seemingly sudden and chaotic burst of rapid motion of the plugs and bubbles, back and forth, and around the bends, followed by a relatively quiescent dwell period, and so on as this cycle is repeated. This rapid motion followed by quiescent periods continues for as long as there is a temperature difference between the heated and cooled ends, hence the name *pulsing heat pipe*, and heat continues to be transferred. This is what may be called a low heat flux operating condition. If the heated section is above the cooled end (called the *top heat mode*), the heat pipe will start to transfer heat, and against gravity, when the temperature difference exceeds about 60°C [71]. This can be called the *high heat flux* operating mode. In this mode the liquid plug and vapour bubble flow pattern develops into a uni-directional annular type of flow reminiscent of the annular convective two-phase flow and heat transfer flow pattern regime.

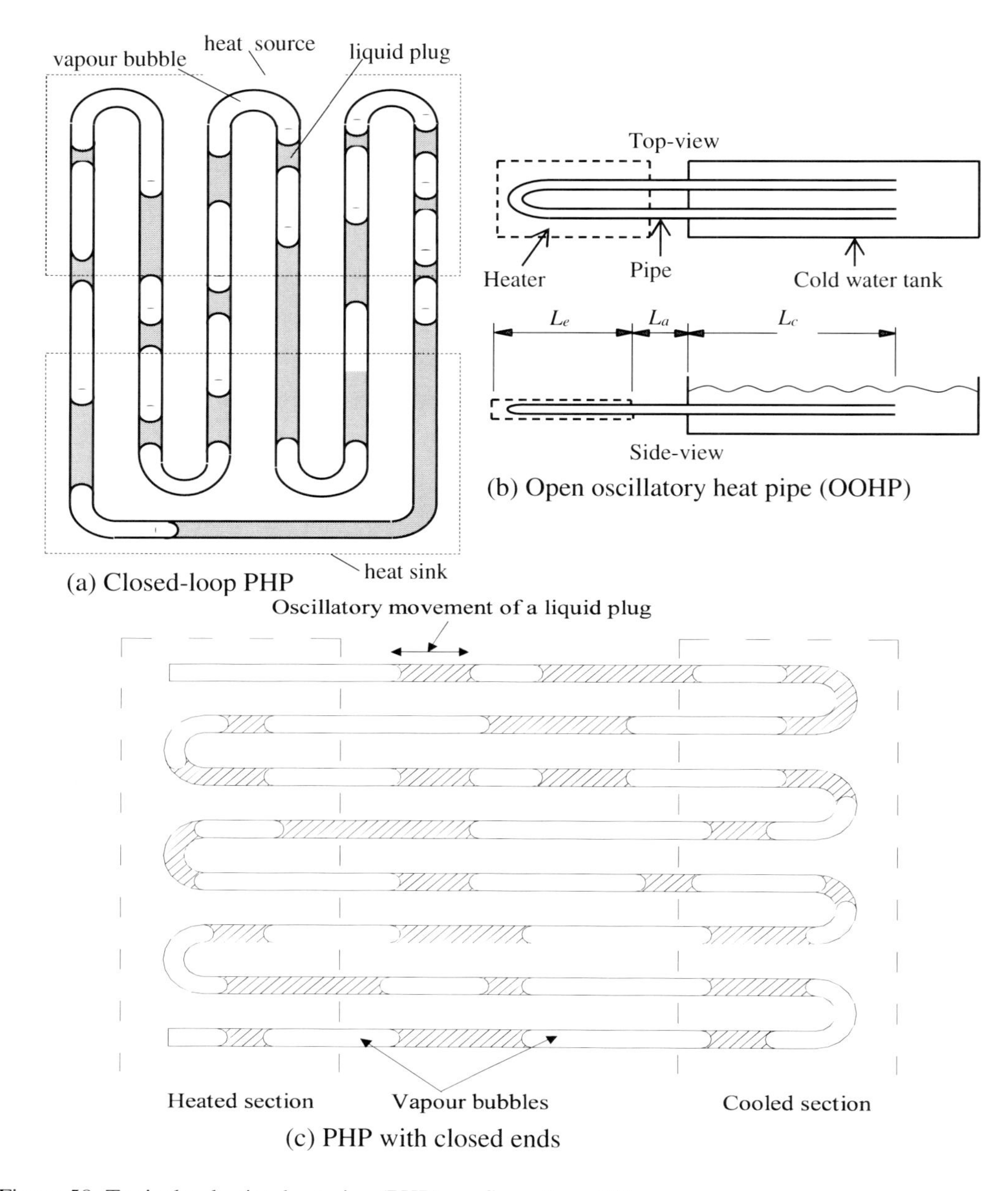

Figure 58. Typical pulsating heat pipe (PHP) configurations [32].

Note that to initiate this high heat flux top heat mode and if the cold source is in the order of say 20°C, then the hotside temperature must be in the order of 80°C. This hot side temperature exceeds the generally accepted maximum allowable casing temperature of 60°C allowed for the cooling of electronics; thus excluding it as a potential electronic component cooling option. Nevertheless, researchers continue to suggest its potential usefulness for the cooling of electronic components.

The relatively simple liquid plug-vapour bubble geometry in low heat flux mode lends itself to theoretical analysis and mathematical formulation by applying the equations of change (the conservation of mass, momentum and energy) to the relatively discrete liquid and vapour control volumes, and using the constituent working fluid property equations for closure. Wong et al. [72] proposed a simple liquid plug and vapour bubble theoretical model. Dobson and Harms [32] however included a film thickness in their simple liquid plug and vapour bubble model and showed that the model tended to capture the oscillatory behaviour of PHPs. This theoretical approach was also applied to a single vapour bubble-two-liquid plug open oscillatory heat pipe and it was able to successfully capture the chaotic behaviour that is observed to occur in this type of PHP [73, 74]. Other papers by Zhang, Faghri [75] Shafii, Faghri and Zhang [76], Yang, Khandekar, Groll [77], Shao and Zhang [78] have also presented theoretical models, and others [79] have experimentally explored the detailed behaviour of PHPs. Tripath, Khandekar and Panigrahi [80] consider the liquid-vapour-solid contact region. Only one other paper by Senjaya, Inoue and Suzuki [81] pertinently considers the liquid film thickness. In this paper no mention is made of what the thickness of the liquid left behind by the trailing edge of the liquid plug might be but it assumes however that the film thickness is the same in both the evaporating and condensing sections.

If the PHP is charged with a 50% fill ratio of working fluid then, correspondingly, about 50% of the inside pipe surface area is in contact with liquid plugs while the rest of the inside surface is in contact with vapour bubbles. Once the PHP starts working (that is transferring heat), the moving liquid plugs leave behind (at their trailing ends) a thin layer of fluid of finite thickness on the pipe surface and at the leading front take up liquid left behind by the preceding liquid plugs. At any one time, 50% of the heat transfer, ignoring any axial heat transfer, is associated with liquid plugs. For the remaining surface, the heat transfer will then be associated with a relatively intact liquid film; increasing in thickness in the condensing sections and decreasing in thickness in the evaporating sections. For the theoretical modelling of pulsating heat pipes, an idea of the thickness of the liquid film deposited at the trailing edge as it moves through a capillary tube is thus needed. Because the boiling and condensation rates of thin liquid films is very much higher than that of convetion/conduction heat transfer rate, it can be expected that the so called *latent* heat transfer rate associated film evaporation and condensation is significantly higher than the heat transfer rate by conduction and convection between the liquid plug and axial conduction between the vapour and the liquid plug and the tube wall, and which is sometimes referred to as *sensible* heat transfer.

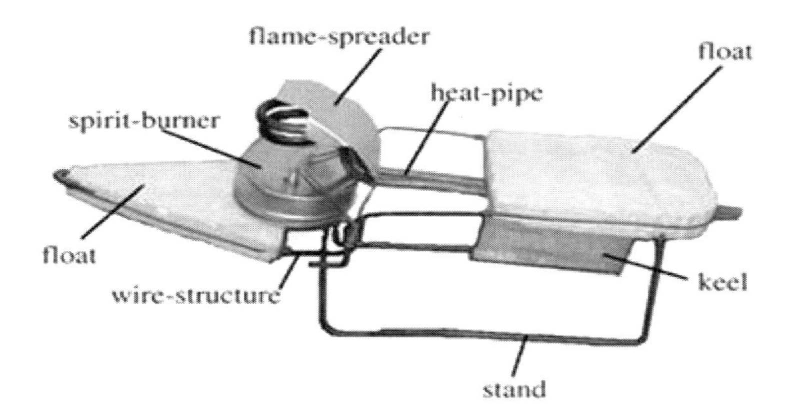

Figure 59. The Great African Lakes Steamer - an example of a toy open oscillatory heat pipe (PHP) out of the water and on a stand [74].

5.1. Theory of Operation

Consider one symmetrical half of the OOHP as shown in Figure 59. The right hand corresponds to the centre of the OOHP while at the left is one of the open ends. The liquid in the hot section evaporates and the vapour bubble pressure increases and forces the liquid plug to the left. As the liquid plug moves to the left water flows out of the pipe and into the cylinder. The liquid plug now tends to oscillate back and forth in the cooled section with a reduced displacement and with the vapour bubble-liquid plug interface in the cooled section. When the liquid plug is moving in this mode liquid continues to evaporate in the heated end and vapour condenses simultaneously in the cooled section. When all the liquid in the heated section has evaporated, the vapour bubble pressure will decrease due to the continued condensation of vapour in the cooled section. The liquid plug now moves back into the heated section; liquid moves into the pipe from the cylinder into the OOHP and liquid is drawn into the heated section. The vapour bubble pressure again increases and the liquid plug moves out of the heated section leaving behind at its trailing end a relatively thin film of liquid in the inside surface of the pipe, and so on.

This characteristic plug-bubble flow behaviour may be simulated by applying the equation of change to three one-dimensional control volumes; a control volume representing the vapour bubble entrapped in the closed end, a control volume representing a liquid plug moving back and forth at the open end, and an annular control volume representing the thin liquid film deposited on the inside surface of the pipe by the trailing end of the liquid plug, as shown in Figure 60.

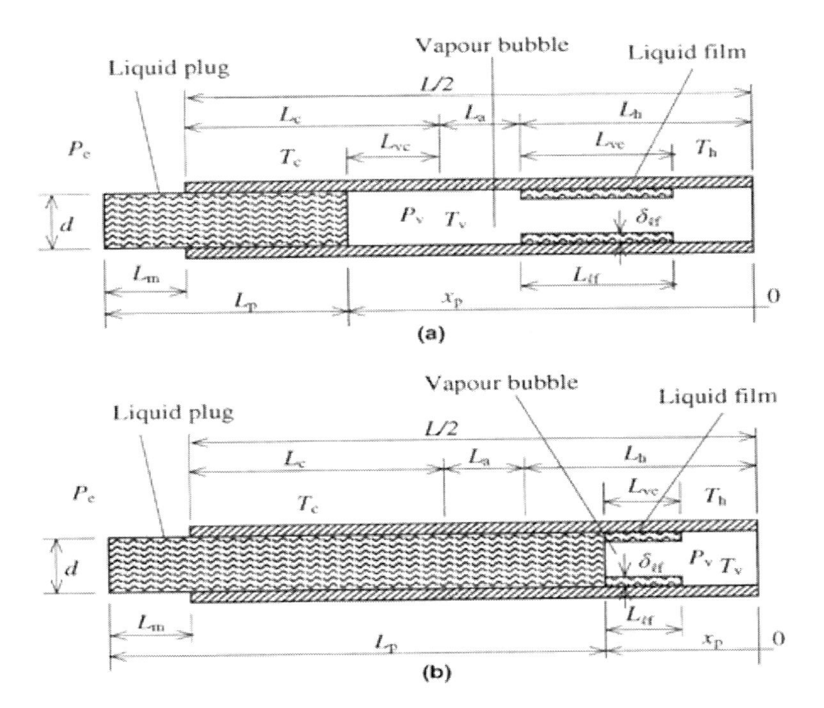

Figure 60. One symmetrical half of a simple open oscillatoty heat pipe (with its centre of symmetry at the origin 0, and with the curved section straitened out); with $x_p > L_h + L_a$ (a) and with $x_p < L_h$ (b) [74].

5.1.1. Conservation of Mass

The rate of change of mass of the vapour bubble with time depends on the net mass flow rate

$$\frac{\Delta m}{\Delta t} = \dot{m}_{\mathrm{vi}} - \dot{m}_{\mathrm{ve}} - \dot{m}_{\mathrm{v,dep}} \tag{70}$$

The mass flow rate of vapour entering the vapour bubble is by evaporation of the liquid film in the heated length

$$\dot{m}_{\mathrm{ve}} = \dot{m}_{lfe} = U_{\mathrm{e}} \pi d L_{\mathrm{ev}} (T_{\mathrm{h}} - T_{\mathrm{v}}) / i_{\mathrm{fg}} \tag{71}$$

L_{ev} is the length of the liquid film in the evaporator that is in contact with the vapour bubble and U_i is a characteristic overall heat transfer coefficient for heat transfer from the heat source into the liquid film. Similarly, the rate at which mass is leaving the vapour bubble by condensation onto the cooled length in contact with the vapour L_{vc} is given by

$$\dot{m}_{\mathrm{vc}} = \dot{m}_{lfc} = U_{\mathrm{c}} \pi d L_{\mathrm{cv}} (T_{\mathrm{v}} - T_{\mathrm{c}}) / i_{\mathrm{fg}} \tag{72}$$

Under certain conditions the temperature of the vapour may rise above the heat source temperature due to the work done by the liquid plug on the vapour as it moves towards the closed end of the pipe. Under this condition

$$\dot{m}_{\mathrm{v,dep}} = U_{\mathrm{ve}}\pi d L_{\mathrm{v,dep}}(T_{\mathrm{v}} - T_{\mathrm{h}})/i_{\mathrm{fg}}$$ $$\dot{m}_{\mathrm{v,dep}} = U_{\mathrm{ve}}\pi d L_{\mathrm{v,dep}}(T_{\mathrm{v}} - T_{\mathrm{h}})/i_{\mathrm{fg}}$$ (73)

where U_{ve} is assumed to be equal to U_{e}.

Eqs. (75), (76) and (77) have assumed that the mass transfer rates for evaporation and condensation are essentially determined by the rate of heat transfer to or from the surface. Typically the total thermal resistance for heat transfer through a pipe wall would depend on three resistances; the thermal resistance for heat transfer from the source/sink to the outside of the pipe wall, across the pipe wall and boiling/condensation. Of these three resistances the last two are relatively small compared to the first and thus the controlling resistance is essentially the heat transfer from the heat source/sink to the outside of the OOHP wall.

The rate of change of mass of the liquid film similarly depends on the net mass flow

$$\frac{\Delta m_{lf}}{\Delta t} = \dot{m}_{lf\mathrm{i}} - \dot{m}_{lf\mathrm{e}} - \dot{m}_{lf,\mathrm{dep}} \tag{74}$$

At the trailing edge, the liquid plug deposits a layer of liquid of thickness $\delta_{\ell\mathrm{f}}$ on the pipe wall as it moves towards the open end of the pipe. This mass rate of liquid deposited is given by:

$$\dot{m}_{lf,\mathrm{dep}} = \rho_l \pi d\, \delta_{lf} v_p \tag{75}$$

An idea of the film-thickness $\delta_{\ell\mathrm{f}}$ can be obtained by experimental observation and evaluating the flow of water through a glass tube. For water at 20°C and a 4 mm diameter glass tube a value of about 0.030 mm, it is a typical thickness of the liquid film deposited on the wall. Applying a multi-linear regression technique [82] to an experimentally obtained data set using distilled de-aerated water as the working fluid was obtained. The experimental variable ranged for inside diameters 1.5 to 2.86 mm, temperatures between 0 and 60°C, and hence with surface tensions of 0.0671 to 0.0786 N/m and dynamic viscosities of 0.00049 to 0.00176 Pas, and average velocities of 0.33 to 1.8 m/s. The results of this analysis is given by way of a power series with a correlating coefficient R^2 = 82% as:

$$\delta_{\ell f} = 603\,1 d_i^{0.485}\sigma^{1.931}v_p^{0.0942}\mu^{-0.130} \tag{76}$$

where $\delta_{\ell f}$ is in μm if d_i is in μm, σ is in N/m, v is in m/s and μ is in Pas. Of the independent variables in Eq. 76, the surface tension is the most important, followed by the diameter whilst the plug velocity and viscosity tend to show a more-or-less negligible effect on the film thickness.

5.1.2. Conservation of Mass

The mass of a liquid plug is given in terms of the bubble length x_p and the pipe half-length $L/2 = L_h + L_a + L_c$ as

$$m_p = \rho_l (L/2 - x_p + L_m) \pi d^2/4 \tag{77}$$

The length L_m is introduced to take into account the mass of the water already outside the open ends that have to be moved to make way for the liquid emerging from the open end as well as the inlet resistance as liquid from the cylinder moves back into the OOHP.

5.1.3. Conservation of Momentum

The forces acting in the direction of motion on the liquid plug are due to shear (friction), gravity, surface tension, and vapour bubble and external water pressures. The equation of motion for the liquid plug is:

$$m_p \frac{\Delta v_p}{\Delta t} = -F_\tau \pm F_g - F_\sigma + F_P \tag{78}$$

The forces are:

$$F_g = \rho_\ell L_p \pi d^2 g \sin\phi / 4 \tag{79}$$

where ϕ is the inclination of the OOHP to the horizontal

$$F_\sigma = \pi d \sigma (\cos\theta_{T,L} - \cos\theta_{L,T}) \tag{80}$$

Eq. (80) is the net capillary force on a liquid plug that has both its ends in a pipe. Referring to Figure 60, it is seen that only one end of the liquid plug is in the pipe. The contact angle for the end out of the pipe is taken as $\theta_{L,T} = 90°$.

$$F_P = \pi d^2 (P_v - P_e) / 4 \tag{81}$$

where P_e is the liquid pressure at the open end of the OOHP

$$F_\tau = \tau\pi d L_p = C_f \rho_\ell \pi d L_p v_p^2/2 \qquad (82)$$

The coefficient of friction is (conveniently) approximated by $C_f = 0.078Re^{-0.25}$ for $Re > 1180$, $16/Re$ for $Re < 1180 < 1$, and $Re = 1$ for $Re < 1$, and $Re = \rho_\ell v_p d/\mu_\ell$. These often-called *slip-stick* parameters have been arbitarilly given here as being more-or-less where the laminar flow friction factor would cut the tube turbulent flow curve on the well known Moody diagram. The liquid plug length L_p includes the additional length L_m to take the exit and entrance resistances as the liquid emerges and re-enters OOHP. It is assumed that the coefficients of friction between the liquid plug and the pipe and the liquid plug and the surrounding liquid are the same.

The vapour bubble and liquid film momentum are both relatively small compared to the liquid plug momentum and are neglected.

5.1.4. Conservation of Energy

The internal energy and hence the temperature of the vapour bubble as a function of time depends on the net convective heat transfer, the net enthalpy, and the rate at which the bubble is doing work on the liquid plug and is given (ignoring the kinetic and potential energies) by:

$$\frac{\Delta E_v}{\Delta t} = \dot{m}_{vi} i_{vi} - \dot{m}_{ve} i_{ve} - P_v A \frac{\Delta x_p}{\Delta t} \qquad (83)$$

where $\Delta E_v = m_v c_{vv} \Delta T_v$, $i_v \approx 2500 + c_{pv} T_v$, $i_{vi} \approx i_{ve}$ and $A = \pi d^2/4$.

Convective heat transfer from the inside surface of the pipe to the vapour and the end of the liquid plug exposed to the vapour bubble are both small compared to the enthalpy and work done and hence have been neglected in Eq. 86. Convective heat transfer to the liquid, although the heat transfer between a moving liquid and the pipe is relative large, the variation of the energy as the liquid plug moves back and forth is small compared to the (latent) heat transfer due to the formation of vapour in the heated section and its subsequent condensation in the cooled section. Convective heat transfer to the liquid plug has hence also been neglected.

5.1.5. Equation of State

The vapour bubble is assumed to be an ideal gas, and hence if the volume and temperature are known, the pressure may be given by:

$$P_v = \frac{m_v R_v (T_v + 273.15)}{\pi d^2 x_p / 4} \qquad (84)$$

In Eq. (84) the effect of the relatively thin liquid film has been ignored when determining the volume of the vapour bubble. The temperature of the vapour has been converted to Kelvin by adding 273.15 to the Celsius value.

5.1.6. Thrust

The net thrust or the force restraining the pipe from moving in the axial direction is calculated by:

$$R = 2[(P_v - P_e)\pi d2/4 \mp F_\tau] \tag{85}$$

There are two open ends and hence the value "2" in Eq. (85). F_τ is given by Eq. (82) and its sign depends on whether the liquid plug is moving out or into the pipe, and F_σ may be neglected because it is relatively small.

5.2. Numerical Solution Procedure

An explicit finite difference scheme numerical solution method to solve the equations of change, developed in section 5.1 is advocated as follows.

Depending on the position of the plug relative to the closed-end (symmetrical centre of the vapour bubble) x_p, the length of the vapour bubble exposed to the heated L_{vi} and the cooled L_{ve} sections of the pipe the volume of the vapour $V_v = x_p\pi d^2/4$ and the mass of the plug $m_p = \rho_\ell (L/2 - x_p + L_m) \pi d^2/4$ are determined.

The liquid film in the evaporator is further divided into a number of smaller liquid film control volumes and the mass of liquid in each of the control volumes accounted for. Liquid is lost by evaporation but gained by the trail of thickness $\delta_{\ell f}$ left by the liquid plug as it moves out of the evaporator. The effect of the liquid film in the cooled section, however, has been neglected.

Assuming appropriate values for the overall heat transfer coefficients between the heated pipe wall and the vapour U_i and U_{ve} and the cooling environment U_c the mass flow rates' $\dot{m}_{vi}$, $\dot{m}_{ve}$, $\dot{m}_{v,dep}$ and $\dot{m}_{lf,dep}$ using the applicable equations.

The *new* values at time step t + Δt are determined from the known *old* values at time t by

$$m_v^{new} = m_v + (\dot{m}_{vi} - \dot{m}_{ve} - \dot{m}_{v,dep})\Delta t \tag{86}$$

$$m_{\ell f}^{new} = m_{\ell f} + (\dot{m}_{\ell fi} - \dot{m}_{\ell fe} + \dot{m}_{\ell f,dep})\Delta t \tag{87}$$

$$T_{\mathrm{v}}^{\mathrm{new}} = T_{\mathrm{v}} + \frac{(\dot{m}_{\mathrm{vi}} - \dot{m}_{\mathrm{ve}})(2500 + c_{\mathrm{pv}}T_{\mathrm{v}}) - P_{\mathrm{v}}(\pi d^2/4)\Delta x_{\mathrm{p}}/\Delta t}{m_{\mathrm{v}}c_{\mathrm{vv}}}\Delta t \quad (88)$$

$$P_{\mathrm{v}}^{\mathrm{new}} = \frac{m_{\mathrm{v}}\mathrm{R}_{\mathrm{v}}(T_{\mathrm{v}} + 273.15)}{V_{\mathrm{v}}} \quad (89)$$

$$v_{\mathrm{p}}^{\mathrm{new}} = v_{\mathrm{p}} + \frac{\sum F_{\mathrm{p}}}{m_{\mathrm{p}}}\Delta t \quad (90)$$

where the sum of the forces acting the liquid plug $\sum F_{\mathrm{p}} = -F_{\tau} \pm F_{\mathrm{g}} - F_{\sigma} + F_{\mathrm{P}}$ are given by Eqs. (78) to (82).

$$x_{\mathrm{p}}^{\mathrm{new}} = x_{\mathrm{p}} + v_{\mathrm{p}}\Delta t \quad (91)$$

$$\Delta x_{\mathrm{p}}^{\mathrm{new}} = x_{\mathrm{p}}^{\mathrm{new}} - x_{\mathrm{p}} \quad (92)$$

The old values are replaced by the new values and the procedure is repeated for the next time step, and so on.

5.3. Example

The simple theory as given in the preceding section was applied to an open oscillatory heat pipe with geometrical, heat transferand operational parameters as shown in Figure 61 and Table 9. The results are given by way of Figures 62, 63 and 64.

Table 9. Geometrical dimensions, operating parameters and initial conditions for the as tested and theoretically evaluated open oscillatory heat pipe

$L_{\mathrm{e}} = 0.18$, $L_{\mathrm{a}} = 0.02$, $L_{\mathrm{c}} = 0.48$, $d = 0.00334$, $L_{\mathrm{m}} = L_{\mathrm{fe}} = 15d$, $\delta_{lf,\mathrm{dep}} = 0.000025$ m
$T_{\mathrm{e}} = 150$°C, $T_{\mathrm{c}} = T_{\mathrm{w}} = 20$°C, $U_{\mathrm{e}} = 1000$ W/m^2°C, $U_{\mathrm{a}} = 0$, $U_{\mathrm{c}} = 600$ W/m^2°C
$h_{\mathrm{e}} = 0$ W/m^2°C, $h_{\mathrm{a}} = 0$ W/m^2°C, $h_{\mathrm{c}} = 0$ W/m^2°C
$\mathrm{R}_{\mathrm{v}} = 461$ J/kgK, $c_{\mathrm{pv}} = 1900$ J/kg°C, $c_{\mathrm{vv}} = c_{\mathrm{pv}}$ - R, $\rho_l = 1000$ kg/m^3, $\sigma_l = \sigma_{@(Th+Tc)/2}$
$h_{\mathrm{fg}} = 2.34$ MJ/kg, $\mu_l = \mu_{@(\mathrm{Th+Tc})/2}$ kg/ms, Initial conditions: $\delta_{\ell\mathrm{f}} = 0.00001$ m
$x_0 = 0.15$ m, $P_{\mathrm{v0}} = 100\,000$ Pa, $T_{\mathrm{v0}} = 20$°C,
and hence $V_{\mathrm{v0}} = x_0\pi d^2/4$, and $m_{\mathrm{v0}} = (P_{\mathrm{v0}}V_{\mathrm{v0}})/(\mathrm{R}_{\mathrm{v}}\,(T_{\mathrm{v0}} + 273.15))$

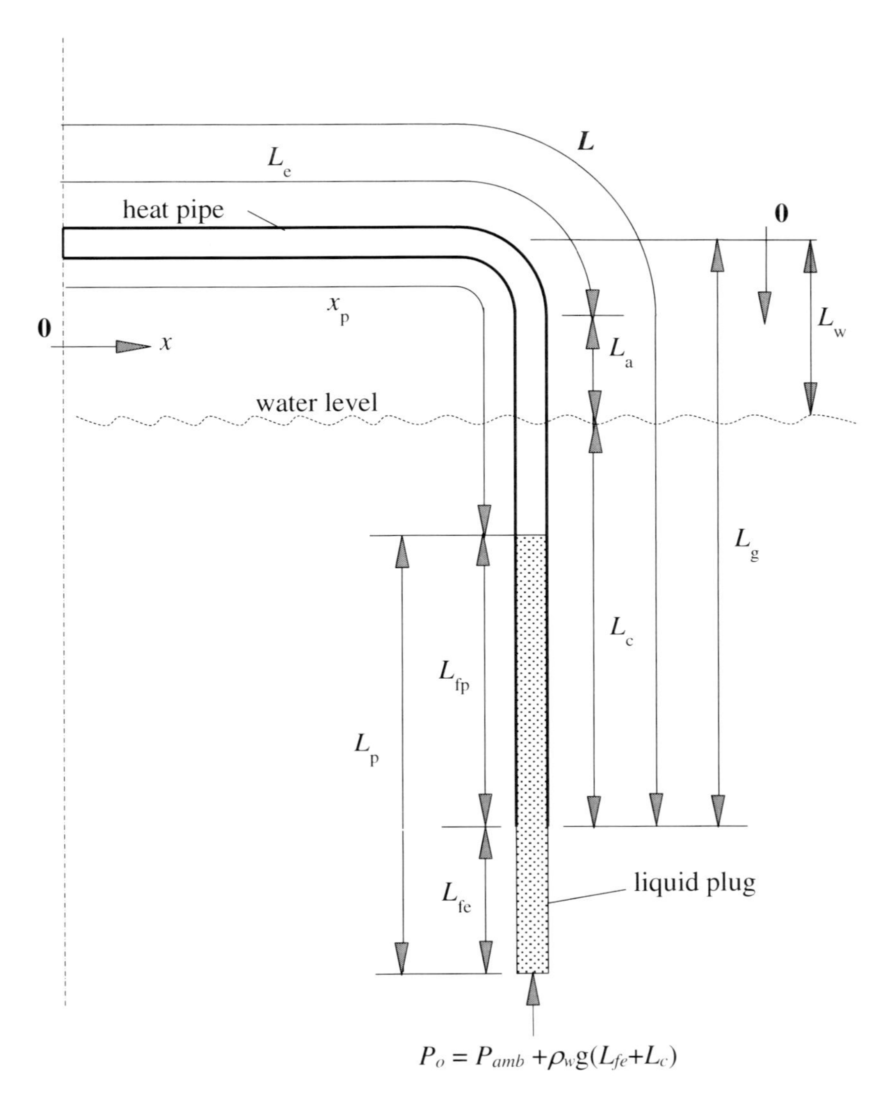

Figure 61. Symmetrical half of an open oscillatory heat pipe [73].

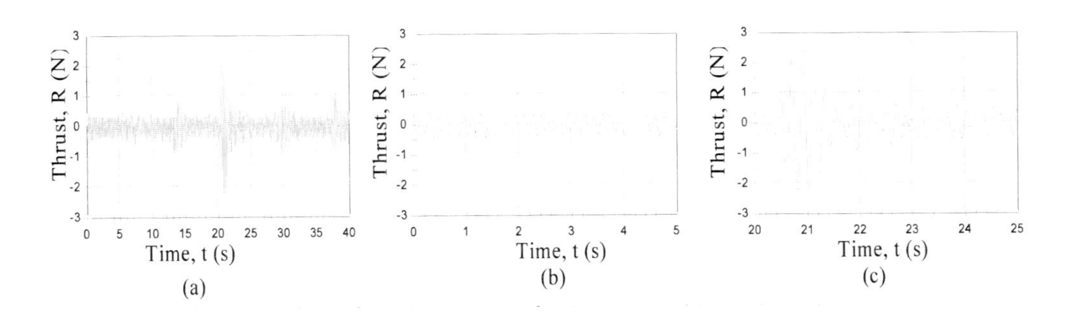

Figure 62. Experimentally determined thust as a function of time for the open oscillatory heat pipe given in Figure 61. [73]

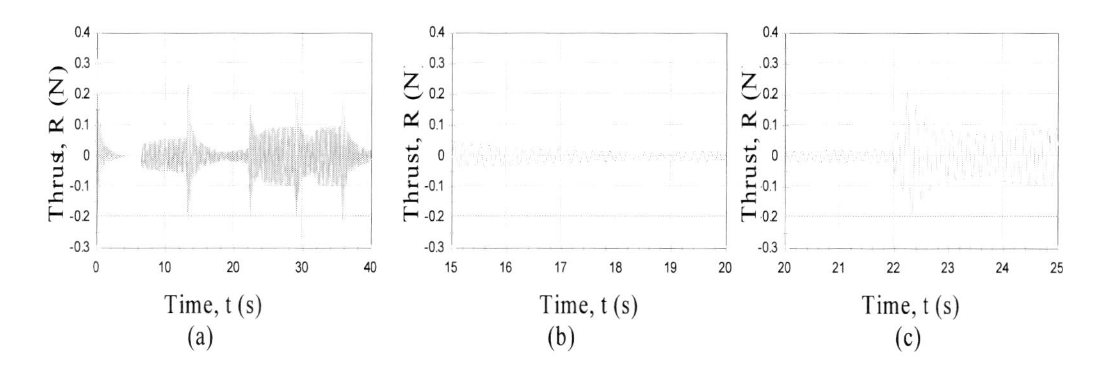

Figure 63. Theoretically determined thust as a function of time for the open oscillatory heat pipe given in Figure 61 [73].

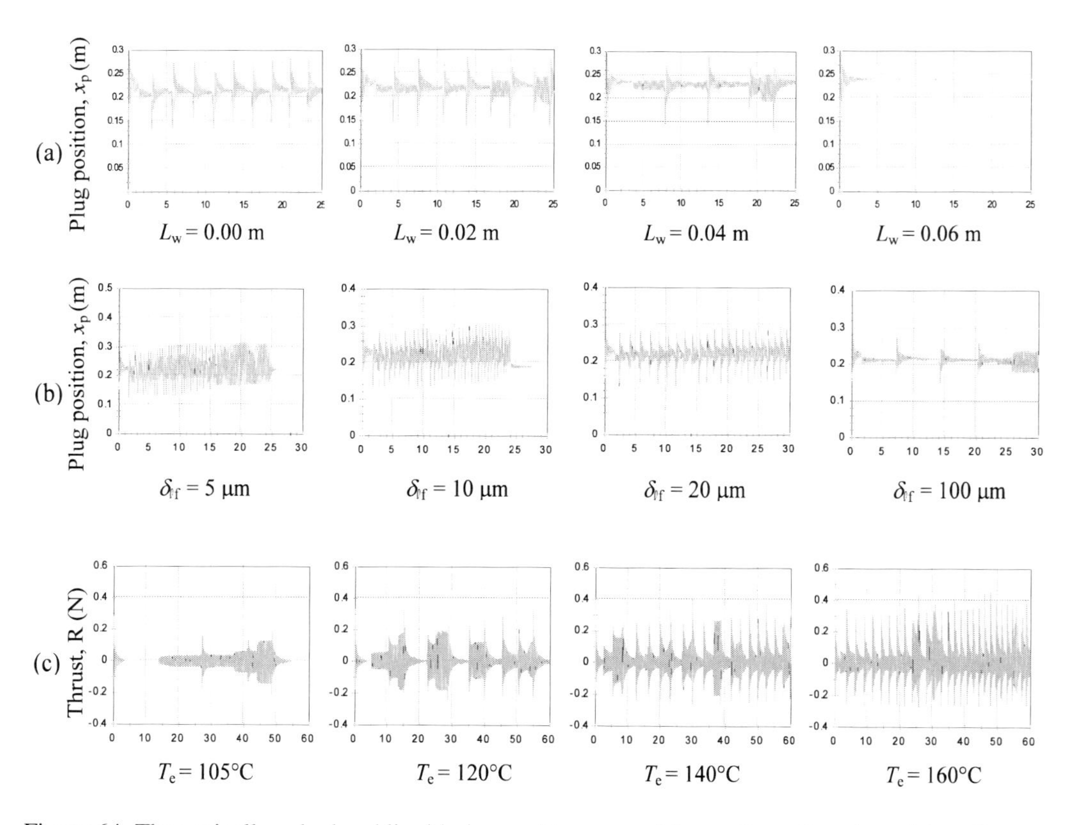

Figure 64. Theoretically calculated liquid plug positions x_p and thrust R as a function of time (in seconds) for different values of: vertical distance between evaporator and water level L_w, film thickness deposited at the trailing end of a liquid plug δ_{lf} and evaporator temperature T_e for the open oscillatory heat pipe defined by way of Figure 61 [73].

6. Sundry Natural Circulation Heat Pipe Applications

In this section a number of widely varying heat pipe applications are considered. These include a number of possible water pumping applications, a night-sky cooling and day-time solar heating system, cooling of electronic and electrical equipment, and a relative performance comparison of supercritical carbon dioxide, water and nitrogen.

6.1. Water Pumping

The need to pump water for irrigation or for humans and livestock remains a primary requirement in developing rural areas [83]. Traditionally, water lifting has been through the use of petroleum fueled internal combustion engines. Recognizing that solar irradiation is a technically appropriate, sustainable and viable source of energy there are a number of non fossil-fuel based small-scale water pump options available. Also, in an attempt to better manage our world's finite energy resources, solar energy is regarded these days as an automatic choice for water pumping. Normally such pumping of water would be to use photovoltaic solar cells to generate electrical energy to drive a submersible pump. In this section however we will we consider a number of relatively novel and innovative water pumping concepts. These concepts considered will be called:

- Drinking bird water pump
- Surface tension driven water pump
- Green building natural air circulation water pump
- Open oscillating heat pipe water pump

6.1.1. Drinking Bird Water Pump

The *happy* drinking bird, as depicted in Figure 65, is a well-known novelty toy. It is actually though, a simple heat engine that includes in some ways a simple closed two-phase heat pipe device [84]. It consists of two bulbs with an interconnecting open-ended pipe and is charged with a working fluid (usually dichloromethane or ether). The working fluid liquid phase partially fills the bottom bulb and its vapour fills the rest of the working space as shown in Figure 65(a). In this position the top bulb, bottom bulb and interconnecting pipe are all at the same temperature, If however the wicking material covering the top bulb is wetted with water, the water evaporates and then cools the *head* and the vapour within the top bulb condenses, the pressure decreases and the now higher pressure in the now hotter bottom bulb pushes liquid up through the interconnecting pipe and into the top bulb. To keep in balance, the device must rotate about its pivot point as shown in Figure 65(b) until the bottom end of the interconnecting pipe is exposed to the vapour allowing a vapour bubble to ascend up the pipe, thereby breaking down the liquid plug flow. Vapour can now flow upwards into the top bulb and thus allowing the liquid to flow back as a stratified rivulet in the lower portion of the pipe. While this is happening the device is held back from further rotation by the water container the wicked beak dips into the water and the evaporated water is now replenished. The bottom is now heavier than the top and the device returns to its starting position allowing the cycle to repeat itself continuously, provided its wick is always wetted. This repeated cycle of operation is shown by way of Figure 65 (c).

The flow of fluid in the working space can be subdivided into six one dimensional control volumes; 1, 2, 5 and 6 as variable volume Legrangian control volumes and 3 and 4 as fixed volume Eulerian control volumes. The fluid motion and device rotation can now be simulated using the conservation equations. If this is done for the happy drinking bird shown in Figure 67(a), a remarkably good correlation can be obtained between the experimentally and theoretically obtained birds, as shown in Figure 67(b).

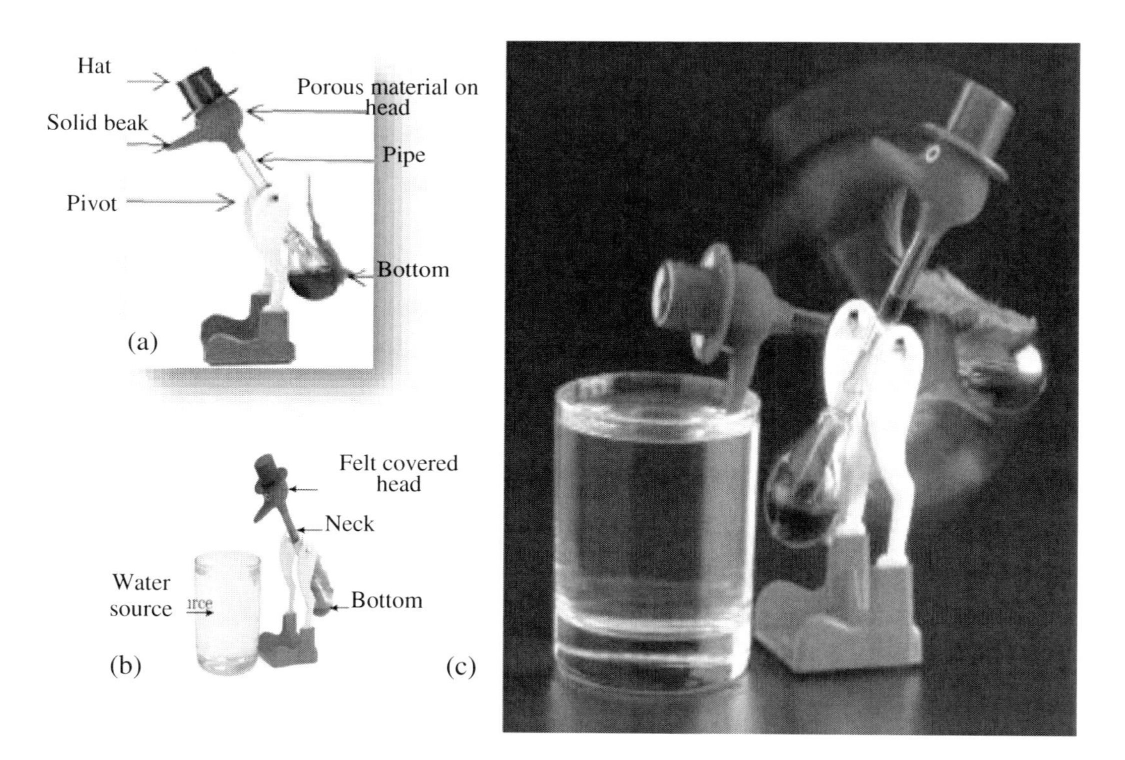

Figure 65. A toy happy drinking bird in action [84].

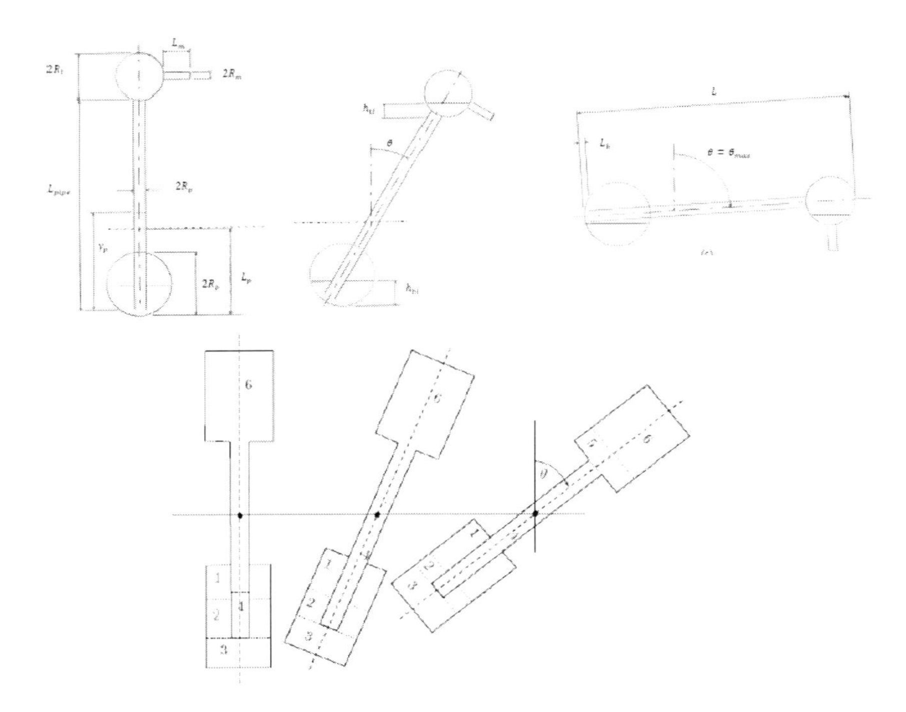

Figure 66. Drinking bird depicted as six interconnected control volumes [83].

To investigate the practical water pumping ability, an experimental prototype was built and instrumented as shown in Figure 68 [83]. The prototype was able to exert a force of 30 N over a period of nearly 2000 s or 0.0085 W of mechanical power. If this force was applied to a hand pump, 0.30 litres of water could be pumped a distance of 2 m and of this amount of water pumped, 0.10 had to be used to cool the wicking structure.

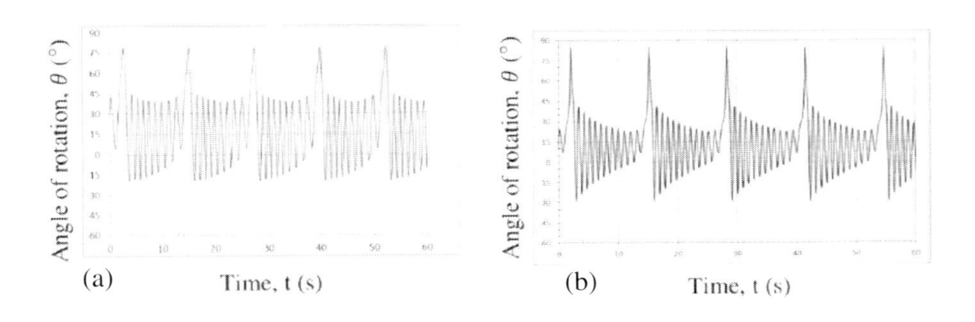

Figure 67. Experimental (a) and theoretical (b) simulations of the toy drinking bird [83, 84].

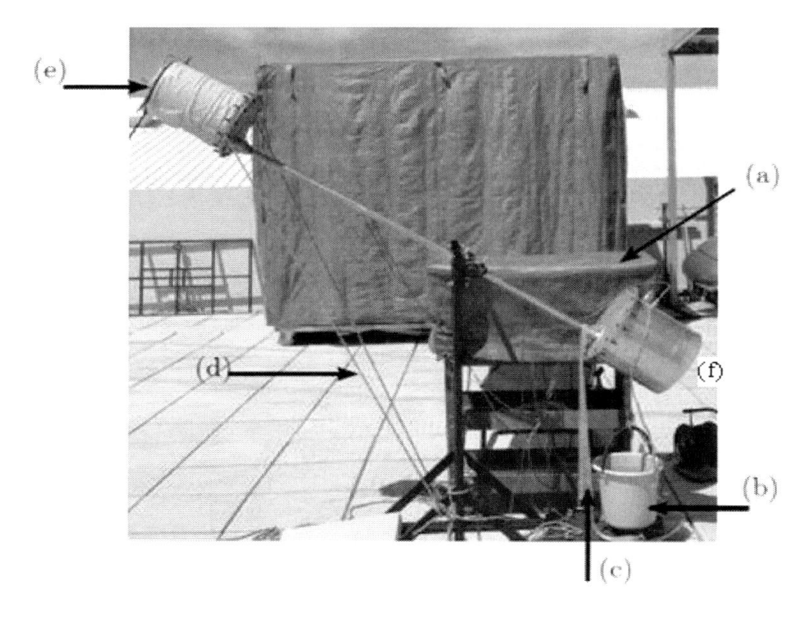

Figure 68. Large scale experimental prototype water pump on the roof-top open-air laboratory, showing the weather proof shelter for instrumentation (a), working load (bucket of water) (b), retaining strap to prevent over-rotation (d),wetted head (e) and bottom (f) [83].

6.1.2. Surface Tension Driven Water Pump

The transport of water (and associated processes) that take place in a plant from the ground up to he leaves and then back down to the very lowest roots is undoubtably a very complex phenomenon. Based on the somewhat simplistic description as depicted in Figure 69, this water pumping process can be likened to a two phase natural circulation heat pipe in so-called top heating mode. As the plant grows it pulls water up with it in small diameter, short but interconnected tubes with water with no-return valves. This water-chain, so to say, is prevented from breaking because molecular cohesion and cavitation is prevented by adhesion between the water and the tube walls. In the "photosynthesis cell", which can to likened to an open-celled sponge with a small diameter (about 1 to 15 μm) capillary tube sized opening; the water is held up by capillary forces. At these capillary sized openings water is exposed to the environment and evaporation takes place. For each molecule of water evaporated, a molecule of water is drawn up from the water surrounding the smallest of the roots exposed to the water in the ground and in this way mineral rich water is drawn up. In the photosynthesis cell energy, water, minerals and carbon dioxide CO_2 come together and carbohydrates and oxygen O_2 are formed and the now somewhat denser carbohydrate laden water can flow back down and carry nourishment to the very lowest roots. CO_2 can now diffuse into the leaf structure from the higher CO_2 concentration in the air, and the higher concentration of O_2 in the leaf allows it to diffuse into the lower concentration of O_2 in the surrounding air. [As such, all animal life as we may know it requires plant-life effluent to survive.]

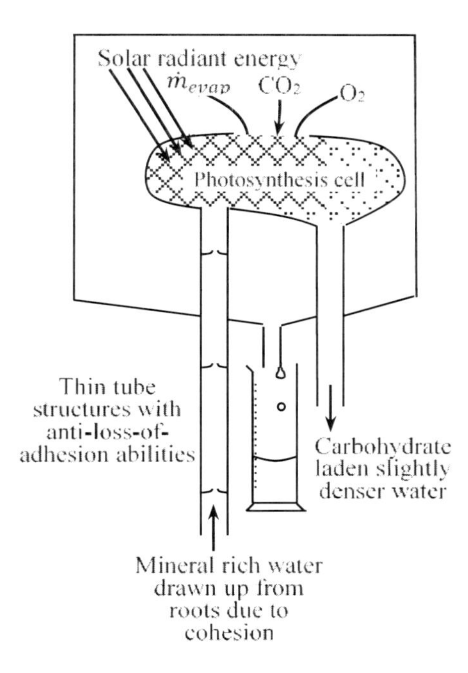

Figure 69. Simplistic depiction of a plant's water transport and food production plant system and showing a leaf photosynthetic cell consisting of open-celled micro capillary structure.

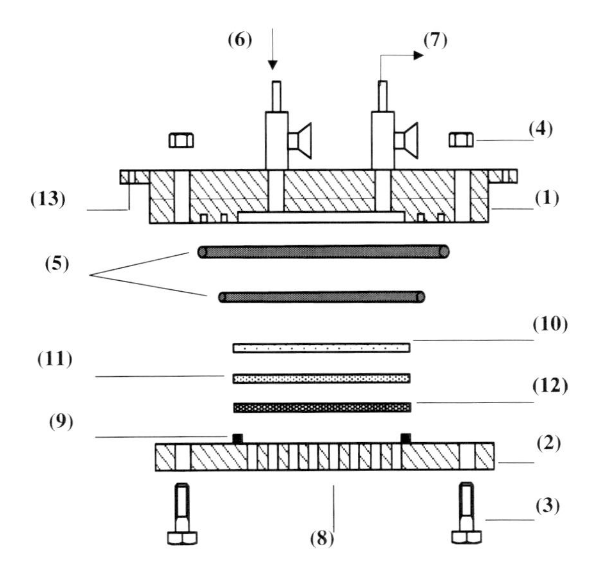

Figure 70. Mechanical construction details of the photosynthesis cell [85].

Mimicking this water pumping action in plants, a single "leaf" may be constructed as shown on Figure 70 and tested in an experimental chamber as shown in Figure 71 at different heights as shown in Figure 72 [85]. The "leaf" consists of an upper (1) and lower disc (2). Both discs were constructed from marine grade aluminium (AL 5083) for its relatively lightweight, strong and corrosion resistant properties. 12 M8 bolts (3) and nuts

(4) were used to ensure that the two discs are firmly held together. A double set of O-rings (5) was used to prevent air entering the system from the outer edges. The upper disc features two brass needle valves. The primary needle valve (6) regulates feed water to the "leaf" while the secondary needle valve (7) is used to alleviate pressure buildup during the initial filling procedure as well as air and bubble removal from the "leaf." Additionally, the upper disc contains a small chamber which houses the internal "leaf" materials. A set of 63 ϕ7 mm holes (8), at varying PCD's up to 80 mm were laser cut into the lower disc. The laser cut holes would provide openings for water vapour to escape from the "leaf" whilst keeping enough material to support the internal "leaf" material. A gasket (9) was placed on the lower disc to prevent air seepage past the internal "leaf" material through minute crevices in the discs. The internal "leaf" material consists of three layers: A super absorbent sponge material (10) made of high density PVA (polyvinyl alcohol) was used as the first layer. The PVA sponge was used as it is hydrophilic with estimated effective capillaries of 200 μm which would help prevent bubble formation in the "leaf." Thereafter, ϕ90 mm Grade 393 (Quantative) Munktell filter paper (11), with 1 to 2 μm pores, was used as the second layer. The second layer is also hydrophilic and also helps prevent bubble formation in the "leaf." Additionally, the second layer has much finer pores than the first layer which traps larger particles before reaching the final layer to prevent clogging of the final layer. For the final layer, Millipore membrane filters (12) were used. Both ϕ90 mm MF-Millipore™ membrane filters (0.22 μm pores, 3.52 bar bubble pressure) as well as ϕ90 mm Durapore® (0.22 μm pores, 3.45 bar bubble pressure) were tested. MF-Millipore™ was found to work much more effectively than the Durapore® for this application. The winglets (13), shown in Figure 70, are used to secure the "leaf" to a frame or supporting structure.

Inside the control chamber (Figure 71) copper cooling coils (3) were used to condense the evaporated water from the "leaf." A collection container (4) inside the control chamber was used to collect the condensate. To increase evaporative rates, a 9 W heating pad (5) was placed on top of the "leaf" and insulated. Further, an axial flow fan (6) was placed inside the chamber to force air flow over the surface of the "leaf" and to promote air circulation.

It was found that the pumping head had little effect on the evaporation rate from the "leaf" and that the water could be pumped at a rate of nearly 400 mL/hrm^2 and with a maximum functional lifespan of 13 days for pumping heads of 1.8 m. Although the pumping head had little effect on the evaporation rate from the "leaf," the pumping head height, however, was found to greatly affect the pump's functional lifespan. Figure 73 shows the time to failure as the pumping head is increased.

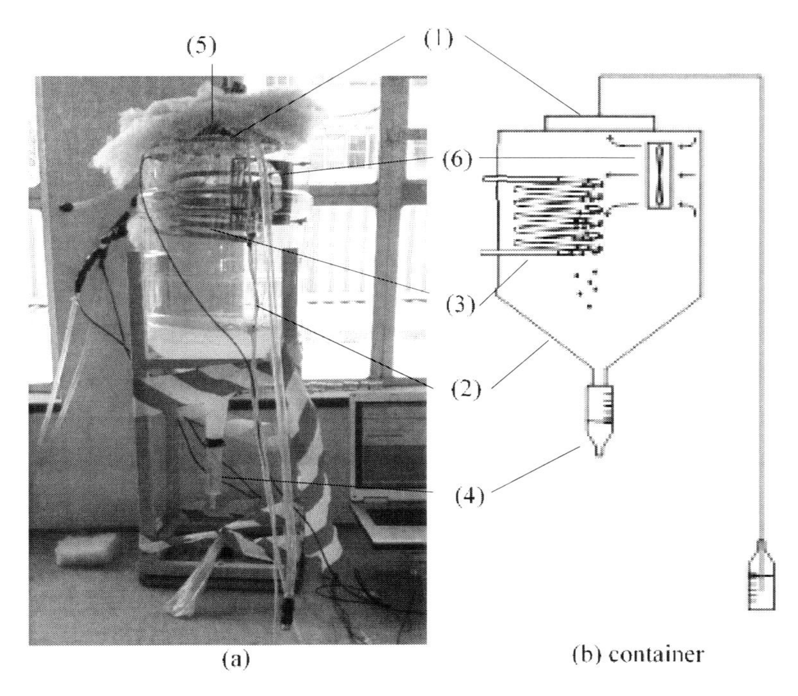

Figure 71. Photgraph of photograph and details of the water collection system [85].

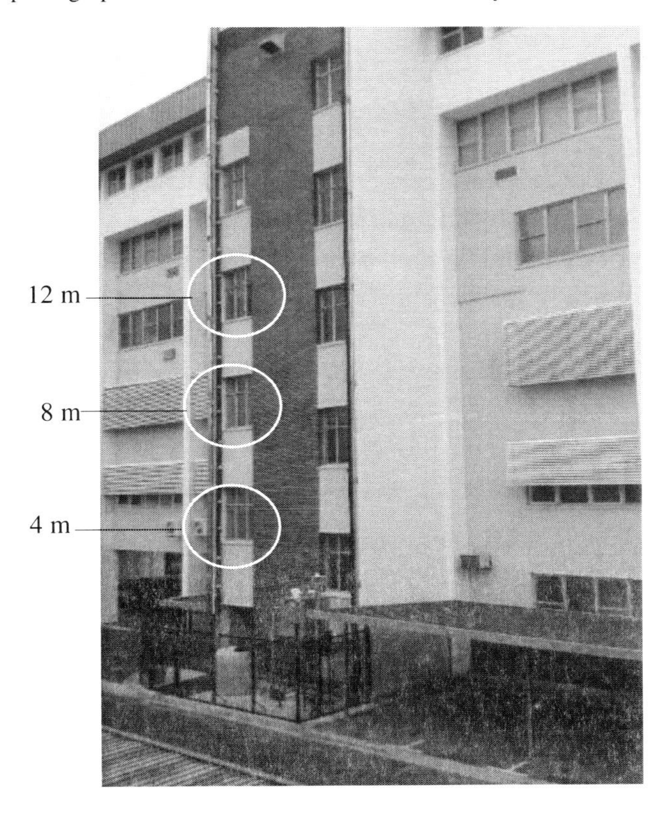

Figure 72. Water pumping ability test positions on side of the facility building.

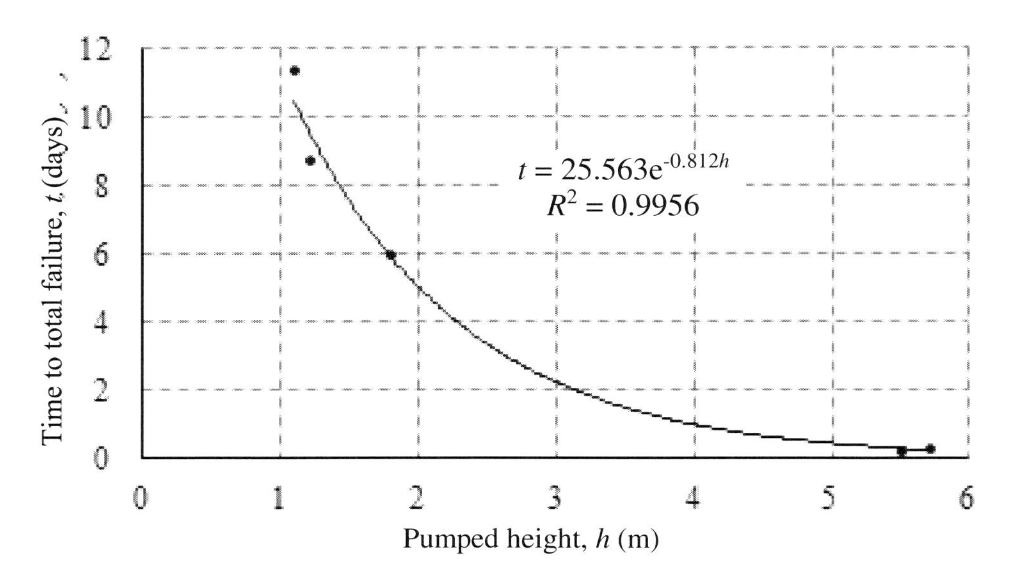

Figure 73. Time to failure of pumping action as a function of pumping height [85].

6.1.3. Natural Air-Circulation Water Pump

Green building design is a manner of designing structures and using operational practices which are energy efficient, resource efficient and environmentally responsible [86]. The design philosophy is to utilize renewable, energy conscious and sustainable design methods for buildings, with the focus on reducing the energy consumption and reducing climate changing effects. An alternative to conventional HVAC systems in buildings is the passive downdraft evaporative cooling (PDEC) system. The Torrent Research Centre in Ahmedabad, India is an example of an installed PDEC system.

A laboratory scale PDEC is shown by way of Figure 74. It consists of a 5.5 m high, 600 mm diameter 100 x 100 mm and 3.1 mm wire-diameter mesh covered with a layer of absorbant natural cotton material. The absorbant material is wetted and to ensure that evaporation of water takes place only from the inside, the ouside of the material is covered with a thin layer of transparent water proof plastic sheeting.

Many of the PDECs that have been investigated and reported in the literature make use of ground level water storage, electrically driven water pumping, and spray nozzles to evenly distribute the water droplets over the absorbent evaporation pads. Such a water pump and droplet distribution system however seriously compromised the energy saving potential of such a designed a PDEC in as much as the energy consumed in the pumping and water atomization process will probably negate its energy-savings potential. It is for this reason that an entirely passive natural system in these applications is, so to say, mandatory and hence the need for a natural circulation pump to get water from a ground level storage tank to the top of the PDEC.

Inlet at the top

At the bottom

Figure 74. Experimental 5.5 m high 600 mm diameter PDEC tower.

To get water to the rooftop level to wet the evaporation pads of a PDEC two experimental natural circulation air water pump set-ups of different cross-sectional areas were built to evaluate the pump design and the theoretical model that was developed [87]. A small scale (2.8 m high and 100 x 100 mm cross-sectional area) natural air circulation loop was constructed as shown in Figure 75. It was assumed that solar heated water at about 60°C was available at ground level to keep the evaporating section water-evaporation tray charged with hot water. A temperature difference of 9 to 12.5°C between the heating and cooling sections induced an average velocity of 0.4 to 0.6 m/s for a duct cross section of about 100 mm square.

Experiments were also conducted using a larger natural circulation air water pump set-up as shown in Figure 76, of larger cross-sectional areas to evaluate the pump design and the theoretical model [87]. For this larger cross section of 400 mm square, a temperature difference of 2 to 5°C induced an average velocity of 0.25 to 0.3 m/s. An asymmetrical velocity profile was observed which varied at different points in the loop. A water delivery rate of 1.2 to 7.5 L/day was experimentally determined. This compares well to the passive air-conditioning water requirements of a small building.

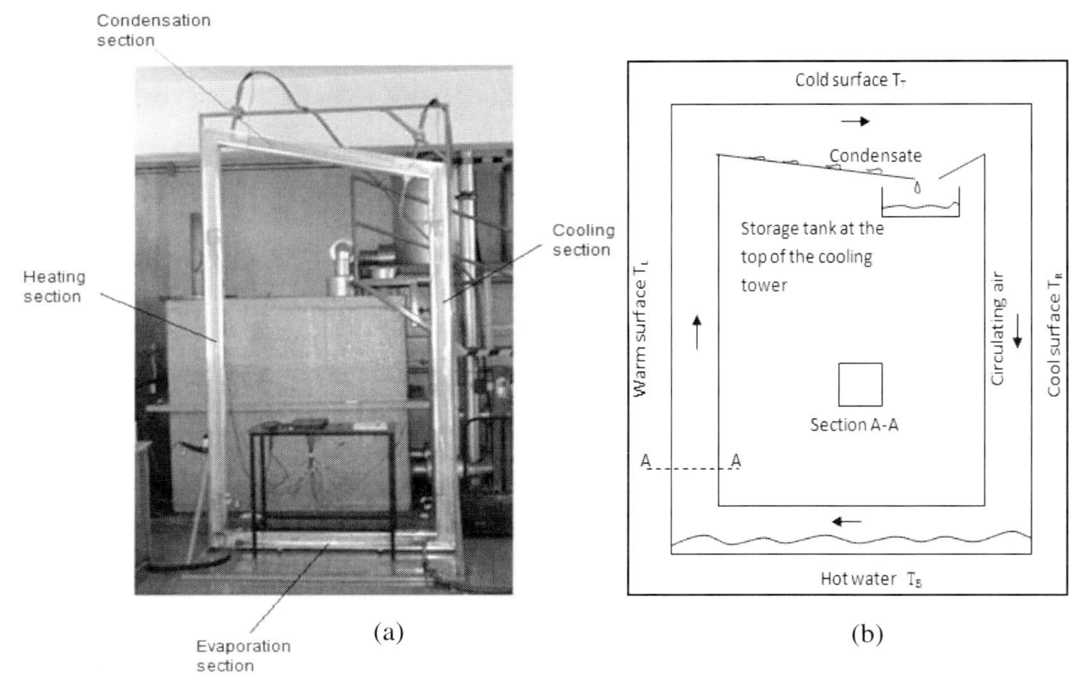

Figure 75. Small scale experimental natural circulation air water pump [87].

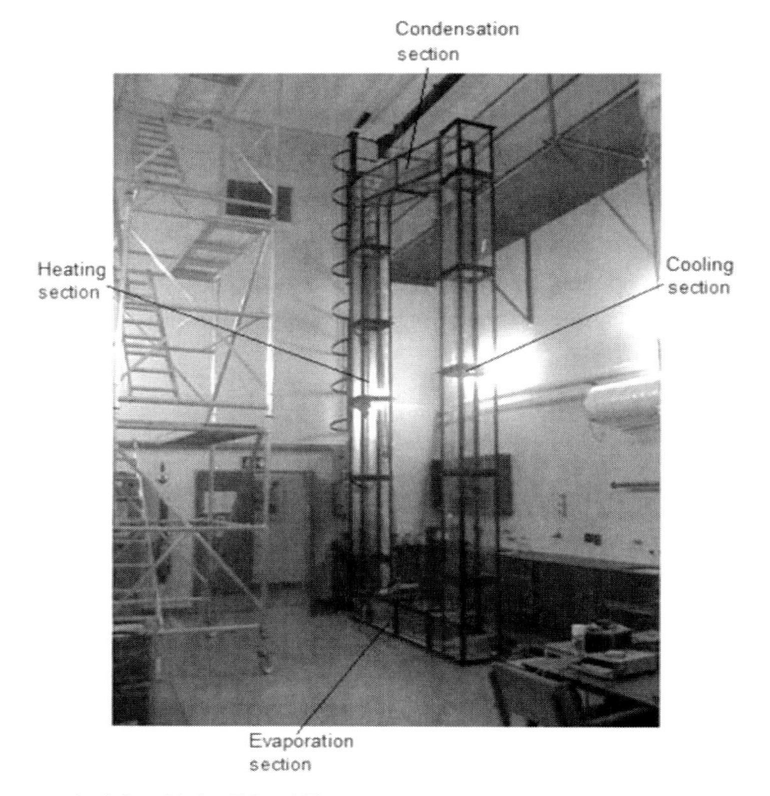

Figure 76. Larger-scale 5.8 m high, 400 x 400 mm cross sectional area experimental natural circulation air water pump [87].

6.1.4. Open Oscillating Heat Pipe Water Pump

The very simplicity of the concept of the open oscillatory heat pipe (OOHP) makes it an attractive option for consideration as a water pump; and for which a theoretical basis of operation suitable for engineering design has, as yet, not been fully established. It is for these reasons that the OOHP is considered as a water pump in this article.

An open oscillatory heat pipe (OOHP) is a simple two-phase flow heat transfer device and may be regarded as a type of a pulsating heat pipe (PHP) [67]. An OOHP consists of a bent pipe and operates, by way of a simple example, as shown in Figure 77. Provided that the temperature of the hot oil is above the boiling point of the water, the two liquid plugs will oscillate back and forth and heat will be transferred from the hot oil to the cold water. The important feature of the device is that it is able to maintain its characteristic liquid plug vapour bubble flow pattern even when the heat source is above (relative to gravity) the heat sink. In this way heat may be effectively transferred from a higher to a lower elevation without the need of any special capillary wick structure or moving mechanical parts [74].

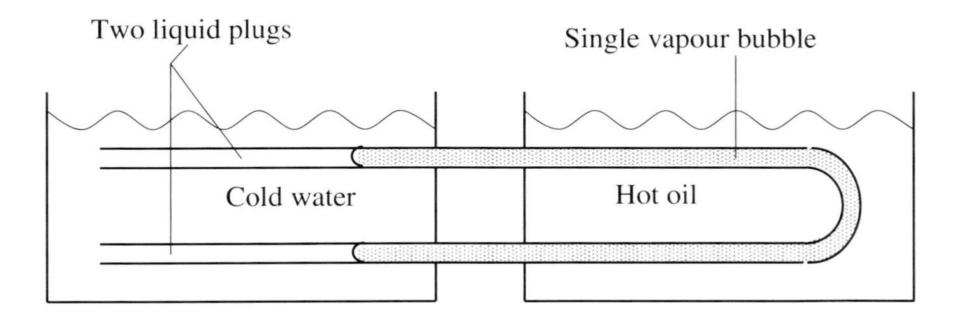

Figure 77. Example of an open oscillatory heat pipe [88].

The diameter of the pipe is important. It must be small enough that under operating conditions a liquid plug and vapour bubble type flow pattern occurs. If the diameter is too large the tendency is for a stratified flow pattern to occur with the liquid flowing in the bottom of the pipe and the vapour in the top. The two competing forces, surface tension and gravity, are related to the diameter by the Eötvös or Bond dimensionless numbers ($Eö = Bo^2$) (Khandekar et al., 2003). It is generally accepted that the diameter of the pipe should be in the region of 0.7 to 1.8 of $[\sigma/g(\rho_\ell-\rho_v)]^{\frac{1}{2}}$ to ensure the formation of liquid plugs and vapour bubbles and hence the desired oscillatory behaviour. This diameter works out to between about 2 and 5 mm for water at 100°C, 1.1 and 2.8 mm for methanol at 64.7°C, and 0.7 and 1.9 mm for Freon 113 at 47.7°C.

A schematic representation of an open oscillatory heat pipe water pump (OOHPWP) is given in Figure 78. The pump operates similarly as a positive displacement piston pump [89] and replaces the need for a mechanical reciprocating mechanism in a positive displacement pump [90]. It consists of two valves, a cylinder and an OOHP with its heated, adiabatic and cooled sections. The cylinder and OOHP are filled with water and the heated section (or evaporator) is exposed to a heat source. When the saturation temperature corresponding to the local pressure is reached, the water begins to boil in the heated section of the OOHP, vapour is formed, the pressure builds up, water is forced out of the OOHP and into the cooled section (or condenser), valve-A closes, valve-B opens and water flows into the discharge line. As the liquid plug moves out of the OOHP, the steam constituting the vapour bubble comes into contact with the cooled section, condenses, the pressure decreases, valve-B closes, valve-A opens and the atmospheric pressure forces water into the cylinder and into the OOHP. The "piston" of this pump can thus be likened to the *liquid plug* moving back and forth in the OOHP.

A small working model was built and was able to pump 0.2 mg/s at a pumping head of 100 mm and mechanical, theoretical and thermal efficiencies of 3%, 0.03% and 0.00003%, respectively. [Letting ones mind wander/wonder/stray, might this not be possible to duplicate trillions pumps in parallel thereby making it more useful using nanotubes and nano technology?].

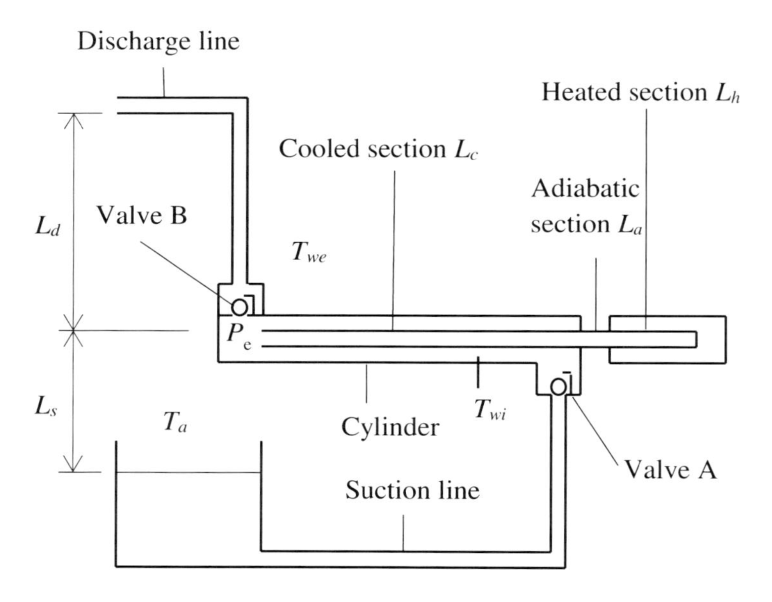

Figure 78. Schematic representation of an OOHP-Pump [88].

6.2. Night-Sky Cooling and Day-Time Solar Heating System

In this section an entirely passive natural circulation combined night-sky room cooling and solar water heating system having no control valves but making use entirely of natural thermosyphonic flow, will be experimentally configured and theoretically simulated [91, 92]. The theoretical simulation model will be compared with an actual experimental model, thereby validating the theoretical simulation model and allowing the cooling and heating performance of such combined cooling and heating systems to be determined. These systems would be ideal for inland desert-like locations where the days are invariably very hot and the nights very cold.

The experimental system that was investigated and theoretically modelled is shown schematically in Figure 79a and photographically in Figure 79b. The modeled system consists essentially of three subsystems: a night cooling cycle (1), day water heating cycle (2) and a day room cooling cycle (3). At night the cooling cycle operates and water flows naturally from the cold water storage through to the radiation panel where it is cooled before returning back to the cold water storage tank. During the day the water heating cycle operates, and the cold water in the cold water storage tank also flows naturally through the convector situated inside the room. The convector extracts heat from the air in the room, the water heats up and flows back to the cold water storage tank. In this way, during the day hot water is supplied as well as the cooling of the room all by entirely natural means and without the use of any active controls and pumps and without the consumption of any fossil fuel generated energy sources.

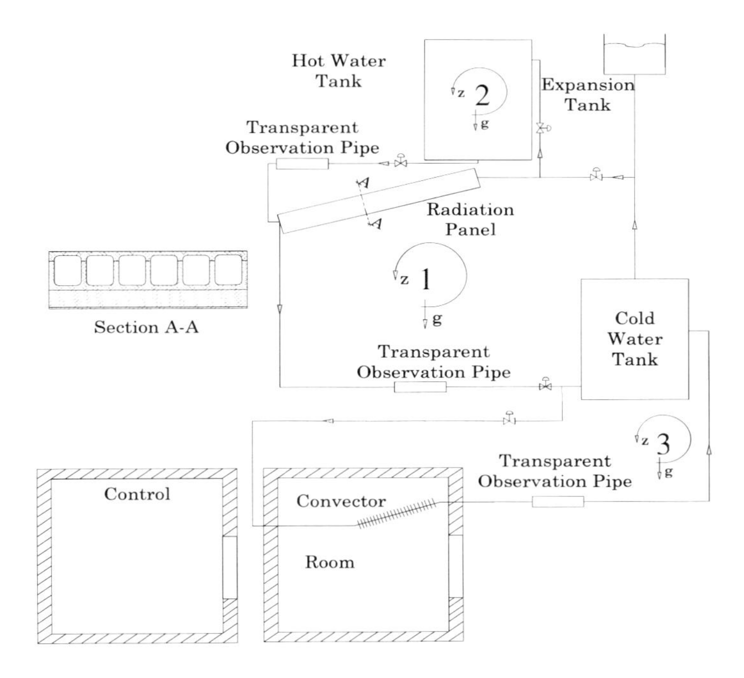

Figure 79a. Schematic layout of the experimental passive night-sky radiation system; indicating the night time cooling cycle 1, day time heating cycle 2, and room convector cooling cycle 3 [92].

The radiation panel used in this case was a swimming-pool solar water-heater, and is similar to a conventional solar water-heater but without glazing and fins. The radiator consists of numerous small channels as depicted by section A-A in Figure 79a. The bottom of the radiation panel is insulated with plywood that also provides structural stiffness. The radiation panel serves as a heat-emitting component at night and as a heat absorber during the day. The placement of the radiation panel is therefore important to ensure that thermosyphoning occurs. The panel should be tilted at an angle equal to the latitude of the experiment location for heating but should be horizontal for better cooling cycle performance. It was therefore decided to tilt the radiator at only 5° (just enough to ensure a downwards-flow through the panel at night and an upwards flow during the day). The small tilt of the angle has negligible effects on the cooling and heating performance [93].

The amount of energy absorbed by the radiation panel is given by Eq. (93) as:

$$\dot{Q}_{absorb} = \alpha I A \tag{93}$$

where α is the absorptivity of the radiator, I the direct solar radiation and A the aperture area. The aperture area is influenced by the sun's azimuth angle $\alpha_{sol,}$ and the tilt angle θ_{tilt} of the radiator. The aperture area is given by Eq. (94) as:

$$A = A\sin\big((90 - \alpha_{sol}) - \theta_{tilt}\big) \tag{94}$$

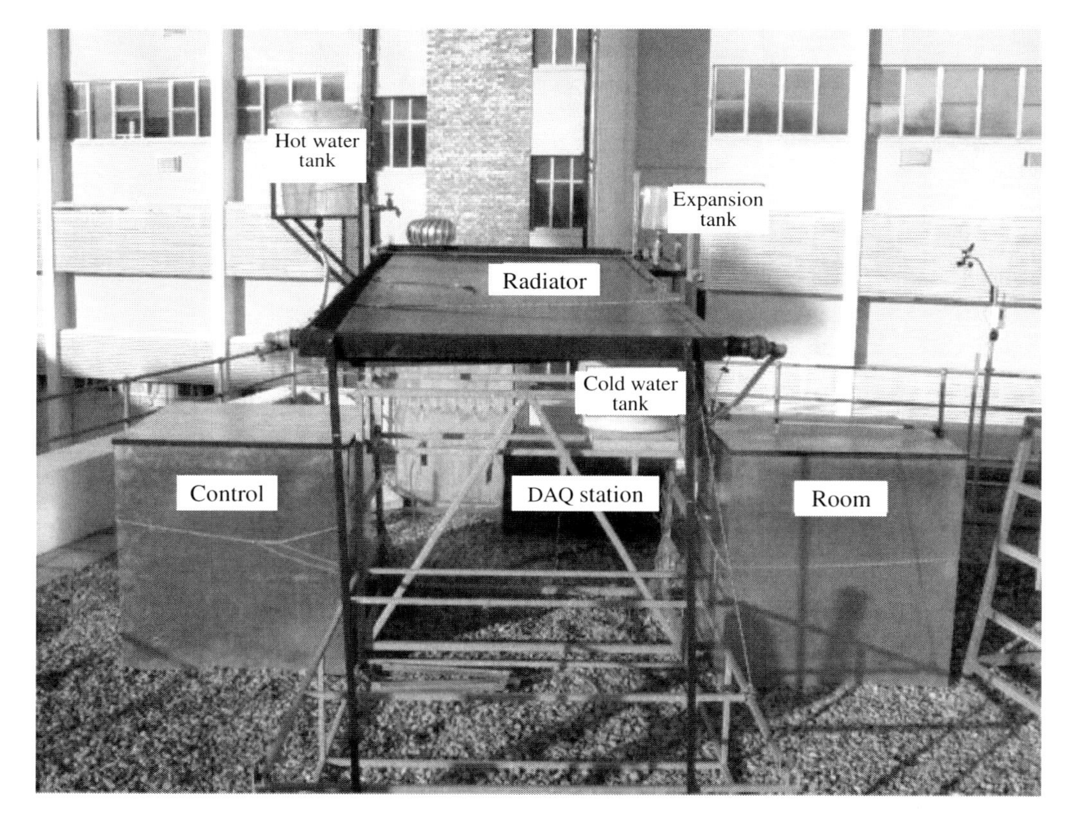

Figure 79b. Photograph of the experimentally evaluated passive night-sky radiation cooling and day time room cooling and solar water heating and system. 3.0 x 1.25 m = 3.75 m^2 radiator surface area, 150 L cold water storage tank, 68 L hot water storage tank, 1.2 x 1.2 x 1.3 m high test and control rooms, and 2.81 m^2 convector fin heat transfer surface area [92].

During both the heating and cooling cycles, energy is either gained or lost to the environment by radiation and convection. The heat radiated to the cold sky, acting as a heat sink, is a function of the sky temperature. A colder sky temperature increases the heat transfer rate. The sky temperature can be calculated using the ambient air temperature and relative humidity (Mills, 2009) as expressed by Eq. (95).

$$T_{sky} = (\varepsilon_{sky}(T_{amb} + 273.15)^4)^{0.25} - 273.15 \tag{95}$$

where $\varepsilon_{sky} = 0.741 + 0.00162T_{dp}$ at night and $\varepsilon_{sky} = 0.727 + 0.00160T_{dp}$ during the day. With the sky temperature known, the heat lost or gained due to radiation can be calculated using Eq. (96).

$$\dot{Q}_{rad} = \frac{T_{hot} - T_{cold}}{\varepsilon\sigma A({T_{hot}}^2 + {T_{cold}}^2)(T_{hot} + T_{cold})} \tag{96}$$

The heat lost or gained by convection is given by Eq. (97).

$$\dot{Q}_{conv} = hA(T_s - T_{amb}) \tag{97}$$

where the convection heat transfer coefficient is a function of the wind speed v_{wind} and is given by [94] as expressed by Eq. (98).

$$h = 18.6v_{wind}^{0.605} \tag{98}$$

The tank and interconnecting pipes lose or gain heat by convection. The external convection heat transfer coefficients $h = Nu\, k / D_h$ of the tank and pipes are calculated in terms of a Nusselt number Nu. For wind speeds greater than zero the Nusselt number is calculated [64] according to Eq. (99).

$$Nu = \left\{0.3 + \frac{0.62Re^{0.5}Pr^{0.33}}{1(1 + (0.4/Pr)^{0.66})^{0.25})}\right\}\{(1 + (Re/282500)^{5/8}))\}^{\frac{4}{5}} \tag{99}$$

where the Reynolds number $Re = \rho vD/\mu$. When there is no wind, natural convection is assumed and the Nusselt number is calculated by Eq. (100).

$$Nu = \{0.6 + \frac{0.387Ra^{\frac{1}{6}}}{\left(1 + (0.559/Pr)^{\frac{9}{16}}\right)^{\frac{8}{27}}}\}^2 \tag{100}$$

The internal convection heat transfer coefficient depends on the cross-section of the component. The analysis assumes internal forced convection. The Nusselt number was taken as the average of the constant surface temperature and the constant heat transfer rate conditions. Using values suggested by Cengel and Ghajar [64], the Nusselt number for a circular section is 4.01. The tank was divided into a number of control volumes. Fluid velocity is low and thus de-stratification in the tank is unlikely since the natural flow being gravity driven is low. The water that enters the tank flows either down or up with very little mixing until it finds the temperature level closest to its inlet temperature and then settles or mixes with the water coinciding with its inlet temperature.

The natural circulation convector removes the energy added to the room. The convector is situated in the top section of the room at an angle of 5° to allow natural circulation of the water. Air flows through the convector, is cooled and finds its way to the bottom, allowing new, hotter air to enter the convector. Natural convective circulation in the room is, consequently, initiated. The convector is tilted slighty at 5°C and with its higherst point some 40 mm from the top of the room. The convector consists of seven copper tubes with aluminium fins attached to the tubes. The normal conduction resistance for a cylinder, $R_{cond} = \ln(r_o/r_i)/2\pi Lk$, is used for the tubes. The convection resistances for the water side and air side are calculated as $R_{conv} = 1/hA$, while the fin resistance includes a fin efficiency η to the equation to get $R_{fin} = 1/\eta hA$. It is necessary to calculate convection heat transfer coefficients for the water, the base of the tube as well as the fins of the convector. The fin heat transfer coefficient is calculated (Mills, 2009) using Eq. (101).

$$h_{fin} = \left(\frac{1.07\Delta T}{L_{fin}}\right)^{1/4} \text{ for } 10^4 < Gr < 10^9 \quad (101a)$$

$$h_{fin} = (1.3\Delta T)^{1/3} \text{ for } 10^9 < Gr < 10^{12} \quad (101b)$$

where $Gr = \beta \Delta T g \rho^2 {L_{fin}}^3/\mu$, and the coefficient of volumetric expansion β, the temperature difference between the fin surface and air ΔT and the length $L_{fin} = 0.165$ m of the fin along which natural convection takes place. It is assumed that the air acts as an ideal gas, which implies that $\beta = 1/T$, where T is in Kelvin.

The solar load is affected by the thermal capacity of the room, the solar irradiation, the position of the sun and the outside weather conditions. As the position of the sun changes, the solar load varies due to the change in the incident solar radiation aperture area of the room. At sunrise, for instance, the western side has no solar load, but as the day progresses this load changes, causing the eastern side to have no load, while the western side is subjected to a solar load. The position of the sun has to be known at any point in time to calculate the aperture area. The position of the sun is affected by the day, time of day and location of the experimental setup. The position of the sun is described by two angles, namely the solar azimuth, describing the sun's position measured clockwise from the south, and the zenith, describing the angle of incidence on a horizontal surface, as indicated in Figure 80.

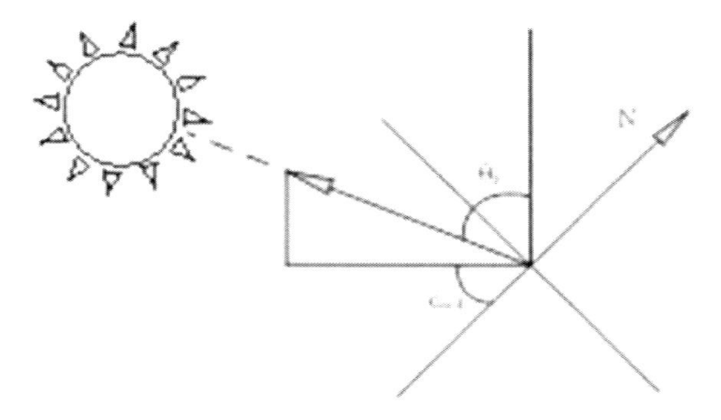

Figure 80. Diagram indicating solar zenith and azimuth angle [91].

The solar heating load can be calculated using $\dot{Q} = \alpha I A$, where I is the direct solar radiation, the aperture area $A = WH\cos\alpha_{sol}\cos\theta_z$, and α_{sol} is the solar absorptivity of the surface. The steel is very thin and has a high conductivity, so the outside and inside temperatures of the galvanized steel sheet are assumed to be equal. The room is also subjected to convection heat losses on the outside. The convection heat transfer coefficient is a function of wind-speed. Loveday and Taki (1996) discussed numerous correlations of wind-speed and heat transfer coefficients on building facades. For the specific case of this study, the seemingly best correlation for a flat surface subjected to varying wind-speeds appears to be given by Eq. (102).

$$h = 1.7v_{wind} + 5.1 \tag{102}$$

The thermofluid time dependent response of the system, as depicted by way of Figure 79a and Figure 79b, is modeled by discretizing the system into a series of one dimensional control volumes as shown in Figure 81, 84 and 85. Each figure representing the night-sky radiation cooling, the solar hot water heating and the room natural convection cooling cycles, respectively. The conservation of mass, energy and momentum are then applied to each of the control volumes. The night-sky cooling theoretical modelling will be shown in detail. The day time solar water heating and room cooling modelling follows in similar fashion and thus will not be considered further in so far as the detailed theoretical modelling is concerned.

Consider the discretized loop shown in Figure 81, which consists of a radiator, cold water tank, manifold and pipes. During the night, water circulates naturally through the radiator. The radiator thermally radiates energy to the sky and extracts heat from the water. The water cools, its density increases and under the influence of gravity flows to a lower position and enters the tank. Hotter water from the tank is, in turn, pushed into the radiation panel to complete the cycle. The theoretical model requires the application of the conservation of mass, energy and momentum to the indicated control volumes.

The mass of the control volume for the next time step can be calculated by applying the conservation of mass. The conservation of mass involves the mass flow rate into and out of the control volume. With the mass at the next time step the conservation of energy can be applied to determine the temperature of the control volume for the next time step $t + \Delta t$. Eq. (103) shows the application of the general statement of conservation of mass to each ith control volume shown in Figure 81.

$$\frac{\Delta m}{\Delta t} = \sum \dot{m}_{in} - \sum \dot{m}_{out} \tag{103}$$

Rearranging of Eq. (103) gives Eq. (104).

$$m^{t+\Delta t} = m^t + \Delta t(\sum \dot{m}_{in} - \sum \dot{m}_{out}) \tag{104}$$

where $m = \rho A_z \Delta_z$, $A_z = \pi d^2/4$ and $\dot{m} = \rho v A_z = \rho G$ since $v = G/A_x$.

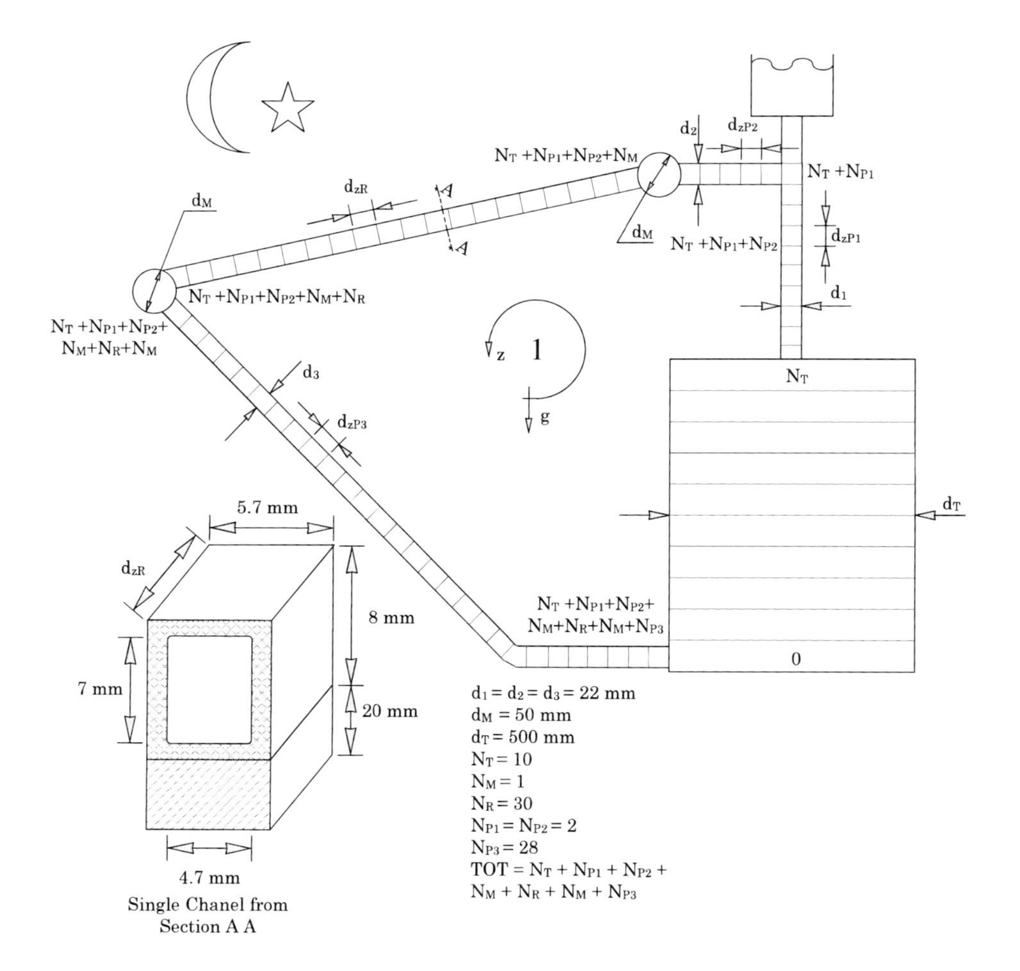

Figure 81. Discretized loop for night cycle operation [92].

In order to predict the temperature of a control volume at each time step, the conservation of energy needs to be applied to the control volume. The conservation of energy considers the flow of energy due to mass flow, conduction, convection and radiation as indicated in Figure 83. With the direction of the flow of energy known, a thermal resistance diagram of the control volumes is drawn as shown in Figure 82, and includes the convection, conduction and radiation modes of heat transfer.

Applying the general statement of conservation of energy to each unique control volume in Figure 81 and ignoring kinetic and potential energy, Eq. (105) is obtained

$$\frac{\Delta mU}{\Delta t} = \sum \dot{m}h_{in} - \sum \dot{m}h_{out} + \dot{Q}_{in} - \dot{Q}_{out} - \frac{P\Delta V}{\Delta t} \tag{105}$$

where enthalpy, $h = u + PV$, therefore, yielding Eq. (106).

$$\frac{\Delta mH}{\Delta t} = \sum \dot{m}h_{in} - \sum \dot{m}h_{out} + \dot{Q}_{in} - \dot{Q}_{out} \tag{106}$$

when enthalpy is expressed as $h = c_pT$, Eq. (107) is produced.

$$\frac{\Delta mc_pT}{\Delta t} = \sum \dot{m}h_{in} - \sum \dot{m}h_{out} + \dot{Q}_{in} - \dot{Q}_{out} \tag{107}$$

And with manipulation, Eq. (108) is obtained.

$$T^{t+\Delta t} = \frac{mc_pT^t}{(mc_p)^{t+\Delta t}} + \frac{(mc_p)^t\Delta t}{(mc_p)^{t+\Delta t}}\left(\sum \dot{m}h_{in} - \sum \dot{m}h_{out} + \dot{Q}_{in} - \dot{Q}_{out}\right) \tag{108}$$

where $m^{t+\Delta t}$ is given by Eq. (104) and c_p the specific heat is essentially constant.

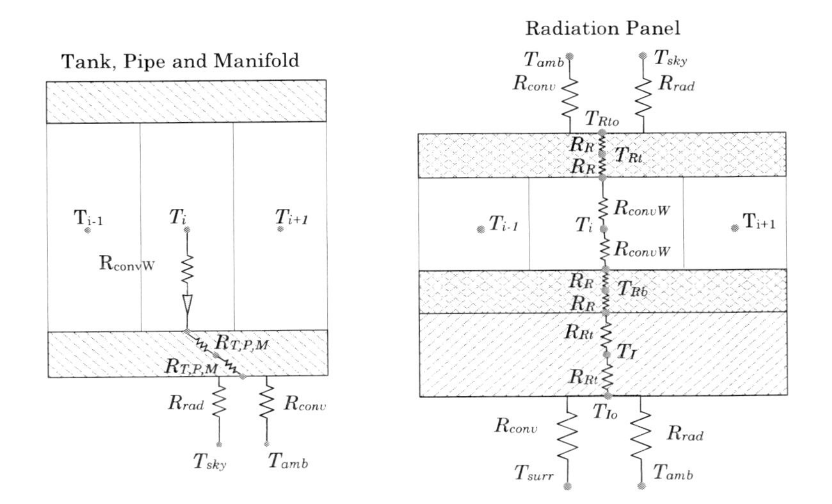

Figure 82. Thermal resistance diagram for the typical ith control volume. (Joubert and Dobson, 2017).

Application of the conservation of momentum to the cycle allows the calculation of the volumetric flow rate at the next time step. The new volumetric flow-rate is used to calculate the mass flow-rate of the control volumes for the next time step. Gravity is only present in the conservation of momentum equation and, therefore, the Boussinesq approximation becomes applicable.

Applying the general statement of the conservation of momentum as in Eq. (109)

$$\frac{\Delta mv}{\Delta t} = \sum \dot{m}v_{in} - \sum \dot{m}v_{out} + (P_{in} - P_{out})A_x - mg - \tau A_z \tag{109}$$

to each control volume shown in Figure 81, and then dividing by $A_x = \pi d^2/4$ and summing around the loop, the pressure terms cancel out to yield Eq. (110).

$$\frac{\Delta}{\Delta t}\sum \frac{mG}{{A_x}^2} = \sum \frac{G}{A_x}^2 (\rho_{in} - \rho_{out}) + \sum \rho g sin(\theta) - \sum \frac{\tau A_z}{A_x} \tag{110}$$

where $\tau = C_f \rho G^2/2A_x^2$, $A_z = \pi d \Delta z$ and $\theta = -\frac{\pi}{2}; \frac{\pi}{2}$ if the gravity acting against the flow θ is negative and with the flow θ is positive. Dividing by Δz and in the limit, as Δz and A_x tend to zero and integrating around the loop Eq. (110) becomes Eq. (111)

$$\frac{\partial}{\partial t}\oint \rho v_z = -\oint \frac{\partial \rho {v_z}^2}{\partial z} - \oint \frac{\partial P}{\partial z} - \oint \rho g_z - \oint \frac{\partial \tau}{\partial z} \tag{111}$$

where $\oint \frac{\partial P}{\partial z} = 0$

or explicitly as in Eq. (112)

$$G^{t+\Delta t} = \sum \frac{(G\rho)^t \Delta z}{(\rho)^{t+\Delta t}\Delta z} + \Delta t \frac{M+B-F}{\sum_1^{N_{TOT}} \rho^{t+\Delta t}\Delta z/A_x} \tag{112}$$

where M, B and F are respectively given by Eqs. (113), (114) and (115).

$$M = \sum \left(\frac{G}{A_x}\right)^2 (\rho_{in} - \rho_{out}) \tag{113}$$

$$B = \sum \rho g sin(\theta) \tag{114}$$

$$F = \sum(\tau \pi D(\Delta z + \Delta z_{minor})/A_x \tag{115}$$

The thermal and thermofluid modelling procedure as described for the night cycle is equally applicable to both the day water-heating and day room-cooling cycles. Discretized representations of these two cycles are given in Figures 84 and 85, but for the sake of brevity, their thermal and thermofluid modelling will not be further described, as they are given by Joubert (2014). An explicit numerical solution method was used to solve for the set of finite difference equations constituting the thermofluid model. Data provided by the weather station of the University of Stellenbosch was used in the theoretical simulation of the system (Stellenbosch-Weather, 2014). The theoretical and experimental results are compared with each other to establish the validity and accuracy of the theoretical model solution. The experiment was carried out over an extended period in April 2014. The weather conditions during this period varied from clear to cloudy skies, with most of the days enjoying sunny summer weather. For winter conditions the experiment was carried out only in the night cycle, and this experiment took place during clear night-sky conditions in June 2014.

In Figure 84 the experimental average temperature (of five equally spaced measurements) of the cold water tank is compared with the theoretically determined average (of ten control volumes) cold water tank temperature. The average temperatures vary from a minimum of about 20°C to a maximum of about 35°C. It is also seen that the theoretical tank temperatures tended to lag the experimental temperatures, but other than that, correspond reasonably well with each other. On the morning of 10 April it is seen that the experimental temperature is not as low as the theoretically predicted value, this being ascribed to the reduction in radiation from the panel to the sky that night due to the presence of cloud coverage, which was not taken into account in the theoretical model.

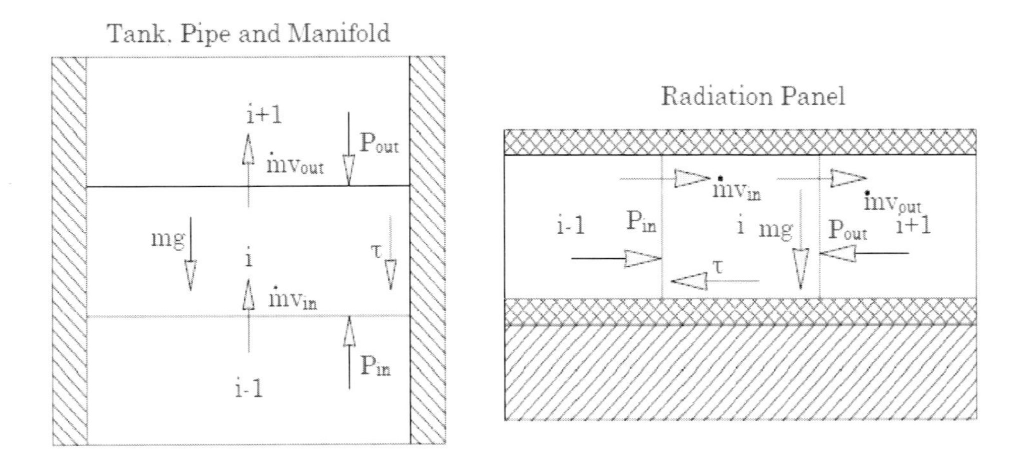

Figure 83. Diagram indicating the forces acting in on the ith control volume [92].

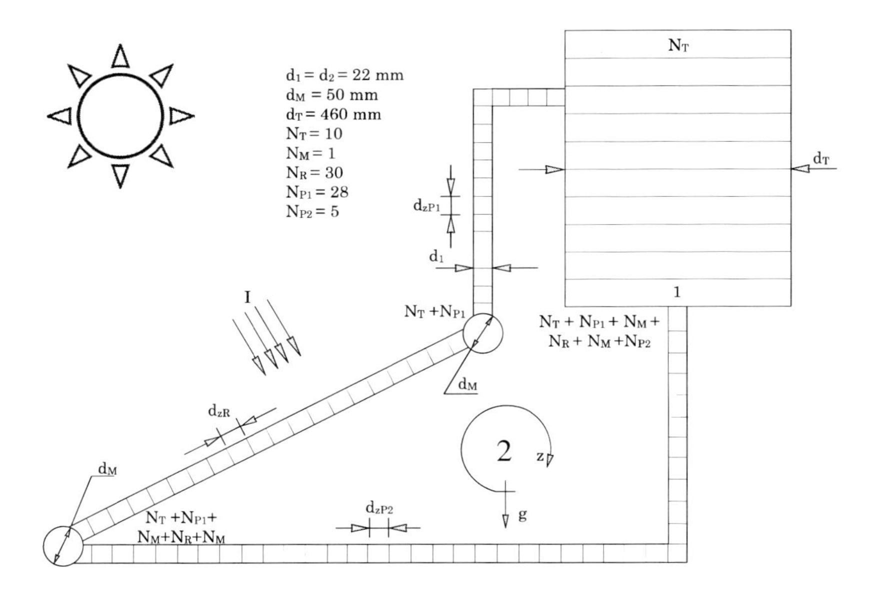

Figure 84. Discretised day (water-heating) cycle [92].

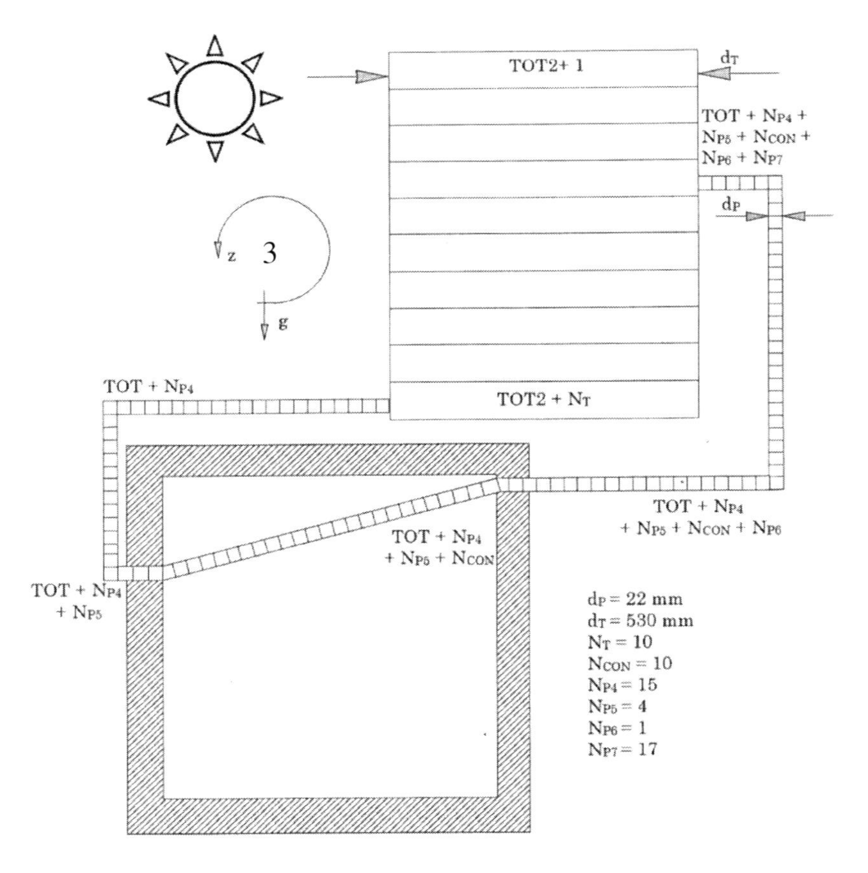

Figure 85. Discretized room convector cooling cycle [92].

The rate of heat removed from the cold water tank is given by Eq. (116).

$$\dot{Q} = mc_p(T_{t+\Delta t} - T_t)/\Delta t \tag{116}$$

where the mass m = ρ_{avg}V, ρavg is the average density water temperature and V is the volume of the cold water tank, which in this case was 150 L. Depending on the weather conditions the heat removal rates from the cold water tank varied between 39 and 75 W/m^2, but on average was 55 W/m^2. This average value corresponds well with the heat removal rates of 60 W/m^2 reported by Dobson [88] and Okoronkwo [95].

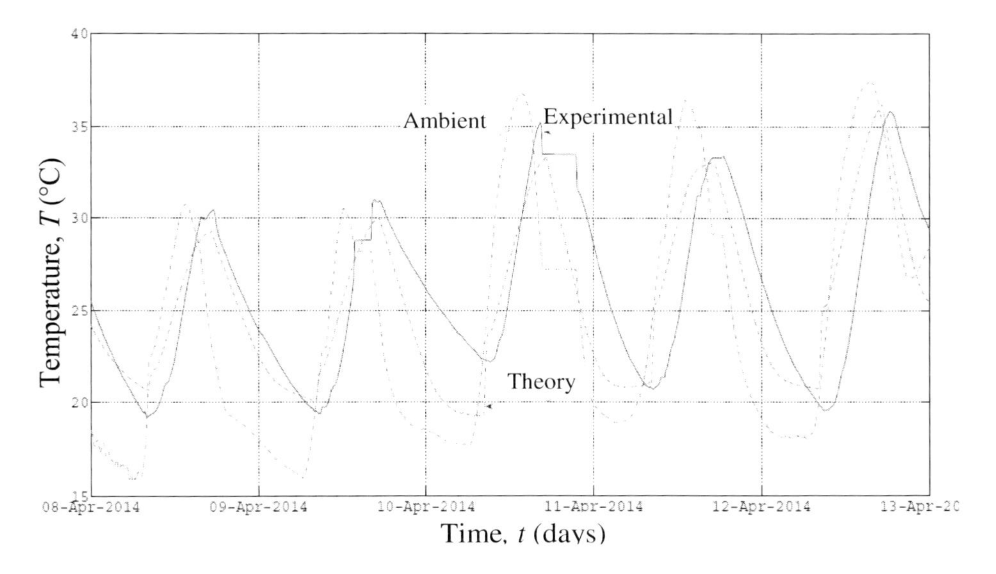

Figure 86. Comparison between the theoretical and experimental water temperatures in the cold water tank [92].

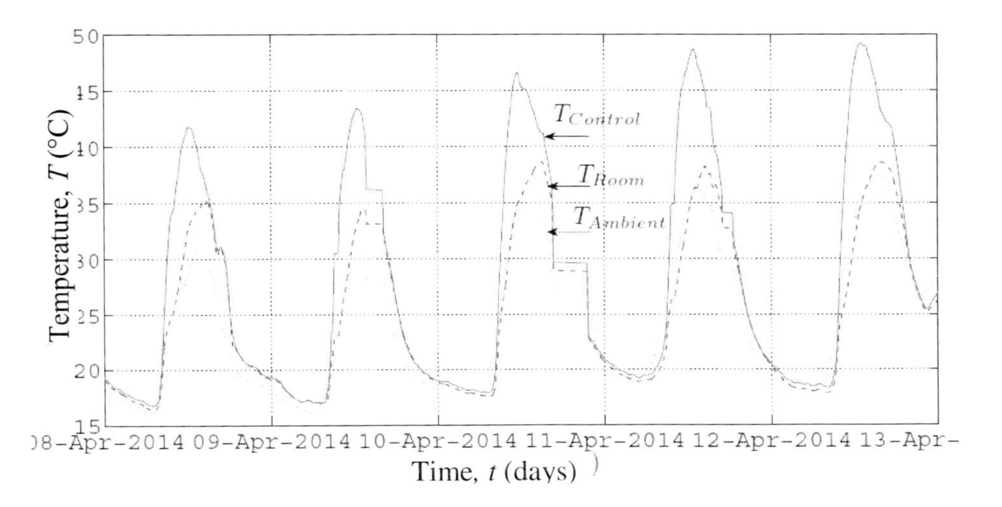

Figure 87. Comparison of the experimentally determined room and control temperatures [92].

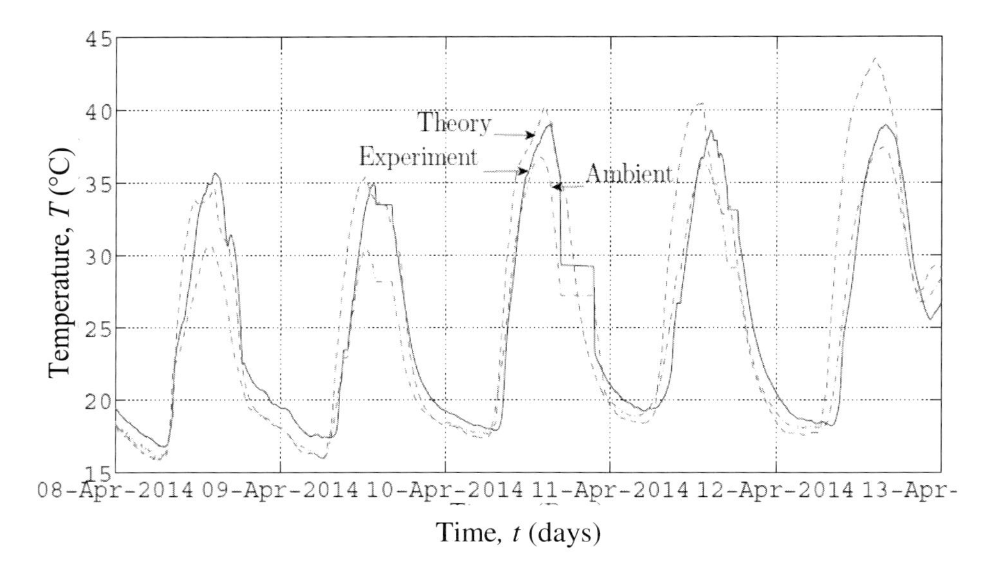

Figure 88. Comparison of the measured room and theoretically calculated temperatures [92].

For comparison, a control room was used. The control room had no cooling. The experimentally determined temperatures of the control and room are shown in Figure 87. The cooling system was able to reduce the temperature in the room by between 7 and 12°C.

In Figure 88 the theoretically calculated and experimentally measured room temperature are compared. It is seen that the daily temperature profiles closely follow each other. The rate of heat removed from the air in the room by the convector and subsequently absorbed by the water in the cold water tank varied from 102 to 150 W/m^3, but on average was 126 W/m^3.

The theoretically calculated and experimentally measured hot-water storage tank temperatures are shown in Figure 89. Note that the water heated in the panel circulates upwards from the panel to the hot water storage tank only during the day and stops circulating at night. At night when the water in the panel is cooled and is more dense it circulates downwards and into the cold water storage tank and the warmer water at the top of the cold water storage tank is displaced upwards and is, in turn, cooled in the panel, and so on. Despite this circulatory flow stopping during the day when the water in the convector in the room is heated by the hot air in the room, its density deceases and it flows upwards and into the cold water storage tank. Subsequently denser cold water in the cold water storage tank, in its turn, now flows downwards and back into the room convector, and so on. Experimental and theoretical temperatures (in Figure 89) compare reasonably with each other. The temperature of the water in the hot water storage (of 68 L) increases by about 30°C to 40°C and the rate at which it is heated varies from about 230 to 448 W. On average it is about 96 W/m^2 of panel area.

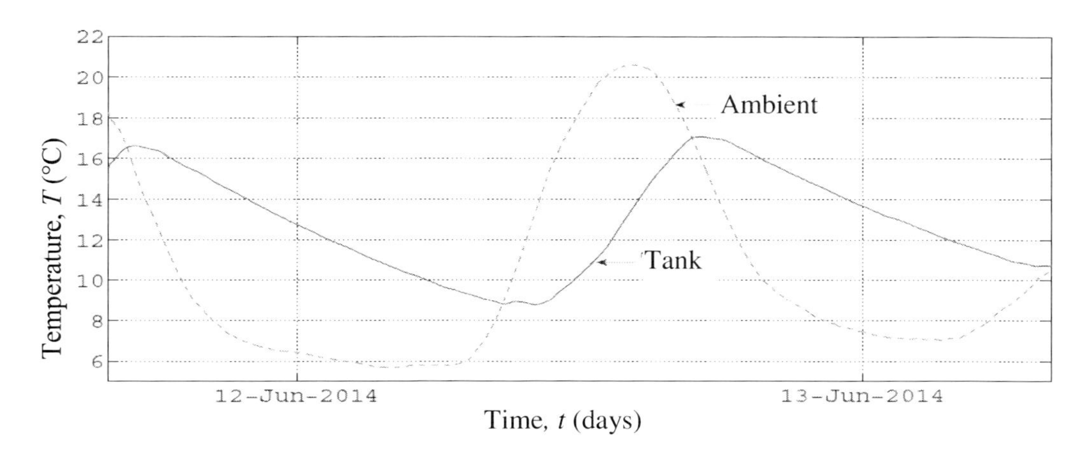

Figure 89. Comparison of the experimentally measured ambient and cold water tank temperatures as a function of time [92].

During typical (but relatively cloudless) winter weather conditions the temperature of the water in the cold water tank is compared with the ambient temperature as in Figure 89. Although the water is only cooled to within 5°C of the now cold ambient temperatures, it was cooled at a rate of 56.6 and 55.8 W/m^2 of panel area, for two days. These values, however, also correspond to the summer panel cooling rates of on average also about 55 W/m^2 of panel area.

In conclusion, the design, construction, experimental testing and theoretical modelling of a night-sky radiation system, comprising of night cooling of water and day time cooling of a room and heating of water, was successfully undertaken. The theoretically determined temperatures of the cold water tank, hot water tank and room were graphically compared with the experimentally measured values. The theoretical model used data supplied by the weather station of the University of Stellenbosch and which is situated close to the experiment location. The theoretical model results compared reasonably well with the experimental results. From a numerical solution procedure point of view, it was also shown that reasonably sized control volumes and time step length were indeed used.

The theoretical model of the night-time cooling of the water in the cold water-storage tank by the radiating panel yielded results that predict the performance of the system favourably well and this is confirmed by way of Figures 86 to 88. During the night the system was able to radiate energy to the cold sky at a rate of 55 W/m^2 of radiating panel. Any cold tank temperature deviations from the experimental results were ascribed to cloud coverage during the night, which was not taken into account by the theoretical model. For the as-sized experimental set-up, during the day, the air in the room was cooled by the water circulating through the convector at a rate of between between 102 to 150 W/m^3, but on average was 126 W/m^3 of the room volume. The theoretical model was able to emulate the experimentally measured values as illustrated in Figure 88. Also during the day, the water in the hot water-storage tank was heated at a rate of 96 W/m^2 of panel area. Based on the foregoing discussion, it is seen that it is entirely plausible to design and operate a passive system, that is, without the use of any moving mechanical equipment such as pumps and active controls, for water-cooling during the night and room-cooling during the day, as well as for water-heating during the day. It is thus concluded that such a passive cooling/heating system (as illustrated schematically by way of Figure 79) is an entirely viable energy-saving option, and also that the theoretical simulation model, as formulated, can be used with confidence as an energy saving system design and evaluation tool.

6.3. Cooling of Electronic and Electrical Equipment

The demands of the electronics industry have dictated that electronic assemblies operate at increased speeds in smaller packages with higher reliability. These requirements create extreme challenges in thermal management for the packaging engineer. Faster speeds generate more heat, while smaller packages increase the difficulty of removing it from sensitive devices which are adversely affected by increased temperature. In this section we will thus consider a number of possible cooling applications of potential interest to a thermal management engineer.

6.3.1. Separated Thermosyphon Heat Pipe

Consider a typical cabinet containing electronic equipment as shown by Jeggles [96] in Figure 90. A fan in the cabinet draws room-temperature air in at the bottom, though the cabinet where its temperature increases as it is heated by the heat dissipated by the electronic equipment, and then expelled out at the top as shown in Figure 91. The room shown in Figure 91 is a control cabin for a mobile radar system and which requires three operating personnel. The heated room air is in turn cooled by a conventional air conditioning unit (ACU).

This cooling option has a number of disadvantages. Firstly, as the air flows through the cabinet it is progressively heated and the equipment has to be cooled by progressively hotter air and hence the equipment in the upper levels will have to operate at higher temperatures, for the same heat removal temperature differences. Secondly, if there is a temperature limit placed on the equipment to keep the electronic equipment within specified operating temperature limits, the air entering the cabinet may have to be kept much cooler than the comfort level for the operating personnel. Thirdly, the air conditioning and cooling systems would need to be over-designed to allow for future upgrades using more and more powerful electronic equipment.

A possible cooling solution, without changing the general layout of the electronic cooling equipment in the cabinet in the room, would be to specially cool the air, either as it enters the cabinet or at a convenient level along the air flow path in the cabinet [97]. This could be done by channeling cold air as it exits the ACU directly to the inlet to the cabinet.

Rather than additional ducting, and an accompanying additional loss of room space, the use of a *separated* two-phase closed loop thermosyphon can be envisaged as shown in Figure 92. The condenser section could be placed at the cold air outlet of the ACU and the separate evaporator section in the cabinet, and to complete the loop, a riser and condensate return pipe.

Two heat exchangers were specially manufactured using commercially available refrigeration industry manufacturing technology and standards [97]. The evaporator heat exchanger as specified in Figure 93 is designed for a horizontal orientation. The condensing heat exchanger is designed for vertical operation and is specified in Figure 94.

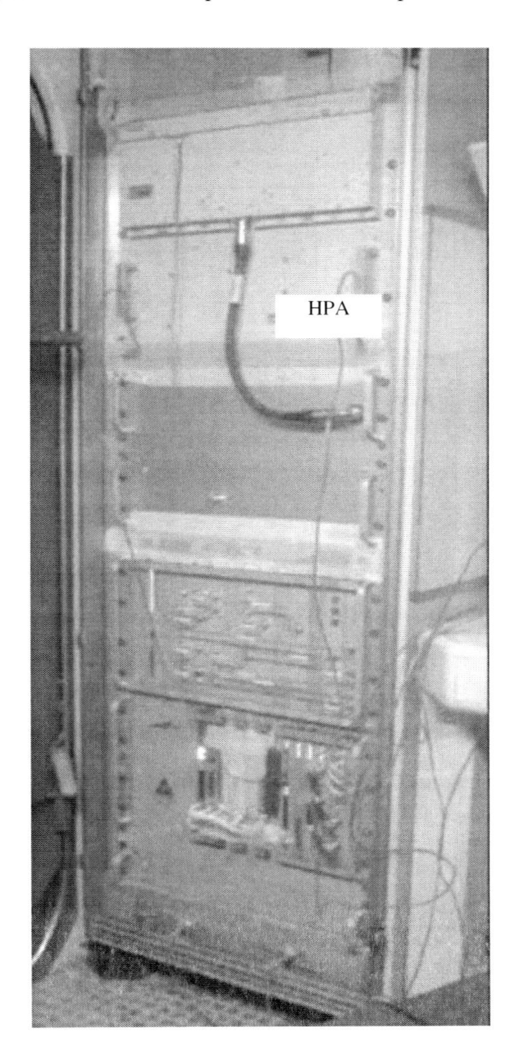

Figure 90. Example of a cabinet (with its door removed) containing high power electronic equipment.

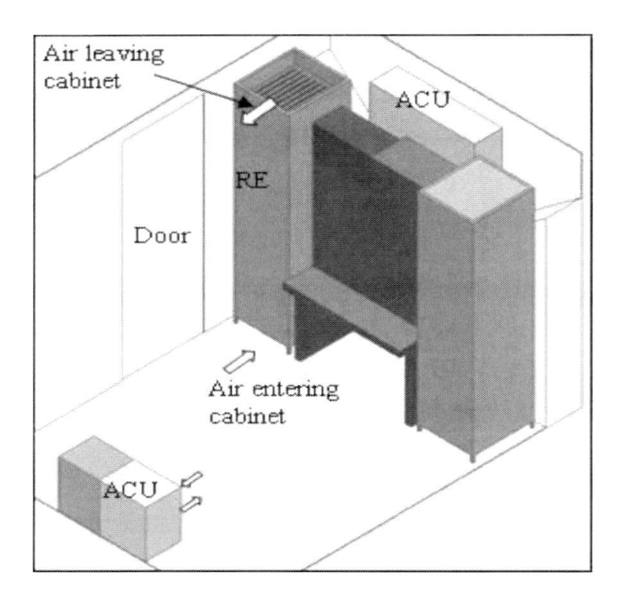

Figure 91. Schematic layout of a room with cabinets containing electronic equipment and also showing position of the air-conditioning units [96].

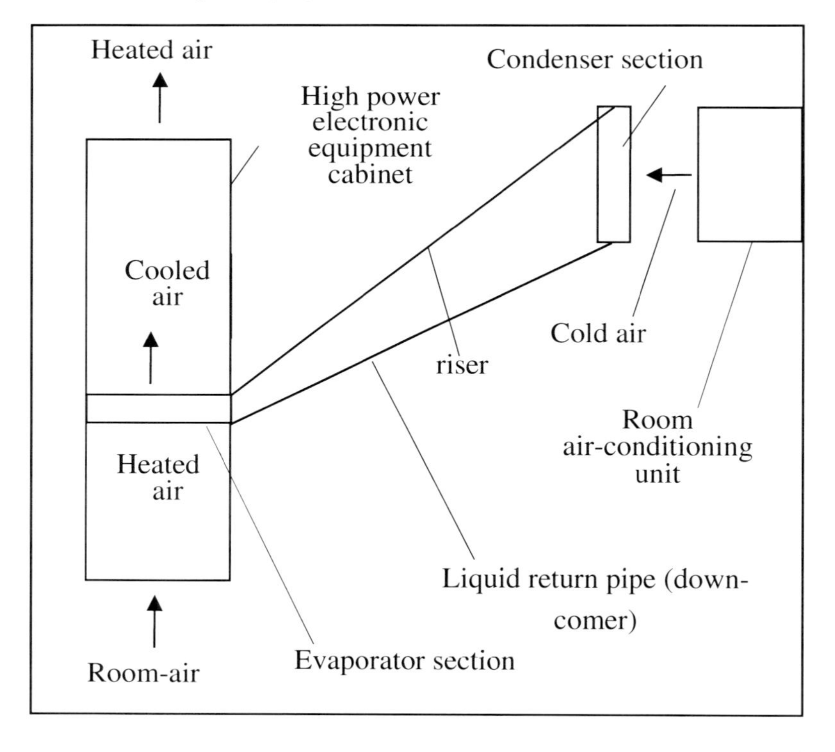

Figure 92. Two-phase closed loop thermosyphon with separated evaporator and condenser sections [96].

The heat transfer and pressure loss characteristics were obtained using a laboratory wind tunnel and hot water supply as shown in Figure 95. Heat exchanger inlet and outlet air and water stream temperatures and pressures were measured, the air flow rate was determined using bell-mouthed nozzles and the water flow rate using a measuring cylinder and a stopwatch. Hot water was passed through the heat exchanger tubes at a relatively high flow rate. This allowed the water side heat transfer coefficient to be relatively accurately calculated using standard heat transfer theory. Also, the water side thermal resistance will be much lower than the air-side thermal resistance. Any inaccuracies in determining the water side thermal resistance will thus not necessarily impact significantly on the much larger and *controlling* air-side thermal resistance. In the determination of the heat transfer coefficients, the small effect on the thermal conduction of the copper tube-wall was neglected. Correlations for the heat exchanger air-side heat transfer coefficients based on the inside tube surface area were in this way experimentally determined.

The evaporator heat transfer coefficient h_{eo} in W/m^{2}K in terms of the face velocity in m/s and with a correlating coefficient $R^2 = 0.964$ was determined as:

$$h_{eo} = 324.9 V_{face}^{0.5530} \tag{117}$$

Similarly, the condenser airside heat transfer coefficient h_{co} and with a correlating coefficient $R^2 = 0.969$ was determined as:

$$h_{co} = 410.5 V_{face}^{0.5972} \tag{118}$$

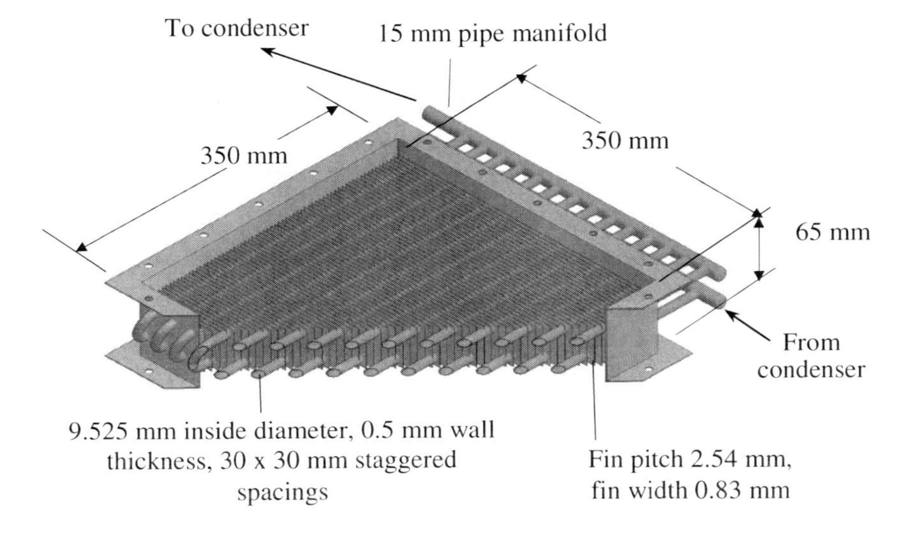

Figure 93. Construction details of the horizontally orientated evaporator, copper tubes and aluminium fins [96].

The frictional pressure loss Δp in Pa across each of the two heat exchanges was correlated in terms of the face velocity V_{face} in m/s for the evaporator with $R^2 = 0.9467$ as:

$$\Delta p = 0 - 0.4080\ V_{face} + 4.436\ V_{face}^2 \tag{119}$$

And similarly for the condenser with $R^2 = 0.9946$ as:

$$\Delta p = 0 + 10.14\ V_{face} + 2.6046\ V_{face}^2 \tag{120}$$

The above correlations are applicable for air face velocities (defined as air mass flow rate divided by heat exchanger height times width and density $V_{face} = \dot{m}_a / \rho_a A_{face}$ and at the linear average temperature through the heat exchanger) of up to about 5 m/s.

The inside-tube evaporator heat transfer coefficients were determined using equations 121, based on the inside tube-surface area of the evaporator as:

$$h_{ei} = \left[\left(\frac{\overline{T}_{aeo} - \overline{T}_{ei}}{\dot{Q}_e} - \frac{1}{h_{eo} A_{eo}}\right) A_{ei}\right]^{-1} \tag{121}$$

The condenser heat transfer coefficient, based on the inside tube-surface area of the condenser, as:

$$h_{ci} = \left[\left(\frac{\overline{T}_{ci} - \overline{T}_{aco}}{\dot{Q}_c} - \frac{1}{h_{co} A_{co}}\right) A_{ci}\right]^{-1} \tag{122}$$

where

$$\dot{Q}_e = \dot{Q}_c = \dot{m}_{ae} c_{pae} (T_{aei} - T_{aeo}) = \dot{m}_{ac} c_{pac} (T_{aco} - T_{aci})$$

$$\overline{T}_{aeo} = (T_{aei} + T_{aeo})/2$$

$$A_{ei} = \pi d_{ei} L_e$$

$$A_{eo} = A_{face\,e} = L_{e1} L_{e2},$$

$$\overline{T}_{aco} = (T_{aci} + T_{aco})/2$$

$$A_{ci} = \pi d_{ci} L_c$$

$$A_{co} = A_{face\,c} = L_{c1} L_{c2}.$$

L_e and L_c are the total lengths of the copper piping over which air flows in the evaporator and condenser, respectively. L_{e1}, L_{e2}, L_{c1} and L_{c2} are the face dimensions (length and breadth) of the evaporator and the condenser, respectively.

The experimentally determined evaporator heat transfer coefficient as determined using Eq. (121) is correlated, essentially in terms of the evaporator wall heat flux, in terms of a Reynolds number and to within a correlation coefficient $R^2 = 0.924$ as

$$h_{ei} = 0.001550\ Re^{1.226} \tag{123}$$

where

$$Re = \frac{4\dot{Q}/h_{fg}}{\pi d \mu_h} \tag{124}$$

The viscosity of the working fluid is assumed as:

$$\mu_h = 1/\left(x/\mu_h + (1-x)/\mu_h\right) \tag{125}$$

and where the two-phase mass fraction x is assumed as being $x = 1$ - FR where the liquid charge fill ratio FR = V_ℓ / V_e, where the liquid charge fill ratio was 50% of the evaporator tube volume.

Similarly, using the same Reynolds number for the condenser but with a correlation coefficient $R^2 = 0.977$ the heat transfer coefficient may be predicted using

$$h_{ci} = 0.000422\ Re^{1.39} \tag{126}$$

The test results may be summarized, as shown in Figure 96, as the heat transferred by the separated thermosyphon as a function of the temperature difference between the average air across the evaporator and the condenser $\bar{T}_{aeo} - \bar{T}_{aco}$ for different air flow face velocities $V_{face} = \dot{m}_a / \rho_a A_{face}$, $\bar{T}_{aeo} = (T_{aei} + T_{aeo})/2$ and $\bar{T}_{aco} = (T_{aei} + T_{aco})/2$ $\bar{T}_{aco} = \left(T_{aci} + T_{aco}\right)/2$. It is seen that the separated thermosyphon heat transfer rate is not only dependent on the temperature difference but also on the face velocity, for instance at a temperature difference of 20°C and at a face velocity of 3.46 m/s the heat transfer rate is about 1 500 W, whereas at 1.94 m/s the heat transfer rate is 1 000 W.

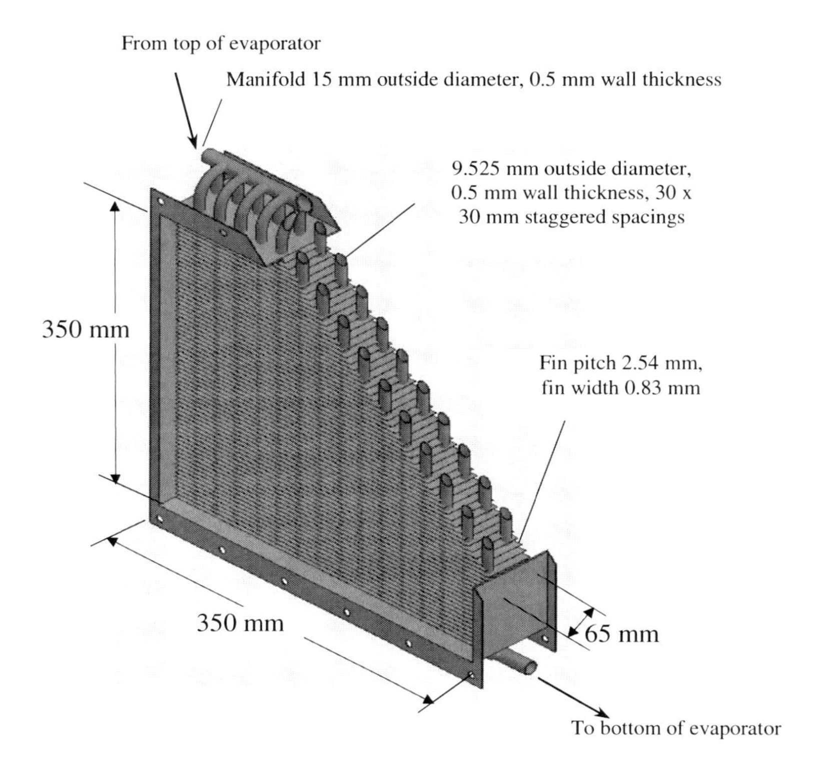

Figure 94. Construction details of the condenser copper tubes and aluminium fins [96].

Butane (with a liquid charge fill ratio of 50% of the evaporator inside-tube volume) was used as the working fluid. If water is used, and after charging, any non-condensable gases that may have to be exhausted to the atmosphere can only exhausted provided the thermosyphon is specially heated to well-above 100°C. With butane this heating procedure is not required; as at room temperature its saturated pressure is suitably higher than atmospheric pressure. Although water has better heat transfer characteristics than butane, the thermal performance test results tend to be less eratic than when water is used. This could be due to the relatively high water-vapour velocities at the same temperatures and heat transfer rates due to its relatively lower density, at the low temperatures up to about 50°C in order to ensure case temperatures not exceeding 60°C needed for cooling of electronics using the ambient air as the ultimate heat sink.

It was not possible to determine the maximum heat transfer rate at which, for any further increase in the temperature difference between the evaporator and the condenser, the heat transfer rate ceases to increase as the available water heat source was only 80°C. In Figure 96 it is seen that the heat transfer rate is clearly still increasing as a function of the temperature difference and shows no sign of leveling off at temperatures of up to 80°C. Based on all the test results it was concluded that for the as-defined separated heat using butane as the working fluid and the heat transfer coefficients presented the individual and system heat transfer performance may be calculated reasonably well to within 10%.

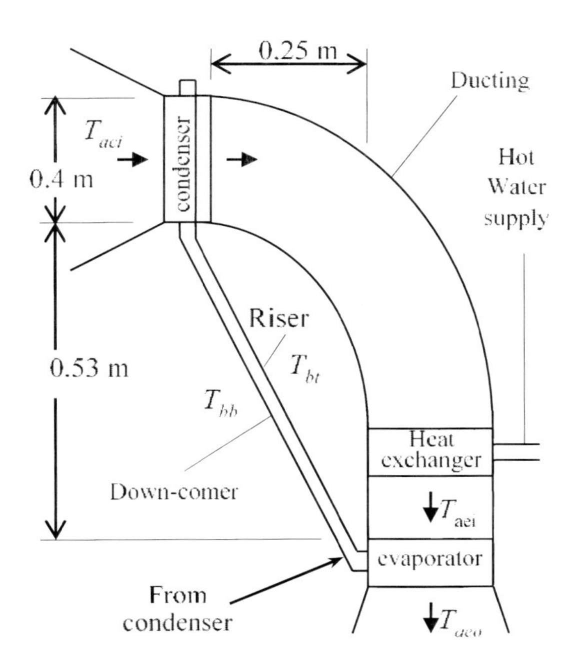

Figure 95. Experimental test set-up of the separated thermosyphon heat exchanger in wing tunnel [96].

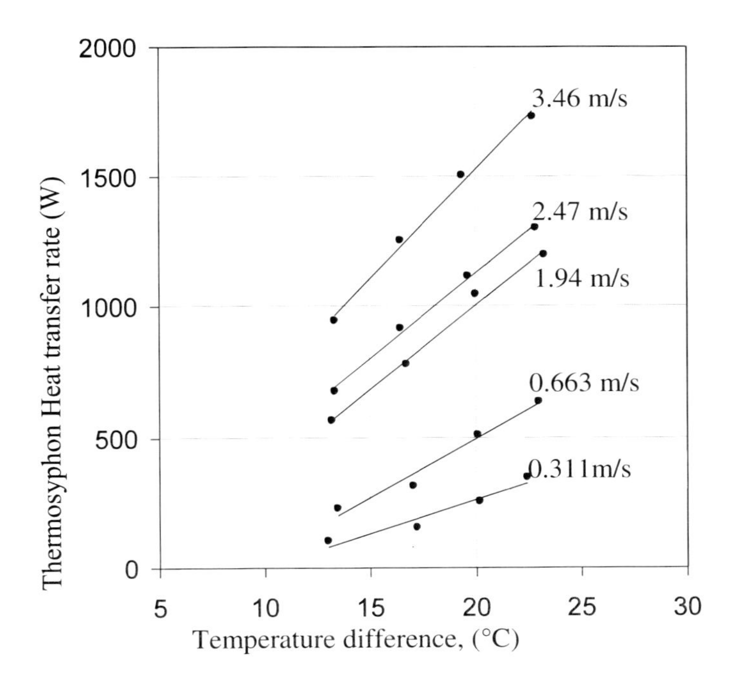

Figure 96. Heat transferrer rate as a function of the average air temperature temperature difference for different air flow face velocities [96].

6.3.2. Bent (BT) and Looped Closed Thermosyphons (CLTs)

The demands of the electronics industry have dictated that electronic assemblies operate at increased speeds, in smaller packages and with higher reliability. In particular consider a high power amplifier for which needing localized cooling of the power transistors is needed in order to prevent overheating. Two heat pipe cooling configuration were considered [98], the one using a bent closed two-phase thermosyphon (BT) as shown in Figure 97 (a), and the other using a closed loop two-phase thermosyphon (CLT) as shown in Figure 97 (b). Both these options are configured to slot into a standard sized rack such as the rack for the high power amplifier (HPA) shown in Figure 90. The rack is inserted then pulled up tightly against the cold plate at the back of the rack using long threaded bolts and nuts.

The as-tested BT was manufactured from 12.7 mm outside diameter cold drawn copper tube incorporating a standard long radius elbow, a vacuum/charge/discharge valve and two thermocouple insertion fittings and thermocouples as shown in Figure 98. The system was evacuated and charged with working fluid (R134a) using standard refrigeration equipment. The liquid charge fill ratio was about 75% of the volume of the horizontal length of the BT. The horizontal length is L_H = 250 mm and the vertical length L_V = 150 mm. Heat is supplied evaporator by 3 x 25 W power resistors mounted on pairs of 50 x 50 square x 10 mm thick aluminium plate heat spreaders that are spaced between 1 and 2 mm apart, which for three pairs of heat spreaders gives an evaporator heated length of about L_e = 3 x 50 + 3 = 153 mm. The heat was removed from the condenser section of the thermosyphon by a tube-in-tube heat exchanger of length L_c = 110 mm.

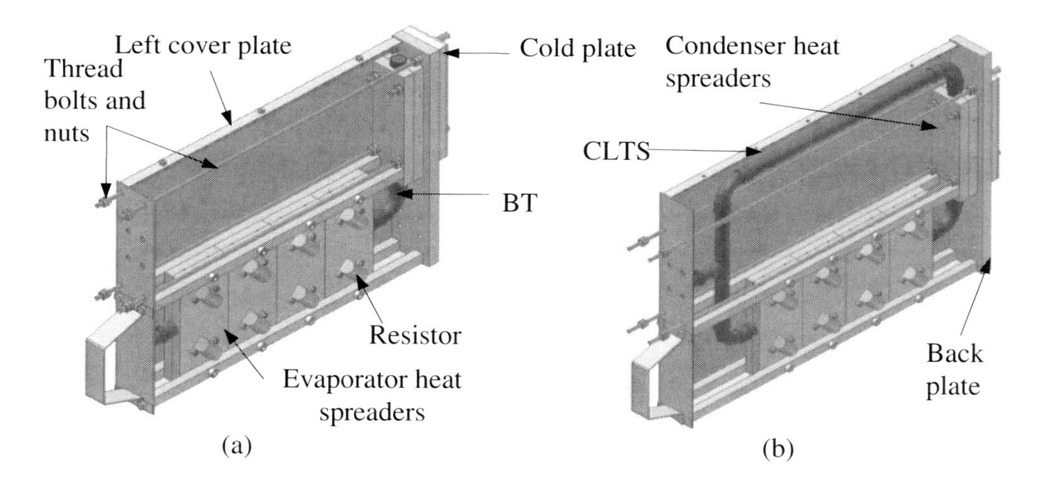

Figure 97. Possible transistor (shown as resistors) cooling solution using a bent thermosyphon (BTS) and a closed loop thermosyphon (CLTS) and an externally cooled cold plate [99].

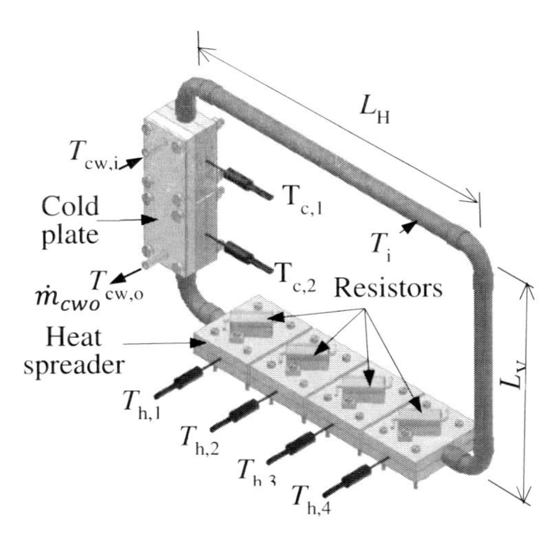

Figure 98. Test set-up for the CLT cooling configuration [99].

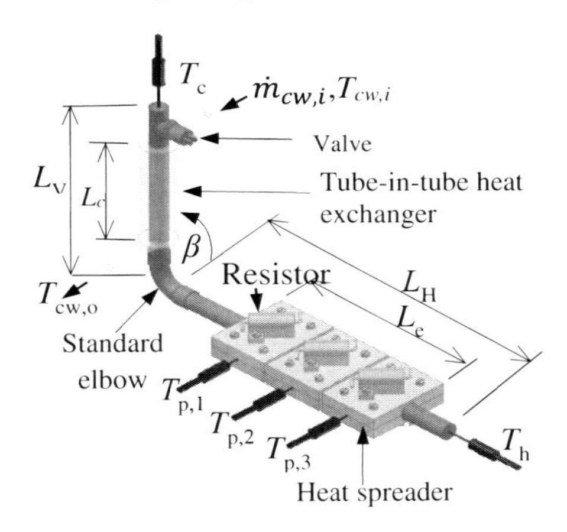

Figure 99. Test set-up for the BT cooling configuration [99].

The shape of the closed loop thermosyphons (CLT) that were tested is shown in Figure 98. Three different pipe outside and inside diameters of sizes (12.7 x 11.0 mm, 9.5 x 8.3 mm and 6.2 x 5.7 mm were used for the loops. Three different working fluids were tested: water, R134a and butane. The liquid charge fill ratio is about 60% of the volume of the horizontal length of the CLT. The heat spreaders are all 50 x 50 mm square x 10 mm thick and bolted tightly to the pipe with a little thermal paste to reduce the contact resistance. Grooves were machined in the heat spreaders allowing for 1.6 mm diameter thermocouple probes to be snugly fitted and inserted all the way up to the pipe outer surface. An idea of the inside temperature of the working fluid in the loop was obtained by measuring the surface temperature of the top (unheated), horizontal portion of the loop. Heat input was achieved using 4 x 25 W power resistors and a variable power supply. The heat was removed by a 100 mm long x 50 mm wide cold plate heat exchanger through which cooling water could be circulated.

Electrical power was supplied to the resistors using a variable voltage power supply and heat was removed by the cooling water flowing through the tube-in-tube heat exchanger. To ensure a more uniform flow of cooling water through the tube-in-tube heat exchanger the annular water channel was supplied with spiralled inserts that forced the water to spiral around the outside surface of the BT condenser. The cooling water flow rate was measured using a stop watch and a measuring cylinder. During testing the heated section was wrapped in 50 mm thick ceramic insulating wool. The BT was tested with the condenser section in a vertical plain but could be inclined (in a vertical plane) at an angle β to the horizontal. The working fluid was R134a and liquid charge fill ratios of about 0.75 of the volume of the evaporator section, cooling water temperatures of between 20 and 40°C and condenser inclinations to the horizontal of 45, 60 and 90° were tested.

The heat supplied to the resistors is reckoned using Ohm's law as:

$$\dot{Q}_{\mathrm{elec}} = V/R_{\mathrm{ei}} \tag{127}$$

where V is the applied voltage in Volts and electrical resistance R, in Ω. Heat loss from the resistors through the insulation to the surrounding air was experimentally determined as about

$$\dot{Q}_{\mathrm{e,loss}} = 0.25(\bar{T}_{\mathrm{h}} - \bar{T}_{\mathrm{amb}}) \tag{128}$$

where the average heat spreader temperature $\bar{T}_{\mathrm{h}}$ is the average temperature of the plate heat spreader thermocouple temperature readings $T_{\mathrm{p,i}}$ shown in Figure 98. The heat into the evaporator $\dot{Q}_{\mathrm{e}}$ is thus:

$$\dot{Q}_{\mathrm{e}} = \dot{Q}_{\mathrm{elec}} - \dot{Q}_{\mathrm{e,loss}} \tag{129}$$

The heat removed by the cold plate is

$$\dot{Q}_{\mathrm{c}} = \dot{m}_{\mathrm{cw}} c_{\mathrm{pcw}} (T_{\mathrm{cwo}} - T_{\mathrm{cwi}}) - \dot{Q}_{\mathrm{c,loss}} \tag{130}$$

where, for the evaporator section, the heat loss/gain to/from the surroundings was experimentally determined as:

$$\dot{Q}_{\mathrm{c,loss}} = 0.10(\bar{T}_{\mathrm{cw}} - \bar{T}_{\mathrm{amb}}) \tag{131}$$

Figure 100(a) shows the heat removed from the condenser section as a function of the heat into the evaporator section of the BTS. Only values of up to 180 W are reflected as above the value the BTS attained its maximum heat transfer rate, it could not remove heat at a greater rate than this value even if the temperature difference between the hot and cold sections was further increased. At heat transfer rates of less than 30 W it is seen that the operation is somewhat erratic, however above 30 W it is seen that an energy balance of less than about ± 15% was maintained; thereby indicating a reasonable level of confidence in the experimental results.

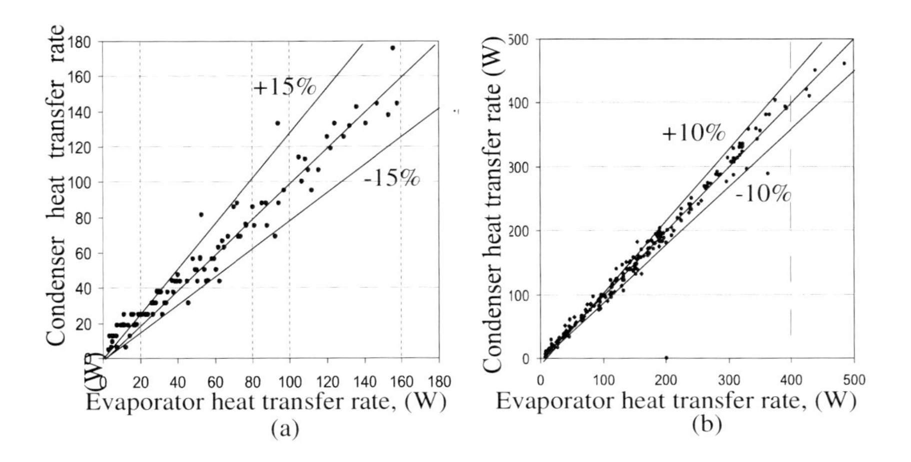

Figure 100. Energy balance for BT (a) and (b) CLTS showing the heat removed by the condenser as a function of the heat input to the evaporator [99].

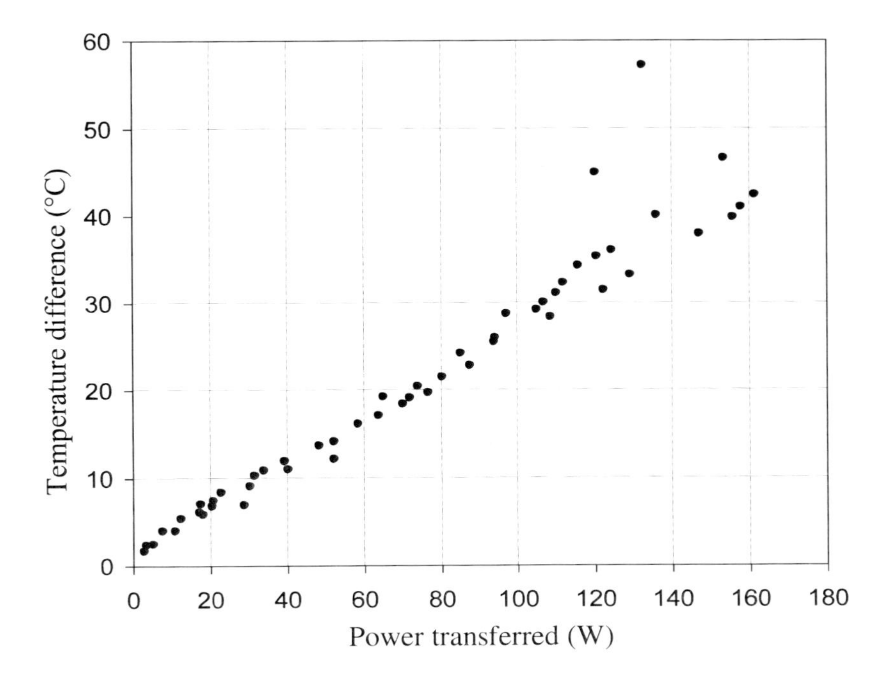

Figure 101. Temperature difference between the average heat spreader plate temperature and the average cooling water temperature for the bent thermosyphon (BT) $\overline{T}_{p} - \overline{T}_{cw}$ [99].

Figure 100(b) shows the heat removed from the condenser section as a function of the heat supplied to the evaporator section of the BT and CLTS. The condenser heat transfer rates are seen to correspond to within about 10% of the heat input rates indicating a reasonable degree of confidence in the validity of the experimental results. It is also seen that a significantly larger maximum heat transfer rate amount (500 W) can be transferred by the CLTS compared with the 180 W with the BT.

Figure 101 gives the temperature difference between the average plate temperature and the average cold plate cooling water temperatures as a function of the heat input into the evaporator section of the BT. The BT was not investigated further because at temperature differences of about 40°C and greater the maximum heat transfer rate was in to order of only about 150 W, that is, the heat transfer rate did not increase further when the temperature difference was increased above 40°C. This value compared to the CLT was deemed too low (see next section). All things being equal it is much lower than for a closed loop thermosyphon and as such no further tests were carried out with the BT.

To enable a designer to determine the thermosyphon heat transfer rate, in general, the heat transfer coefficient between the working fluid and the inside wall temperature must be known. It was not feasible to measure the wall temperature but rather the temperature of the heat plate spreaders as measured by temperatures $T_{h,1,2,3,4}$ and $T_{c,1,2}$. The working fluid temperatures were not measured directly but rather the outside wall temperature in the top horizontal leg at a position T_i, as indicated in Figure 98. As the wall resistance of the copper pipe is essentially negligible, this temperature is assumed to give a reasonable estimate of the average temperature of the working fluid in the thermosyphon. The heat transfer coefficient was, accordingly, thus defined as:

$$h_{ei} = \frac{\dot{Q}_c}{(\bar{T}_{hp} - \bar{T}_i)A_e} \tag{132}$$

and

$$h_{ci} = \frac{\dot{Q}_c}{(\bar{T}_i - \bar{T}_{cp})A_e} \tag{133}$$

where A is defined as the inside wall surface area of the pipe for a length covered by the heat spreaders, $\bar{T}_{hp}$ is the arithmetic average of temperature of $T_{h,1\to4}$. The evaporator heat transfer coefficient was then assumed to be dependent on the heat flux using the Kutateladze number Ku, a Jacob number Ja to capture the temperature dependence of the working fluid properties, and the diameter d and fill ratio FR. The condenser heat transfer coefficient was assumed to depend on the Reynolds number Re and the film condensation coefficient FC, and also the diameter d and the fill ratio FR, defined, as for the BTS, as the ratio of the volume of the liquid working fluid to the volume of the lower horizontal pipe of length L_H.

The heat transfer coefficients were then experimentally correlated [98] using multi-linear regression and assuming a power law dependence for the evaporator heat transfer h_{ei} coefficient and condenser heat transfer coefficient h_{ci} as:

$$h_{ei} = a\, Ku^{b1}\, Ja^{b2}\, d^{b3}\, FR^{b4} \quad (134)$$

$$h_{ci} = a\, Re^{b1}\, FC^{b2}\, d^{b3}\, FR^{b4} \quad (135)$$

and where the coefficients for the as-tested CLTs and working fluids are given in Table 10, and the dimensionless variable are:

$$Ku = \frac{\dot{Q}''}{\rho_v h_{fg}[\sigma g(\rho_l - \rho_v)/\rho_v^2]^{\frac{1}{4}}},\ Ja = c_{pl}(T_{wall} - T_{sat})/\ h_{fg},\ Re_l = \frac{\dot{Q}/h_{fg}}{\mu_l \pi d_i/4}$$

and

$$FC = \frac{g\rho_v(\rho_l - \rho_v)d_i^3 h_{fg}}{\mu_l k_l(T_{sat} - T_{wall})}$$

The coefficient of determination R^2 for these correlations was significantly better than 90%. Also the experimental heat input agreed to the experimental heat output to within 10% as shown in Figure 102. As such the correlations may be used with reasonable confidence.

Table 10. Regression coefficients for Eq. (134) and Eq. (135)

Evaporator heat transfer coefficient h_{ei}						
	R^2	a	Ku	Ja	d	FR
Bu	0.933	169889	0.770	-0.704	-0.669	0.678
R134	0.922	30568	0.685	-0.306	-0.818	0.183
H_2O	0.870	5.18×10^{-3}	1.360	-0.881	-3.052	0.822
Condensation heat transfer coefficient h_{ci}						
	R^2	a	Re	FC	d	FR
Bu	0.454	1.59×10^{-3}	0.406	1.190	-3.353	-0.139
R134	0.651	5.16×10^{-6}	0.164	0.594	-1.873	0.034
H_2O	0.995	5.19×10^{-90}	0.822	6.488	-20.19	0.123

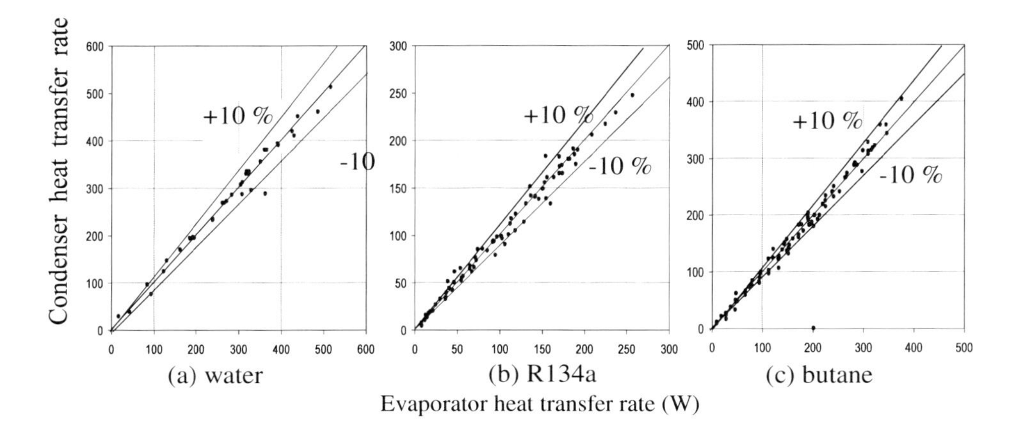

Figure 102. Comparison of CLTS condenser and evaporator heat transfer rates for (a) water, (b) R134a and (c) butane working fluids [99].

6.3.3. Plate and Thermosyphon Heat Transfer Comparison

Aluminium plates are commonly used to transfer heat from the inside to the outside of electronic equipment enclosures because aluminium has a low density, reasonable strength and high conductivity. However, with the rapid increase of board and chip heat generation rates, aluminium plates have certain heat transfer limitations.

A typical application in electronic equipment cooling is the removal of the heat from high-power dissipating modules in a sealed enclosure. Originally this type of cooling was done by solid metal (aluminium) heat conducting plates to finned heat sinks that are exposed to a stream of outside air. Due to increased electronic component density and power dissipation rates, overheating may occur and the equipment would not necessarily meet its temperature operating specifications. To lower the operating temperatures without any significant electronic re-design, use of better heat conducting devices such as heat pipes, is necessary.

The cooling heat transfer performance of two different aluminium conductor plates of 3 mm and 9.6 mm thickness are compared with two different heat pipes of outside diameters 8.0 and 12.7 mm [99]. The relative cooling performance of these various cooling options will be evaluated by means of experimental testing and theoretical simulations using a specially developed simple theoretical model. Based on the result of these evaluations, conclusions will be drawn and recommendations made as to the relative advantages and disadvantages of using heat pipes and solid plates.

The basic experimental configuration tested using aluminium plate conductor plates is shown in Figure 103(a), whilst the test configuration for the heat pipes is shown in Figure 103(b). The heat sources (power resistors) are attached to the aluminium plate conductor and heat spreaders. At the other end, the plates or heat pipes are attached to the finned heat sinks, taking care to correctly apply thermal paste and to exert the maximum possible pressure between the heat transfer surfaces using screws. Air is channelled through the fins using an air flow rate calibration unit consisting essentially of a fan and bell-mouth type flow meter. The air flow through the fins was kept more or less constant at 40-42 m^3/hour. A single thermocouple measures the inlet air temperature $T_{m,i}$ whilst four thermocouples are used to measure the average temperature $T_{m,o}$ of the air flowing through the fins. Two temperatures are measured along the conductor/heat pipe T_t and T_b. The heat sink temperatures $T_{f\text{-}1}$, $T_{f\text{-}2}$, $T_{f\text{-}3}$ are measured at the bottom of 1.7 mm diameter holes drilled half-way down the base of the heat sink at locations as shown in Figures 98 and 99. The conductor plate and heat pipes are insulated using foam rubber.

To compare the time that it takes to achieve steady-state temperatures power was supplied only to the bottom pair of resistors. Figure 104 shows the difference between the bottom temperature and the ambient air temperature ($T_b - T_{amb}$) as a function of time for a 35 W heat load. The heat pipe has a significantly faster response time, having reached steady state in less than 5 minutes. On the other hand, the aluminium conductor configurations only tend to a steady state condition after about 30 to 50 minutes.

Table 11. Experimentally determined temperatures T_b and T_t at positions as indicated in Figure 103 for a ϕ12.5 x 240 mm closed thermosyphon and a 3 and 9.6 mm thick aluminium conductors of width 50 mm and lenth 240 mm conductors

Load	sensor	Heat pipe	Al 3 mm	Al 9.6 mm
W	position	°C	°C	°C
25	T_t	38.9	48.3	38.7
25	T_b	39.9	57.7	44.1
40	T_t	46.7	61.1	47.0
40	T_b	48.0	76.3	55.1

The cooling heat transfer performance of two different aluminium conductor plates of 3 mm and 9.6 mm thickness are compared with two different heat pipes of outside diameters 8 and 12.7 mm. The relative cooling performance of these various cooling options have been summarized by way of Table 11; where the lower the temperature difference, the more heat is transfered.

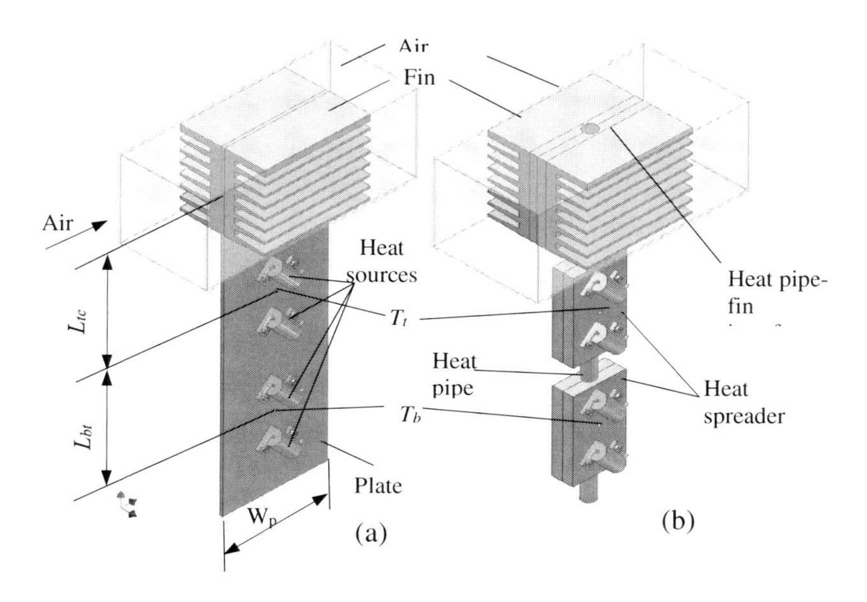

Figure 103. Schematic representation of an aluminium plate conductor onfiguration (a) and a heat pipe configuration (b). Note that the insulation around heat pipe and plate is not shown [99].

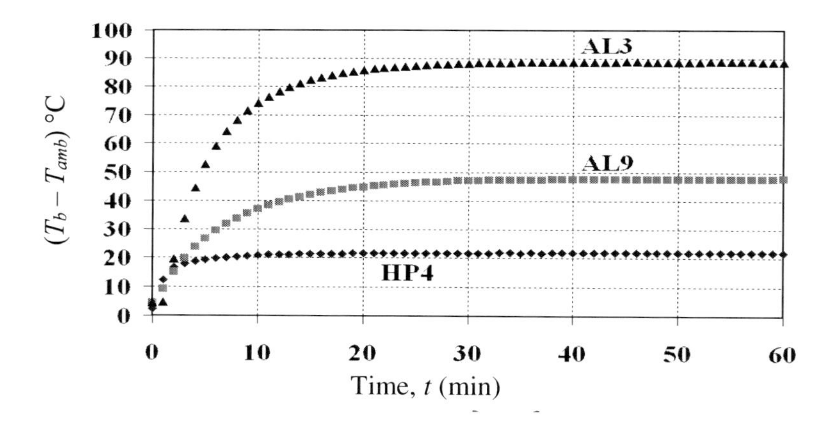

Figure 104. Experimentaly determined heat pipe and aluminium conductors' configuration temperature response (Tb – Tamb) curves for a heat load of 35 W [99].

From the results of this comparison between the use of heat pipes and solid conductors, the following conclusions may be drawn.

- The heat transfer performance of aluminium conductors and heat pipes was investigated for their use in electronic equipment cooling. They were analyzed using experimental and theoretical methods. Heat pipes were found to have a thermal conductivity of around 7200 W/mK, while the aluminium conductors had a conductivity of 200 W/mK. The use of heat pipes, instead of aluminium conductors, are recommended if the distance between the thermal source and thermal sink is significant or there are multiple thermal sources along the heat transfer path.

- Steady state temperature of a constant heat dissipating thermal source cooled by a heat pipe was achieved in around 1-5 min, while the thermal source cooled by an aluminium conductor took between 30 and 50 minutes to achieve steady state temperature. This shows that if electronic response time to reach rated operating conditions is important, then the heat pipe cooling option has a distinct advantage over conductor plates.
- Heat pipes were found to be better than conductor plates to transfer the heat from the heat source. However, once the heat has been transferred away from the heat source, an important, if not a more important consideration, is the transfer of this heat to the surrounding air. This is because of the relatively large thermal resistance associated with surface to air cooling. The fin's area has thus to be designed large enough so as not to negate the superior heat transfer characteristics of using heat pipes.

6.4. Supercritical Closed Loop Thermosyphon Heat Transfer

There are a number of reasons why a supercritical closed loop thermosyphon heat transfer device is deemed important:

- Technology has in general been steadily progressing and heat loads and temperatures have been increasing and hence better heat transfer devices are needed. Heat pipes are already extensively used in the cooling of electronic equipment; modern Rankine cycles operate in the region of 20 MPa and 650°C. However, using CO_2 at these conditions, efficiencies of 43 to 46% are possible (Singer, 2011). These efficiencies are thus obtained at a much lower temperature than the 950°C operating temperatures necessary for comparable closed Brayton cycle power generation equipment. Dostal (2004), in his supercritical CO_2 cycle study for next generation nuclear reactors shows that the size of the equipment can be significantly decreased using CO_2. For instance, a conventional 250 MW steam turbine would typically have a linear length of 35 m; for the same generating power the length can be reduced to 5 m for a helium turbine; for a 300 MW supercritical CO_2 turbine however the length of only 1.5 m!
- Although the excessive generation of CO_2 by industrial activity is considered as being a major contributing factor towards global warming; CO_2 with its so-called *global warming potential* of "1" is nevertheless the lowest of all known substances. Furthermore, because of its higher density and high specific heat at these high temperatures and pressures, under these conditions it is the working fluid of choice

for power and refrigeration systems. Sandia Laboratories [100] claimed that the supercritical-CO_2 (S-CO2) Brayton cycle, compared with fossil fuelled Rankine cycles, could increase electrical power produced per unit of fuel consumed by 43%!

- A requirement of all next-generation nuclear reactors (so-called GEN IV reactor technology) is that the use of passive cooling, because of its acceptably high reliability, needs to be considered a priori, over mechanically pumped and actively controlled devices [101]. Not only can heat pipes transfer large quantities of heat over relatively long distances, they are also self-controlling – as the temperature difference increases, so too does the heat transfer rate.
- Although a plethora of published literature regarding CO_2 is available, it is generally overly complicated and does not specifically apply to closed loop natural circulation heat pipes, often called *gravity assisted* heat pipes.

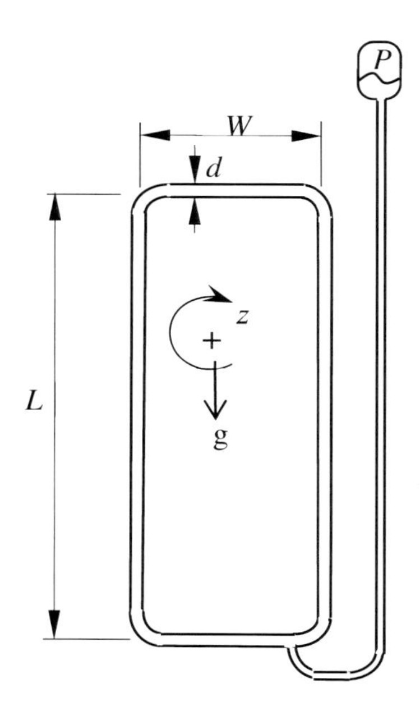

Figure 105. Constant pressure closed loop natural-circulation thermosyphon-type heat pipe.

Considering the loop shown in Figure 105 with $L = W = 1.0$ m and $d = 0.01$ m but with different heating and cooling wall temperatures T_{hw} and T_{cw} (heating mode HM = 1), respectively. The following is a theoretical analysis based on Dobson [102]. The rate at which heat is transferred by the loop for CO_2, H_2O and N_2 at a loop pressure of 2 MPa is given in Watts in Table 12. It is seen that at typical superheated conditions of about 2 MPa and $(T_{hw} + T_{cw})/2 = 600$°C that the CO_2 loop transfers some 5.64 W at a temperature difference of 50°C. This is some 94.5% better than the 2.90 W for H_2O. At a higher pressure of 10 MPa and at the same average temperature of 600°C it is 96.6% better.

The numerical procedure outlined in section 4 is also able to compare the heat transfer for different geometries. For instance, the rate of heat transfered at 10 MPa for different CO_2 loop diameters are given in Table 13, for a constant heating and cooling wall temperatures T_{hw} and T_{cw} (heating mode HM = 1). It is seen that the heat transfer rate does not increase proportionally with increasing pipe diameter. This is ascribed to the highly non-linearity and interdependence of the various terms such as the Reynoltds number *Re*, the heat transfer coefficient α the coefficient of friction friction C_f and the shear stress τ of the governing equations of change.

Table 12. Heat transferred in W for $L = W = 1$ m and $d = 0.01$ m at different temperatures and temperature differences between the constant tenperature hot and cold sides for CO_2, H_2O and N_2

P	T_{hw}	T_{cw}	ΔT	CO_2	H_2O	N_2
MPa	°C	°C	°C	W	W	W
2	610	590	20	1.11	0.41	0.25
2	625	575	50	5.64	2.90	2.75
2	650	550	100	16.7	10.6	9.46
2	700	500	200	44.0	34.3	29.2
10	610	590	20	5.89	6.8	4.17
10	625	575	50	40.5	20.6	13.0
10	650	550	100	111	44.7	28.4
10	700	500	200	305	94.0	158
25	650	550	100	216	184	-

Table 13. Heat transferred in W for different CO_2 loop diameters *d* for a constant hot and cold side wall temperature differences of $(T_{hw} - T_{cw}) = 100°C$

					diamet	Er	
P	T_{hw}	T_{cwd}	ΔT	0.005	0.01	0.02	0.03
MPa	°C	°C	°C	m	m	M	m
10	650	550	100	22.5	111	216	283

REFERENCES

[1] P. B. Whalley, *Boiling Condensation and Gas-Liquid Flow.* Clarendon Press, Oxford, 1990.

[2] G. S. H. Lock, *The Tubular Thermosyphon: Variations on a Theme.* Oxford University Press, New York, 1992.

[3] V. P. Carey, *Liquid-Vapor Phase-Change Phenomena, Hemisphere,* 1992.

[4] A. Faghri, *Heat Pipe Science and Technology.* Taylor and Francis, London, 1995.

[5] I. Groenewald, A. H. Basson, Dobson, R. T. *Theoretical modelling and numerical simulation of a two-phase thermosyphon considering splashing and geysering*, Sixth International Heat Pipe Symposium, Chiang Mai, 2000, 209-219.

[6] R. T. Dobson, Transient response of a closed loop thermosyphon *R and D Journal,* 9 (1) (1993) 32-38.

[7] L. S. Pioro, I. L. Pioro, I*ndustrial Two-phase Thermosyphons*, Begell House, (1997).

[8] D. Reay, P. Kew, *Heat Pipes Theory, Design and Applications,* fifth ed., Butterworth-Heinemann, Oxford, MA, USA, 2006.

[9] S. Bhattacharyya, Two-phase natural circulation loops: a review of the recent advances, *Heat Transfer Engineering* 45 (2012) 461-482.

[10] D. Jafari, A. Franco, S. Filippeschi, P. Di Marco, Two-phase closed thermosypyons: areview of studies and solar applications, *Renewable and Sustainable Energy Reviews* (53) (2016) 575-593.

[11] R. T. Dobson, D. G. Kröger, Thermal characteristics of an ammonia-charged two-phase closed thermosyphon, *Tenth International Air Conditioning, Refrigeration and Ventilation Congress*, Midrand, South Africa, 8-10 March (2000).

[12] A. Meyer, *Development of a range of air-to-air heat pipe heat recovery heat exchangers.* MSc thesis, Stellenbosch University, Stellenbosch, South Africa, 2004.

[13] NEI, Nuclear Energy Institue. *World Statistics* - Nuclear Energy Institute, 1997, www.nei.org/Knowledge-Center/Nuclear-Statistics/World-Statistics.

[14] G. Vécseyy, P. G. K. Doroszlai, GEYSER, A simple, new heating reactor of high inherent safety, *Nuclear Engineering and Design* 109 1988 141-145.

[15] IAEA, International Atomic Energy Agency, *Advances in small modular reactor technology developments, 2014*. IAEA Advanced Reactors Information System (ARIS). http://aris.iaea.org.

[16] I. Marcel, *CAREM: Argentina's innovative SMR, Nuclear, Engineering International, 2014*. [http://www.neimagazine.com/features/featurecarem-Argentinas-innovative-smr-4266787/featurecarem-argentinas-innovative-smr-4266787-2.html].

[17] W. G. Halsey, Gen IV Nuclear energy Ssstems interim status report on pre-conceptual LFR design studies and evaluations, UCRL, UCRL-TR-209718, 2005.

[18] M. Taube, The inherently-safe power reactor DYONISOS (Dynamic Nuclear Inherently-Safe Reactor Operating with Spheres), *Annals of Nuclear Energy* 13 (12) (1986) 641-648.

[19] G. Yadigaroglu, M. Zeller, Fluid-to-fluid scaling for a gravity- and flashing- driven natural circulation loop, *Nuclear Engineering and Design* (151) (1994) 49-64.

[20] K. A. Loubser, *An experimental study of an inherently-safe, natural circulating, flash-tube type system for a nuclear reactor steam Supply Concept.* MSc thesis, Stellenbosch University, Stellenbosch, South Africa, 2014.

[21] F. M. Slabber, *Technical description of the PBMR demonstration power plant*, doc. No. 016956, rev 4, PBMR (Pty) LTD, Centurion, South Africa, 2006.

[22] R. Greif, Y. Zvirin, A. Mertol, The transient and stability behavior of a natural circulation loop, *Journal of Heat Transfer* 101 (1979) 684-688.

[23] M. Gordon, E. Ramos, M. Sen, A one-dimensional model of a thermosyphon with known wall temperature, *Heat and Fluid Flow* 8 (3) (1987) 177-181.

[24] M, Sen, D. A. Pruzan, K. E. Torrence, Analytical and experimental study of steay-state convection in a double loop thermosyphon, *International Journal of Heat and Mass Tranfer* 31 (4) (1988) 709-722.

[25] O. Salazar, M. Sen, E. Ramos, Flow in naturnal circulation loops, *Journal of Thermophysics* 2 (2) (1988) 180-183.

[26] J. C. Chatto, Natural convection flow in parallel-channel systems, *Journal of Heat Transfer,* (November 1963) 339-345.

[27] K. P. Halliman, R. Viskanta, Dynamics of natural circulation loop: analysis and experiments. *Heat Transfer Engineering,* 7 (3-4) (1991) 43-52.

[28] P. K. Vijayan, S. K. Metha, A. W. Date, On the steady-state performance of natural circulation loops, *International Journal of Heat and Mass Transfer* 34 (9) (1991) 2219-2230.

[29] R. K. Sinha, A. Kakodkar, Design and development of the AHWR – the Indian thorium fuelled innovative nuclear reactor, *Nuclear Engineering and Design* 236 (2006) 683-700.

[30] I. Sutharshan, M. Mutyala, R. P. Vijuk, A. Mishra, The AP1000TM reactor passive safety and modular design, *Energy Procedia* 7 (2011) pp 293-302.

[31] J. B. Joshi, Computational flow modelling in and design of bubble column reactors, Che*mical Engineering Science* 56 (21-22) 2001 5893-5933.

[32] R. T. Dobson, T. M. Harms, Lumped parameter analysis of closed and open oscillatory heat pipes *Eleventh International Heat Pipe Conference,* Tokyo, 12-16 September 1999.

[33] M. V. Sardeshpande, V. V. Ranade, *Two-phase flow boiling in small channels: A brief review,* Sãdhanã, 38(Part 6) (2013) 1083-1126.

[34] A. Manera, *Experimental and analytical investigations on flashing-induced instabilities in natural circulation two-phase systems,* PhD thesis, Delft University of Technology, Delft, The Netherlands, 2003.

[35] X. Guo, Z. Sun, J. Wang, S. Yu, L. Gao, Numerical simulation of the transient behaviors in an open natural circulation system with a large scale, *Annals of Nuclear Energy* 77 (2015) 83-93.

[36] G. Su, J. Dounan, J. I. A. Kenji, Fukuda, Y. Guo, Theoretical study on density wave oscillations of two-phase natural circulation under low quality conditions, *Journal of Science and Technology* 38 (8) 607-613.

[37] F. Matovu, *Drift flux models, TKP11,* Department of Chemical Engineering, Norwegian University of Science and Tecnology, 1996.

[38] G. Yadigaroglu, R. T. Lahey, On the various forms of the conservation equations in two-phase flow, *International Journal of Multiphase Flow* 2 (1975) 477-494.

[39] P. K. Vijayan, A. K. Nayak, D. Saha, M. R. Gartia, Effect of loop diameter on the steady stae and stability behaviour of single-phase and two-phase natural circulation loop, *Science and Technology of Nuclear Installations* (2008). Article ID 672704. [Doi:10.1155/2008/672704].

[40] K. Yoneda, A. Yasuo, X. Okawa, Bubble characteristics of steam-water two-phase flow in a large-diameter pipe, *Experimental Thermal and Fluid Science,* 26 (2002) 669-676.

[41] K. Yoneda, A. Yasuo, X. Okawa Flow structure and bubble characteristics of steam-water two-phase flow in a large-diameter pipe *Nuclear Engineering and Design,* 217 (2002) 267-281.

[42] A. H. Stenning, C. B. Martin, An analytical and experimental study of air-lift performance, Transactions of the ASME *Journal of Engineering for Power,* (April 1968) 106-110.

[43] J. C. Ruppersberg, *Transient modelling of a loop thermosyphon: transient effects in single and two phase natural circulation thermosyphon loops suitable for the reactor cavity cooling of a pebble bed modular reactor,* MSc Eng thesis, University of Stellenbosch, Stellenbosch, South Africa, 2008.

[44] Sittmann, *Inside heat transfer characterisation of a loop-type heat pipe suitable for the reactor cavity cooling system of a pebble bed modular reactor,* MSc Eng thesis, University of Stellenbosch, Stellenbosch, South Africa, 2010.

[45] H. A. White, *Investigating instabilities in a multi-channel thermosyphon loop*. BSc Eng project, University of Stellenbosch, Stellenbosch, South Africa, 2010.

[46] L. S. Sangweni, *Experimental and numerical modelling of the response of a two-phase natural circulation multi-parallel channel system.* MEng thesis, University of Stellenbosch, Stellenbosch, South Africa, 2015.

[47] F. M. Senda, *Aspects of Waste Heat Recovery and Utilization (WHR&U) in Pebble Bed Modular Reactor (PBMR) Technology*. MSc Eng thesis, Stellenbosch University, Stellenbosch, South Africa, 2012.

[48] F. M. Senda, R. T. Dobson, A Natural circulation waste heat recovery system for high temperature gas-cooled reactor used and/or spent fuel tanks: PART I - design considerations and theoretical simulation, *Proceedings of the ASME 2013 Power Conference Power 2013,* Power2013-98132, July 29-August 1, Boston, Massachusetts, 2013.

[49] F. M. Senda, R. T. Dobson, A Natural circulation waste heat recovery system for high temperature gas-cooled reactor used and/or spent fuel tanks: PART II - theoretical and experimental validation, *Proceedings of the ASME 2013 Power*

Conference Power 2013, Power2013-98135, July 29-August 1, Boston, Massachusetts, 2013.

[50] R. T. Dobson, R. Laubscher, Heat pipe heat exchanger for high temperature nuclear reactor technology, *Frontiers in Heat Pipes* (FHP) 4 (2) (2013) 1-7, DOI: 10.5098/fhp.v4.2.3002, Global Digital Central, ISSN: 2155-658X.

[51] R. T. Dobson, R. Laubscher, Heat pipe heat exchanger for high temperature nuclear reactor technology, *Eleventh International Heat Pipe Symposium,* Beijing, China, June 2013, 9-12.

[52] R. Laubscher and R. T. Dobson, Theoretical and experimental modelling of a heat pipe heat exchanger for high temperature nuclear reactor technology, *Applied Thermal Engineering* 61(2) (2013) 259-267, doi: 10.1016/j.applthermaleng. 2013.06.063.

[53] R. Laubscher, R. T. Dobson, Boiling and condensation heat transfer coefficients for a Dowtherm-A charged heat pipe heat exchanger, *Frontiers in Heat Pipes* (FHP) 4 (2) (2013) 1-6, 023003. DOI: 10.5098/fhp.v4.2.3003, Global Digital Central, ISSN: 2155-658X.

[54] R. Laubscher and R. T. Dobson, Boiling and condensation heat transfer coefficients for a Dowtherm-A charged heat pipe heat exchanger, *Eleventh International Heat Pipe Symposium,* Beijing, China, June 9-12, 2013.

[55] R. Laubscher, *Development aspects of a high temperature heat pipe heat exchanger for high-temperature gas cooled Nuclear Reactor Systems,* MSc Eng thesis, University of Stellenbosch, Stellenbosch, 2013.

[56] A. E. Russwurm, Q-pipes add a new dimension to waste heat recovery, recycling energy Part 1, *Heating, Air Conditioning and Refrigeration* (January 1980) 27-39.

[57] E. Russwurm, Recovering waste heat with Q-pipes, recycling energy Part 2 Recycling energy. *Heating, Air Conditioning and Refrigeration*, (March 1980) 45-49.

[58] R. T. Dobson, S. A. Pakkies, Development of an air-to-air Refrigerant R134a charged two-phase closed thermosyphon heat exchanger. *Eleventh International Air Conditioning, Refrigeration and Ventilation Congress,* Midrand, South Africa (March 2002) 13-15.

[59] R. T. Dobson, S. A. Pakkies, Development of a heat pipe (two-phase closed thermosyphon) heat recovery heat exchanger for a spray drier, *Journal of Energy in Southern Africa.* 13 (4) (2002) 130-138.

[60] A. Meyer, R. T. Dobson, Thermal performance characterization of R134a and Butane charged two-phase closed thermosyphons, *R and D Journal* 11 (2) (2005).

[61] Meyer, A. and Dobson, R. T. (2006). A heat pipe heat recovery heat exchanger for a mini-drier. *Journal of Energy in Southern Africa*, 17(1): 50-57.

[62] Mills, A. F., *Heat and Mass Transfer.* Pearson, 2009.

[63] G. D. Joubert, R. T. Dobson, Modelling and testing of a passive night-sky radiation system, *Journal of Energy in Southern Africa* 28 (1) (2017) 76-90.

[64] Y. Cengel, A., Ghajar, *Heat and Mass Transfer: Fundamentals and Applications*, McGraw-Hill Education, 2011. ISBN 9780077366643.

[65] R. T. Dobson, Theoretical modelling of an open oscillatory heat pipe. *Sixth International Heat Pipe Symposium,* Chiang Mai, 5-9 November 2000.

[66] N. S. Thomas, N. S. *Performance characterisation of a separated heat-pipe heat-recovery heat-exchanger for the food drying industry*, MEng thesis, University of Stellenbosch, Stellenbosch, South Africa, 2016.

[67] H. Akachi, *US Patent No. 4921041*, 1992.

[68] H. Akachi, *US Patent No. 219020*, 1993.

[69] H. Akachi, F. Polasek, Pulsating heat pipe: review of the present state of the art, *Technical report to Industrial Technology Research Institute Energy and Resources Laboratory*, Chutung, Taiwan, 1995.

[70] H. Akachi, F. Polášek Pulsating Heat Pipes, *Fifth International Heat Pipe Symposium,* Melbourne, Australia, 17-20 November 1886.

[71] R. T. Dobson, G. Graf, Thermal characterisation of an ammonia-charged pulsating heat pipe, *Proceedings of the Seventh International Heat Pipe Symposium IHPS*, Jeju, Korea, October 12-16, 2003, 325-330.

[72] T. N. Wong, Theoretical modelling of pulsating heat pipe, *Eleventh International Heat Pipe Conference,* Tokyo, 12-16 September 1999.

[73] R. T. Dobson, Theoretical and experimental modelling of an open oscillatory heat pipe including gravity. *Twelth International Heat Pipe Conference,* Moscow-Kostroma-Moscow, 19-24 May 2002.

[74] R. T. Dobson, An open oscillatory heat pipe steam-powered boat, *International Journal of Mechanical Engineering Education,* 31 (4) (2003) 339-358.

[75] Y. Zhang, A. Faghri, Heat transfer in a pulsating heat pipe with open end, *International Journal of Heat and Mass Transfer* 45 (2002) 755-764.

[76] M.B. Shafii, A. Faghri, Y. Zhang, Analysis of heat transfer in unlooped and looped pulsating heat pipes, *International Journal of Numerical Methods for Heat and Fluid Flow* 12 (5) (2002). 585-609, https://doi.org/10.1108/09615530210434304.

[77] H. *Yang*, S. *Khandekar*, M. *Groll*, Operational limit of closed loop *pulsating heat pipes*, *Applied Thermal Engineering* 28 (1) (*2008*) 49-59.

[78] W. Shao, Y. Zhang, Effects of film evaporation and condensation on oscillatory flow and heat transfer in an oscillating heat pipe, *Journal of Heat Transfer*, 133 (*2008*) / 042901-3. DOI: 10.1115/1.4002780.

[79] S. Khandekar, M. Groll, P. Charoensawan, S. Rittidech, P Terdtoon, Closed and Open Loop Pulsating Heat Pipes, *Proceedings of the Thirteenth International Heat Pipe Conference,* Shanghai, China, 2004, pp.1-14.

[80] A. Tripathi, S. Khandekar, P. K. Panigrahi, Osillatory contact line motion inside capilliares, *Fifteenth International Heat Pipe Conference*, Clemson, USA, April 2010, pp. 25-30.

[81] R. Senjaya, T. Inoue, Y. Suzuki, Oscillating heat pipe simulation with bubble generation, *Fifteenth International Heat Pipe Conference,* Clemson, USA, April, 2010, 25-30.

[82] R. T. Dobson, G. Swanepoel, An experimental investigation of the thickness of the liquid-film deposited at the trailing end of a liquid plug moving in the capillary tube of a pulsating heat pipe, *Frontiers in Heat Pipes*, 1 (1) (2010).

[83] R. J. Craig, *Modelling of a thermodynamically driven heat engine with application intended for water pumping,* MSc Eng thesis, Stellenbosch University, Stellenbosch, South Africa, 2014.

[84] Oostenbrink, H. Natural evaporation gravity powered water pumping. *BSc Mechanical Engineering Project Report*, Stellenbosch University, Stellenbosch, South Africa, 2012.

[85] Fraser, *Surface tension driven water pumping: a bio-inspired passive water pump.* MSc Eng thesis, Stellenbosch University, Stellenbosch, South Africa, 2015.

[86] Green Building Council of South Africa. (2014). *About green building* [Online]. Available: https://www.gbcsa.org.za/about/about-green-building/14 September 2014.

[87] Hobbs, *Thermally driven natural circulation water pump.* MSc Eng thesis, Stellenbosch University, Stellenbosch, South Africa, 2015.

[88] R. T. Dobson, An open oscillatory heat pipe water pump, *Applied Thermal Engineering* 25 (2005) 603-621.

[89] Van Amerongen, *How Things Work The Universal Encyclopaedia of Machines.* Paladin, London, 1972.

[90] J. F. Douglas, J. M. Gasiorek, J. A. Swaffield, *Fluid Mechanics,* Longman, 1995.

[91] G. D. Joubert, *Investigation of a passive night-sky radiation system.* MSc Eng thesis, University of Stellenbosch, Stellenbosch, South Africa, 2014.

[92] G. D. Joubert, R.T. Dobson, Modelling and testing a passive night-sky radiation system, *Journal of Energy in Southern Africa* 28 (1) (2017).

[93] Meir, J. Rekstad, O. LØvvik, O. A study of a polymer-based radiative cooling system. *Solar Energy,* 73(6) (2002) 403-417.

[94] Loveday, Taki, A. Convective heat transfer coefficients at a plane surface on a full-scale building façade, *International Journal of Heat and Mass Transfer*, 39(8) (1996) 1729-1742.

[95] Okoronkwo, K. Nwigwe, N. Ogueke, E. Anyanwu, D. Onyejekwe, P. Ugwuoke, An experimental investigation of the passive cooling of a building using nighttime radiant cooling. *International Journal of Green Energy*, 11(10) (2014) 1072-1083.

[96] Jeggels, Y. (2008). *Thermal management and temperature control of a containerised rapid deployment radar system.* MSc Eng thesis, University of Stellenbosch, Stellenbosch, South Africa.

[97] O. O. Kritzinger, R. T. Dobson, Electronic equipment cabinet cooling enhancement using a two-phase thermosyphon with separated evaporator and condenser sections. *9th International Heat Pipe Symposium*, Kuala Lampur, Malaysia, 20-24 November 2008.

[98] R. T. Dobson, Y. U. Jeggels, Cooling of electronic equipment using bent and closed loop two-phase thermosyphons. *Tenth International Heat Pipe Symposium*, Kuala Lumpur, 17-20 November 2008.

[99] Y. U. Jeggels, R. T. Dobson, D. H. Jeggels, Comparison of the cooling performance between heat pipes and aluminium conductors for electronic equipment enclosures. *Fourteenth International Heat Pipe Conference*, Florianoplis, Brazil, April 2007, 22-27.

[100] Singer, *Supercritical carbon dioxide Brayton Cycle turbines promise giant leap in thermal-to-electric conversion efficiency*, Sandia news media release March 4, Sandia Corporation, Albuquerque, 2011.

[101] IAEA, International Atomic Energy Agency. *Progress in design, research and development, and testing of safety systems for advanced water cooled reactors.* Vienna, 1996. IAEA-TECDOC-872.

[102] R. T. Dobson. Relative performance of supercritical CO_2, H_2O, N_2 and He charged closed-loop thermosyphon-type heat pipes, *Heat Pipe Science and Technology An International Journal,* 3 (2-4) (2012) 169-185.

In: Heat Pipes: Design, Applications and Technology
Editor: Yuwen Zhang
ISBN: 978-1-53613-908-2

Chapter 2

HIGH HEAT FLUX PHASE CHANGE AND WICK STRUCTURES

Steve Cai*

Climate, Controls, & Security, the United Technologies,
Syracuse, NY, US

ABSTRACT

High-powered 3D microelectronics devices, such as RF power amplifier and laser diodes, have brought emerging demands for removing highly concentrated heat from small device areas. As a passive cooling method, heat pipe heat spreaders that are capable of high density phase change are highly desired. In the past twenty years, this technical demand has constantly driven development of advanced wick structures to provide more effective heat and mass transfer. As a result of these efforts, cooling heat flux phase change > 1kW/cm^2 has been demonstrated in lab studies. Heat pipe platforms capable of a few hundred W/cm^2 cooling have also been developed for industry uses. In these advances, more tolerance is given to phase change superheat and operating temperature, with less emphasis on isothermal characteristics stressed in the conventional heat pipes. This mission shift renders designers to design wick structures with large particles to allow occurrence of nucleate boiling and fast vapor ventilation from the wick structures. Accordingly, some traditional heat pipe limits (e.g., boiling limit) are either inapplicable or not considered significant in high heat flux thermal management. Heat and mass transfer in the heated wick structures becomes the major concern. In addition, high heat flux phase change requires high density heat conduction from the wick subtract to phase change interfaces, dramatically increasing superheat and complexing evaluation of the heat transfer coefficient. All these unique characteristics, attributing to high heat flux phase change, expand our knowledge basis and steer industrial practices in the new era.

* Corresponding Author Email: steve.q.cai@carrier.utc.com.

Keywords: porous wick structure, capillary limit, heat flux, heat pipe heat spreader

NOMENCLATURE

C	constant
d	diameter, m
D_h	wick hydraulic diameter, m
f	dynamic function of vapor phase
g	dynamic function of liquid phase
h	heat transfer coefficient, $W/m^2 \cdot K$
h_{fg}	liquid latent heat, kJ/kg
K	permeability, m^2
k	thermal conductivity, $W/m \cdot K$
L	wick dimension along flow direction, m
l	width of heated wick area, m
Ge	Wick geometrical number
Me	dimensionless Merit number
P	pressure, Pa
$\dot{q}$	heat flux, W/m^2
r	radius
T	temperature, ^oC
t	thickness of wick structure, m
v	velocity, m/s

Greek Symbols

α	liquid volume fraction
β	dimensional ratio between heating area width and wick thickness
γ	tortuosity
θ	contact angle
σ	surface tension, N/m
ρ	density, kg/m^3
μ	dynamic viscosity, $N.s/m^2$
ϕ	porosity
ε	thin-film evaporation area ratio
η	dimensionless ratio of $\Delta R/R$
φ	wick geometrical number of liquid phase

τ	wick geometrical number of vapor phase
Δ	difference
Ψ_{CK}	dimensionless Kozeny factor

Subscripts

bi	bi-porous wick structure
c	critical
cap	the maximum capillary
cha	characteristic
CK	dimensionless Kozeny factor
CNT	carbon nanotube
cluster	cluster
eff	effective
h	hydraulic
l	liquid
lf	liquid film
max	maximum
q	quality
ε	thin-film evaporation enhancement
sat	saturation
s	splash
sup	superheat
sub	wick substrate
super	superheat
v	vapor
wire	wire resistor
wick	wick structure

1. CHALLENGES OF HIGH HEAT FLUX THERMAL MANAGEMENT

High density heat dissipation occurs in many cutting-edged electronics, aerospace, optical weapon and nuclear power systems, such as the RF power amplifiers, phase array radars, diodes of quantum cascade lasers, high power converters, high-performance computer CPUs, hypervelocity gliders, and nuclear fusion and fission reactors. In some cases, heat dissipation flux of the compact packing systems could reach or exceed 1,000 W/cm^2. For example, heat dissipation of the GaN power amplifier is beyond $50,000 W/cm^2$

at the gate level. After heat spreading through its package substrate, the die level heat flux is still above 400W/cm^2. Even with large surface area of ~ 1.0cm^2, chips of the SiC power converter dissipate heat up to 500W/cm^2. Similar situation is also known in the high-energy laser systems, waste heat dissipation of the laser diodes is projected to be over 1,000W/cm^2. Such high heat dissipation exceeds the cooling capabilities of most convection cooling systems, in single-phase or two-phase, with requirement of dielectrical coolant. In industrial practices, high density thermal management becomes one of the major constrains that limits system performance and design. This technology demands associated with abroad application platforms call research efforts to develop more powerful cooling solutions.

In history, heat pipes are widely used in electronics and aerospace industries [1]. Benefitted from highly efficient evaporation and condensation on the wick meniscus interfaces, phase change heat transfer in heat pipe gains an ultrahigh heat transfer coefficient [2, 3]. In theory, operating range or the maximum phase change heat flux is well-defined by several heat and mass transport limits, such as viscous, sonic, boiling, entrainment, and capillary limits, which have been introduced in previous chapters and other heat pipe books. To obtain high heat flux capability, most of limits can be overcome by optimizing the heat pipe dimensions, elevating operating temperature, or selecting appropriate operating fluid. For instance, high operating temperature increases vapor density, which significantly reduce or eliminate likelihood of viscos and sonic limits.

The other limits, including the boiling and capillary limits, typically occur at higher operating temperature and heat flux. A review of the boiling limit routes back to temperature differences across the wick structure. Between the saturated vapor and the hot substrate, increased heat flux and temperature difference trigger the onset of nucleate boiling (ONB) at certain point [7, 8]. Depending on the wick thickness and the wick pore size, vapor generated from nucleate boiling could be trapped in the wick structures, forming a vapor blanket to block the return of liquid, or they rise and burst at the liquid-vapor interface to disrupt the menisci for circulating operating fluid. Both the situations could lead to the dryout. Thus, nucleate boiling/the boiling limit is inhibited in designing a classic heat pipe system. In contrast, the capillary limit describes the relation between the flow viscous loss and capillary pressure. The maximum mass transport capability is directly dictated by flow/wick cross-section area, transport distance, as well as the effective wick pore size. As a result, wick geometry plays the most critical role for these two limits. Because of existence of these limits, tradition wick structures, such as microgroove and sintered powder wicks, only provide limited phase change capability < 20W/cm^2. Even with surface enhancement technologies, the maximum heat fluxes relying on meniscus interfacial evaporation are reported to be less than 80W/cm^2 [4, 5], which is lower than the critical heat flux (CHF) of nucleate boiling of water that could reach 150W/cm^2 [6].

Wick structures capable of high heat flux phase change are designed to contain uninterrupted interfacial menisci and efficient vapor ventilation paths. Not only does not the occurrence of nucleate boiling result in the consequence of the dryout, but the boiling in wick structures is utilized as a booster to enhance phase change. Unlike the identical flow friction coefficient in all the wick structures of the traditional heat pipe, liquid and vapor flows at high heat flux have strong interaction in the phase change area. This interaction significantly increases flow pressure drop in the evaporator wick and its weight in the whole circulation loop. Consequently, local mass transport on the heated wick structures becomes a major concern for maximizing phase change capability.

2. ADVANCED WICK STRUCTURES CAPABLE OF HIGH HEAT FLUX PHASE CHANGE

Over the last decade, various advanced porous wick structures capable of high heat flux phase change have been proposed and developed. One of the common features of these wick structures is to have different sizes of pores, consisted of either mono or multiscale particles. As shown in Figure 1, the larger pores have less capillarity but provide low viscous flow paths for both the heated and non-heated wick structures. More importantly, they also provide vapor ventilation paths when nucleate boiling occurs in the evaporator wick. The smaller pores enhance the thin film evaporation and maintain interfacial menisci for circulating the two-phase flow. As heat flux increases, escaping vapor bubble joints together and forms vapor jets/wells. The perpendicular vapor jet flow from the wick structures creates the expanded interfacial surface for evaporation to enhance phase change capability, as well as the heat transfer coefficient.

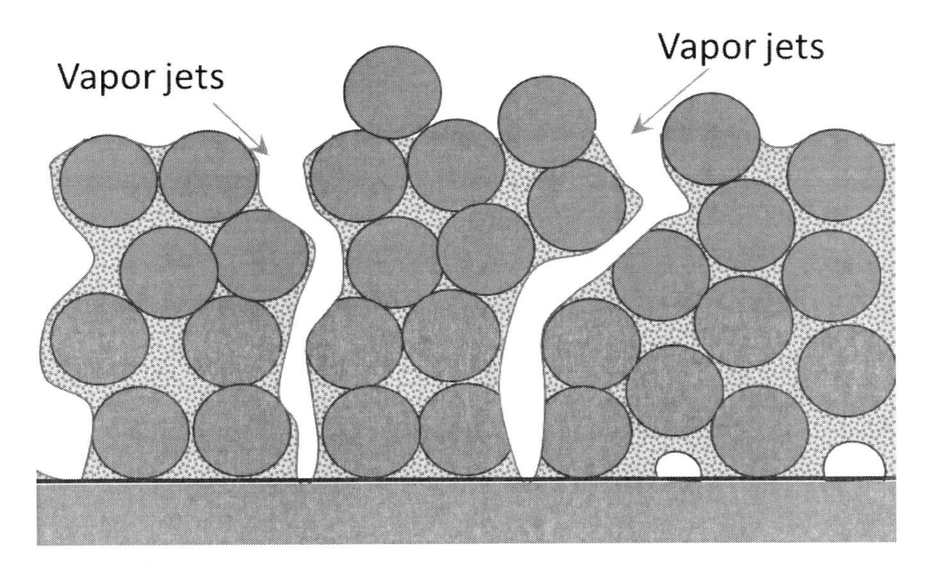

Figure 1. Phase change on the multiscale porous wick structure: Small pores retain liquid for phase change, and large pores provide vapor ventilation paths.

Up to today, wick structures reported to be capable of high heat flux phase change covers three major categories, which are bi-dispersed porous wick structures/bi-wick structures, mono porous wick structures, and multiscale nanoporous wick structures. In the next sections, we will discuss in details about design, development, characteristics and high heat flux lab tests of these advanced wick structures.

2.1. Bi-Dispersed Porous Wick Structures

Initial research on high heat flux wick structures was started from bi-dispersed porous wick. The concept of bi-porous wick structure (also called bi-wick) was proposed by Vityaz et al. in 1990's [9]. A brief review was later conducted by Rosenfeld et al. [10]. This novel bi-porous wick structures were made from a duel sintering process. In the first stage of the process, elementary metal particles with diameter of ~80μm are sintered to create numerous sintered clusters, working as large particles in the bi-porous structure. A simple way to achieve this goal is to break a bulky mono sintered structure to many small pieces/clusters with equivalent size of 200 to 600μm in diameter. Through mesh screening, the sintered clusters can be selected within the designated range. The second stage of the process is to apply another sintering process to bond these clusters and form bi-porous wick structures. Because of different sizes, both large and small pores were formed between the clusters and elementary particles. This wick configuration is able to rapidly vent the vapor generated in the wick structures, therefore allowing occurrence of nucleation boiling and a higher superheat at high heat flux phase change.

A broad research of the bi-porous wick structures has been conducted by Catton's team in the University of California at Los Angeles. Wang et al. observed phase change on/in bi-porous wicks and found that evaporation interface consists of two categories: menisci in small and large pores [11]. When the menisci in large pores recede or vapor jets form in the bi-porous wicks, the meniscus evaporation area from the small pores significantly increases. Using bi-porous wick structures sintered by spherical copper powders (average cluster diameter is 302 μm), Semenic et al. reported the maximal CHF at 520 W/cm^2 at superheat of 50°C on 1mm thick bi-porous wick structures [12]. With increase of cluster size to 455 μm and wick thickness to 3mm, high heat flux of 990 W/cm^2 was achieved at a high superheat of 147°C. To understand this result, attention should be switched to porosity of the bi-porous wick structures. Because of two different scales of pores, porosity of the bi-porous wick structure is expressed as

$$\phi = \phi_{cluster} \times \phi_{bi} \quad (1)$$

Here, ϕ, $\phi_{cluster}$, and ϕ_{bi} are porosity of the bi-porous wick, porosity of clusters, and wick porosity when the clusters are assumed to be solid, respectively. Compared with one tier/mono wick structure, the effective thermal conductivity of the bi-porous wick is reduced by a factor $\phi_{cluster}$. As a result of this change, the bi-porous wicks made through the dual sintering process are much less thermally conductive and have 3-5X higher superheat in high heat flux phase change. This drawback reduced the wick applicability in sensitive electronics systems.

Continuing with the Semenic's work, Reilly et al. added a mono-porous layer underneath the bi-porous layer to increase the effective thermal conductivity and to delay the dryout [13]. They demonstrated over 600W/cm^2 phase change with the wick effective thermal conductivity >60W/m.K. In spite of these efforts, superheat was not obviously reduced when compared their results with Semenic's.

In addition to the bi-porous wicks developed for heat spreading and vapor chamber systems, the concept was also presented and tested in loop heat pipe systems. By adding Na_2O_3 pore formers, Yeh et al. developed one-step sintering process to sinter nickel powder based bi-porous wicks [14]. The pore formers (32–48 µm or 74–88 µm) were controlled to be 20% or 25% by volume so that large pores were connected but avoid vapor penetration to the liquid core of the compensation chamber. Because of the unique on-top heating approach of the loop heat pipe wick, increase of evaporator menisci overweighs the need of vapor ventilation. With higher flow resistance in the less permeable wick structure, the maximum heat flux was limited below 300W/cm^2 in this study.

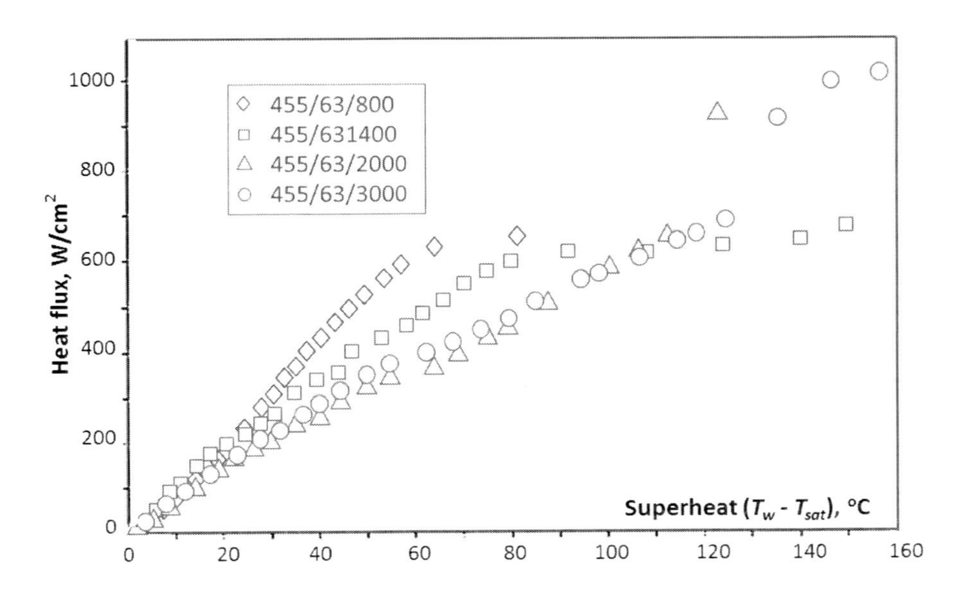

Figure 2. Heat flux vs. superheat at different bi-porous wick thickness (Cluster diameter/particle diameter/wick thickness), by Semenic et al. [12].

Deng et al. investigated a bi-porous wick fabrication process in micro V-sharp grooves [15]. Their test results show that the optimal groove geometry has the groove depth of 0.85 mm and pitch of 0.45 mm to achieve the maximum capillary performance. With copper particle size in the range of 75–110 μm, their sintering temperature at 950°C with 30-minute duration can also be used for other sintering wick developments.

Teledyne Scientific developed bi-porous wick structures with a one-step process in which the large pores were created by a stainless steel pre-molding [16]. The perpendicular molding generates vertically straight vapor escape paths/large pores, reducing vapor phase flow resistance within the wick structures. Using a 5cm×5cm copper heater, the maximum heat flux over 500W/cm^2 has been reported. However, large vertical open pores on the wick structures reduced the total meniscus evaporation area. As a result, phase change superheat tends to be higher at low heat flux.

2.2. Microfabricated Mono Wick Structures

Wick structures using mono particles can be properly placed to form different size pores. For the mono wick structures illustrated in Figure 3, the transverse liquid flow (in the wick plane) and the axial vapor flow (perpendicular to the wick plane) co-exist within the wick cavities between perpendicular pillars. Since the radius of interfacial meniscus varies between two solid points, capillarity is a function of locations. The maximum capillary force appears between two adjacent pillars with the smallest gaps. Cavities in the middle of the triangle area have the lowest capillarity at which the interface recedes into wick structures to form vapor wells/jets at high input heat flux. Like the bi-porous wick structures discussed previously, this flow configuration alters the traditional heat and mass transport mechanism from two aspects. First, at high heat flux, liquid phase recedes into the wick cavity while vapor phase advances into wick structures, creating more surface area for interfacial evaporation. The heat conduction distance from the hot substrate to the vapor-liquid interface is dramatically reduced. The heat transfer coefficient of phase change increases at high heat flux. Second, vapor bubbles or jets share the wick cavities with liquid flow. The presence of vapor jets/walls reduces wick permeability, debilitating liquid supply.

Using sintered spherical copper particles, Semenic, et al. investigated phase change capability of thick mono wick structures with a thickness range from 0.5 to 2mm and large particle diameter varying from 107 to 586μm. They reported approximate 300W/cm^2 maximal heat flux with superheat of 21°C. This study demonstrated high heat flux capability on/in mono wick structures after nucleation boiling started. Compared with their previous studies on the bi-porous wick structures, superheat was significantly reduced due to increase of the wick solid fraction. However, the thick sintered structures created tortuous ventilation path, increased vapor traveling distances, and reduced liquid volume fraction in the wick cavities. All these impacts limited phase change to reach higher heat flux.

Weibel et al. studied phase change in wick structures made through similar sintered process with spherical mono copper particles. They reported heat flux were close to 600 W/cm^2 on a 25 mm^2 heating area without occurrence of dryout, as shown in Figure 4 [17]. The particle sizes that they tested ranges from 45 to 355μm in diameter and wick structure thickness varies between 0.6 and 1.2mm. In their experiments, authors perpendicularly set mono wick structure sample. A liquid pool submerges the bottom part of the test wick sample so that, against gravity, liquid is automatically supplied to the heated wick area by its capillary force. Since there was no dryout reported in this study, the maximal heat flux versus the mono wick parameters was not correlated.

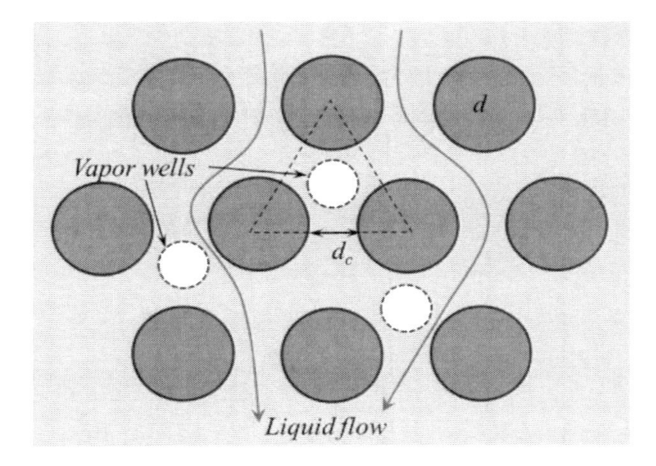

Figure 3. A schematic top view of flow in the wick structure composed of perpendicular pillars: large cavities in the middle of triangle area are used for venting vapor.

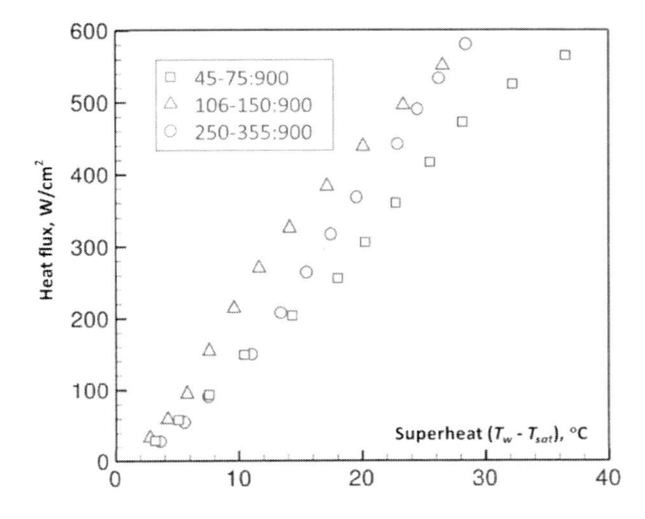

Figure 4. Steady-state thermal performance results for the 900 μm thickness test samples. The maximum heat flux test point does not correspond to the dryout (Weibel et al.).

Weibel et al. also visualized phase change at various heat flux. They found that boiling occurred within the wick structures that led to recoverable temperature overshoots. A reduction in the overall evaporator thermal resistance was also identified, while phase change interfacial evaporation is transitioned to the boiling regime, with effective reduction of the wick conduction resistance.

Mono wick structures with precisely defined dimensions of wick particle and the heating area were developed in the research of Cai et al. [18]. They utilized MEMS (MicroElectroMechanical System) microfabrication technologies to lithographically define the wick pattern and dry-etch silicon to form perpendicular pillars on a wafer. Because there were no sintering interfaces between wick particles, the pillar wick structures gain high effective thermal conductivities close to 70W/m.K with wick porosity of ~ 0.5. Among a few mono wick structures they developed, a representative of wick *A* (shown in Figure 5a) has pillar diameter of 100μm and height/wick thickness of 320μm. With a 4×4 platinum film heater deposited on the backside of the silicon substrate, the ratio of particle radius over the heating area width, *r:l*, approached to 1/40. This wick design slightly sacrificed capillarity, but stressed on vapor ventilation through reducing wick thickness and straightening the vapor escape paths.

In addition to precisely define wick parameters, efforts were also spent to monitor substrate heat spreading and precisely quantify input heat loss [19]. As shown in Figure 5b, the Z sharp circuit is a platinum film resistor fabricated through e-beam evaporation process. Wire bonds between the resistor and electrode pads defined heat generation area of 4mm×4mm. As a platinum resistance thermometer (PRT), a thin straight platinum wire was deposited 100μm apart from the right side of the resistor. It can monitor temperature variation and substrate heat spreading before and after dryout occurs. Since gravity effect is negligible in mass transport, silicon mono wick structure was perpendicularly set to allow liquid supply from its bottom reservoir.

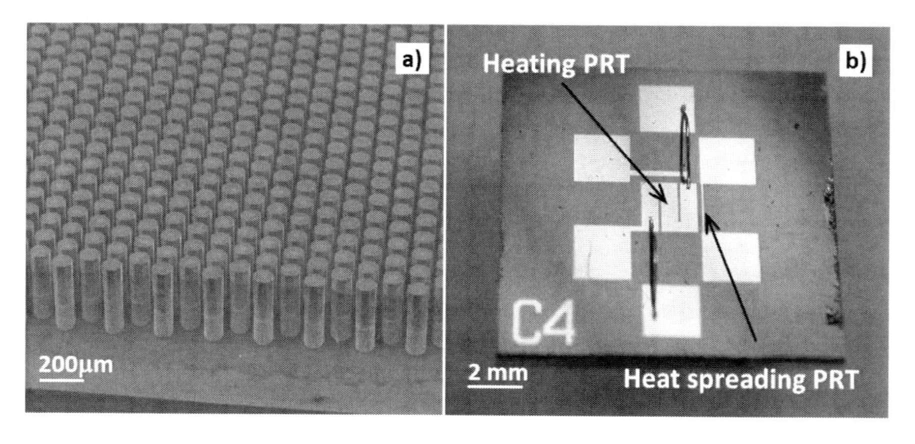

Figure 5. Mono wick pillars made through silicon MEMS processes: a) the representative wick structures *A* has the ratio of radius over the heating area width of 1/40, b) platinum film heater deposited on the backside of the mono wick.

Using water as operating fluid, the maximum heat flux over 1,000W/cm^2 has been reached in this study. As shown of the right curve in Figure 6, from point S to M, phase change proceeds with a fast increase of superheat. Wick conduction thermal resistance plays a critical role at this stage. As input heat flux exceeded ~ 400W/cm^2 (starting from point M), the heat transfer coefficient or the curve slope quickly increased. The maximal heat flux was reached at 1,130W/cm^2 (point O) with superheat measured at 54°C, followed by dryout and the rapid increase of the substrate temperature, T_{sub}. T_{sub} is defined as the substrate surface temperature on the opposite side of the heater.

$$T_{sub} = T_{heater} - \Delta T_{sub} \tag{2}$$

The resistor heater temperature, T_{heater}, was directly measured through the effect of temperature on resistance. Temperature drop across the silicon substrate, ΔT_{sub}, must be subtracted when converting T_{heater} to T_{sub}. A 1D conduction model is used to estimate ΔT_{sub} after the sample backside is well thermally insulated and in-plane heat loss of the substrate was estimated based on the measurement of the wire PRT. As shown in Figure 6, at the maximum phase change heat flux, T_{wire} is below 120°C and the heater temperature was measured at ~ 204°C. A large temperature gradient between the heater and the wire PRT indicated very effective phase change over the heated wick area so that heat conduction through the substrate becomes thermally expensive. However, as the dryout occurred, expansion of phase change area led to a fast rise of T_{sub} (as the left curve shown in Figure 6).

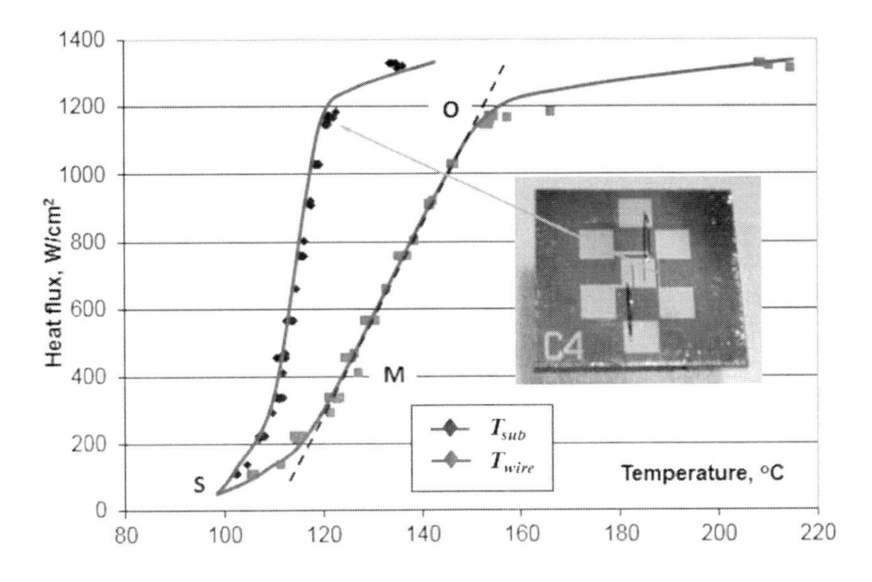

Figure 6. Input heat fluxes vs. substrate temperature of the mono silicon wick structure *A*, Cai et al. [18].

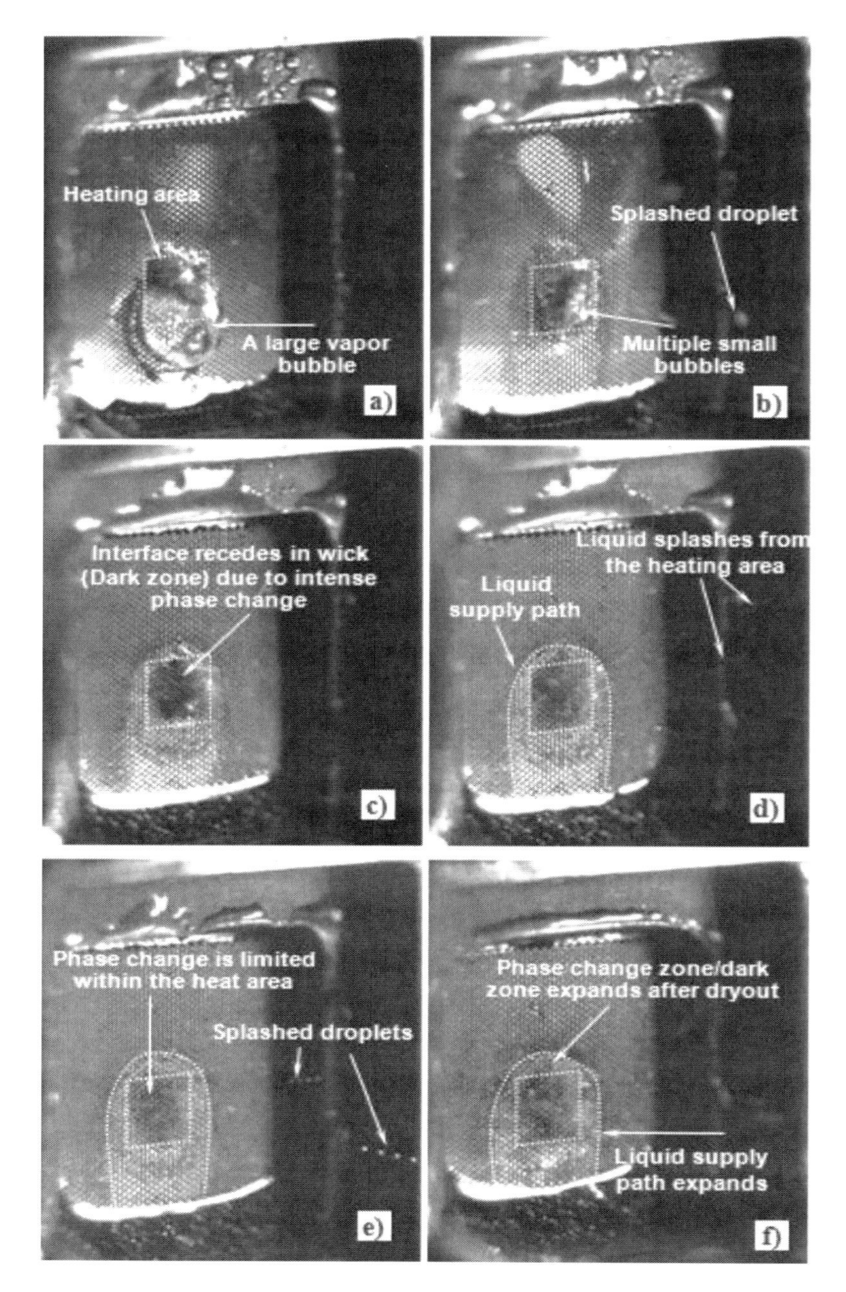

Figure 7. Phase change images of the silicon pillar wick structure at: a) 180W/cm^2, b) 300W/cm^2, c) 450W/cm^2, d) 650W/cm^2, e) 1,130W/cm^2, f) 1,300W/cm^2.

Phase change visualization was also conducted in this study at each input heat flux level to cross-validate heat transfer characterization curves. White dash lines in Figure 7 outline the actual heated wick area. With increase of input heat flux, increase of superheat initiated nucleate boiling. These bubbles expanded in volume as it raised from the bottom of the wick structure. When heat flux reached 300W/cm^2, increase of vapor flow speed reduced bubble growth time and size (Figure 7b). Scattered liquid splashes were observed from the bubble disruptions. When heat flux exceeds 450W/cm^2, liquid splashing took place of the role of the emerging bubbles. Vapor escaping from the wick formed vapor wells/jets (Figure7c), corresponding to the point M in Figure 6. Because operating liquid was a few degrees cooler than the saturated vapor, vapor condensation on the top of silicon pillars reflected light and highlighted liquid supply paths around the heating area (outlined in Figure 7d). At the maximum heat flux at 1,130W/cm^2, intense phase change area spread to the entire heated wick area, as shown in Figure7e. Phase change zone expanded out of the heated wick area as the dryout occurred (Figure 7f).

Phase change capability studies were also performed on the metallic wick made through a 3D additive fabrication process, electroplating technology. Electroplating technology was originally developed to create stacked metallic RF networks, which uses electrical current to reduce dissolved metal cations so that they form a coherent metal coating on lithographically defined electrodes through iterated process cycles. Compared with lithographic and dry-etch approach on silicon, electroplating could generate equally good quality on precisely defining the wick particle dimensions, as shown in Figure 8. Moreover, wick structures made through this approach (e. g. copper wick) has higher effective thermal conductivities and are more thermomechanically compatible with its metallic heat sink. Nam et al. first evaluated wick properties made through this process [20]. In his later study, the maximum heat flux exceeded 800W/cm^2 was demonstrated with a small 2mm×2mm heating area [21]. Benefited from the highly conductive wick structure, phase change superheat was reduced to ~ 30°C at the highest heat flux.

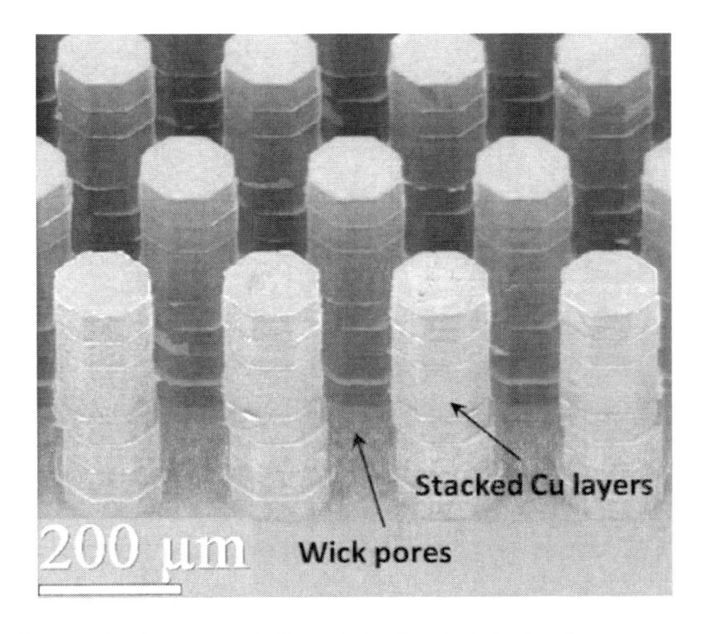

Figure 8. Copper pillars and substrate made through the iterated electroplating process cycles.

2.3. Multiscale Nanoporous Wick Structures

Nanoscale structures, such as carbon nanotube (CNT) clusters with SP2 atomic structures, are highly thermally conductive and porous. These structures contain nanoscale pores that are able to generate over 10X capillarity for flow circulation. These features make nanoporous wick structures potentially capable of high heat flux and high efficient phase change [22].

Depending on the growth height and approach, porosity of the CNT clusters/forests varies from 0.98 to 0.95, leaving only 2-5 percent in solid fraction. A thicker CNT wick structure generally has higher porosity after a number of catalyst seeds defunctionalized in their synthesis process. Unlike the straight pillars made through microfabrication processes, the actual CNTs are highly tortuous. As shown in Figure 9, depending on growth temperature and height, CNT tortuosity may vary between 1.0 and over 3, leaving a large uncertainty for estimating the wick effective thermal conductivity. Ideally, a CNT with perfect SP2 crystalized atomic structures have ultrahigh axial thermal conductivity >2,000W/m.K. However, low growth temperature and imperfect crystallization increase formation of amorphous carbon that embedded in the structures and the phonon-to-phonon scattering. Considering all these factors, the effective thermal conductivity can be expressed as

$$k_{eff} = C_q \frac{(1-\phi)}{\gamma^2} k_{CNT} \tag{3}$$

Here, γ is tortuosity, C_q is the CNT quality factor with a range from 0 to 1.0, and k_{CNT} is the CNT theoretical thermal conductivity with perfect SP2 atomic structures. Literature reported thermal conductivity data of CNT varies in a very large scope. The term of $C_q \cdot k_{CNT}$ is between 350 and 6,600W/m.K [24-28]. Convert to the CNT cluster through Eq.(3), the wick effective thermal conductivity varies between 0.8 and ~180W/m.K., consistent with reported test data from 0.7 to 262W/m.K [29-31].

The effective pore size of the CNT clusters defines its maximum capillarity. Thus, the minimal meniscus radius between the two adjacent CNT fibers is used to characterize the parameter.

$$r_{eff} = R\left(\sqrt{\frac{\pi\gamma}{(1-\phi)}} - 2\right) \tag{4}$$

Here, r_{eff} is the effective porous radius and R is radius of a single CNT. For CNTs with diameters between a few nanometers to 20 nanometers, the effective pore radius varies from 50nm to 180nm, which can generate capillary pressure one or two orders of magnitude higher than traditional microporous wick structures.

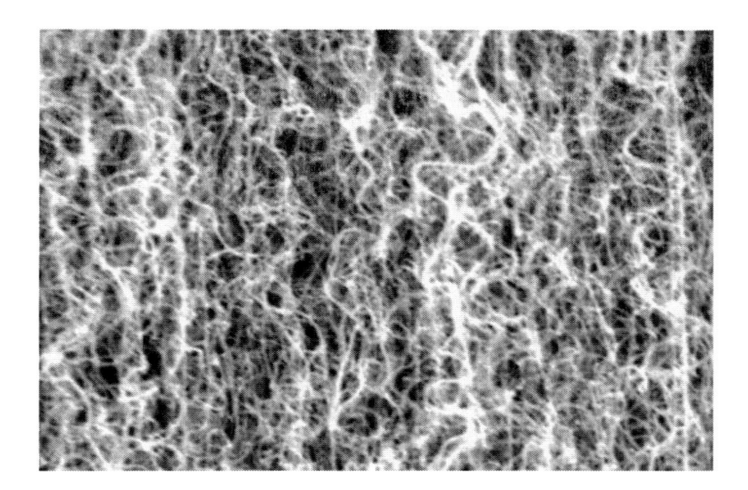

Figure 9. Scanning Electron Microscope (SEM) images of tortuous CNTs.

Permeability of the CNT forest based on the correlation of multiple wire mesh screens (plain or sintered) provides the best approximation.

$$K = \frac{d^2 \cdot \phi^3}{122(1-\phi)^2} \quad (5)$$

Here d is the wick particle diameter. Permeability of the porous CNT forest wick falls in a range between 1.1×10^{-16} and $5.8 \times 10^{-15} m^2$, which is 3-4 orders lower than microporous wick structures used in conventional heat pipes. High flow resistance in such the dense nanoporous structures make high density heat removal impossible.

To minimizing the flow viscous loss but utilize high capillarity and the nanoscale evaporation interfacial area, multiscale CNT wick structures are presented and developed. As shown in Figure 10, CNTs were synthesized on selective areas of a silicon wafer. Large cavities are formed between these CNT clusters. Besides providing vapor ventilation when nucleate boiling starts, these microscale large cavities function as low viscous liquid transport path, pumping liquid from the condenser to the evaporator area. When phase change occurs on the nanoporous evaporation interfaces, liquid is supplied from surrounding liquid arteries/micro cavities. The liquid transport length in the dense mono nanoporous structures is reduced to the characteristic dimension of the CNT clusters at the scale of 50 - 100μm, 2-4 orders of magnitude shorter than the flow path through the single-tie and uniform CNT forest. With aid of 1-2 orders of magnitude higher capillarity, high heat flux phase change with efficient heat and mass transport becomes achievable on the multiscale CNT wick.

Development of multiscale nanoporous structure involves MEMS process. On the silicon substrate, CNT growth areas are lithographically defined (shown in Figure 10a). Following with a sputtering catalyst coating (Figure 11b) and photoresist lift-off (Figure 11c), wafers were then sent to furnace to grow CNTs. At elevated temperature, an annealing process was applied to create nano catalyst particles for the CNT growth. The patterned CNTs were synthesized at 745°C environment with flowing gases of hydrogen and C_2H_4 at ambient pressure (Figure 11d). After the CNT growth, the UV or Ozone dry surface treatment is required to convert the hydrophobic surface to hydrophilic.

Figure 10. CNT multiscale wick structures: a) large cavities between the CNT clusters create liquid transport paths with low viscous loss; b) nanoscale pores in the CNT clusters increase phase change surface area.

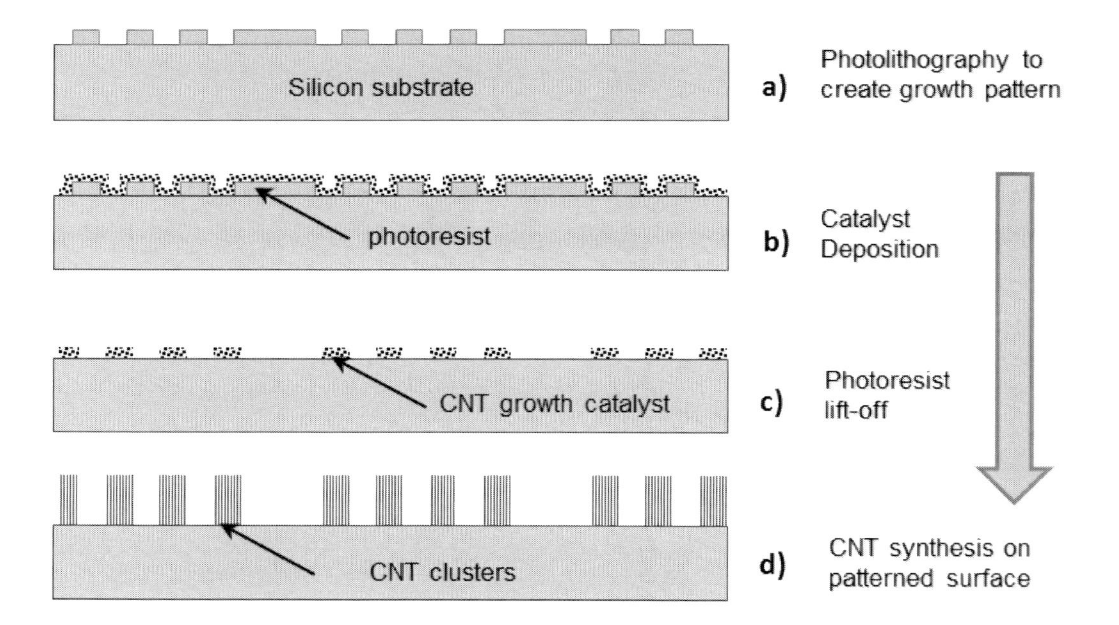

Figure 11. Development process of the multiscale CNT wick structures.

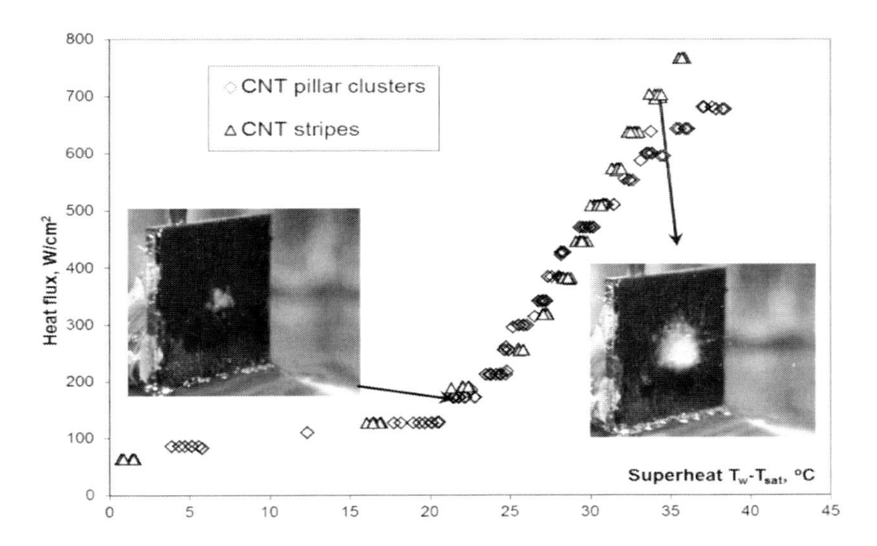

Figure 12. Phase change heat flux vs. superheat of multiscale CNT pillar and stripe wick structures with the heating area of 2mm×2mm.

Cai et al. developed multiscale CNT structures composed of continuous stripe and isolated pillar clusters to investigate phase change capability. On the stripe wick structures, the maximum heat flux of 770W/cm^2 was reported with superheat less than 40°C, as shown in Figure 12. Large pores of the patterned CNT pillar clusters offered lower capillarity for liquid transport, therefore reducing phase change heat flux to 600W/cm^2. A visualization system was also set to closely observe phase change zone and occurrence of the dryout. In Figure 12, the left zoom-in image shows the initial stage of phase change. Nanoscale pores suppressed generation of nucleation bubbles in the CNT clusters. As a result, small vapor bubbles can only be seen between CNT strip clusters, raising from the bottom of the microscale grooves/pores. Similar to bi-porous wick structures and mono porous wick structures, the in-plane liquid flow and perpendicular venting vapor flow shared the same wick flow passages. While input heat flux approaches to its CHF at 690W/cm^2, intense phase change covered the entire heating area, as shown in the right image of Figure 12. The rapid vapor ventilation from the microscale pores carried liquid phase to splatter as liquid droplets and films. It created a phase change column constantly suspended over the heated wick area. As further increase of power density, strong interaction of vapor and liquid phases reduced liquid supply, leading to the dryout.

Nam et al. investigated oxide nanostructures formed on the copper micropost surfaces with post diameter of 50μm. Nanostructures created through the chemical surface oxidation are ~ 0.2μm thick. They reported that Cu_2O nanostructures on the microposts enhanced the critical heat flux by over 70%, reaching >500 W/cm^2 on a 2 mm×2 mm heating area. Because nanostructures were only applied on the surface of micro wick particles, the effective thermal conductivity of the wick structure was not sacrificed for generating high capillarity. As a result of this effort, the heat transfer coefficient of phase change is maintained >10 W/cm^2 K, which justified the low superheat of < 30°C at high heat flux.

With consideration of the superior thermal and physical properties of CNTs, Weibel et al. developed a numerical capillary limitation fluid flow model to assess the potential reduction in evaporation thermal resistance [32][33]. Their research had consistent findings with Cai et al. that the low permeability limits the uses of mono CNT forest as the wick structure over a large heating area. They proposed new evaporator wick structures that the nanostructured areas were surrounded by interspersed traditional wick structures (e.g., sintered copper particles) with high permeability for liquid phase mass transport. An optimization that balanced between the high permeability of the sintered materials and the greater capillary pressure was reported to minimize the evaporator thermal resistance while using interspersed nanostructure to provide phase change interfaces.

When taking advantages of merits of nanoscale phase change heat transfer, all researchers considered the multiscale structures as feasible solutions. The multiscale structures fully utilized nanoscale pores to generate high capillarity and increase thin film evaporation efficiency, and meanwhile avoided paying the penalty by transporting liquid and venting vapor between the dense nanostructure clusters.

3. High Heat Flux Phase Change Modes and Transitions

In most high heat flux applications, input power may rapidly vary in a large range, leading unsteady state phase change heat transfer, or fast phase change mode transitions. In responding to variation of input heat flux, phase change in/on the wick structure manages itself to alter its heat transfer mechanism to achieve the most efficient heat and mass transfer. In the effort of creating a diagnose tool for understanding this dynamic process, a direct link between phase change images and heat transfer characterization data establishes the most effective method.

Using the copper wick structures made through the 3D additive electroplating processes, a comprehensive research of phase change transitions was conducted by Cai et al. [34]. They used water as the operating fluid, running experimental at saturation temperature of 100°C. Visualization images, input heat flux, and the evaporator substrate temperature were simultaneously recorded, providing real time cross-validations between phase change phenomena and temperature responses. Their results were summarized and shown in Figure 13. At lower heat flux (shown in Figure 13a), the meniscus interfacial evaporation was the only phase change mechanism. Due to the recession of the center of the menisci, reflected light was steered off from the observation direction. Thus, the evaporation zone was slightly darker than the surrounding area. At this mode, heat must conduct through the wick structure to reach the liquid-vapor interfaces. The wick effective thermal conductivity dictated the phase change superheat. While changing the heat flux to the range of 50 to 100W/cm^2, a large vapor bubble growing from nucleate site emerged over the wick structure, covering the entire heating area (shown in Figure 13b). Phase change occurred at both the bottom of the wick structure as nucleate boiling and on the top interfacial meniscus as evaporation. With involvements of both mechanisms, this phase change mode was called meniscus interfacial evaporation associated with nucleate boiling. At this mode, wick structures below the emerging bubble were maintained to be saturated. Before eruption, the large on-top vapor bubble kept increasing in volume while vapor mass was added through the interfacial evaporation.

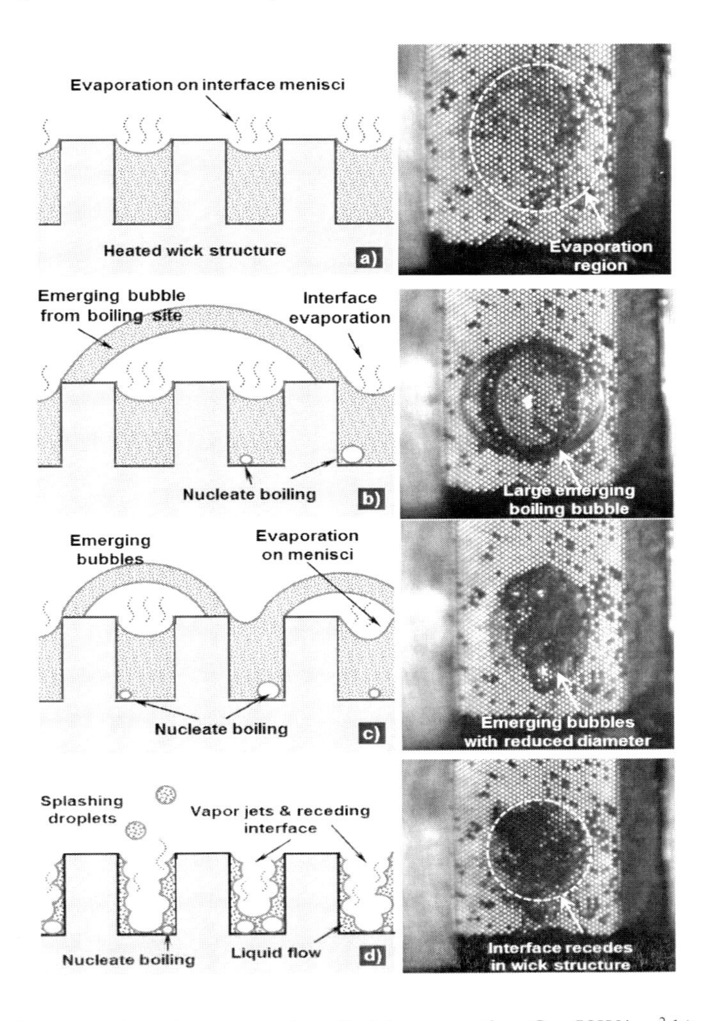

Figure 13. Phase change modes: a) meniscus interfacial evaporation @ <50W/cm^2 b) meniscus interfacial evaporation associated with nucleate boiling @ 50-100W/cm^2, c) meniscus interfacial evaporation enhanced by nucleate boiling @ 175W/cm^2, d) evaporation enhanced by interfacial receding @ >250W/cm^2.

With increasing heat flux to a higher level at 175W/cm^2, phase change entered the mode of meniscus interfacial evaporation enhanced by nucleate boiling. High shear-force of vapor flow reduced the survival size of the bubbles and multiplied bubbling spots. The breakup of the vapor bubbles is attributed to the Rayleigh-Taylor instability at which a liquid film accelerated by a vapor column breaks when the ratio of inertia to surface tension forces exceeds a critical value. The non-dimensional Weber number is used to characterize this competition.

$$We = \frac{\rho v_v^2 d_c}{\sigma} \tag{6}$$

where ρ is fluid density, v_v is vapor velocity, d_c is the maximum vapor bubble diameter at the critical *We* number, and σ is fluid surface tension. Compared with water, fluids with low latent heat and surface tension (e.g., ethanol) generate higher vapor flow speed at given heat flux. With the same critical/breakup *We* number, the maximum survival diameter of ethanol is much smaller than that of water.

As heat flux was increased above 250W/cm^2, even small size vapor bubbles cannot be found over the wick structure. Small droplets were entrained by high speed vapor flow, splashing from the wick structure. Liquid-vapor interfaces, through vapor jets, spatially extend from open vapor ambient to the bottom of the wick structure (shown in Figure 4d). The stretched vapor volume significantly increasing the total surface area for evaporation phase change, the heat transfer coefficient, as well as energy removal density. This fully developed phase change mechanism proceeds until the input heat generated enough vapor that the expanded vapor jets cut down liquid supply, leading to the dryout at the critical heat flux.

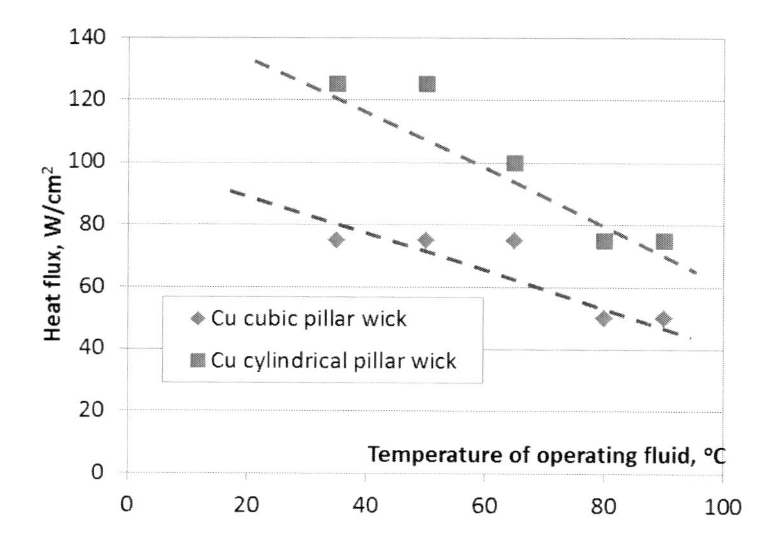

Figure 14. Onset of nucleation boiling vs. fluid temperature.

Within the wick structure, ONB is a landmark that distinguished the traditional interfacial evaporation from the rest modes capable of high heat flux phase change. Since

ONB is wick parameters and fluid property dependent, phase change mode transitions and heat transfer characterization curves may vary between different operating systems [35]. In Figure 14, the cubic and cylindrical wicks refer to their cross-section shapes of the copper wick particles. Their corresponding porosities of these two wick structures are 0.75 and 0.71, with the solid volume fractions of 0.25 to 0.29 for conduction heat transfer, respectively. The effective thermal conductivity of the cubic wick is about 20% lower. Thus, for the same amount of heat flux input, the conduction thermal resistance/temperature difference across this wick structure is proportionally higher, starting ONB at lower heat fluxes. For all the wick structure with high porosity, the earlier ONB advances intense phase change mode and interface receding in the wick structure, creating an opposite effect to suppress the rapid superheat rise as increase of heat flux. This self-adjustable mechanism is particularly useful for the wick structures with the low effective thermal conductivity. Similar to high porosity wick, ONB starts earlier at lower heat flux for the less conductive wick structures. The advanced mode transition hedges the negative impact of more conduction heat transfer losses in the wick structure.

4. Parameter Analysis of High Heat Flux Wick Structures

4.1. Equilibrium of Heat and Mass Transport in Wick Structures

The capability of phase change heat transfer reflects the limit of the mass transport or liquid delivery to wet the heated wick structure. The maximum heat flux is strongly dependent on the wick geometric and operating fluid parameters, such as wick porosity, thickness, particle diameter, character length of the heated wick area, latent heat, viscosity, etc. The mono CNT wick is an example in which liquid supply, nucleate boiling and vapor ventilation are suppressed by the dense and nanoscale structures. As a result, in spite of over 100 times higher capillarity, the maximum heat flux is less than $100 W/cm^2$.

An analytical model was developed to understand wick geometry and fluid property contributions to the complicated physical process [36]. This study assumes the wick structure is composed of perpendicular cylindrical pillars with precisely defined particle geometry, similar to the etched silicon and the 3D additive copper wick structures discussed previously. The wick has a characteristic heating length of l and thickness of t, as shown in Figure 15. Liquid is fed into the wick structure from the left side. Vapor vents perpendicularly to open vapor ambient. To simplify the analysis, hydrostatic pressure and flow pressure drops out of the heated wick structures are not considered. The heated wick dimension perpendicular to the l - t plane is assumed to be infinite long so that only the 1D mass transport needs to be considered.

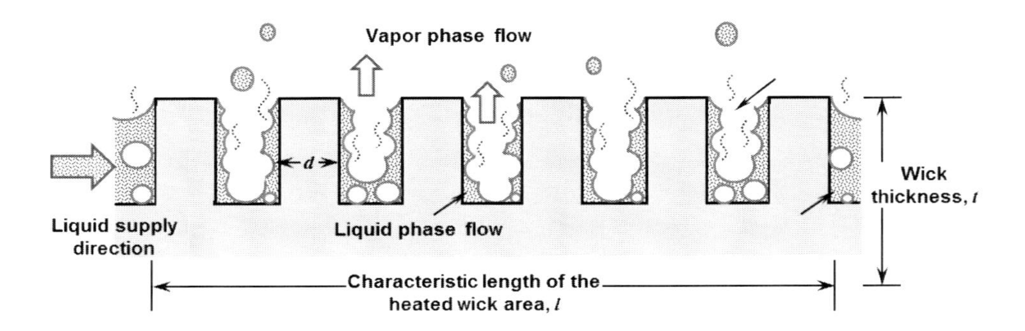

Figure 15. Schematic diagram of phase change analytical model.

As we mentioned in the previous sections, with implementation of nucleate boiling as part of phase change mechanism, capillary limit is the major constrain for high heat flux mass transfer. To achieve effective phase change, the maximum capillary pressure generated by the wick structure must be able to overcome the total flow resistance within the heated wicking area, including pressure drops of both the liquid and vapor phase flows.

$$P_{cap,\max} = \frac{4\sigma}{d_{eff}} > \Delta P_l + \Delta P_v \tag{7}$$

Here, d_{eff} is the wick effective pore size, and σ is liquid surface tension. ΔP_l is the pressure drops of the liquid flow to wet the wick structure. ΔP_v is the pressure drop to vent vapor from the wick structure to open vapor ambient. At this case, liquid flow in the transverse direction is perpendicular to the vapor flow. For the periodic cylindrical pillar structures, d_{eff} is expressed as [2]

$$d_{eff} = \frac{d}{1-\phi} \tag{8}$$

Here d is the pillar diameter and ϕ is the structural porosity. In the porous wick structures, pressured drops of both liquid and vapor phase flows are dictated by the Darcy's law.

$$\Delta P = \frac{\mu L v}{K} \tag{9}$$

where, K is wick permeability, μ is dynamic viscosity and v is flow velocity. L is the characteristic length along the flow direction. For liquid phase, L is replaced by the characteristic length of the heated wick area of l; for vapor flow, L is the wick thickness of t. An analytical correlation of wick permeability, K, applicable to both the transverse liquid flow and axial/perpendicular vapor flow, is given by Carman [37].

$$K = \frac{\phi D_h^{\;2}}{\psi_{CK}} \tag{10}$$

where, Ψ_{CK} is the dimensionless Kozeny factor, characteristic of the wick particle geometry or shape. D_h is hydraulic diameter, defined by the ratios between the cavities available for flow and the total wetted surface area [38].

$$D_h = \frac{\phi d}{(1-\phi)} \tag{11}$$

At high heat flux, vapor and liquid phases share the flow passages/cavities in the wick structure. If α is defined as the liquid volume fraction in the wick cavities, the wick porosity available for the liquid and vapor phase flows becomes

$$\phi_l = \alpha\phi,\ \phi_v = (1-\alpha)\phi \tag{12}$$

Hydraulic diameter of the vapor flow passage is the function of input heat flux. However, growth of the jet hydraulic diameter is constrained by the surrounding wick particles/adjacent pillars, as well as capillary forces that tend to recover meniscus interfaces at where has the smaller pore radius. Before phase change reaches its dryout, these vapor jet flows behavior as isolated vapor wells/jets. This flow pattern allows liquid turn around the jets to wet all the heated wick structures through the cavity gaps formed between high-speed vapor jets and solid wick particles. For typical mono wick structures with a porosity range between 0.5 - 0.7, the size of the fully developed vapor jets is considered to be equivalent to that of wick particle. Thus, the hydraulic diameter of the vapor flow passage, $D_{h,v}$, is approximated to the pillar diameter, d.

Combine the above Eq.(10)-(12), the transverse and axial permeability for liquid and vapor phase flows can be written as

$$K_l = \frac{1}{\psi_{CK,l}} \frac{\alpha^3\phi^3 d^2}{(1-\alpha\phi)^2} \tag{13}$$

$$K_v = \frac{1}{\psi_{CK,v}} (1-\alpha)\phi d^2 \tag{14}$$

Considering phase change occurs in saturated environment, the mass conservation is written as

$$h_{fg}\rho_v v_v = h_{fg}\rho_l v_l \frac{t}{l} = \dot{q} \tag{15}$$

Here, $\dot{q}$ is phase change heat flux. Substituting Eq. (8),(9), (13), (14) and (15) into Eq.(7),

$$\dot{q}_{max} = \frac{4\sigma h_{fg}(1-\phi)}{\frac{\mu_l l^2}{\rho_l t d}\cdot\frac{\psi_{CK,l}(1-\alpha\phi)^2}{\alpha^3\phi^3}+\frac{\mu_v t}{\rho_v d}\cdot\frac{\psi_{CK,v}}{(1-\alpha)\phi}} \tag{16}$$

$\Psi_{CK,l}$ *and* $\Psi_{CK,v}$, are Kozeny factors for liquid and vapor flows, respectively. If both vapor and liquid phases play equivalent roles in phase change, the maximum heat flux is the function of wick geometrical parameters, and local thermal properties of liquid and vapor flows. Although this wick analysis is based on micro cylindrical pillars, the correlation of Eq.(16) is also applicable for other types of wick structures, such as sintered metallic powders, micro meshes and micro grooves, by adjusting the Kozeny factors.

4.2. Wick Geometrical Effect

For the thick wick structure with a small characteristic heating dimension, liquid phase flow has relatively low pressure loss and liquid supply is sufficient due to the increased flow cross section area and the shorter transport distance. However, vapor becomes easier to be trapped in the wick structure, referring to the boiling limit [2]. Between the two pressure drop terms in Eq.(7), the vapor phase loss is dominant. Therefore, Eq.(16) can be simplified as

$$\dot{q}_{max} = \frac{1}{\psi_{CK,v}}\cdot\frac{d}{t}\cdot\frac{\rho_v h_{fg}\sigma}{\mu_v}\cdot f(\alpha,\phi) \tag{17}$$

Here, $f(\alpha,\phi) = 4\phi(1-\phi)/(1-\alpha)$. Function *f* reflects dynamic variations of the wick cavity available for the liquid flow. For different types of wick particles, the Poiseuille flow model is applicable for the vapor jets. Thus, $\Psi_{CK,v}$, is a constant of 32 [38]. A dimensionless number, Ge_v, called vapor phase geometrical number, is defined to summarize wick geometrical effect on the maximum phase change capability.

$$Ge_v = \frac{d}{t} \tag{18}$$

Wick geometrical contribution to the phase change capability is correlated to the ratio of the wick particle diameter of *d* over thickness of *t*, but is irrelevant to the characteristic length of the heating area of *l*.

In most phase change applications, the characteristic dimension of the heat source is much larger than the wick thickness. Meanwhile, wick structures are usually designed to be as thin as possible to reduce phase change superheat. At this case, vapor ventilation becomes much easier. The major mass transfer constrain is switched to the liquid phase flow due to the reduced cross section area and the extended transport distance. Moreover, the vapor jet diameter and vapor jet zone expand at high heat flux, further reducing wick cavities/permeability for liquid supply. As a result, liquid flow resistance dictates the heat and mass transfer limit. By ignoring the vapor phase term in Eq.(16), the maximum heat flux can be expressed as

$$\dot{q}_{max} = \frac{1}{\psi_{CK,l}} \cdot \frac{td}{l^2} \cdot \frac{\rho_l h_{fg} \sigma}{\mu_l} \cdot g(\alpha, \phi) \tag{19}$$

Here, function $g(\alpha,\phi) = 4\alpha^3\phi^3(1-\phi)/(1-\alpha\phi)^2$ reflects dynamic vapor phase expansion impact on the maximum phase change capability. $\Psi_{CK,l}$, is 90 for the periodical cylindrical pillars after a tortuosity correction of the shape factor is made [37] [38]. Correspondingly, a liquid phase geometrical number is defined as

$$Ge_l = \frac{t \cdot d}{l^2} \tag{20}$$

The wick geometrical influence on the maximum heat flux is correlated to a product of the wick thickness of t and the particle diameter of d over square of the characteristic length of the heating area of l^2.

In practice, applicability of the geometrical numbers relies on a clear definition between these two extreme scenarios. For this purpose, a dimensionless parameter, β, is defined as the ratio of characteristic length of the heated wick area over the wick thickness, l/t. By introducing β into Eq.(20), the liquid phase geometrical number is rewritten as

$$Ge_l = \frac{1}{\beta^2} \cdot \frac{d}{t} = \frac{1}{\beta^2} Ge_v \tag{21}$$

The difference between the liquid and vapor phase geometrical numbers is reflected through the dimensional ratio of β. Since the heating area size of l and wick thickness of t represent resistance characters of the liquid and vapor flows, respectively, the dimensionless β can be treated as a corresponding characteristic non-dimensional number to justify the applicability of the previous analytical model.

Based on Eq.(16), the ratio of between the liquid and vapor phase pressure drops can be expressed as

$$\frac{\Delta P_l}{\Delta P_v} = \beta^2 \frac{\psi_{CK,l}}{\psi_{CK,v}} \cdot \frac{Me_v}{Me_l} \cdot \frac{f(\alpha,\phi)}{g(\alpha,\phi)} \tag{22}$$

Here, $Me_v = \rho_v h_{fg} \sigma / \mu_v$ and $Me_l = \rho_l h_{fg} \sigma / \mu_l$ are vapor and liquid phase merit numbers, representing contributions of vapor/liquid properties to phase change. For most operating fluids, typical wick particles and liquid fraction α ranging from 0.6 to 0.9, liquid phase flow resistance is at least one order of magnitude higher than the vapor flow resistance when β is >1.0. In contrast, the vapor flow resistance may become dominant when β is <0.1. In most practical cases (e.g., SiC die cooling) where the heated wick characteristic length is far larger than the wick thickness, the liquid phase geometrical number takes the dominant role. However, some particular applications (e.g., GaN power amplifier) may have very tiny heat source much smaller than the wick thickness dimension at which the vapor phase geometrical number may apply. When the dimension ratio is in the range of $0.1< \beta <1.0$, both the liquid and vapor flow resistances are important in determining the maximum phase change capability. The full correlation of Eq.(16) should be considered. In our previous discussion, the etched silicon and the 3D additive copper wick structures have β ranging from 6.25 to 12.5. Within this range, the liquid flow resistance is three orders of magnitude higher than that of the vapor flow, exclusively supporting the liquid phase geometrical number Ge_l.

Model applicability between vapor and liquid phase geometrical numbers can be also validated through comparing experimental data with analytical prediction. A group of etched silicon mono wick structures with precisely defined wick particle geometry, summarized in Table 1, were tested within a closed chamber with water as operating fluid. For all test samples, the maximum phase change heat flux/CHF was recorded vs. their geometrical numbers.

Using the vapor phase geometrical numbers, no clear relation can be found to prediction variations of the maximum heat flux. A chaotic distribution of experimental data implies that the vapor phase geometrical number is not the dictating parameter for qualifying the maximum phase change capability. The inference can be verified by comparing wick structures of d30, t320, l2000 and d30, t320, l4000. They have the identical vapor phase geometrical number of 0.09 but more than three times difference in the maximum phase change capability. The only difference between these two wicks is the uncaptured characteristic length of the heating area. Another opposite case is, with increase of vapor phase geometrical number Ge_v from 0.31 to 0.45 (wick d100, t320, l2000 and wick d100, t220, l2000), the maximum heat flux reduces, exhibiting reverse variations.

In contrast, experimental data well aligned to the analytical prediction while using the liquid phase geometrical number as the index. As plotted in Figure 15, the curve of the thin wick model is based on experimental data to assume that the liquid volume fraction α linearly reduces from 0.8 to 0.65 when increasing heat flux from 50 to 1,200W/cm^2 (from ONB to the dryout) [36]. As increase of heat flux, vapor ventilation speed and jet diameter increase. The liquid phase fraction in the shared flow passages, α, is squeezed to reduce. On the other hand, improved wick geometry reflected by the geometrical number allows more liquid being delivered to wet the wick structure. Interactions between the improved wick geometry and the reduced wick permeability steer the curve to be asymptotic. The continuous and monotonic increase of the maximum heat flux reflects a unique way that wick geometry contributes to mass transport. The maximum phase change heat flux falls between 1,000 to 1,200W/cm^2 as the geometrical number approaching to 1.0×10^{-2}. This upper limit confines the wick geometrical effect. Beyond this limit, the wick geometrical effect vanishes (e.g., keep increasing wick thickness) and vapor phase flow starts to play critical roles in the mass transport.

Given porosity of a wick structure, the effective pore size is considered to be proportional to diameter of wick particles. It is worthy to mention that increase of particle diameter or pore size reduces the resistance for both liquid and vapor phase flows, but ends by encountering the gravity effect on the interfacial meniscus [39]. Beyond the capillary length, a flat zone may appear at the bottom of the meniscus, which may fail the local capillary mechanism for circulating operating liquid.

The aforementioned research was also experimentally validated by Chen et al. who published two-part papers to investigate effects of the wick thickness, porosity and mesh pore size [40, 41]. Their observation of the evaporation and boiling phenomenon on the mesh wick structures (wire diameter ~ 56μm) was divided into four modes as: conduction-convection, receding liquid, nucleate boiling, and film boiling. With varying the wick thickness from 0.21 to 0.74 mm, the maximum heat flux over 350W/cm^2 was reached, as shown in Figure 17. They discovered that the heat transfer coefficient of phase change is independent on wick thickness. However, increases of the wick thickness helped to increase the maximum heat flux/CHF. Their study indicated that CHF was strongly dependent on the mesh size and the volumetric porosity. Both these two parameters work together to determine the effective pore size (from 119.3 to 232.8μm), which is equally reflected by the wick particle diameter, *d*, with a given wick porosity.

Table 1. Geometrical parameters of tested wick structures, μm
(*d*, *t* and *l* are wick particle diameter, thickness and characteristic length)

Wick structures	*d*	*t*	*l*
d100, t320, l2000	100	320	2000
d75, t320, l2000	75	320	2000
d60, t320, l2000	60	320	2000
d30, t320, l2000	30	320	2000
d30, t320, l4000	30	320	4000
d20, t320, l4000	20	320	4000
d10, t320, l4000	10	320	4000
d100, t220, l2000	100	220	2000

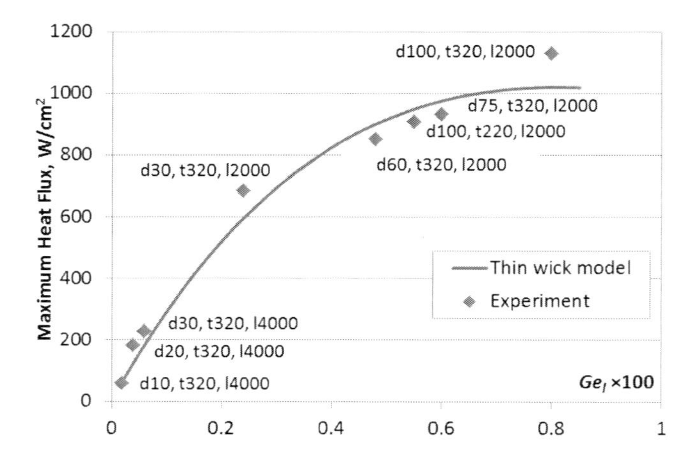

Figure 16. The maximum heat fluxes vs. the liquid phase geometrical numbers: α reduces from 0.8 to 0.65 as input heat flux increases from 50 to 1,200W/cm^2.

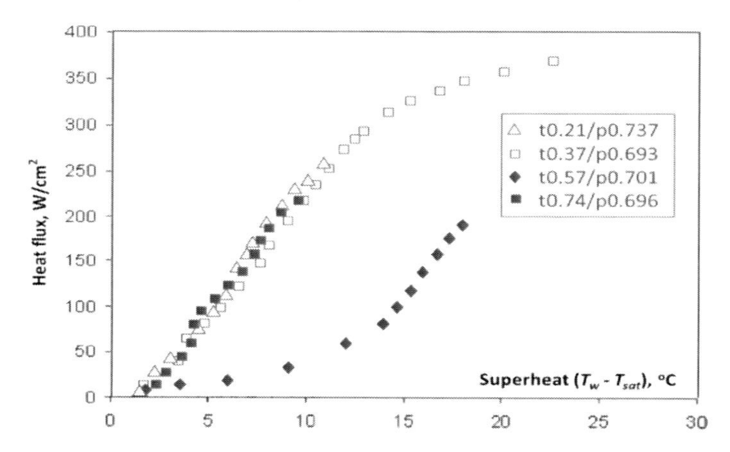

Figure 17. Heat flux as a function of superheat T_w-T_{sat} (Chen et al., 2006): labels of t and p represent thickness in millimeter and wick porosity, respectively.

4.3. Liquid Properties Contribution

Similar to the geometrical numbers, property contributions of operating fluid on phase change capability can be extracted from Eq.(17) and (19) to be defined as dimensional Merit numbers [42]. For vapor or liquid flow dominated wick structures, the Merit numbers are respectively written as:

$$Me_v = \frac{\rho_v h_{fg} \sigma}{\mu_v} \tag{23}$$

$$Me_l = \frac{\rho_l h_{fg} \sigma}{\mu_l} \tag{24}$$

As mentioned before, for most wick structures in practice, liquid flow dominated phase change capability. Thus, the Merit number, *Me*, is referred to Me_l in the following discussion.

Corresponding research of liquid property contributions on the maximum phase change heat flux has been reported through comparing *Me* number prediction with the lab experimental results [43]. Three different operating fluids, water, ethanol and Novec 7200, were investigated, with an order of magnitude difference in the Merit number, as shown in Table 2.

In the early discussion, the maximum heat flux > 1,000W/cm^2 was reported while using water as the operating fluid [18]. After switching the operating fluid to ethanol, the maximum heat flux was reduced to ~ 275W/cm^2 with the same wick design, as shown in Figure 18. Using Novec 7200 as the operating fluid, the maximum heat flux was further reduced to ~ 50W/cm^2. Both ethanol and Novec 7200 have low vaporization latent heat that enhances volumetric speed of vapor generation, amplifying impact of vapor flow on liquid phase within the limited wick cavities. At the same time, low surface tensions of these two fluids reduce capillarity, leading to insufficient liquid supply. These experimental results, showing orders of magnitude difference in the CHF values, validated the Merit number as the fluid property indicator in determining the maximum phase change capability.

Table 2. Merit Numbers of Operating Fluids

Operating fluid	Water	Ethanol	Novec 7200
Me, $\times 10^{10}$	45.5	3.4	0.5

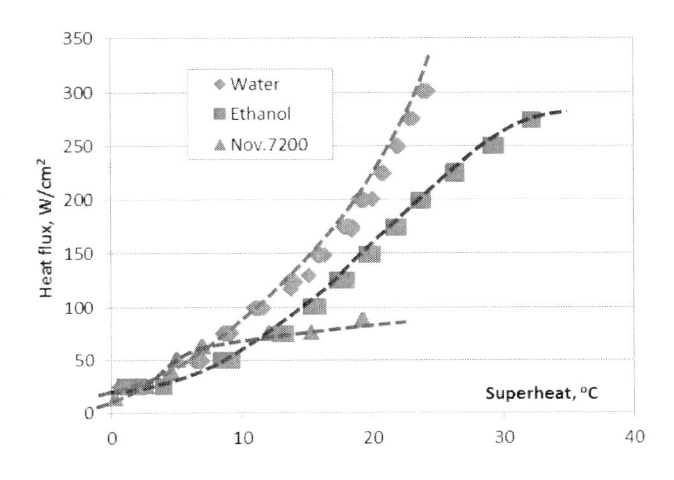

Figure 18. Early dryout/the lower CHF when using ethanol and Novec 7200 as operating fluids.

Besides the maximum heat flux, liquid properties are also playing critical roles in determining the heat transfer coefficient of phase change, h, defined by the input heat flux divided by the superheat, also the temperature difference between saturated vapor and the substrate. Heat comes from the bottom wick substrate. Prior to reaching the liquid-vapor interfaces, it must conduct through the layers of the porous wick structure and the thin liquid film of the meniscus (shown in Figure 19a). At the 1D conduction model, phase change superheat is dictated by thermal resistance of the conduction heat transfer.

$$\Delta T_{sup} = \frac{\dot{q}}{h} = \Delta T_{wick} + \Delta T_{lf} \tag{25}$$

where, $\dot{q}$ is the local heat flux, and ΔT_{wick} and ΔT_{lf} are the temperature drops across the solid wick layer and the liquid film, respectively. Equating the conduction and phase change heat transfer, the heat transfer coefficient can be expressed as:

$$h \approx \frac{k_l k_{wick}}{l_{lf} k_{wick} + l_{wick} k_l} \tag{26}$$

where, k_{wick} and k_l are the thermal conductivity of the wick material and operating fluid. Corresponding to ΔT_l and ΔT_{wick}, l_{lf} and l_{wick} are the average heat conduction distances through these two different materials.

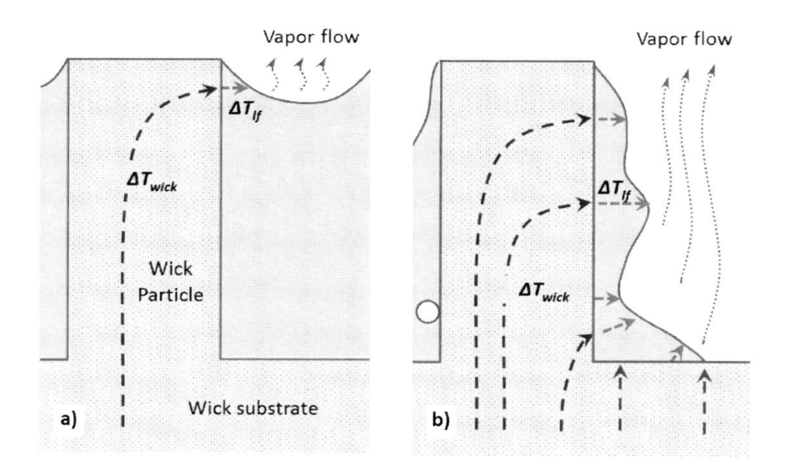

Figure 19. Temperature drops within the wick structure: a) mode of meniscus interfacial evaporation, b) mode of evaporation enhanced by interfacial receding.

Table 3. Comparisons of the thermal conductivity of operating liquids

Operating fluid	Water	Ethanol	Novec 7200
k_l, W/m.K	0.68	0.17	0.07

For highly conductive wick materials, such as silicon and copper, have the thermal conductivity 2 - 4 orders of magnitude higher than the conventional operating fluids (e. g. 400 vs. 0.07 W/m.K between copper and Novec 7200). Liquid thickness for highly effective thin film evaporation (10s nm to a few micrometers) is 2 – 4 orders of magnitude lower than the wick thickness (100s μm). When the meniscus interfacial evaporation is the dominant mode of heat transfer, as shown in the schematic of Figure 18a, the terms of $l_{lf}k_{wick}$ and $l_{wick}k_l$ are equivalent. However, with increasing heat flux, the phase change interface recedes into the wick structure. The conduction heat transfer distance l_{wick} in the solid wick material is significantly reduced, particularly when the receding phase change interfaces approach the bottom of the wick structure at which l_{wick} is less than one-tenth of the original thickness (shown in Figure 19b). The term $l_{lf}k_{wick}$ becomes one order of magnitude higher than $l_{wick}k_l$. As a result, the heat transfer coefficient shown in Eq. (26) can be simplified as

$$h \approx k_l/l_{lf} \tag{27}$$

In addition to the liquid film thickness, the thermal conductivity of the operating fluid plays the dominant role in dictating the phase change heat transfer coefficient and superheat. Table 3 shows a summary of the thermal conductivities of three different operating fluids.

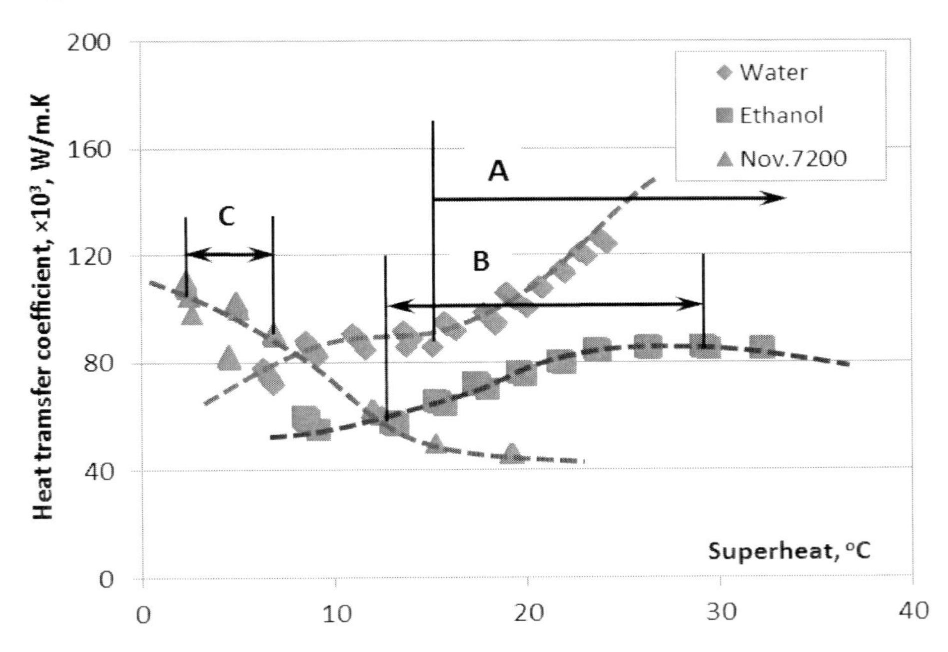

Figure 20. Comparisons of the heat transfer coefficients of different operating fluids vs. superheat: Zone A, B and C illustrated the mode of evaporation enhanced by interfacial receding.

Following this clue, the heat transfer coefficients of three different operating fluids vs. superheat are plotted and compared in Figure 20. Zone A, B and C marked in this chart illustrated the fully developed phase change mode of evaporation enhanced by interfacial receding for water, ethanol and Novec 7200, respectively. Between water and ethanol, the heat transfer coefficient of water is consistently higher than that of ethanol at the overlapped region between A and B. Novec 7200 does not have any overlapped region in the mode of evaporation enhanced by interfacial receding. Despite having an order of magnitude lower thermal conductivity than water, this fluid demonstrated a higher heat transfer coefficient at the low superheat. At the region C, phase change of Novec 7200 early enters the mode of evaporation enhanced by interfacial receding while the other two fluids are still in the mode of meniscus interfacial evaporation with large heat conduction resistance across the wick structures. As discussed in the previous section, the transition of the phase change mode itself is related to the fluid properties (e.g., non-dimensional Weber number). Thus, a comprehensive consideration combining phase change modes with the fluid thermal conductivity is required when evaluating fluid property contribution on the phase change heat transfer coefficient.

4.4. Nanostructure Enhanced Phase Change

For all phase change modes, the high heat transfer coefficient or phase change efficiency relies on formation of thin-film evaporation regions on liquid-vapor interfaces. The thin-film evaporation region is referred to the portion of the interfacial meniscus where liquid film thickness is less than a few micrometers. At this thickness region, evaporation proceeds effectively with a very small temperature difference across the liquid layer. Research studies show that more than 99 percent of evaporation phase change occurs on the region where the liquid film is less than 2.5μm [44] [45].

Nanoscale porous structures, in addition to their high capillarity and thermal conductivity, have a large quantity of nanoscale pores. Compared with microscale porous structures, these nanoscale pores can dramatically increase the total thin-film evaporation surface area and enhance the heat transfer coefficient of phase change. An analytical model, illustrating the thin film evaporation on both micro and nanoscale porous structures, is shown in Figure 21 [23]. In this model, porous wick particles are assumed to be perpendicular solid cylinders/pillars with radius of R.

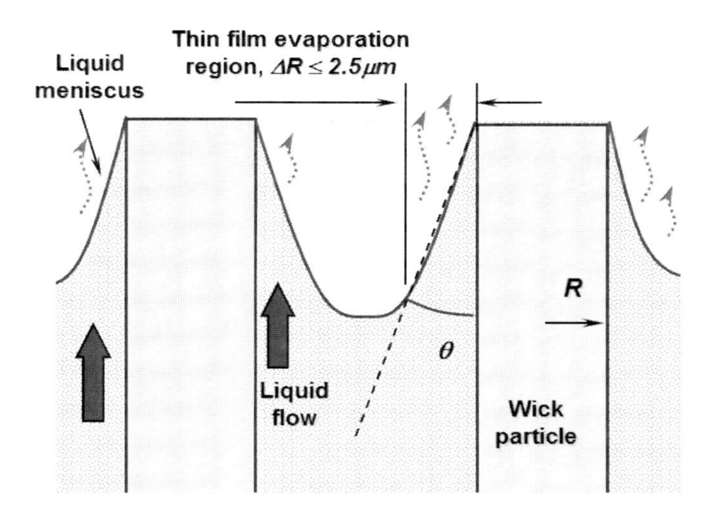

Figure 21. Cylindrical wick particles/units with interfacial meniscus. Effective evaporation region is defined in the film region that $\Delta R \leq 2.5\mu m$.

Based on definition of wick porosity, a characteristic surface area of the wick, A_{cha}, can be expressed as:

$$A_{cha} = \frac{n\pi R^2}{(1-\phi)} \tag{28}$$

where, n is the quantity of the cylindrical wick particles in the characteristic area. The thin-film evaporation area around each pillar can be approximated as a truncated prim. Thus, the total thin-film evaporation area, S_{tf}, can be written as:

$$S_{tf} = \frac{n\pi \cdot \Delta R(2R + \Delta R)}{\sin\theta} \tag{29}$$

Here, ΔR is the maximum film thickness with effective evaporation phase change. θ is the contact angle. Dividing Eq.(29) by Eq.(28), the ratio of the thin-film evaporation area over the characteristic wick area, ε, is

$$\varepsilon = \frac{S_{tf}}{A_{cha}} = \frac{\eta(2+\eta)\cdot(1-\phi)}{sin\theta} \tag{30}$$

where, η is the dimensionless ratio of $\Delta R/R$. Referring to the previous research, ΔR can be approximated to $2.5\mu m$. When interface recedes to the bottom of the wick structure, as discussed, temperature drop in the solid wick layer becomes negligible. The ratio of the thin-film evaporation area, ε, turns to be one of key factors to dictate phase change superheat, or the heat transfer coefficient. A large ε tends to reduce the local thin film evaporation heat flux, therefore increasing the heat transfer coefficient.

A classic microscale wick structure has wick particle radius of 10-100μm and porosity ~ 0.5, varying ε in a range of 0.0125/sinθ to 0.125/sinθ. Comparatively, nanoporous wick structures (e.g., CNT wick structures) have particle radius three orders of magnitude smaller, in a range of 5-25nm. Despite of high porosity, the fine and dense nanostructures make ε approach to 1.0/sinθ. As a result, the thin-film evaporation area in the nanoporous structures is at least ~ 10 times larger than that of the microscale porous structures, which can significantly increase the phase change efficiency and the heat transfer coefficient.

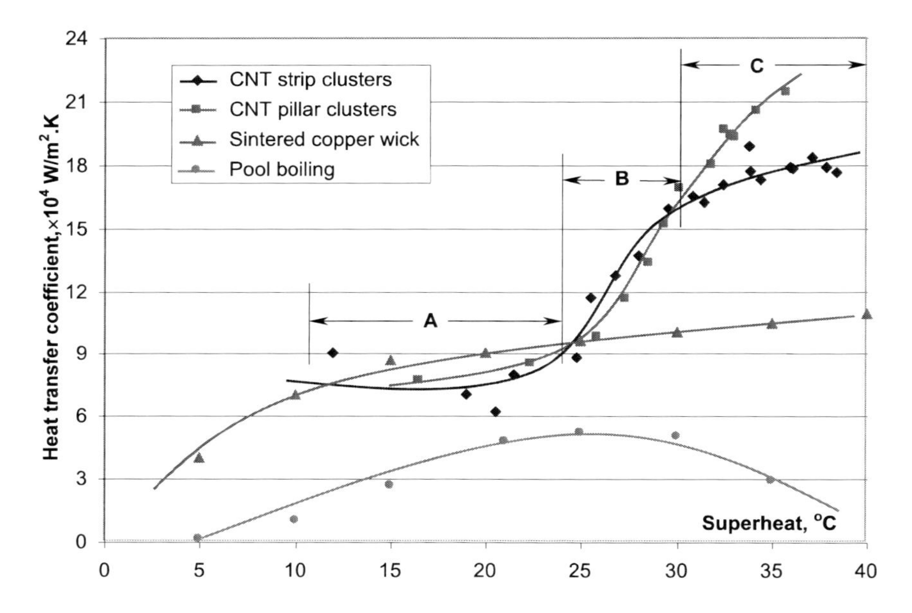

Figure 22. Comparisons of the heat transfer coefficients between the nanoporous wick structures, microporous wick structures, and the pool boiling.

Experimentally, the heat transfer coefficients of two different multiscale porous CNT structures were compared with the sintered microporous wick structures and the pool boiling reported in the literatures [13, 6]. Because of highly efficient interfacial evaporation over the enlarged thin-film region, both the bi-porous CNT structures demonstrated over 100% increases of their heat transfer coefficient in contrast to nucleate pool boiling, as shown in Figure 22. At the superheat region A < 23°C where input heat flux is less than 200W/cm^2 (also shown in Figure 12), phase change occurred interfacial meniscus located on the top of the wick structures. At this region, the wick thermal conductivity still played an important role in dictating the heat transfer coefficient. As a result, the heat transfer coefficients of nanoporous CNT wick structures were equivalent to that of the microporous wicks. A rapid increase in the heat transfer coefficients of the nanoporous CNT structures starts when entering the region B with superheat >23°C. As the superheat increased to 35°C, vapor jets form in the large pores and phase change interfaces recedes into the wick structures. Thermal resistance of the thin film evaporation becomes the governing factor s $\Delta T_{lf} >> \Delta T_{wick}$. The phase change heat transfer coefficients show 70~100% higher than the microporous sintered wick structures.

In spite of theoretically having high impact, advantages of nanoporous wick on increasing the heat transfer coefficient is partially cancelled by efforts to enhance the wick permeability. In all multiscale CNT wick structures, the large microscale flow passages must be reserved as liquid arteries, which reduce the nanostructure covered area. Moreover, the transported liquid inevitably floods or covers a large number of nano pores, further reducing the actual exposed thin film evaporation area. As a result, the enhancement of the heat transfer coefficient show in experimental data is less than the analysis indicated by the ratio of ε. With consideration of the nanoscale enhanced thin-film evaporation area, Eq. (27) can be updated as.

$$h \approx C_\varepsilon \cdot k_l / l_{lf} \tag{31}$$

where C_ε represents the enhancement ratio of thin film evaporation area of the nanoporous wick over the microporous wick.

5. Dryout at High Heat Flux

Dryout occurs when the input heat flux exceeds the maximum phase change capability, at which capillary force is not able to overcome the increased resistance to transport sufficient liquid to wet the wick structures. For a 2D isotropic wick structure, the dryout appears first in the furthest place for liquid delivery. Experimentally, the dryout was recognized as a rapid increase of evaporator temperature after reaching CHF. However, this phenomenon may not be very pronounced when a thick and conductive substrate/package layer was applied and the serious conjugated heat transfer was involved.

Cai's team based on the multiscale CNT wick structures to investigate dryout at high heat flux [46]. The wick structure is ~210μm thick with varying the size of the heated surface areas. By combining visualization images and temperature measurement data at different input heat flux, cross-validation of occurrence and development of dryout can be performed. On the wick structure with a large heating area of 10mm×10mm, the test point A, as shown in Figure 23, show an active phase change area covering the entire heated wick structures. As heat flux was increased to 125W/cm^2 at point B, the curve of heat flux vs. the evaporator temperature tends to quickly steer right. The fast increase of the evaporator temperature accompanied with appearance of a small dryout spot, as highlighted over the heated wick area. The inflexion point of the characterization curve was identified and defined as the start of dryout. The dryout area expanded out of the heating area when kept increasing heat flux (at point C). Temperature hysteresis was observed when reducing heat flux from its maximum from the point D (~ 195W/cm²), which reflected the different wetting status when advancing and receding the vapor-liquid interfaces in the wick structure.

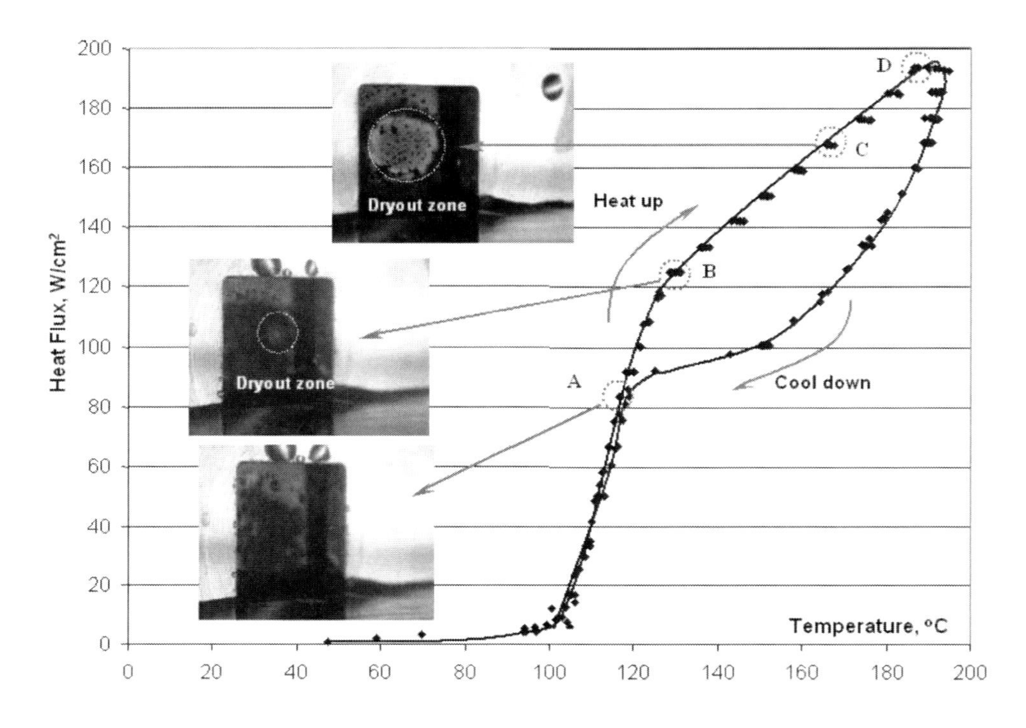

Figure 23. Heat flux vs. the evaporator temperature with a 10mm×10mm heated CNT wick structure: The first inflexion point of the characterization curve was identified as the start of dryout. Hysteresis appeared when the wick structure dries and rewets.

With a smaller heated wick area, the maximum heat flux on the nanoporous wick reached over 700W/cm². Because of a large ratio between substrate thickness and the heated wick characteristic dimension, the conjugated heat transfer effect became more pronounced after dryout occurred. The evaporator temperature variations were smoothened to veil the curve inflexion point. Temperature hysteresis was significantly reduced when drying and rewetting the small heating area. On the experimental images, high heat flux created fluctuating phase change interfaces and strong droplet splashes, which blocked the visual detection of the dried wick area. As a result, both approaches encountered difficulty to identify the dryout spot.

To determine the early stage of dryout, a supplemental image analysis approach was developed. After phase change area exceeded its maximum capability, the increased local temperature amplifies thermal conduction heat transfer in the substrate, creating extended phase change areas or spots out of the heated wick area. By analyzing the phase change image, the start of dryout can be found by identifying boundary of the phase change zone.

As shown in Figure 24, phase change images were divided into many sectional areas, equal to the heated wick area at the center. At high heat flux close to the dryout, liquid splashes and fluctuated phase change interfaces cover the heating area. Under gravitational force, the phase change volume slightly dips down when departing from the wick structure. At the same time, small nucleate bubbles raise along the large microscale grooves under buoyance force, appearing over the heater defined area (shown in Figure 24a). Both the top and bottom sides cannot be used for finding the evidence of the dryout. Limited by visualization angle, only the left side of the wick structure was undisturbed for judging expansion of phase change zone. In Figure 24, the major phase change zone is still restrained within the heated wick area at $370W/cm^2$. However, after increasing to dryout point of $770W/cm^2$, phase change region moved out of the left boundary (shown in Figure 24b). The increased in-plane heat transfer is a direct indicator of the dryout in the wick structure.

Figure 24. Phase change image analysis on the CNT bi-wick structure: a) intensive phase change at $370W/cm^2$, b) dryout occurs at $770W/cm^2$.

At high heat flux, not all pumped liquid was used for phase change. Because of high shear-force, liquid in the form of droplets was entrained with high-speed vapor flow to form splashes. In this study, experiments were designed to quantitatively measure the amount of splash liquid that was transferred to the wick but not utilized for phase change. At the low heat flux of $100W/cm^2$, liquid splash is close to zero with interfacial evaporation. Nearly 5 percent liquid was carried by vapor stream when heat flux was increased to $400W/cm^2$. Over 12 percent liquid in volume ratio left the wick structure without phase change before occurrence of the dryout. To consider the effect of splash liquid in phase change, the function g in Eq.(19) can be rewritten as

$$g(\alpha, \phi) = \frac{4\alpha^3\phi^3(1-\phi)}{C_s(1-\alpha\phi)^2} \tag{32}$$

where, splash constant C_s is larger than 1.0, varying with the input heat flux. This constant tends to reduce the maximum phase change capability, particularly at high heat flux.

Conclusion

Evaporation of interfacial menisci associated with nucleate boiling in the wick structure is one of the most effective ways to achieve high energy density liquid-to-vapor phase change. As heat and mass transports reached their limits, intense interactions between vapor & liquid flows and the wick structures constructed a complex figure for understanding the physics. This chapter reviewed and summarized the latest research progress made in recent ten years. Through diligent efforts of scholars and nationwide enormous investment in this field, advanced wick designs, mechanisms of high heat flux phase change, and relevant theoretical analyses are available for engineers to utilize them in industrial practices.

To enable high heat flux phase change capability, multiscale pores in the wick structures were developed for the purpose of venting vapor and meanwhile providing sufficient liquid supply. With an approximate 50°C superheat, high heat flux phase change over 1kW/cm^2 was demonstrated in lab environment. The entire procedure was divided four phase change modes that end at the mode of evaporation enhanced by interfacial receding. Weber number was used to interpolate transitions of the phase change modes. To correlate wick geometry to the maximum phase change capability, thorough theoretical analysis and experimental validation were conducted. Both vapor and liquid phase geometrical numbers were defined to quantify wick dimension contribution and predict the phase change limit. In contrast, Merit number was experimentally validated in determining fluid property contributions in the maximum heat flux/CHF. By analyzing superheat variations, both wick particle size and the fluid thermal conductivity were correlated to the phase change efficiency. Nanoscale porous structures and highly conductive fluid were able to boost the heat transfer coefficient for phase change. At the end of the chapter, the major wick/heat pipe failure mechanism, dryout, was investigated as a part of theory. An explicit definition of dryout at high heat flux was presented to further implementation of the high heat flux technologies.

References

[1] G. P. Peterson, *An Introduction to Heat Pipes: Modeling, Testing, and Applications*, John Wiley & Son, Inc., New York, 1994.

[2] P. C. Wayner Jr., Y. K. Kao, L. V. LaCroix, The interline heat transfer coefficient of an evaporating wetting film, *Int. J. Heat Mass Transfer* 199 (1998) 487–492.

[3] H. B. Ma, G. P. Peterson, Temperature Variation and Heat Transfer in Triangular Grooves with an Evaporating Film, *J. Thermophysics and Heat Transfer* 11 (1997) 90-97.

[4] C. C. Yeh, C. N. Chen, Y. M. Chen, Heat transfer analysis of a loop heat pipe with biporous wicks, *International Journal Heat Mass Transfer* 52 (2009) 4426–4434.

[5] X. L. Cao, P. Cheng, T. S. Zhao, Experimental Study of Evaporative Heat Transfer in Sintered Copper Bidispersed Wick Structure, *Journal of Thermophysics and Heat Transfer* 16 (2002) 547-552.

[6] F. P. Incropera, D. P. Dewitt, T. L. Bergman and A. S. *Lavine, Fundamentals of Heat and Mass Transfer*, Sixth Ed., John Wiley & Sons, Hoboken, 2006.

[7] N. Basu, G. R. Warrier, V. K. Dhir, Onset of nucleate boiling and active nucleation site density during subcooled flow boiling, *Journal of Heat Transfer* 124 (2002) 771-728.

[8] A. Faghri, *Heat Pipe Science and Technology*, Taylor & Francis, New York, 1995.

[9] P. A. Vityaz, S. K. Konev, V. B. Medvedev, V. K. Sheleg, Heat Pipes with Bidispersed Capillary Structures, *Proceedings of 5th International Heat Pipe Conference, Tsukuba Science City*, Japan, 1984, pp. 127-135.

[10] J. H. Rosenfeld, M. T. North, Porous Media Heat Exchangers for Cooling of High-power, *Optical Engineering* 34 (1995) 335-341.

[11] J. Wang, I. Catton, Vaporization Heat Transfer in Biporous Wicks of Heat Pipe Evaporators, *Proceedings of 13th International Heat Pipe Conference,* Shanghai, China, 2004, pp. 76-86.

[12] T. Semenic, I. Catton, Experimental Study of Biporous Wicks for High Heat Flux Applications, *International Journal of Heat Mass Transfer* 52 (2009) 5113-5121.

[13] S. W. Reilly, I. Catton, Improving Biporous Heat Transfer by Addition of Monoporous Interface Layer, *Proceedings of the ASME 2009 Heat Transfer Summer Conference,* San Francisco, 2009, HT2009-88257.

[14] Yeh, C. C. Chen, C. N., Y. M. Chen, Heat Transfer Analysis of a Loop Heat Pipe with Biporous Wicks, *International Journal of Heat and Mass Transfer* 52 (2009) 4426–4434.

[15] D. Deng, Y. Tang, G. Huang, L. Lu, D. Yuan, Characterization of Capillary Performance of Composite Wicks for Two-phase Heat Transfer Devices, *International Journal of Heat and Mass Transfer* 56 (2013) 283–293.

[16] Y. Zhao, C. L. Chen, Development of A high Performance Vapor Chamber for High Heat Flux Applications, *Proceedings of Micro/Nanoscale Heat Transfer International Conference,* Pingtung, 2008, MNHT2008-52363.

[17] J. A. Weibel, S. V. Garimella, M. T. North, Characterization of Evaporation and Boiling from Sintered Powder Wicks Fed by Capillary Action, *International Journal of Heat and Mass Transfer* 53 (2010) 4204-4215.

[18] S. Q. Cai, A. Bhunia, Characterization of Phase Change Heat and Mass Transfers in Monoporous Silicon Wick Structures, *Journal of Heat Transfer* 136 (2014) 072001(1-8).

[19] Q. Cai, Y. C. Chen, Development of Platinum Resistance Thermometer on Silicon Substrate for Phase Change Studies, *Journal of Micromechanics and Microengineering* 22 (2012), 085012 (1-8).

[20] Y. Nam, S. Sharratt, C. Byon, S. J. Kim, Y. S. Ju, Fabrication and Characterization of the Capillary Performance of Superhydrophilic Cu Micropost Arrays, *Journal of Microelectromechanical Systems* 19 (2010) 375–383.

[21] Y. Nam, S. Sharratt, G. Cha, Y. S. Ju, Characterization and Modeling of the Heat Transfer Performance of Nanostructured Cu Micropost Wicks, *Journal of Heat Transfer* 133 (2011) 101502.

[22] Q. Cai, C. L. Chen, Design and Test of Carbon Nanotube Biwick Structure for High-Heat-Flux Phase Change Heat Transfer, *Journal of Heat Transfer* 132 (2010) 052403 (1-8).

[23] Q. Cai, A. Bhunia, High Heat Flux Phase Change on Porous Carbon Nanotube Structures, *International Journal of Heat and Mass Transfer* 55 (2012) 5327-5335.

[24] P. Kim, L. Shi, A. Majumdar, P. L. McEuen, Thermal Transport Measurement of Individual Multiwalled Nanotubes, *Physics Review Letters* 8 (2001) 215502(1-4).

[25] C. Yu, L. Shi, Z. Yao, D. Li, A. Majumdar, Thermal Conductance and Thermopower of an Individual Signle-Walled Carbon Nanotube, *Nano Letters* 5(9) (2005) 1842-1846.

[26] S. Berber, Y. K. Kwon, D. Tomanek, Unusually High Thermal Conductivity of Carbon Nanotubes, *Physics Review Letters* 84(20), (2000) 4613-4617.

[27] W. Yi, L. Lu, D. L. Zhang, Z. W. Pan, S. S. Xie, Linear Specific Heat of Carbon Nanotubes, *Physics Review B* 59, (1999) 9015-9018.

[28] S. L. W. Xie, Z. Pan, B. Chang, L. Sun, Mechanical and Physical Properties on Carbon Nanotube, *Journal of Physics and Chemistry of Solid* 61 (2000) 1153-1158.

[29] H. Huang, C. Liu, Y. Wu, S. Fan, Aligned Carbon Nanotube Composite Films for Thermal Management, *Advanced Materials* **17** (2005) 1652-1656.

[30] S. Ganguli, A. K. Roy, L. Dai, L. Qu, *Aligned Carbon nanotube to Enhance Through Thickness Thermal Conductivity in Adhesive Joints, AFRL-ML-Wp-Tp-2007-446*, Dayton, OH, 2007.

[31] D. Lucio, D. Laurent, G. Roger, S. Yasushi, Y. Noriko, KOH Activated Carbon Multiwall Nanotubes, *Carbon: Science and Technology*, 2(3), (2009) 120-124.

[32] J. A. Weibel, S. V. Garimella, J. Y. Murthy, D. H. Altman, Optimization of Mass Transport in Integrated Nanostructured Wicking Surfaces for the Reduction of Evaporative Thermal Resistance, *Proceeding of Thermal and Thermomechanical Phenomena in Electronic Systems,* Las Vegas, NV, 2010, 978-1-4244-5343-6.

[33] J. A. Weibel, S. V. Garimella, J. Y. Murthy, D. H. Altman, Design of Integrated Nanostructured Wicks for High-Performance Vapor Chambers, *IEEE Transactions on Components, Packaging and Manufacturing Technology* 1(6) (2011) 859 – 867.

[34] S. Q. Cai, A. Bhunia, A. Transitions of Heat Transfer Modes on Microfabricated Copper Wick Structures, *Journal of Thermophysics and Heat Transfer* 29(4) (2015) 820-825.

[35] N. Basu, G. R. Warrier, V. K. Dhir, Onset of Nucleate Boiling and Active Nucleation Site Density During Subcooled Flow Boiling, *Journal of Heat Transfer* 124 (2002) 717-728.

[36] S. Q. Cai, A. Bhunia, Geometrical Effects of Wick Structures on Maximum Phase Change Heat Flux Capability, *International Journal of Heat and Mass Transfer* 79 (2014) 981-988.

[37] P. C. Carman, Fluid Flow through Granular Beds, *Transactions of the Institution of Chemical Engineers* 15 (1937) 150-166.

[38] K. Yazdchi, S. Srivastava, S. Luding, Microstructural Effects on the Permeability of Periodic Fibrous Porous Media, *International Journal of Multiphase Flow* 37 (2011) 956-966.

[39] B. Lautrup, *Physics of Continuous Matter,* Talor & Francis, Boca Raton, FL, 2011.

[40] L. Chen, G. P. Peterson, Y. Wang, *Evaporation/Boiling in Thin Capillary Wicks I — Wick Thickness Effects* 128 (2006) 1312-1310.

[41] L. Chen, G. P. Peterson, *Evaporation/Boiling in Thin Capillary Wicks II—Effects of Volumetric Porosity and Mesh Size* 128 (2006) 1320-1328.

[42] D. Reay, P. Kew, R. McGlen, *Heat Pipes: Theory, Design and Applications,* Sixth Edi., Elserver, UK, 2014, pp.66-67.

[43] S. Q. Cai, A. Bhunia, Liquid Property Impacts of Liquid-To-Vapor Phase Change on Porous Wick Structure, *Journal of Heat Transfer* 137(12) (2015) 072001.

[44] H. K. Dhavaleswarapu, P. Chamarthy, S. V. Garimella, J. Y. Murthy, Experimental investigation of steady buoyant-thermocapillary convection near an evaporating meniscus, *Physics of Fluids* 19 (2007) 082103.

[45] H. Wang, S. V. Garimella, J. Y. Murthy, Characteristics of an Evaporating Thin Film in a Microchannel, *International Journal of Heat and Mass Transfer* 50 (2007) 3933-3942.

[46] Q. Cai, Y. C. Chen, Investigations of Bi-porous Wick Structure Dryout, *Journal of Heat Transfer* 134(2) (2012) 021503 (1-8).

In: Heat Pipes: Design, Applications and Technology
Editor: Yuwen Zhang
ISBN: 978-1-53613-908-2

Chapter 3

THERMAL MANAGEMENT OF ELECTRONIC DEVICES USING HEAT PIPES

Boris I. Basok[1], Yurii E. Nikolaenko[2,*] and Roman S. Melnyk[2]
[1]Department of Thermophysical Basics of Energy-Saving Technologies, Institute of Technical Thermophysics, National Academy of Sciences of Ukraine, Kyiv, Ukraine
[2]Heat-and-Power Engineering Department, National Technical University of Ukraine "Igor Sikorsky Kyiv Polytechnic Institute", Kyiv, Ukraine

ABSTRACT

The chapter is devoted to the solution of an important scientific and applied problem of creating highly efficient thermal management techniques for advanced electronic devices based on scientifically based methods of design and technological realization of two-phase technologies. The techniques were developed for the major types of electronic devices built on modern components, taking into account their design features. This allows providing a more reliable performance of complex electronic systems due to a decrease in the temperature of the components, as well as a decrease in energy consumption of water-based thermal management devices based on heat pipes (HP) due to reducing the hydraulic resistance of the water tract.

Theoretical basis was given to improvement of existing and development new design solutions for HPs, heat transfer devices and thermal management devices based on two-phase technologies for the major types of electronic equipment: complex modular computing systems in basic supporting structures with water cooling with a thermal power of up to 30 kW per one instrument cabinet; computing machinery and computer control devices; infrared electronics; microlaser devices for information systems and controls; transmitting units; LED lighting units.

New experimental data was obtained on the thermal characteristics of the proposed design solutions for HP-based thermal management elements, devices and techniques.

* Corresponding Author Email: yunikola@ukr.net.

The data obtained confirms high efficiency of these design solutions. The newly-developed technologies make it possible to industrially produce HPs of the proposed designs. The obtained scientific and practical results allow expanding the functionality of advanced electronic devices built on modern components with an increased specific and total heat release.

Keywords: electronic devices, electronic components, base supporting structure, thermal management, two-phase technologies, heat pipe, thermosyphon, thermal performance

1. Introduction

The global trend in the development of electronics is expanding equipment functionality, enhancing technical parameters, improving efficiency and data processing speed. At the same time, enhancing technical parameters of electronic systems leads to significant growth of electric power consumption P_e. Only small amount of the consumed energy (5 to 25%) is converted into energy of useful signals E_u, while the rest of the power (75 to 95%) is converted into the heat power Q_t (see Figure 1), which leads to an increase in the temperature of electronic components. This process (temperature increase) negatively affects performance and reliability of electronic elements. All these factors considerably reduce the performance of the devices based on such components. For example, when the temperature of semiconductor crystals increases, the clock frequency and the computing performance of microprocessors decrease [1]. Temperature variation in a range from 25 to 75°C in a semiconductor crystal of a laser diode based on InGaAs/AlGaAs/GaAs heterostructures leads to a 27% decrease in the average power of pulsed radiation [2]. An increase in the temperature of semiconductor luminophore crystals of white and ultraviolet emitting diodes (LEDs) causes accelerated aging of the phosphor, a change in color parameters and a decrease in the light flux [3, 4]. As for the RGB LEDs, their maximum emitting wave length, spectral width and output power strongly depend on the temperature level. The temperature of three-colored light sources affects their parameters thus: when the transition temperature increases from 20 to 80°C, the chromaticity point shifts, both color rendering index and light output decrease, and the color temperature increases [5].

To ensure the required temperature range of electronic components, different thermal management devices (TMD) are used. They take the heat generated in the electronic components out to the heat-absorbing environment – outer space, water or air, depending on the operating conditions and function of a particular electronic device or system (Figure 1).

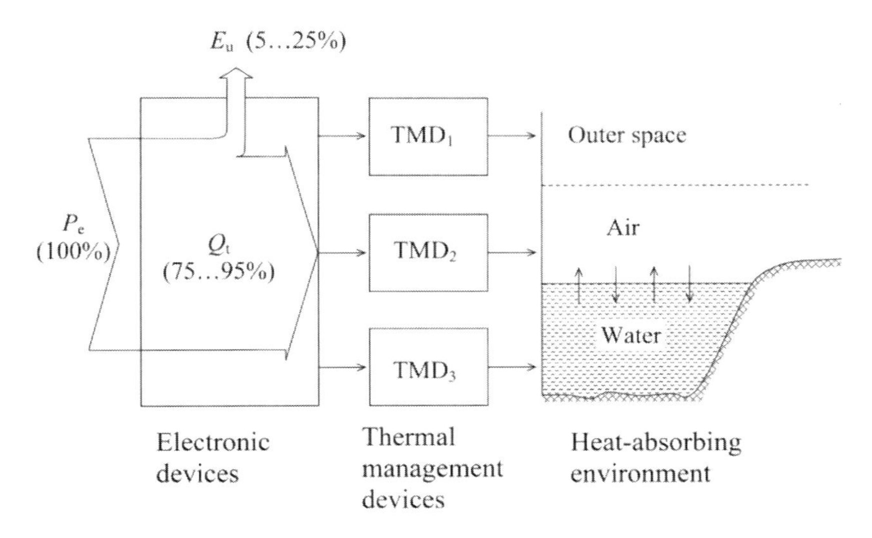

Figure 1. The function of thermal management devices for electronics.

The most effective systems for thermal management of electronic devices and systems developed in the late 20th century in the former USSR (air–water, water, conductive water, conductive evaporation and liquid evaporation systems) allowed up to 10–15 kW of heat output from one instrument cabinet by pumping cooling water through zigzag channels of heat exchangers built into the instrument cabinet, and by building in compressors, air conditioners and powerful refrigeration units. Increased hydraulic resistance of water channels led to significant power consumption for pumping the coolant.

Over the past 20–25 years, the components of all the levels of packaging hierarchy of electronic devices has been significantly upgraded (see Figure 2): instead of low-power integrated circuits, powerful semiconductor electronic components appeared, such as large-scale and very-large-scale integrated circuits (LSI and VLSI) multichannel microprocessors (MP), light-emitting diodes (LED) and laser diodes (LDs), as well as strips, matrices and modules on their basis, matrix crystals for photodetecting devices, etc.

The power of the main electronic component (microprocessor) is already on the 130 W mark, and in the near future (before 2020) we may expect its growth up to 360 W [6].

The global trend is that the number of transistors on a crystal steadily increases (22-core Xeon E5 Intel processor based on Broadwell-EP cores has a 456 mm^2 crystal area and contains 7.2 billion transistors [7]), the size of electronic components is reduced and their number in a modern device increases (Sunway TaihuLight supercomputer contains 40960 Sunway SW26010 processors, each containing 256 cores, National Supercomputing Center in Wuxi, China, 2016, [8]), which causes a further increase in the specific heat release of electronic components. By 2020, the maximum specific heat flux from microprocessor-based high-performance chips is expected to be about 190 W/cm^2 [6]. The total heat release of complex electronic devices per instrument cabinet may be up to 20–30 kW.

Rapid increase in specific and total heat generation values of electronic components raises a problem of finding a new conceptual approach to organizing effective heat removal for the major electronic devices based on modern hardware components. More effective TMTs need to be developed, which would allow decreasing total thermal resistance of the pair electronic component – heat sink, and, as a result, significantly enhance the heat output (20–30 kW instead of 10–15 kW).

Thermal management using two-phase technologies is a promising technique at all hierarchical design levels. Theoretical foundations of thermal and hydrodynamic processes in closed evaporation-condensation systems of heat removal have now been thoroughly studied. At the same time, insufficient attention has been paid to the problem of the scientific substantiation of the ways of constructive and technological realization of two-phase technologies in electronic devices with the modern components. The known separate studies in this direction were not systemic. This hinders the further development of promising electronic systems.

The said above shows the relevance of the development of high-efficiency TMTs for electronic devices with an increased heat release using science-based ways of constructive and technological implementation of two-phase technologies in electronic devices on modern components, taking into account the specificity of their hardware design.

I level Electronic components	II level Cassettes, modules, units	III level Cabinets, racks	IV level Devices, complexes	Location of the objects / Heat absorbing environment
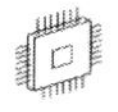	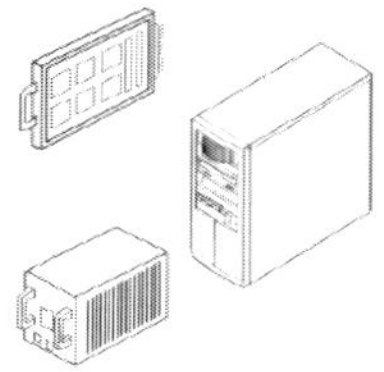	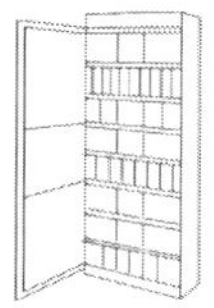	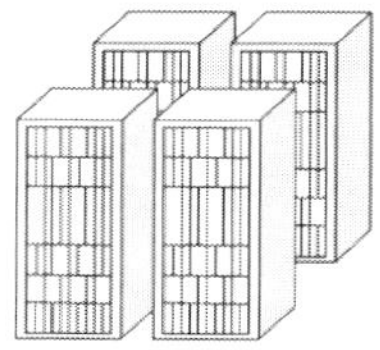	• Spacecraft / Outer space • Aircraft, helicopters, missiles / Air • Ships, buoys, seabed / Water, Air • Stationary ground facilities, railway platforms, cars / Air
MP, VLSI, LSI, LD, LED, photosensitive elements, transformers, etc.	Functional modules, units, PCs, servers, etc.	Information processing devices, generators, power supplies, etc.	Supercomputers, hydroacoustic complexes, radar stations, communication and telecommunication systems, etc.	
Up to 130 W / cm^2 (26 times ↑); Up to 80-130 W → 200 W (16–26 times ↑)	Up to 150-500 W (2–5 times ↑)	Up to 20-30 kW (2 times ↑)	K Computer (supercomputer, 2011) 68544 microprocessors 548352 cores 12.66 MW 672 cabinets	

Figure 2. Packaging hierarchy of electronic devices and their heat generation trends.

2. Heat Pipes and Thermosyphons as Thermal Management Elements. Field for Improvement

There are many different types of heat pipes (HP). Below we describe the most suitable types of HPs from the point of view of thermal management of electronic devices.

Thermosyphon (TS) is the simplest, no-wick type of HP. It is a sealed vacuum case made of a heat-conducting material and partially filled with working fluid. The principle of the TS operation is the heat transfer Q along its length by means of a closed evaporation-condensation cycle. The advantage of the TS as a heat-transfer element, in comparison with other types of HPs, is the simplicity of the design, while its major drawback is that the TS can not operate in the absence of a gravitational field and that its operation depends on the orientation of the TS in space. To return the condensate to the evaporation zone, it is necessary for the condensation zone to be positioned above the evaporation zone. Moreover, the work of the TS is often accompanied by instability of heat exchange processes and temperature pulsations in the evaporation zone.

For thermal management of electronic devices, more suitable are the TSs with dimensions close to the sizes of the basic supporting structures (BSS) for which they are designed (boards, frames, cassettes, modules and block frames, chassis, instrument cabinet sections, etc.). In the case of block construction of complex electronic systems, it is of particular interest to use thermosyphons in the form of functionally complete blocks with the heat-generating elements placed inside the thermosyphon. It is thus necessary to solve the problem of increasing the reliability and service life of thermosyphons under the harsh operating conditions inherent in such electronic systems.

Half way through between the wick heat pipe and the smooth-walled thermosyphon is the gravitational heat pipe (GHP). Heat transfer processes in GHPs with a layer of wick (capillary structure) in the evaporation zone run more stably and vigorously than in the smooth-walled HPs. However, adding a layer of a capillary structure (CS) complicates the manufacturing technology. Therefore, an important task is to find the simplest CS for gravitational HPs, from the point of view of design and technology.

The presence of a capillary structure (wick) in all zones of the HP allows it to operate outside of gravitational field, e.g., in the outer space, regardless of the orientation of the HP in space. In the gravitational field, the wick HPs can work reliably even when the evaporation zone is positioned higher than the condensation zone by an amount determined by the capillary properties of the wick. In this case, the available capillary pressure must exceed the sum of the hydrostatic pressure and the pressure loss when the coolant moves along the wick and the steam moves through the vapor space.

The wick-wire HPs of the simplest forms (cylindrical and flat) [9–12] became the most common types of HPs. However, due to some design features, not every electronic device would allow using the HP of the simplest geometric form as a thermal management device.

In most cases, especially considering the increased heat release caused by functional upgrade of devices, it is necessary to create HPs of an original shape for the equipment in complex electronic systems. Designing HPs of an original shape influences hydrodynamic and heat transfer processes in them and requires experimental studies of their thermal characteristics. Therefore, an important task here is to develop and experimentally test new structural forms of HPs, including the miniature HPs that are constituent elements of functional modules, blocks or devices of electronic systems, parts of their BSS, which would ensure efficient heat transfer to the heat-absorbing medium.

Gas regulated heat pipes (GRHP) are characterized by high accuracy of automatic maintenance of temperature in the evaporation zone with changing perturbing factors (heat input, cooling medium temperature, cooling medium velocity, etc.). Autonomy and high reliability of GRHPs, absence of special sensors, movable parts, control devices, no need in additional energy consumption for temperature control make these HPs quite fitting for use in thermal management of electronic devices where it is necessary to maintain the required level of temperature of the heat-generating element at a wide range of parameters of internal and external impacts.

The most commonly used and studied GRHPs are those with insoluble non-condensible gas (NCG). However, they have a limited operating temperature range, a limited heat transfer capacity in the forward direction (due to partial blocking of the condensation surface by the remnants of the non-condensable gas in the vapor space caused by the smearing of the vapor-gas interface), and a limited heat stabilization accuracy.

It is possible to expand the operating temperature range and increase the accuracy of thermal stabilization by using GRHPs with an NCG that is readily soluble in the working fluid (e.g., ammonia–water). Thermal stabilization in a GRHP with a soluble NCG is achieved due to a more rapid change in the boundary of the transition region between the high-boiling and low-boiling components in the condensation zone. However, it is impossible to fully realize all the advantages of GRHPs due to the imperfection of the known structures, which allow for a moving hot vapor core and a working fluid in the CS of a gas tank to interact.

In this regard, the most important task of improving existing GRHP designs with a soluble NCG is to increase their heat transfer capacity, expand the operating temperature range and thermal regulation quality by eliminating the temperature interaction of the high-boiling component vapor core with the low-boiling one, located in the CS of a gas tank.

Heat pipes with diode properties (HPDP) are characterized by a low thermal resistance in the forward direction and a high thermal resistance in the opposite direction (thermodiode effect). HPDPs can be used as thermal management elements, provided the undesirable or unforeseeable negative temperature effects of the environment or other heat sources are excluded.

According to the principle of achieving the termodiode effect, HPDPs can be divided into four main types, distinguished by how the flow of vapor or heat transfer fluid is controlled in the HP (by blocking the evaporation zone with liquid coolant or non-condensable gas, interruption of steam flow or coolant condensate) in order to change the heat exchange intensity in of evaporation and condensation zones of an HP. The type of HPDPs with the condensate flow interruption is the most widespread and has the greatest variety of designs. It is considered to be the most recommended for thermal management of electronic devices.

At the same time, the known designs of the HPDPs of this type have a certain disadvantage, consisting of a low heat transfer capacity in the forward direction, which is caused by a significant thermal resistance of the condensate film in the annular gap between the case and the CS layer, or of the insert in the condensation zone. Eliminating these lacks is what should be the first and foremost task while improving such HPDP designs.

What distinguishes contour heat pipes (CHP) from the other types of HPs is that the transport zone in the CHP contains separate channels for vapor and liquid. This eliminates pressure loss from friction in counter flows of vapor and liquid and allows large heat flows (up to 1000 W) over a distance of several meters. Low hydraulic resistance to the movement of the coolant in the liquid and vapor phases in combination with the high capillary potential of the finely porous CS of the evaporator makes the CHP operable in any orientation in the gravitational field. Such advantages allow using CHPs for thermal management of electronic devices on aircrafts.

The research areas of scientific and practical interest, when improving CHP designs, are searching for new combinations of structural materials and coolants that provide a reduction in the mass of CHP, developing new types of CS, improving the reliability and manufacturability of their fabrication.

A very important task is to develop new technological ways of implementing new HP designs.

In order to solve the set tasks, new constructive and technological solutions have been developed (described below) improving the main types of HPs.

3. Improving Design of Heat Pipes and Thermosyphons for Thermal Management

3.1. Increasing Heat Removal Efficiency of Thermosyphons under Mechanical Stress

HPs without a wick (thermosyphons) can be used as heat transfer elements in complex electronic systems for thermal management of powerful devices for power supply, generator cells and heat-loaded microelectronic devices for digital signal processing requiring efficient cooling and assembled in water-cooled instrument cabinets. Functional units of such devices are separate hermetic blocks partially filled with a liquid dielectric with heat-generating elements completely immersed in it. In fact, such units represent thermosyphons with internal heat supply. The major sources of heat are the active elements of the unit (transformers, thyristors, inductors, power integrated circuits, microprocessors, etc.). Thermal management of these elements is provided by evaporation of the liquid dielectric from their surface. The height of the unit is determined by the distance between adjacent intersectional partitions of the cabinet. When installing the unit into the section of the instrument cabinet, the top cover, which functions as both the condensation surface and the heat exchanger, is brought into thermal contact with the partition wall (shelf) of the cabinet through which the cooling water is pumped.

Particular design features of a typical electronic unit with a closed evaporation–condensation cooling system determine the specifics of the heat transfer and hydrodynamic processes running in the unit, which must be taken into account when developing a particular thermal management device. The desire to make maximum use of the internal space of the unit when arranging electronic components, and the height of the electronic unit limited by the intersection space of the cabinet lead to the fact that the distance between the level of the liquid dielectric and the condensation surface in the block is chosen to be minimal (of the order of 15 to 25 mm). Thus, if the device operates under vibrations and shocks, the level of the liquid dielectric in the unit fluctuates and the fluid partially overflows the condensation surface of the heat exchanger, preventing vapor condensation. Between the liquid dielectric and this part of the heat exchanger surface, convective heat exchange is carried out, which is less efficient than the steam condensation heat exchange. This leads to an increase in the temperature of vapor and of the electronic components, and thus to a decrease in the reliability of the unit operation.

In order to improve the reliability and cooling efficiency of electronic components under mechanical stress (vibrations and shocks), a design solution has been proposed (see Figure 3) [13]. The solution consists in installing the partition wall of porous material with one or more apertures between the condensation surface and the liquid dielectric of the electronic unit at a distance Δ from the liquid level. This partition prevents the liquid dielectric from getting to the heat exchanger surface while splashing. The distance Δ between the barrier and the liquid dielectric is chosen experimentally taking into account that at the maximum temperature the volume of the liquid dielectric increases and the vapor bubbles are formed when the liquid boils.

During the operation of the unit, the heat released by the electronic components is transferred via the closed evaporation–condensation cycle to the upper lid heat exchanger, from whence it is removed by the cooling water panel. The flow of water in the cooling panel is shown by arrows in Figure 3. Under the influence of gravity, drops of condensate flow into the evaporation zone. Part of the condensate enters the evaporation zone directly through the aperture in the porous partition, while the other part first gets to the surface of the partition, and then through its pores into the evaporation zone. Viscosity of the liquid dielectric is over an order of magnitude higher than that of its vapor, thus splashes on the surface of the liquid do not pass through the pores of the partition and do not flood the condensation surface. The largest amplitude of the splashes is observed near the side walls of the unit (near the edges of the partition). Therefore, the apertures should be located in the central part of the partition.

Experimental studies of the unit without a porous partition under mechanical stress has shown that the liquid dielectric can cover up to 50% of the condensation surface, i.e., the condensation area can be reduced by a factor of 2, and thus the temperature of vapor and electronic components would increase.

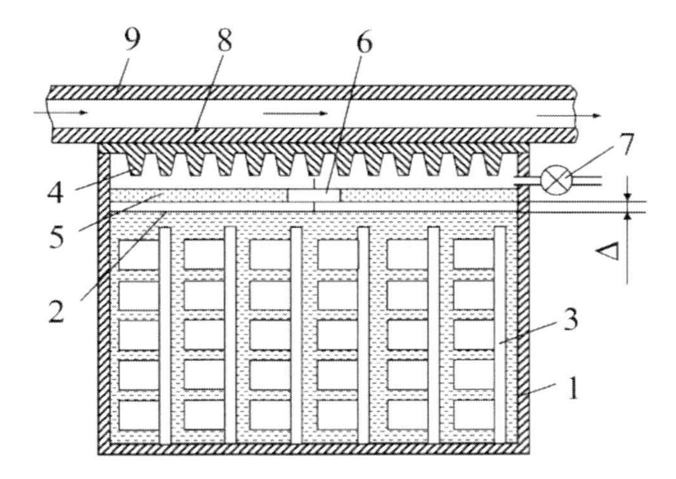

Figure 3. Schematic diagram of a computing unit with an improved thermosyphon: 1 – unit case; 2 – liquid dielectric level; 3 – switching circuit with power electronic components; 4 – condensation surface of the heat exchanger lid; 5 – porous partition; 6 – aperture; 7 – valve; 8 – the contact surface of the heat exchanger lid with the cooling panel; 9 – cooling panel.

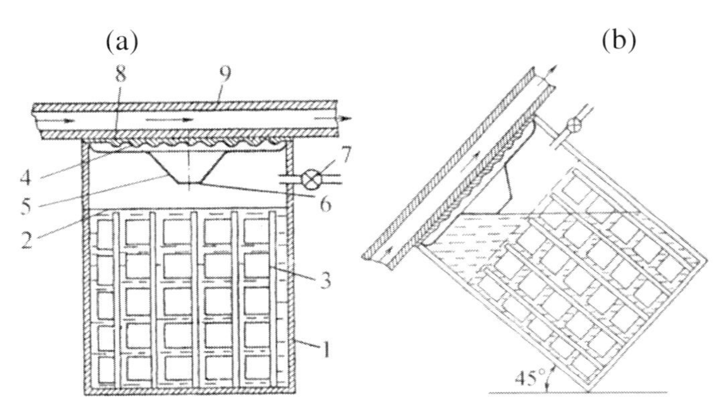

Figure 4. Schematic diagram of a TS with a cone-shaped nozzle: (a) Vertical position; (b) Tilted 45 degrees.
1 – block case 2 – level of liquid dielectric; 3 – board with powerful electronic components; 4 – condensation surface; 5 – cone-shaped nozzle; 6 – hole; 7 – valve; 8 – contact area between the unit and the heat exchanger; 9 – water heat exchanger.

The calculations showed that the use of the proposed design with a porous partition, in comparison with the typical design of the unit (R-113 coolant, power of electronic components of 300 W, condensation area of 0.147 m^2, heat transfer coefficient of 120 $W/(m^{2\circ}C)$ at steam condensation, temperature of the heat exchanger lid of 35°C), allows increasing thermal management efficiency and reducing the temperature of the electronic components by a factor of 1.3 (from 80 to 61°C). This would improve the reliability of the unit operation under mechanical stress (vibrations and shocks).

In the electronic shipborne unit, cooled by a thermosyphon operating under the impact of rolling motion and long slopes, performance reliability and cooling efficiency were increased by adding a cone-shaped nozzle into the construction (see Figure 4) [14]. Arrows in the Figure show the flow of water in the cooling panel.

The cone-shaped nozzle prevents the part of the condensation surface of the upper heat exchanger lid from flooding by the liquid dielectric and reduces the temperature of the electronic components from 80 to 67°C (compared to the unit without a nozzle).

3.2. Extending Service Life of Thermosyphons with Horizontal Condensation Surface

Heat transfer and hydrodynamics in HPs with a horizontal condensation surface are charachterized by the presence of a small amount of NCG, which may appear in the HP during operation as a result of electrochemical processes or during the manufacturing of the HP. This NCG presence blocks the condensation zone with a large thermal resistance interlayer, which leads to an increase in the temperature of the HP heating zone and of the cooled electronic components.

In order to increase the efficiency, extend service life of such thermal management devices and to expand the possible combinations of the materials used for the case, wick and working fluid, new design solutions for HPs and TSs have been developed. Systems of channels forming a reservoir for collecting the NCG are made inside the upper plate. The ends of the channels are connected to the steam space of the HP (see Figure 5) [15]. The gas, which is released as a result of slow chemical reactions during the HP or TS operation, is expelled by the heat carrier vapor into the transverse grooves of the plate and further into the cavity of the channels, thereby freeing the condensation surface from the NCG and ensuring the effective vapor condensation.

Calculations show that during the operation of an HP from, for example, aluminum alloy AMg with acetone as a working fluid, 10.9 ml of hydrogen can be released during 10 000 hours. Such amount of hydrogen would require 555 mm of cylindrical channels 5 mm in diameter, which could amount to 3–4 channels, depending on the geometric dimensions of the plate. These channels with the NCG inside the upper horizontal plate has virtually no effect on the process of heat transfer through the plate by thermal conductivity, since in percentage terms the channels do not significantly decrease the heat transfer area of the cooled plate, while the heat flow envelops the channels along adjacent sections of the solid material of the plate.

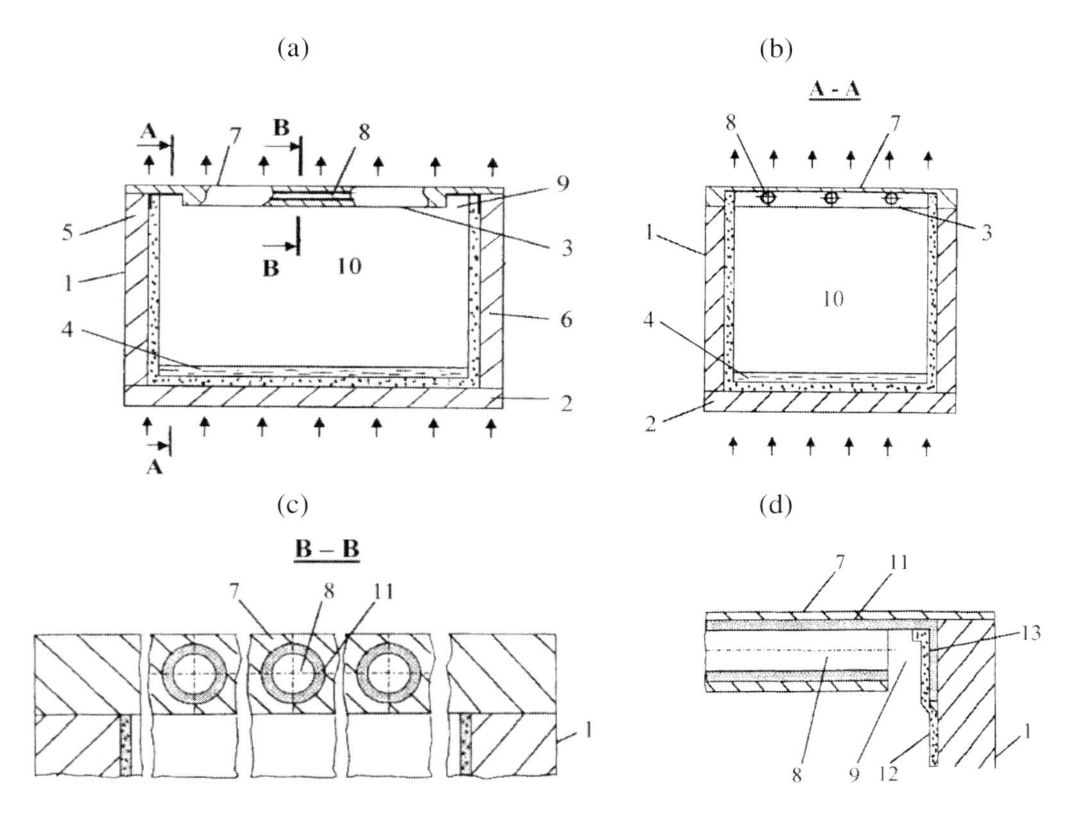

Figure 5. HP with extended service life: (a) Longitudinal section; (b) Cross section along line A-A; (c), (d) Transverse and longitudinal section of the NCG channel, respectively (enlarged scale).
1 – hermetic case; 2 – evaporation zone; 3 – condensation zone; 4 – liquid coolant; 5, 6 – side walls; 7 – horizontal plate; 8 – NCG channel; 9 – transverse groove; 10 – vapor space; 11, 12 – CS (wick) of the channel and case, respectively; 13 – connection of the CSs of channels and the case.

To eliminate possible flooding of small diameter channels (less than 4.5 mm) caused by surface tension forces that keep the condensate inside the channels, the inner surface of the channels is covered with a thin (0.2 – 0.4 mm) layer of wick (see Figure 5 (c), (d)) or lined with wick-like capillary grooves connected to the wick in the case.

Capillary forces cause the condensate formed in the channels to be pumped by the wick on the channel walls into the evaporation zone of the HP, freeing up the volume of the channels for the NCG, which makes it possible to ensure the calculated thermal characteristics of the thermal management device during long-term operation.

3.3. Gravitational Heat Pipe with Threaded Capillary Structure

The known designs of gravitational heat pipes (GHP) with a metal-fiber capillary structure in the evaporation zone [16] are difficult to fabricate. GHPs with threaded capillary structure are easier to produce [17] (see Figure 6).

For experimental studies, a copper GHP was made with the following parameters: length – 830 mm, outer shell diameter – 12 mm, inner diameter – 10 mm, working fluid – R141b. The thermal characteristics of the GHP were compared with the thermal characteristics of a thermosyphon of the same size with a smooth wall surface in the evaporation zone (see Figure 7) [18, 19].

It can be seen from Figure 7, that the thermal resistance of a GHP with threaded capillary structure is 4.4 to 2.0 times (with natural convection) and 1.2 to 1.3 times (with forced convection) less than that of a thermosyphon of the same size with a smooth wall surface in the evaporation zone.

Figure 6. A photograph of a threaded capillary structure with a thread pitch of 0.5 mm in the evaporation zone of a gravitational heat pipe with an external diameter of 12 mm (enlarged scale).

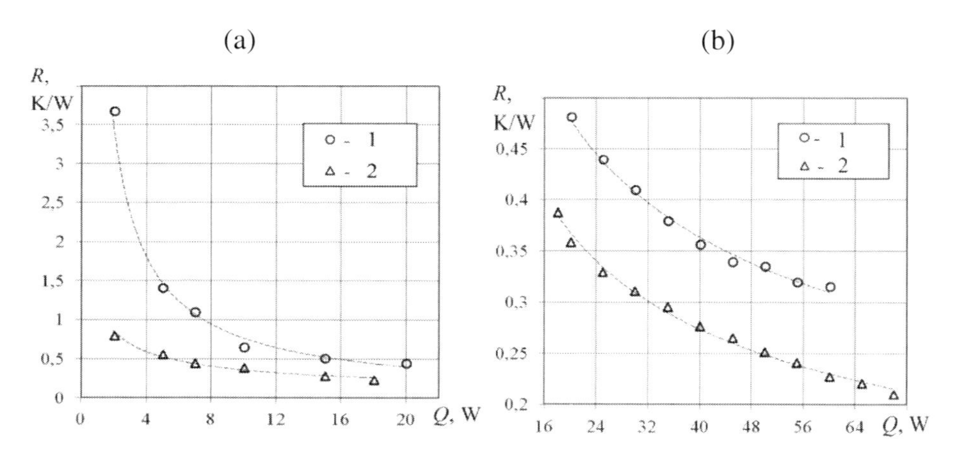

Figure 7. Dependence of thermal resistance R of the TS (1) and GHP (2) from the heat transfer Q: (a) In the cooling zone of the natural condensation zone; (b) When the condensation zone is cooled by forced convection of air.

Thus, due to lower thermal resistance values, gravitational heat pipes are efficient heat transfer elements for thermal management of electronic devices.

3.4. Improved Designs of Heat Pipes with a Wick for the Modernized Electronic Modules

While working on the improvement of heat pipes with a wick, the main task to consider is the development of specific design and technological solutions that would allow creating HPs of the original form. This is important when developing new (or performing functional upgrade of existing) complex electronic systems in order to improve cooling efficiency of the modern components with increased specific heat. Since upgrade of existing electronic systems, as a rule, does not allow increasing the overall dimensions of electronic modules, the possibility of installing additional traditional thermal management devices (radiators and fans) is out of question.

Functional modernization of complex electronic systems and complexes (radar stations, control systems for rocket and space equipment, etc.) goes with a significant increase in the heat release of electronic modules. The problem of increasing heat removal efficiency in this case can be solved by organizing additional parallel channels for removing heat from the most critical electronic components to the case of the module using two-phase technologies based on HPs with branched transport and condensation zones, and bus bars made of high-conductivity material.

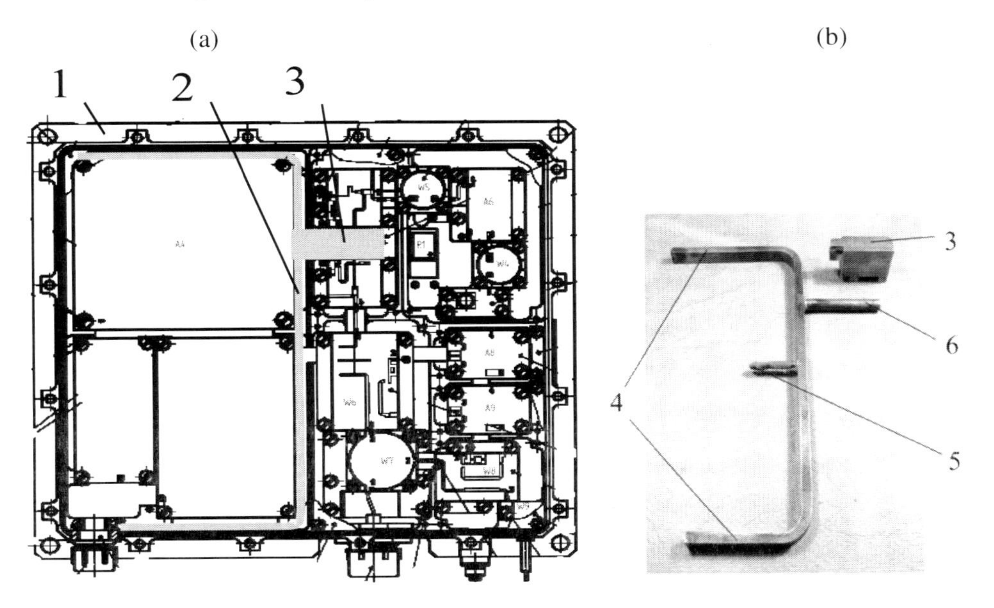

Figure 8. Thermal management of an electronic module based on a complex form of a heat pipe: (a) Schematic diagram of an HP installed in the module; (b) Photograph of the HP and the bus.
1 – module case; 2 – HP of a complex shape; 3 – heat conducting bus bar; 4 – condensation zones of the HP; 5 – exhaust tube; 6 – evaporation zone.

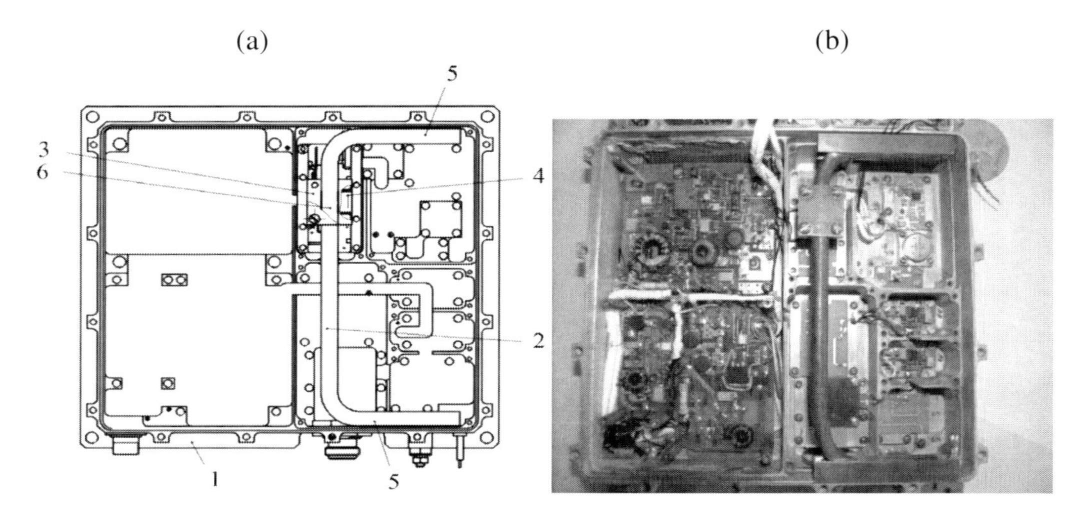

Figure 9. Upgraded electronic module with an advanced heat sink system based on a U-shaped HP with a wick with improved manufacturability, assembly and installation: (a) Drawing; (b) Photograph. 1 – module casing; 2 – U-shaped HP; 3 – heat conducting bus; 4 – a powerful transistor; 5 – HP condensation zones; 6 – zone of evaporation.

The effectiveness of this approach has been confirmed experimentally. The approach was tested on a hermetically sealed electronic module (196 × 181 × 64 mm). Temperature in the module was maintained using improved original HP designs of complex shapes [20] (see Figure 8) and HPs of a similar, but more manufacturable, design (see Figure 9). Figure 8 (c) shows, that in the power range of the transistor from 27 to 51.5 W (the range of increase in the power of the transistor due to modernization of the electronic module), the thermal resistance of the horizontally-oriented HP decreases from 0.2 to 0.1 K/W, which is equivalent to a decrease in the thermal gradient on the HP from 5.4 to 5.15 K, respectively.

The developed HPs allow maintaining the required thermal regime of the most critical electronic component (power transistor, maximum value of the case temperature +66°C). When the power of the transistor was almost doubled (from 27 to 51.5 W) the temperature of the case remained +40°C. This allowed preserving the overall dimensions and the design of the basic version of the module without changing the design of the complex electronic system in which it was installed.

Being less complicated in manufacturing, the improved thermal management device based on a U-shaped HP of a circular cross section (see Figure 9) was chosen to be the basis for modernization of a complicated electronic system.

3.5. Improved Miniature Heat Pipes

Miniature heat pipes (MHP) are promising heat sink elements for electronic modules. An improved MHP design with a capillary structure in the form of flagella, sintered with a micro-wire case has been proposed in [22] (Figure 10). Such design improves performance of the MHP by increasing the porosity of the wick up to 45 – 50%.

The following MHPs were experimentally studied: 1) MHPs with a diameter of 2 and 4 mm (inner diameter 1.3 and 3 mm, respectively) with a wick in the form of a flagellum made up of 60 copper wires 30 μm in diameter (45% porosity); 2) MHPs 6 mm in diameter (inner diameter 5 mm) with a 50 to 250 mm long and 0.5 mm thick metal-fibrous wick (fiber length 3 mm, diameter 30 μm, 90% porosity). Ethanol was used as working fluid. The study showed that the heat transfer characteristics of the MHPs are significantly influenced by the diameter of the vapor channel, the type and parameters of the wick, orientation in space, and the total length of the MHP [23]. It is shown that a decrease in the size of the vapor channel in the MHP leads to an increase in the thermal resistance, a decrease in the maximum heat flux, an increase in the temperature difference in the HP. Thus, when the MHP diameter decreases from 6 to 2 mm, the thermal resistance of the MHP increases from 2 to 12°C/W. The heat transfer coefficients in the evaporation and condensation zones of the horizontally-oriented MHP with a diameter of 2 mm and a flagellum-formed wick reach values up to 1950 and 1100 W/(m^2°C), respectively, at heat flux densities of 2.9–3.4 W/cm^2. The experimental results of the rest of MHP samples are given in [24].

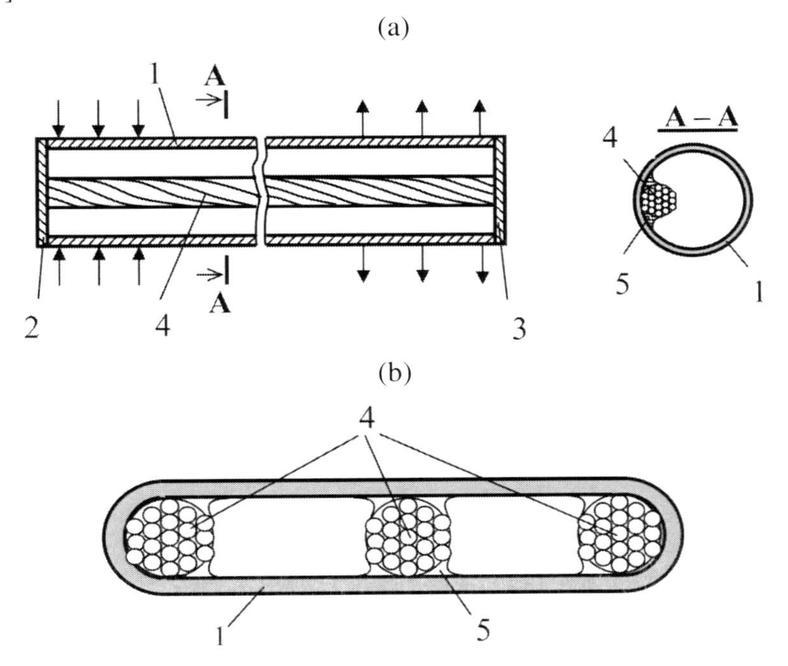

Figure 10. Schematic diagram of the improved MHP design: (a) Longitudinal and transverse sections; (b) Design version of the capillary structure of a flat MHP.
1 – case; 2, 3 – left and right butt ends; 4 – flagellum of micro-wires; 5 – working fluid.

In order to increase the performance of the MHP, it is recommended to use wicks of sintered metal fibers (if the internal section of the MHP allows it) with a high capillary pressure and a higher permeability.

3.6. Improving Heat Pipes with Diode Properties

A new design of a heat pipe with diode properties is presented in [25]. This design has a wick in the evaporation and transport zones, an circular groove and a specifically-shaped insertion in the condensation zone, and a plug with a vapor core divider forming a narrow gap of a particular shape (a 180º bend). The latter accelerates vapor movement in the condensation zone, thus accelerating the two-phase flow of heat transfer medium and intensifying the heat transfer process. Compared with the known analogues, such design gives the following advantages: a safer performance, regardless of orientation in the gravity field, as well as in zero gravity; a larger distance of the heat flow transfer in the forward direction; a wider range of coolants that can be used.

3.7. Improving the Design of Gas-Regulated Heat Pipes

While working on the improvement of the existing designs of gas-regulated heat pipes (GRHP) with NCG, the main tasks are to increase its heat transfer capacity, expand the operating temperature range and improving the quality of thermal regulation. The way to improve the design of the GRHP is to use a gas that dissolves in the coolant as NCG. A new GRHP design with a soluble gas has been developed in [26]. In this design, the gas reservoir with perforated capillary-porous inserts is separated from the condensation zone by a transverse partition (vapor flow core separator). There is a gap between the thin disk of the partition and the case. Upon colliding with the partition, the vapor core of the high-boiling component is directed to the HP wall in the condensation zone and condenses on this wall. The hot vapor core does not reach the perforated inserts directly, and thus does not heat them up. The solubility of the low-boiling component in the high-boiling one (saturating the perforated inserts) is not reduced. This lowers the thermal resistance of this type of GRHPs as compared to the GRHPs without the vapor core separator.

The influence of the of the vapor core separator on the thermal resistance of the GRHP with soluble gas was tested on a specimen of water-ammonia GRHP made of stainless steel. The GRHP had a 95 mm long evaporation zone, a 30 mm transport zone, and a 200 mm condensation zone. The outer diameter of the HP was 24 mm, the inner diameter was 22 mm. The wick with 61.5% porosity was 0.8 mm thick. The volume of CS in the reservoir was 17 cm^3, the porosity of the CS was 90%. The gap between the HP case and the separator was 1.5 mm.

Experiments have shown that when the heat flux value is greater than 150 W, thermal resistance of the GRHP with vapor core separator is 2–3 times lower than the one of the GRHP without the separator. Moreover, the greater the heat flow, the grater the positive effect. The thermal sensitivity coefficient of the developed GRHP is 1.3–2.8 times higher than of the one with insoluble gas. This allows recommending this type of RGHP to be used for thermal management of electronic devices, when it is necessary to regulate the temperature of the components in a given range.

3.8. Improving the Design of Loop Heat Pipes

In order to improve the loop heat pipes (LHP), it is particularly important to reduce their weight, size and cost. Another important task is to expand the range of structural materials and coolants used in such HPs, maintaining corrosive compatibility of the elements. The possibility to create an LHP with reduced weight and size due to the use of aluminum-based materials was studied and the results were presented in [27]. This study differed from the others: the LHP model had transparent insertions which allowed the hydrodynamic processes to be visually observed. This made it possible to observe the processes occurring in the condensation zone and the liquid channel at different positions of the evaporator relative to the condenser, from the launch of the aluminum LHP.

Experiments have shown that for the developed aluminum LHP with acetone as the heat carrier, there minimum value of the heat flux is 6–18 W (when the evaporation zone is positioned 0 to 0.8 m above the condensation zone). Below this minimum value a stable start is not possible. In these conditions, pulsations of the vapor-liquid boundary in the zones of the contour were visually observed, which indicates the dynamic nature of the LHP operation. The stable launch of the LHP was carried out for 5–6 minutes in the range of heat fluxes of 30–70 W. At the same time, the temperature of the evaporator after the start did not exceed +65ºC. The weak dependence of the temperature of the evaporator on the change in the gravity force (the evaporator being positioned higher than the condenser) is experimentally confirmed. Thus, when the difference between positions of the evaporator and the condenser changed from 0 to 0.75 m at a constant nominal heat load on the evaporator, the average temperature on its surface changed only by 5–20%.

At the same time, it was found that the difference between positions of the evaporation zone and the condensation zone significantly affects the heat transfer capacity of aluminum LHPs: at the evaporator temperature of +65°C, the heat flux transferred by the LHP decreased from 170 to 100 W when the position difference changed from 0 to 0.75 m. The aluminum LHP is characterized by its ability to self-regulate. Thus, when one value of the heat flux changed to another, a higher one, the change in the temperature level of the HP was negligible.

Further improvement of the LHP is accomplished by developing a capillary pump made of alumina-based powder [28].

4. Thermal Management of Electronic Devices Using Heat Pipes in Basic Supporting Structures of Various Levels

4.1. Thermal Management in Multi-Channel Secondary Power Supply Units in the Basic Supporting Structures of the Second Level with Air Cooling

Figure 11 shows the original design of eight-channel secondary power supply units with continuous voltage regulators and flat finned HPs with overall dimensions of 26 × 60 × 156 mm (see Figure 12) for thermal management [29, 30]. The power supply units are based on 142EN5 integrated circuits (IC) and are arranged in basic supporting structures of the second level (cassettes) with overall dimensions of 26 × 180 × 320 mm. The units are ventilated by air blowing from the bottom upwards at a speed of 0.31 m/s and an inlet temperature of +40°C. The thermal resistance of the HP does not exceed 0.2°C/W.

When developing the flat finned HP design (see Figure 12), a number of unorthodox design and technological solutions were realized allowing the structure to be much easier to fabricate, making it possible to mechanically attach the IC to the HP, to decouple of the planar pins of the IC on the printed circuit board, to conveniently attach the printed circuit boards and to perform electrical wiring. Among the advantages of the design are high serviceability and a long lifetime of the HP under mechanical stress.

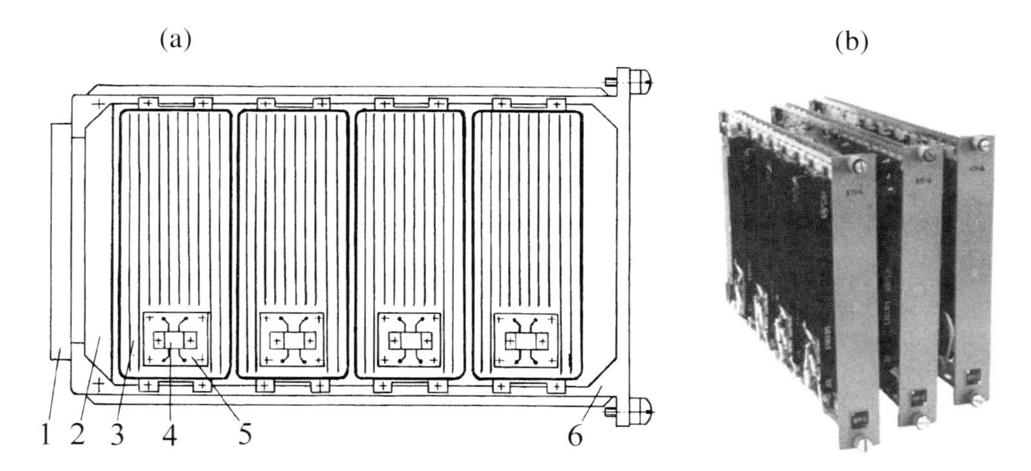

Figure 11. Thermal management of multi-channel secondary power supply units: (a) Power unit arrangement with flat finned HPs; (b) Photograph of three power units.
1 – connector; 2 – circuit board with passive electronic components; 3 – heat pipe; 4 – integrated circuit (9.5 W); 5 – board with the integrated circuit; 6 – cassette.

Unlike traditional finned aluminum radiators of the same dimensions, flat finned HPs allow maintaining the specified temperature of ICs (+78ºC), which made it possible to create functionally complete power supplies assembled in standard cassettes of limited overall dimensions.

Figure 12. Flat finned HP for two microcircuits.

4.2. Thermal Management for Multi-Layer Ceramic Switching Circuits in the Basic Supporting Structures of the Second Level with Water Cooling

Multi-layer ceramic switching boards are widely used in modern electronic systems with a high circuitry density. The switching circuits are installed in the basic supporting structures of the second level with air or water cooling. When using the advanced circuitry with increased specific heat generation, basic water-cooled supporting structures are preferable.

A new design principle was proposed [31] for heat removal devices for multi-layer ceramic switching circuits (MLCSC) with microprocessors (MP) and with very-large-scale integration (VLSI) multichip modules. The new principle consists in embedding the collector thermosyphons directly into the body of the ceramic base of the circuit board [32]. Heat transfer characteristics were tested on glass-ceramic models (see Figure 13) with different shapes of evaporation parallel-plate channels (see Figure 14) [33, 34].

The model contains a copper supply header 5 × 25 mm in cross-section at the bottom, a copper collecting header of the same cross section with a water cooling unit at the top. The dimensions of the ceramic circuit board base are 100 × 120 × 2.5 mm, the dimensions of the return channel are 90 × 10 × 1.5 mm. Distilled water is used as a heat carrier. Flat ohmic heaters (3 heaters per channel) installed on the opposite side of the ceramic base are used as heat flow simulators. A glass insert in the model made it possible to visually observe how processes of hydrodynamics proceed during operation of the heat sink system.

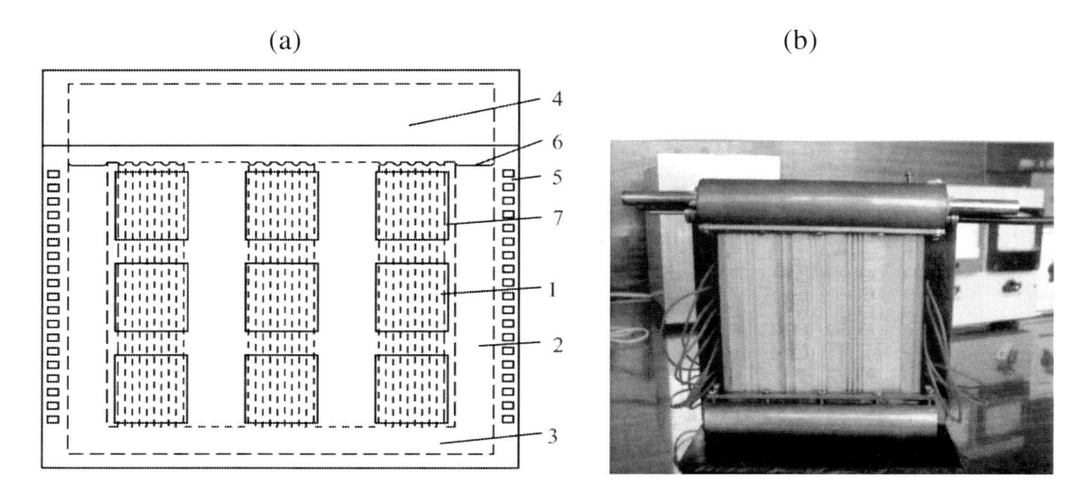

Figure 13. Glass-ceramic model of a thermal management device for the MLCSC based on the TS collector: (a) Schematic diagram; (b) Photograph.
1 – group of evaporation channels; 2 – return channel; 3 – supply header; 4 – collecting header (condensation zone); 5 – MLCSC base; 6 – working liquid level; 7 – heat-generating element (HGE).

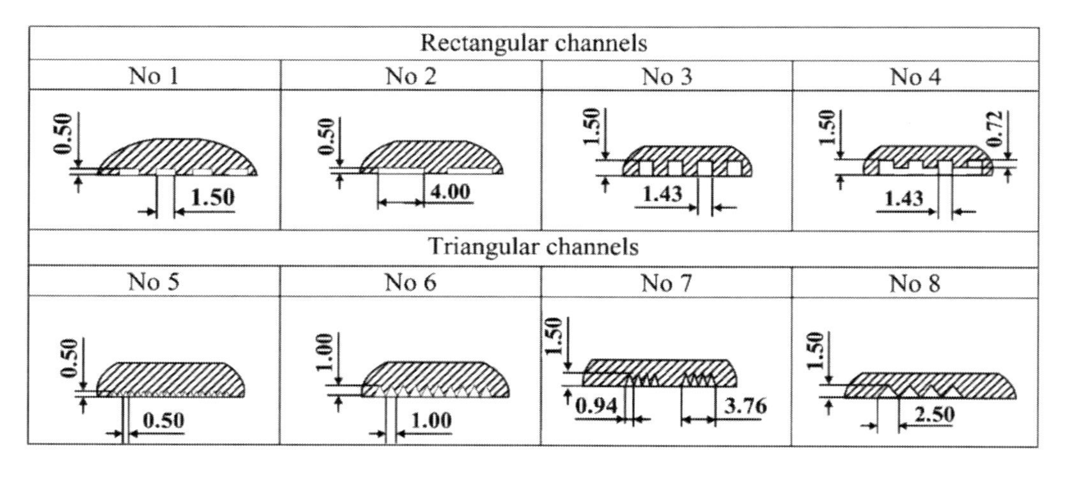

Figure 14. Geometry of the studied parallel-plate channels.

As a result of thermophysical studies and visual observations, it was established that the processes of evaporation and heat transfer occur most intensely and stably in the group of parallel-plate channels No 4 (see Figure 15). This group of channels has a complex the cross-section shape. The equivalent diameter of the channel group is 1.74 mm, the width is 10 mm, and the distance between the rows of channel groups on the board is 11 mm. The general part of the cross-section of the channel with dimensions of 10 × 0.78 mm efficiently replenishes with the liquid coolant the smaller evaporative parallel-plate channels (1.43 × 0.72 mm) adjacent to the heating surface. In evaporation parallel-plate

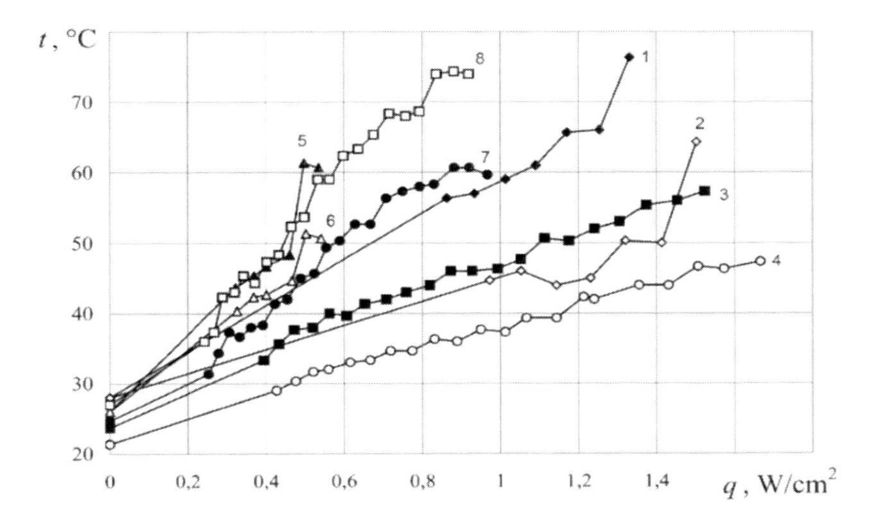

Figure 15. Dependence of temperature in the heating zone t of the circuit on the heat flux density q for evaporation channels with different geometry (the channel numbers correspond to those indicated in Figure 14).

channels, an intensive vaporization occurs. The evaporation channels develop vapor–liquid clusters, a thin fluid film appears on a significant part of the surface, and a movable vapor–liquid boundary is formed along the entire channel section. Rapid replenishment of the dried sections of small evaporative channels with liquid from the general part of the parallel-plate channel stabilizes the boiling process. As the heat flux density increases within the investigated range from 0.3 to 3.0 W/cm^2, there is a tendency for the vapor–liquid interface to move down both in the evaporation and return channels.

The thermal characteristics of a glass-ceramic model with four rows of parallel-plate channel groups are shown in Figure 16.

Investigation of heat transfer intensity showed that as the heat flux density in the evaporation zone increases from 0.3 to 1.0 W/cm^2, the heat exchange rate increases, and starting with the values of 1.0 – 1.2 W/cm^2 and up to 3 W/cm^2, the heat exchange rate is practically independent of the density of the heat flux. The values of the heat transfer coefficient in the evaporation zone of the channels in this regime reached values of 5000 W/(m^2°C).

Curves of average temperature distribution along the height of each of the rows of channels depending on the amount of heat input (see Figure 16) indicate a high cooling efficiency of the ceramic base of the board. Thus, if the total heat input power to the circuit board is 160 W (40 watts per each row of channels), the maximum temperature on the surface of the board does not exceed +60°C (1st row +58°C, 2nd row +56°C, 3rd row +60°C,

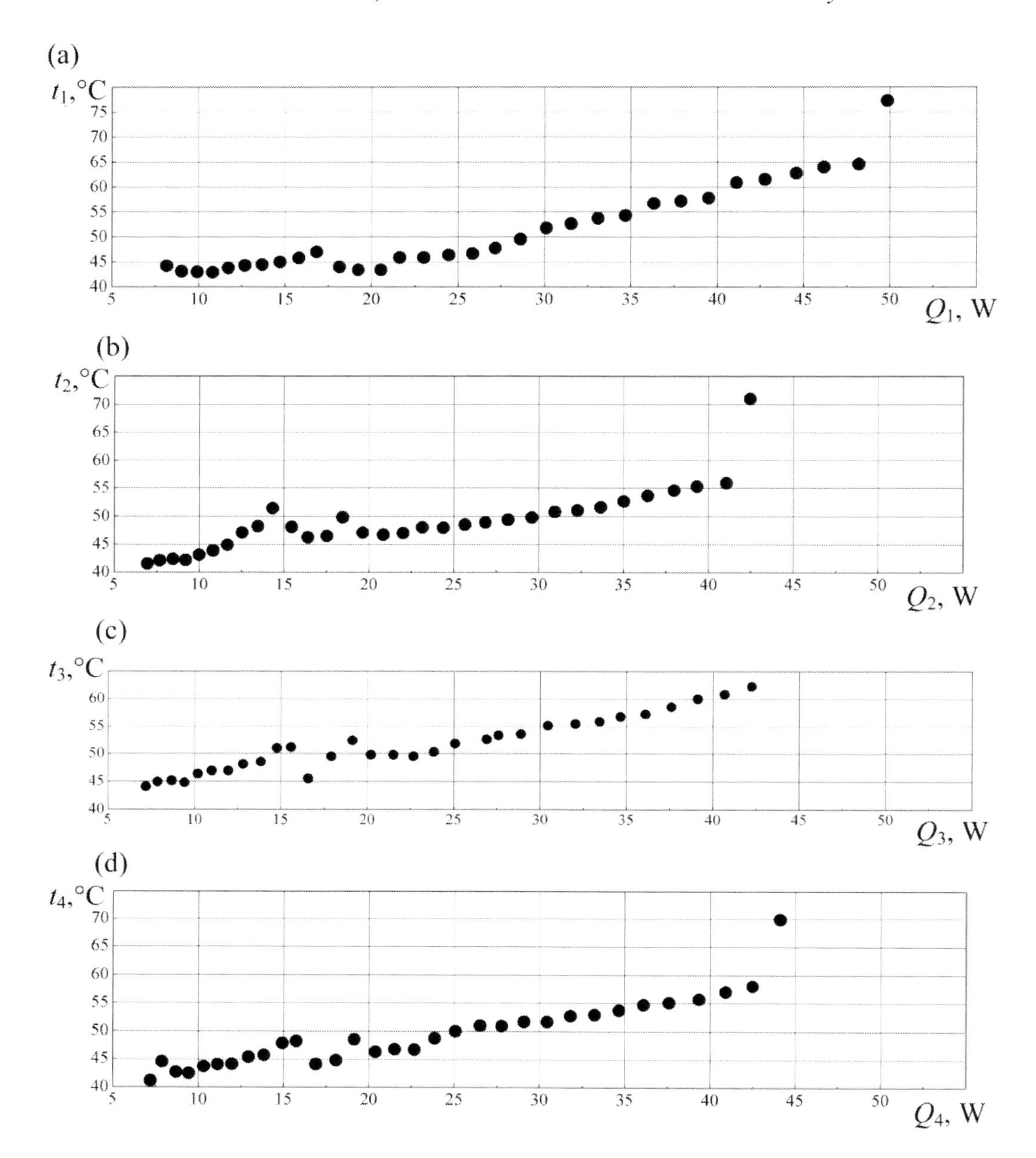

Figure 16. Dependence of the average temperature $t_1 - t_4$ of each row of evaporation channels on the input heat flux Q: (a) For the first row, counting from the return channel; (b) For the second row of channels; (c) For the third row of channels; (d) For the fourth row of channels.

4th row +57°C) at a water temperature of +19°C at the inlet of the water-cooling unit and water flow through a water heat exchanger of 2.1 l/min. When the heat input power exceeds 42 W, vaporization of the evaporation channels become filled with vapor in the second and fourth rows, which led to a sharp increase in the temperature of the surface of the circuit board near the thermal simulators.

The results of physical modeling of a thermal management device for the MLCSC with an internal channel collector TS demonstrate that the effective heat removal is possible due to the closed evaporation-condensation cycle inside the ceramic base of the switching circuit.

4.3. Improved Manifold Heat Pipe for Basic Second Level Supporting Structures in the Form of Removable Electronic Modules with Water Cooling

A sectional HP was designed for a removable electronic module of a water-cooled instrument cabinet. The HP is also part of the basic supporting structure of the module (see Figure 17).

According to the experimental results on the thermal characteristics of the model of the sectional HP [35], if the temperature of the wall of the evaporation channel is +57ºC, the permissible power of the electronic components is 102 W in a module with one section, 204 W with two sections and 306 W with three sections.

HPs of the proposed design can operate under such mechanical stress as shocks, long slopes, oscillation and vibration, and can thus be used in shipborne electronic equipment.

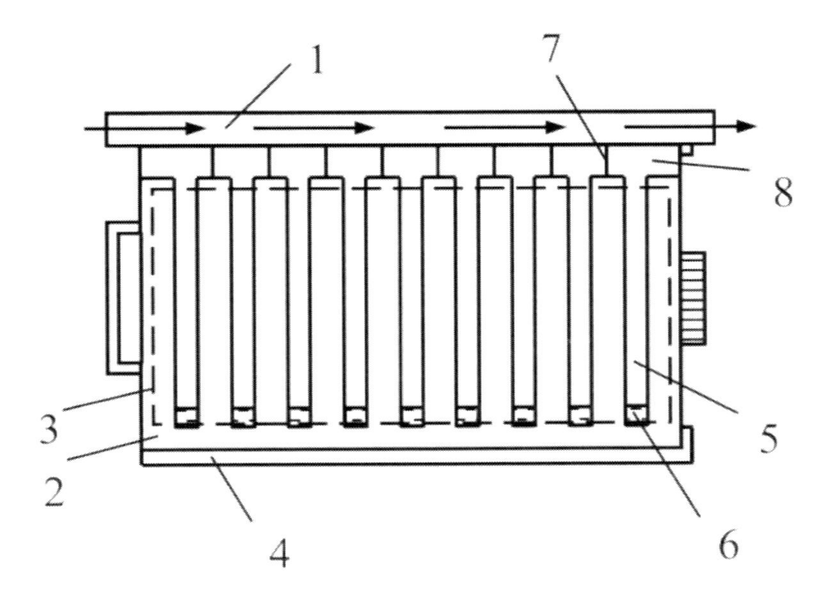

Figure 17. Schematic diagram of the functional module based on the sectional HP.
1 – upper cooling shelf of the instrument cabinet; 2 – switching board; 3 – heat supply zone (where MP and IC are installed); 4 – support base; 5 – HP evaporation channel; 6 – working fluid; 7 – partition in the condensation zone of the HP; 8 – HP section.

4.4. Basic Second-Level Supporting Structure in the Form of a Removable Unit with a Metal Plate with Built-in Evaporation Minichannels and Water Cooling

For more powerful electronic devices, another design of the electronic unit with water cooling has been developed [36]. This design consists of a metal board with an internal system of vertical evaporation minichannels, connected by a distributing collector at the bottom, and a gathering collector (acting as a condenser) at the top. The return channel is designed as a hollow handle of the unit and returns the condensate back to the distributing collector. The surface of the condenser is in thermal contact with the heat sink, which is cooled by water (see Figure 18). The dimensions of the metal board are 170 × 280 mm. Made of aluminum alloy, the metal board contains six vertical evaporative minichannels (1.5x20 mm each). Evaporative minichannels are filled with the working fluid – acetone.

Experiments have shown [37] that the thermal resistance of the basic support structure is 1.4 to 0.1°C/W when the total power of the 24 simulators of the heat-generating electronic components is changed from 1.25 to 570 W. The thermal resistance was defined as the ratio of the difference in the average temperature on the surface of the metal plate in the installation area of the simulators of electronic components and the temperature of the cooling water at the inlet to the heat exchanger to the total power of the heat flow simulators. At a flow rate of 1 l/min with an inlet temperature of 12 ± 1°C, this basic second-level support structure allows removing the heat power of up to 570 W from the unit at an average temperature t of the board surface not exceeding 74.8°C (see Figure 19).

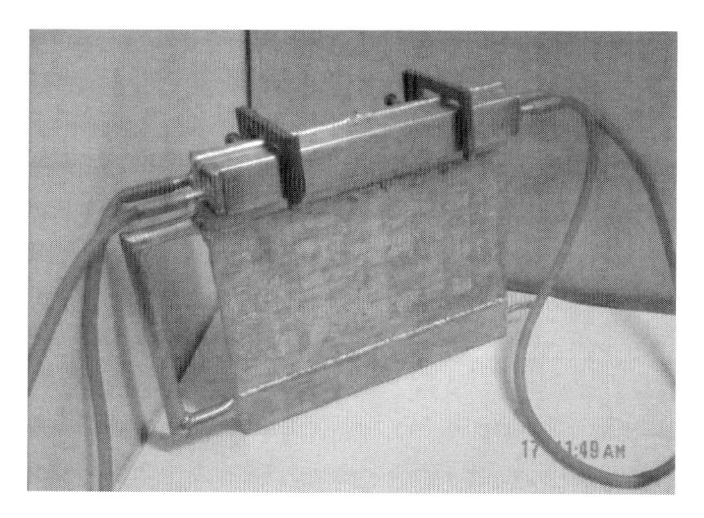

Figure 18. Photograph of the basic water-cooled supporting structure with built-in evaporation minichannels.

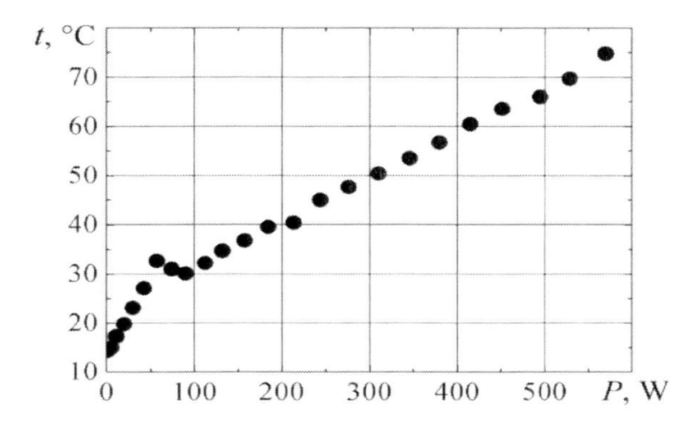

Figure 19. Thermal characteristics of the basic supporting structure with evaporative minichannels.

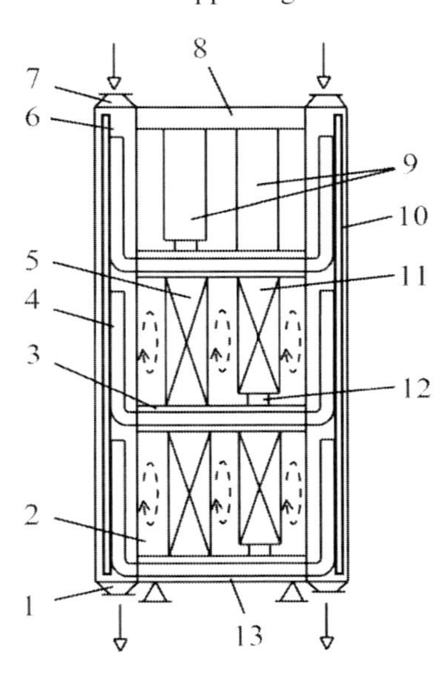

Figure 20. Shematic diagram of a third level BSS with U-shaped HPs: 1 – inlet flange; 2 – cabinet section; 3 – intersectional panel (shelf); 4 – HP; 5, 11 – second level heat generation modules; 6 – water channel; 7 – outlet flange; 8 – top base; 9 – non heat-generating units; 10 – vapor chamber; 12 – hold-down device; 13 – bottom base.

4.5. Using HPs in Third-Level Basic Support Structures with Water Cooling

Structurally complex electronic systems with increased heat release (shipborne equipment, workflow control systems, supercomputers, etc.) are designed according to a functional-modular principle. Functionally completed blocks and modules are placed into water-cooled third level basic supporting structures (BSS), e.g., instrument cabinets, towers, racks, etc. As a rule, cooling water is pumped through numerous channels of complex zigzag shape, which leads to significant energy costs.

The essential point of a new approach to the design of a third-level water-cooled BSC is the use of heat-pipe panel collectors (HPPC) based on U-shaped HPs used as BSS shelves

[38]. The horizontal part of the U-shaped HP serves as an evaporation zone and is mounted into the shelf of the instrument cabinet. The vertical parts of the U-shaped HP are condensation zones and are located in the cavities of the side walls of the cabinet, behind which cooling water is pumped (see Figure 20) [39]. This design of the BSS provides a significant reduction in the hydraulic resistance of the cooling line and a reduction in energy consumption for pumping cooling water, while at the same time increasing the amount of heat removed from the instrument cabinet.

Heat transfer and hydrodynamics in the proposed BSS were studied using computer simulation [40] and experimentally on two thermal models [41, 42]. The first one (see Figure 21) was a model of one BSS section (width – 561 mm, depth – 300 mm, height – 435 mm), and the second one (see Figure 22) was a model of an eight-section rack (width – 561 mm, depth – 303 mm, height – 2035 mm).

Each thermal model has seven cylindrical aluminum U-shaped HPs (30 × 54 × 220 mm each) with Ω-shaped grooves inside and two 30 mm wide shelves outside. The distance between the outer surfaces of the shelves is 14 mm.

The thermal resistance of the HP varies from 0.036 to 0.004°C/W in the range of heat flux values from 100 to 500 W. The temperature gradient along the HP does not exceed 4.5°C.

Experimental studies of the heat transfer characteristics of the model of the BSS section with seven U-shaped HPs (see Figure 21) showed that the set maximum temperature value t_{max} = +60°C in the heating zone of the HP was reached at the total heat flux P values of 3050 and 3600 W, with water travelling downward and upward, respectively. The flow rate is 3 l/min in each side wall of the model (see Figure 23) [42].

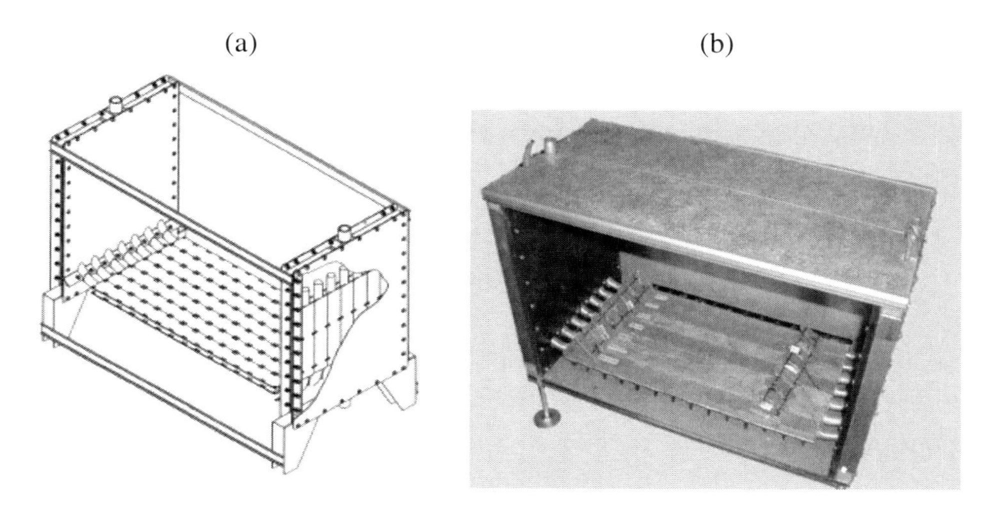

Figure 21. BSS experimental model: (a) Schematic diagram; (b) Photograph.

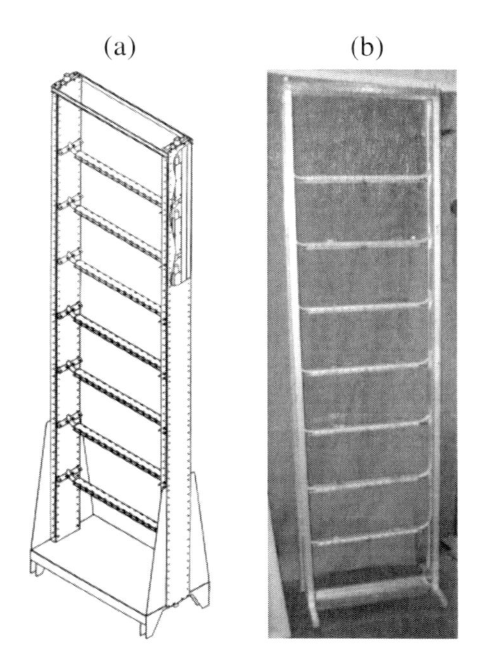

Figure 22. Eight sectioned rack experimental model: (a) Schematic diagram; (b) Photograph.

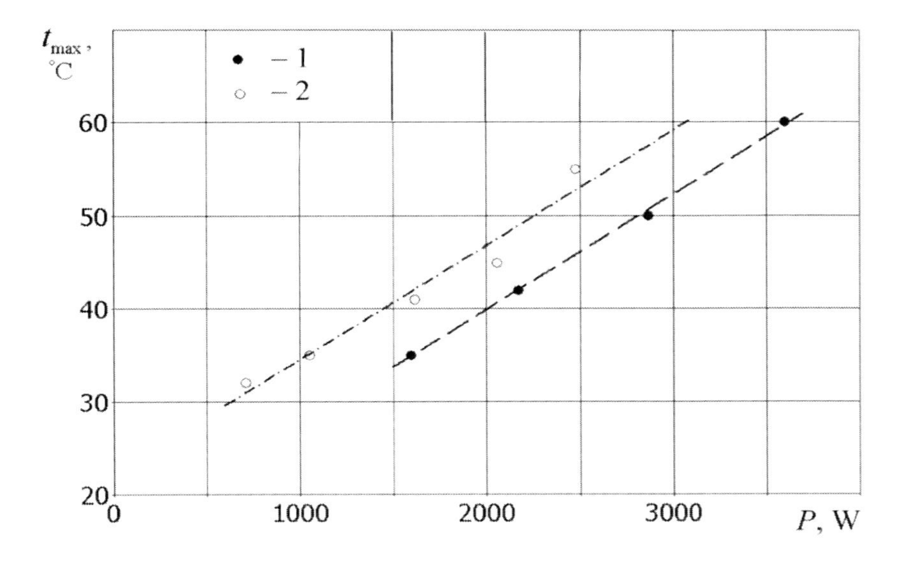

Figure 23. Thermal characteristics of the model of the BSS section: 1 – upward water flow at 6 l/min flow rate; 2 – downward water flow at 6 l/min flow rate.

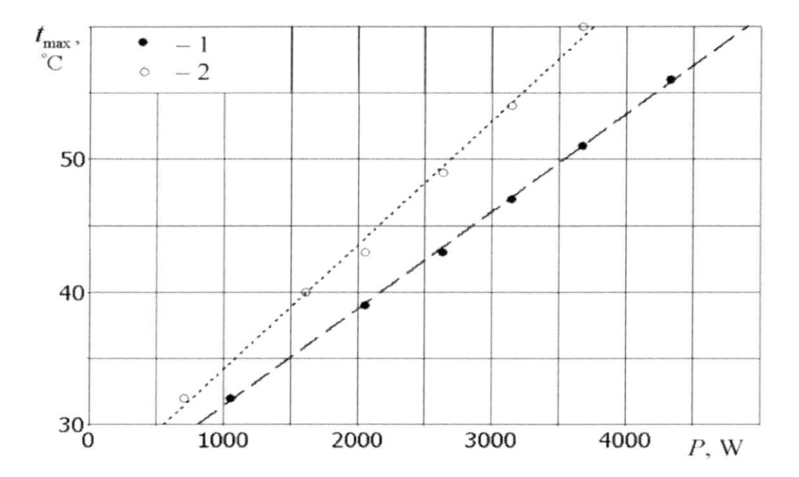

Figure 24. Thermal characteristics of the model of the eight-section rack: 1 – upward water flow at 4 l/min flow rate; 2 – downward water flow at 6 l/min flow rate.

In the eight-section rack model (see Figure 22), each intersection panel consists of one U-shaped HP). The maximum temperature in the heating zone t_{max} = +60°C was reached at the total heat flux *P* value of 3672 W with the downward water flow (at 3 l/min flow rate) in each sidewall, and over 4330 W (power limit for heaters) with the upward water flow (at 2 l/min flow rate) in each sidewall (see Figure 24) [42].

Thus, it was experimentally shown that when constructing a water thermal management device for complex electronic systems in a third-level BSS based on seven heat-pipe panel collectors of U-shaped HPs (7 HPs in each collector), it is possible to remove 25.2–30.3 kW of thermal power from an eight-compartment instrument cabinet at maximum temperature of the heating zone of the HP above +60°C.

Comparing the results of numerical modeling of heat transfer processes in the studied models with different cooling water supply modes and at different levels of thermal load with the results of full-scale experimental studies showed that the difference between the calculated and experimental values of the heat flux did not exceed 25–30% [40], which is acceptable in engineering calculations.

5. Thermal Management of Mobile Infrared Devices Based on Two-Phase Technologies

The electronic component in control systems for mobile infrared devices that requires efficient heat removal the most is the matrix of photosensitive elements (MPE). It is intended that MPEs would be the base for microelectronic photosensitive devices in the leading-edge infrared electronic control systems. To provide the required sensitivity level of the MPEs, they need to be cooled to cryogenic and low temperatures [43, 44].

5.1. Thermal Management of Photosensitive Devices for Control Systems of Infrared Electronics Using Cryocooling

One of the methods of cryogenic cooling of microelectronic photosensitive devices is using an open-loop throttle (gas cylinder) cooling system [44–46]. The disadvantage of such cooling device is the limited period of service, which is determined by the volume of the gas cylinder. An increase in the volume of the cylinder leads to an increase in the mass of the device, which is not always permissible. Reducing the flow rate of the coolant and increasing the period of service of the system is possible by increasing the heat exchange efficiency during boiling of the heat transfer fluid on the cooled surface of the matrix of photosensitive elements.

With this thought in mind, a new cooling device has been developed (see Figure 25). On the heat-dissipating surface of the MPE matrix substrate, the device has fins that form channels for circulation of the coolant. The surface of the channels is covered with a thin layer of capillary-porous material. The layer of the capillary-porous material is sintered from 3–6 mm long copper fibers 20 to 130 μm in diameter and directly adhered to the substrate of the MPE matrix. The porosity of this layer can reach 96%.

Vapor bubbles are rapidly carried away by the coolant through separate channels, which prevents a vapor plug from forming on the porous surface of the channels. Due to capillary forces, the heat-carrying fluid entering the channels is absorbed by the porous structure of the layer throughout the entire inner surface of the substrate, regardless of its size. This increases the number of active vaporization centers and ensures a rational flow rate of the coolant. In such design of the device, the capillary forces retain heat-carrying fluid at the heat-generating surface, which prevents the liquid coolant from ejecting from the boiling surface (as it would be on a smooth surface). This results in a more stable heat transfer process in the developed device.

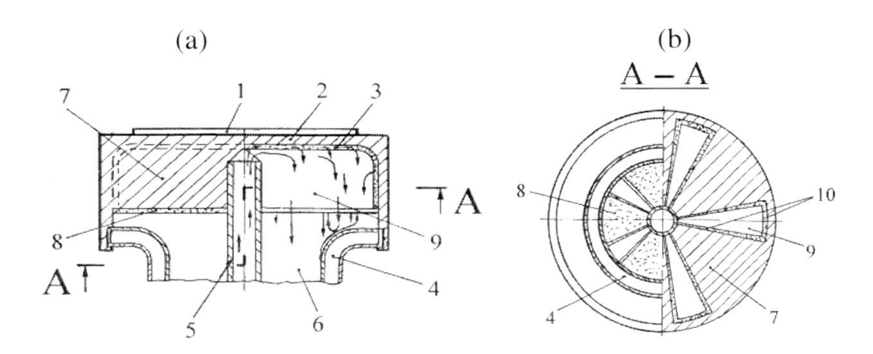

Figure 25. Schematic diagram of the cooling device for the MPE: (a) Longitudinal section view; (b) Cross-sectional view.
1 – MPE; 2 – matrix substrate; 3 – porous layer on the substrate; 4 – Dewar vessel; 5 – nitrogen supply tube; 6 – outlet line; 7 – fin; 8 – porous layer on the end-face surface of the fin; 9 – branch channel; 10 – porous layer on the lateral surface of the fin.

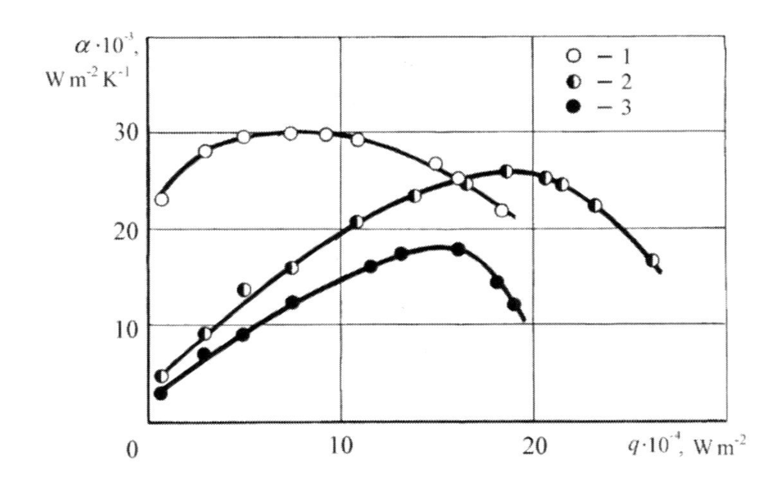

Figure 26. Dependence of the heat transfer coefficient α on the heat flux density q at liquid nitrogen boiling on the capillary-porous (1), (2) and smooth (3) substrate surfaces of the photosensitive element matrix: 1 – 0.5 mm porous layer, 62% porosity; 2 – 0.7 mm porous layer, 87% porosity; 3 – 3 mm copper substrate without porous coating.

In order to experimentally test how the coating of the substrate with a thin layer of capillary-porous material affects the heat transfer efficiency, produced and studied were experimental samples of the matrix substrates of a multi-element photosensitive device with a capillary-porous coating sintered from 3 mm long copper fibers 50 μm in diameter with different thickness and different porosity. Experimental results are given in Figure 18. For comparison, the figure also shows experimental data for a smooth surface without a porous coating.

The curves in Figure 26 indicate, that the maximum value of the heat transfer coefficient of the samples was 30000 W/(m^{2}K) and 26000 W/(m^{2}K), respectively, which is 66.7% and 38.9% more than on a smooth surface.

It was also found that low porosity coating (curve 1) allows high values of heat intensity (over 20000 W/(m^{2}K)) even at low heat flux densities (0.5–1.0 W/cm^2), while with high-porosity coating such an intensity of heat exchange can be achieved only at heat flux densities higher than 10 W/cm^2 (curve 2). Such effect of porosity on the heat exchange process is caused by the high carcass thermal conductivity of the low porosity layer, which leads to the lack of so-called low-intensity undeveloped bubble boiling. In the case of high-porosity coating or smooth surface, such undeveloped boiling occurs and the developed bubble boiling begins only at high heat flux densities.

Experimental studies confirm the high efficiency of 0.5–0.7 mm thick metal-fibrous capillary-porous coating with a porosity of 62–87% on the surface of the fins and channels of the substrate matrix. This allows enhancing the heat exchange efficiency while cooling multielement photosensitive devices of infrared control systems.

Thus, the developed design allows increasing the heat transfer coefficient on the surface of the matrix substrate of photosensitive elements and thereby reducing the flow rate of coolant liquid and increasing the period of service of the cooling system.

5.2. Thermal Management of Infrared Photosensitive Devices of Medium-Temperature Range

Medium-temperature microelectronic photodetector devices (MPD) based on PbSe, CdHgTe or PbStTe are used to register the coordinates of aircrafts and ships, as well as in studying the thermal state of arterial pipelines and determining the epicenter and boundaries of forest fires, etc. [46]. As a rule, creating infrared devices requires taking into account field and power performance, and size of cooling systems.

MPE matrices of mid-temperature range microelectronic photodetectors need to be maintained in the operating temperature of less than 213 K. In order to do so, the matrix is mounted on a cascade thermoelectric cooler (TEC). The heat flow from the cascade TEC is transferred to cooling surfaces that remove the heat into the high-temperature environment. The ambient temperature (of the inhibited air flow) is much higher than the permissible operating temperature of the MPE.

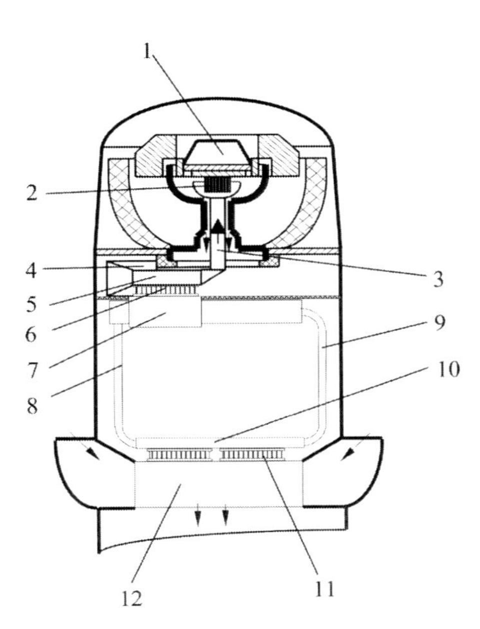

Figure 27. Schematic diagram of a combined thermal management device with convective heat transfer from MPD to LHP: 1 – MPD with a cascade TEM; 2 – radiator of the cascade TEC; 3 – gas circulation channels; 4 – supercharger; 5 – TEC-1 radiator; 6 – TEC-1; 7 – LHP evaporator; 8 – LHP steam pipeline; 9 – condensate pipeline; 10 – LHP condenser; 11 – TEC-2; 12 – heat sink; 13 – heat pipe; 14 – gas gap.

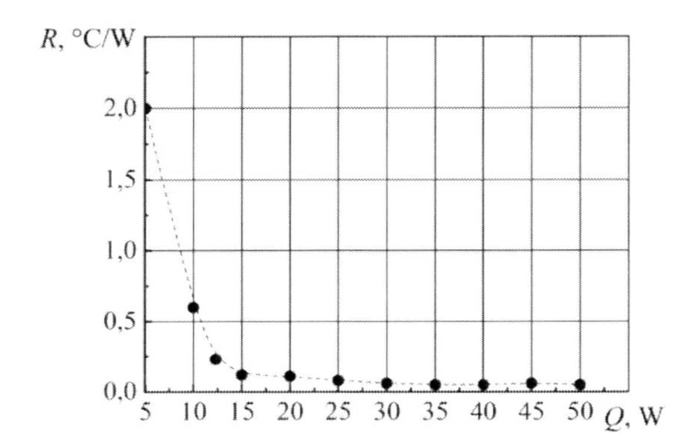

Figure 28. Dependence of the total thermal resistance of the LHP on the thermal load when the evaporator is positioned 125 mm above the condenser.

The idea of the thermal management device for the MPE is the following (see Figure 27): 1) the required MPE temperature is provided with a multistage TEC; 2) the heat generated by the hot junctions of the TEC is then transferred via non-contact scheme to the evaporator of the loop heat pipe (LHP); 3) the LHP removes the heat into the environment; 4) in order to improving of LHP performance, two additional TECs used, which act as a heat pump: TEC–1 located on the evaporator cools the convective circuit radiator and stabilizes the LHP operation; TEC–2 located on the LHP condenser cools the condenser and boosts the removed heat potential to the level that would exceed the ambient temperature (400 K).

The developed calculation algorithms and the calculation results have confirmed the possibility to create a thermal management device based on this concept. Physical simulation was performed using an LHP with a capillary pump made of oxide-based, highly porous ceramic based on a fine powder of alumina with a total porosity of 68%, an open porosity of 60%, an average pore diameter of 2 μm, a permeability of 0.8×10^{-12} m^2, stainless steel case and ammonia as a working fluid.

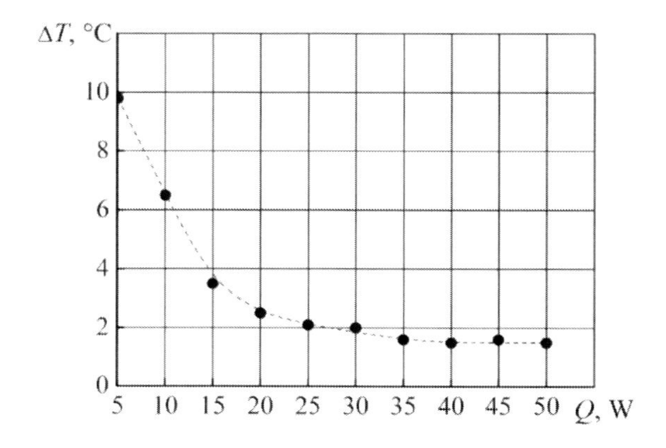

Figure 29. Dependence of the temperature gradient ΔT of the LHP on the heat flux Q when the evaporator is positioned 125 mm above the condenser.

Experimental studies have shown that the developed LHP meets the requirements. The total thermal resistance (see Figure 28) and the driving thermal pressure (see Figure 29) of the ammonia LHP in the investigated range of the transmitted heat flux allow using it in thermal management of advanced small-sized infrared devices. The temperature of the evaporator during the tests remained in the range of 10–70°C.

Thus, the proposed concept of a thermal management device using an LHP and a two-level thermoelectric system makes it possible to overcome the existing temperature difference between the heat-transfer plate of the cascade TEC and the environment, and thus to maintain the MPE temperature above 213 K when the heat is discharged into the high-temperature (400 K) surrounding environment.

6. New Thermal Management Device Based on Two-Phase Technology for Advanced Laser Devices and LED Modules

6.1. Thermal Management Using Two-Phase Technology and Thermoelectric Coolers for Advanced Microlaser Devices

The heat flux structure, the temperature regime of the laser crystal and the features of thermal processes in microlaser devices were analyzed to show that using miniature heat pipes as heat-transfer elements is a promising way to ensure a given thermal regime of a laser crystal and to reduce the total thermal resistance [47].

Thermophysical modeling of the temperature field in a semiconductor crystal of a 500 × 300 × 200 μm laser diode (LD) has shown that the maximum crystal temperature can reach 177.8°C. It is shown that the maximum LD temperature is to a great extent determined by the density of the heat flux of the LD and the conditions of heat removal from the base of the laser module (LM), LD being a constructive part of the LM. Intensification of heat transfer from the lower edge of the LM base allows improving thermal management of the LD.

The new design solutions for the thermal management of LMs are based on the use of small-size heat exchange surfaces and the MHP (first version) and MHP and TEC combination (second version) for heat removal from hot LM sections. Using MHPs allows, in the first case, to effectively remove heat from powerful LMs, e.g., from optical transmission modules (see Figure 30) [47], and in the second case it ensures heat removal from and thermal stabilization of a group of LMs (see Figure 31) [48]. Thermal stabilization of microlaser devices is especially relevant in connection with the trend of using a gigabit speed range and optical multiplexing of signals.

In cases where there is a need to package a significant number of identical heat-loaded LMs inside one device for information or communication systems (particularly of fiber-optic ones), to connect each individual LM with a corresponding TEC or with a fan becomes a challenge. The problem is easier solved by combining several LMs into one group and then to use MHP and TEC to remove heat from the whole group (see Figure 31). With this approach, it is possible to either use the HP to blow the heat from all the LMs with an airflow cooled by the TEC, with a branched air duct (Figure 31, a), or use an HP to remove the heat from all LMs to a cold TEC plate (Figure 31, b). Preliminary evaluation shows that such thermal management techniques allow providing a normal thermal regime for each LM with a thermal power of up to 0.3 W. It is also worth noting, that such thermal management device is comparatively easy to manufacture to install.

The proposed design solutions can be used as the basis for developing thermal management techniques for both separate LDs/LMs and for microlaser devices for advanced information and control systems.

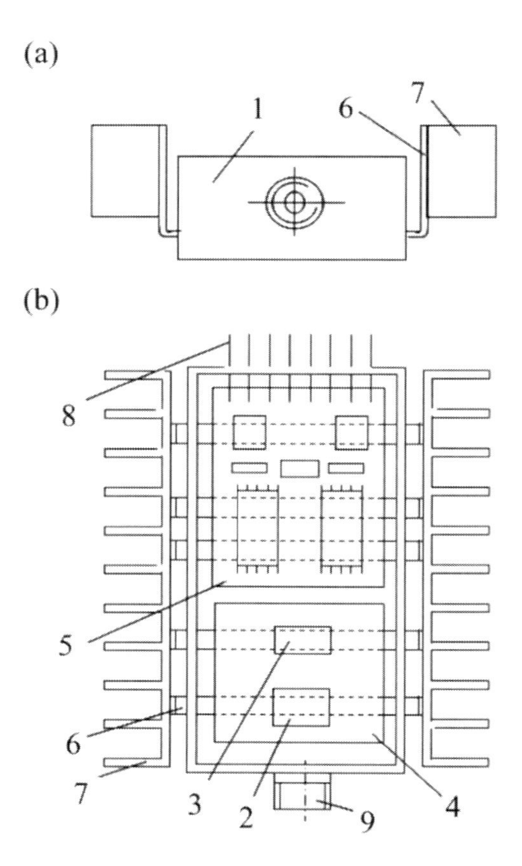

Figure 30. Schematic diagram of thermal management of the optical transmitter using an MHP: (a) Front view; (b) Top view (with the lid removed).
1 – case; 2 – laser diode; 3 – control photodiode; 4 – pedestal; 5 – heat conducting plate with heat-loaded elements; 6 – flat MHPs; 7 – fins; 8 – electrical terminals; 9 – optical connector.

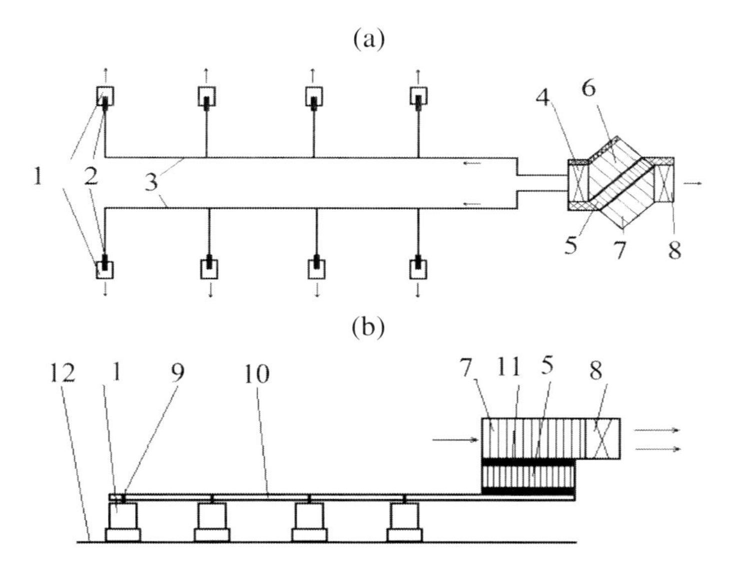

Figure 31. Versions of a combined thermal management device based on MHP and TEC for a group of LMs: (a) Convective scheme; (b) A mixed scheme.
1 – LM; 2 – heat exchanger; 3 – air ducts; 4, 8 – fans; 5 – TEC; 6, 7 – heat exchangers on cold and hot plates of the TEC, respectively; 9 – LM interface; 10 – MHP; 11 – thermal contact zone between MHP and TEC; 12 – board.

6.2. Thermal Management Using Two-Phase Technology and Thermoelectric Coolers for Advanced High-Power Laser Devices

For cooling and thermal stabilization of high-power laser modules, a method has been developed using a closed evaporation-condensation cycle and a device for its realization [49, 50]. Figure 32 shows the schematic diagram of such a device in action: the heat is transferred from the LD semiconductor crystal to the cold plate of the TEC by means of a closed evaporation-condensation cycle in the vapor chamber.

The vapor chamber is a flat HP with one or several semiconductor laser crystals placed on its external surface. Such a design allows lowering the temperature of LD semiconductor crystals and makes their performance more reliable due to effective heat removal using an evaporation-condensation cycle with low thermal resistance at the most heat-loaded area. The low inertia of the heat transfer processes through the evaporation-condensation cycle reduces the time the device needs to reach the operating mode and expands its functional capabilities.

Unlike the existing designs, the proposed device has a vapor chamber with a liquid heat carrier installed into the most heat-loaded part of the heat transfer path (between the LD semiconductor chip and the TEM cold plate) and the external heat exchanger is installed in thermal contact directly to the hot TEM plate. In the vapor space of the vapor chamber, there is a temperature sensor for the feedback of the control system.

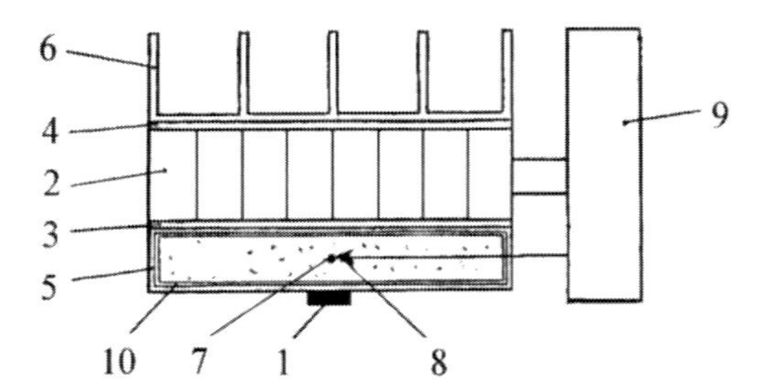

Figure 32. Schematic diagram of a thermal management device for LDs using an HP-shaped steam chamber: 1 – LD crystal; 2 – TEC; 3 – cold plate; 4 – hot plate; 5 – vapor chamber; 6 – external heat exchanger (radiator); 7 – vapor temperature control zone; 8 – temperature sensor; 9 – control unit; 10 – CS layer of the HP.

Heat transfer in the most heat-loaded area by means of a highly effective closed evaporation–condensation cycle allows instantaneous setting of the required temperature level of the saturated vapor and the LD crystal, and the temperature is stabilized within the given limits, regardless of the type of destabilizing factors – external or internal. Thus, the thermal inertia of the device in time is significantly reduced, which makes it possible to use it in devices with a pulsed mode of operation. This advantage expands the functionality of the device.

Heat removal using evaporation-condensation cycle with low thermal resistance at the most heat-loaded area reduces the temperature of semiconductor crystals and makes their performance more reliable.

6.3. Thermal Management Using Heat Pipes and Large-Size Thermoelectric Coolers

When performing thermal management of a powerful modern and next-generation electronic devices based on laser diodes, microprocessors, etc. with an increased density of heat flow, it is also advisable to use large-size TECs with heat plates made in the form of flat heat pipes. Such plates operate in a closed evaporation–condensation heat transfer cycle [51] (Figure 33).

Heat transfer from LD to cold junctions of thermoelements (the most heat-loaded area) is carried out by a highly efficient closed evaporation-condensation cycle allows reducing the thermal resistance and thermal gradient in the heating zone. This leads to a smaller required thermal gradient between the cold and hot plates, which makes it possible to increase the cooling capacity of the TEC and to lower the temperature and reliability of the LD.

Thus, the proposed design of the large-size TEM ensures an increase in the cooling capacity and expands the scope of possible design solutions, including the increase in size (length and width) of the carrier plates without increasing the temperature gradient along the plates. The TEC design allows one or several cooled LDs to be placed in both the center of the surface of the cold plate and in any other place on its surface. The size of the hot plate and the radiator on its surface can also be significantly increased compared to the known TECs. This expands its design scope and possibilities to use it in combined thermal management devices.

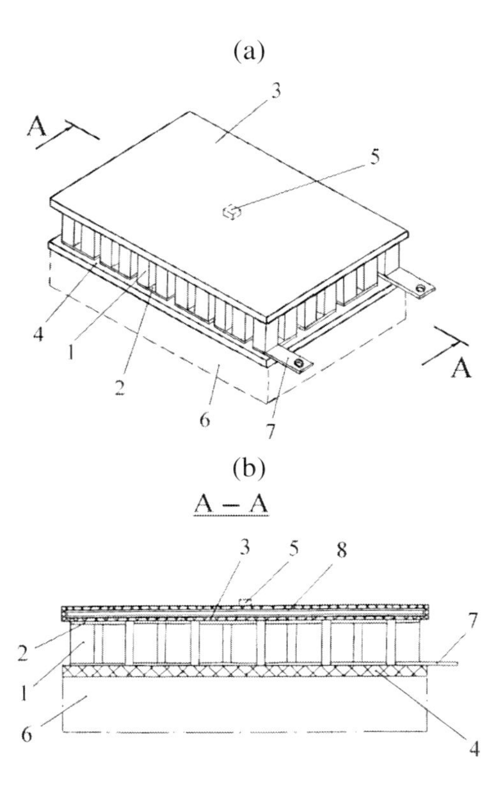

Figure 33. Large-size thermoelectric cooler: (a) General view; (b) Section along the line A-A.
1 – thermoelement (a branch); 2 – contact plate; 3 – cold plate; 4 – hot plate; 5 – LD; 6 – radiator; 7 – output; 8 – CS layer.

Figure 34. Photographs of HPs for a TEC large-size unit: (a) For the cold side of the TEC; (b) For the hot side of the TEC.

In a number of practical applications, it is promising to make cold and hot plates of a TEC of the largest possible sizes, intensifying the heat transfer from the plates in order to increase the refrigerating coefficient and cooling capacity. At the same time, when designing such TECs, one must face a number of constructive and technological difficulties, which urges to search for new design and technological solutions for large-size TECs.

One of the constructive and technological ways of solving this problem is to connect several standard single-stage TECs into a single unit and place them between two flat heat pipes (see Figure 34) [52]. This design of the TEC block makes it possible to significantly intensify the heat transfer with the plates, to increase the size of the heat-transfer and heat-dissipating surfaces of the plates and TEC radiators. Experimental study of large-size TECs based on heat pipes [52, 53] (see Figure 34) showed that the use of aluminum ammonia heat pipes 100 × 100 × 6 mm (to be aligned with the cold plate) and 100 × 300 × 7 mm (to be aligned with the hot plate) provides a uniform temperature over the entire surface regardless of shape and location of the cooled electronic component on the outer surface of the 100 × 100 × 6 mm HP.

The calculated values of the maximum transmitted heat flux, critical heat flux density and thermal resistance of the developed HPs have shown that they are suitable for building large-size TEC units. Calculated values of the critical heat flux density during boiling of ammonia in the temperature range from +20 to –40°C are from 140 W/cm^2 (at a temperature of +20°C) to 60 W/cm^2 (at a temperature of –40°C).

Experimental studies led by Rassamakin B. M. [52, 53] showed the following:

1) For the cold side of the TEC unit, thermal resistance of the HP was 0.17 °C/W with a heat flux density of 5.22 W/cm^2 and a vapor temperature of 37.0°C;
2) For the hot side of the TEC unit, thermal resistance of the HP is 0.03 °C/W at a heat flux density of 200 W and a vapor temperature of 49.0°C.

The research results indicate that using large-size TECs and flat aluminum HPs with ammonia as a base for thermal management devices for electronics has good prospects.

6.4. Improving TEC as Elements for Combined Thermal Management Devices for Electronics

Another aspect of improving TECs as elements of combined HP-based thermal management units for laser and infrared devices is forming optimal load parameters and optimizing the design parameters of heat transitions. The load parameters for TECs at hot plate temperatures T_h = 50°C and T_h = 30°C, developed for thermal management of LMs and photoelectric sensors (PES) are given in [54]. At intermediate temperature values of the heat-generating surface of the TEC, its characteristics can be determined by extrapolation.

Using the load characteristics obtained for different T_h values, it is possible to estimate the change of cooling temperature of the photodetector when changing its thermal capacity Q_0 or supply current I of the TEC to determine the electrical power consumption needed for cooling in each case. These characteristics also allow comparing the parameters of the calculated (only optimal for any one operating mode) TEC design with the parameters of the existing TECs, which are commercially available, and answer the question: whether producing a newly designed unit is worth the effort.

A study of the temperature distribution in interstage thermal bridges of different thickness made of beryllium oxide, aluminum nitride and aluminum oxide (VK96 ceramics) [55] showed that at a typical thickness of ceramic plates of 1–2 mm, thermal gradient on heat transitions from beryllium oxide and aluminum nitride do not exceed 1 K, while on VK96 ceramics the gradient reaches 2 K. These values can be recommended as input data for keeping record of thermal gradient loss in ceramic plates when designing optimal cascade TECs.

Increasing the height of thermal bridges makes the longitudinal gradient to grow and the transverse gradient to decrease, thus there is an optimum altitude at which the total thermal gradient provided by the TEC of this design reaches its maximum value. The optimum height of the thermal bridge is approximately 2 mm. The results show that when thermal bridges are made of materials with high thermal conductivity (beryllium oxide, aluminum nitride), a change in height in the interval of 0.5–3 mm does not lead to a significant decrease in the total TEC thermal gradient. In aluminum oxide ceramic plates (BK96) this interval decreases to 1.0–2.5 mm. The cooling depth depends significantly on the thermal conductivity of the thermal bridge material. If the thermal bridge is made of beryllium oxide, the calculated TEC cooling temperature is 189 K. Replacing this type of ceramics with aluminum nitride raises the temperature by approximately 1 K, and by 7 K in case of replacing it with VK96 ceramics.

The quantitative analysis of the effect of thermal bridges on the cooling temperature can be useful in the development of new designs of cascade TECs for thermal management of electronic devices.

6.5. Using Gravitational Heat Pipes for Thermal Management of High-Power LED Modules

Powerful electronic LED modules are widely used in power-efficient LED lighting fixtures and are characterized by significant heat release. During operation of a LED lighting fixture, about 75% of the consumed electric power is released as heat and leads to overheating of the module's LEDs. In order to ensure the normal thermal conditions of the

LED modules, an effective way is to use gravity heat pipes with a threadlike capillary structure.

In order to experimentally test this, models of high-power volumetric LED modules and a LED chandelier were developed and manufactured (see Figure 35). The frame of the LED chandelier is also a heat sink system made of five curved gravity heat pipes with a thread-like capillary structure and R-*141b* as a working fluid. The HPs carry LED light sources located inside the light-diffusing shades. The light sources (volumetric LED modules) were installed in the heating zone of HPs. Plate radiators were fixed in the condensation zone of heat pipes. The heat emitted by the LED modules is effectively removed by the heat pipes to the plate radiators which dissipate the heat into the surrounding air. The surface area of each radiator is 0.304 m^2.

LED matrices were installed in thermal contact with the flat faces of the module base, made in the form of a polyhedral prism with a through axial hole (see Figure 35, a).

The investigation of thermal characteristics of a gravitational heat pipe with a threadlike capillary structure showed that when the heat flow value is kept in the range of 2 to 27 W, the thermal resistance of a gravitational HP with thread-like capillary structure is 4.5–2.0 times lower, and the average temperature value in the heating zone is respectively 8–6°C lower than the corresponding values for a thermosyphon of the same size.

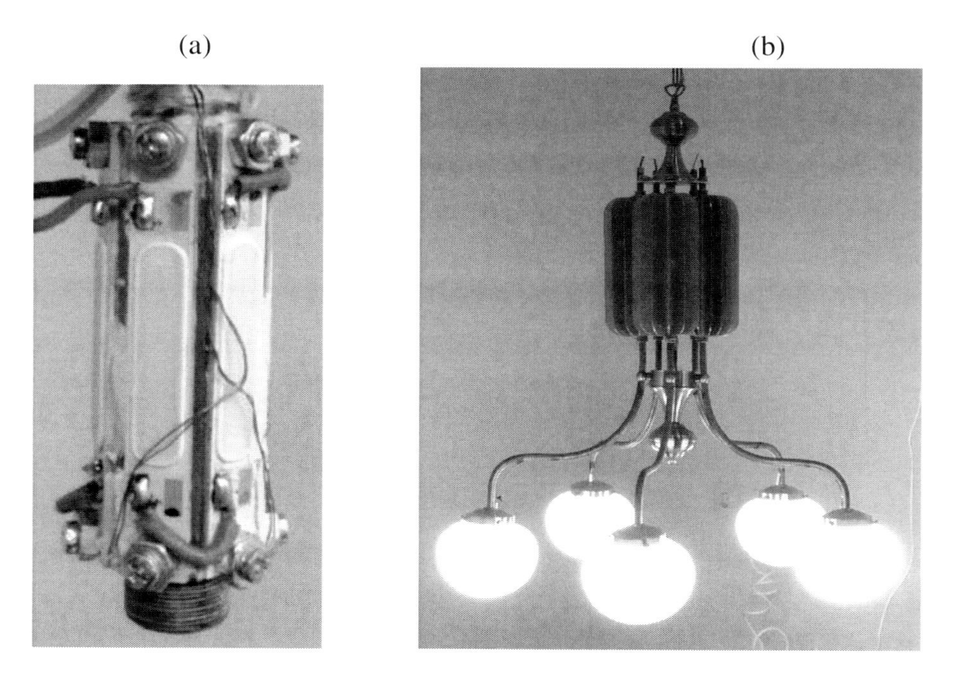

Figure 35. Photographs of a volumetric LED module and an LED chandelier with heat pipes: (a) Hexagonal LED module mounted on a heat pipe; (b) LED chandelier containing a framework of heat pipes with a threaded capillary structure and three-dimensional LED modules as light sources.

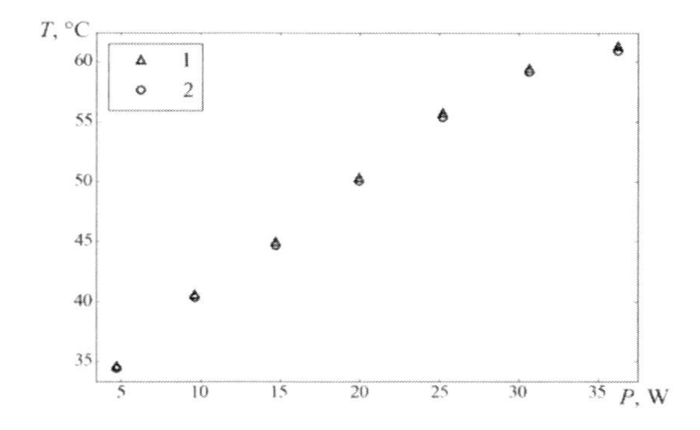

Figure 36. Temperature dependence of power consumption of the LED module on two faces.

Experimental study of the LED modules established (see Figure 36) that at a predetermined maximum value of the electric power LED modules, e.g., 24 W (which is twice the capacity of most modern LED drop-in replacement lamps), the substrate temperature of LED modules under natural convection of ambient air with a temperature of 25 ± 1°C does not exceed +55°C. This proves that such technical solution may ensure long-term reliable performance of LED modules.

Figure 36 presents the experimentally obtained graph indicating the relationship between the temperature of the LED module and the consumed electric power in the range of 5 to 35 W. The Figure shows, that the temperature of the LED matrices does not exceed +60°C when the power consumed by the LED module is 30 W and +55°C when the power is 24 W.

Thus, the experiment confirmed that the proposed design of the LED chandelier with gravitational heat pipes allows maintaining a normal temperature of volumetric LED modules with a power consumption 2.5 times (from 12 to 30 W) higher and thus with twice the emitted light flux than the currently used LED drop-in replacement lamps can afford.

7. New and Advanced Technological Solutions for Manufacturing HP-Based Thermal Management Devices for Electronics

7.1. Technological Solutions for Manufacturing Flat Finned Heat Pipes

The developed technology for manufacturing flat finned HPs (see Figure 37) for thermal management of multichannel sources of secondary power supply of electronic systems allows assembling them in standard cassettes with limited overall dimensions, which could not be done with traditional heat sinks.

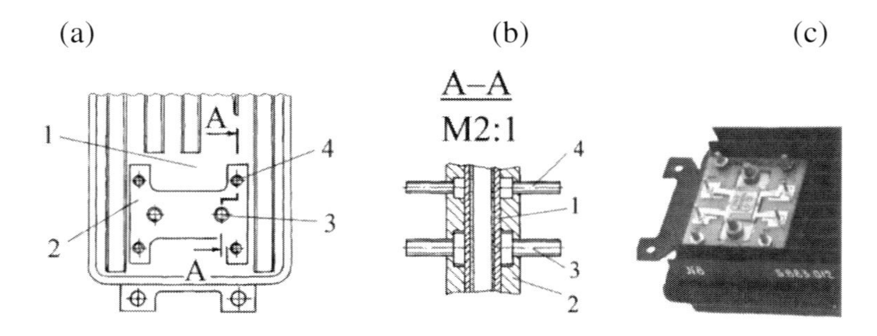

Figure 37. Mounting integrated circuits on a flat finned heat pipe using a special intermediate plate: (a) Position of the intermediate plate on the HP; (b) Cross-section of fasteners mounted into the intermediate plate; (c) Photograph of the installed IC;
1 – HP wall; 2 – intermediate plate; 3 – screw for IC fixing; 4 – screw for PCB fixing.

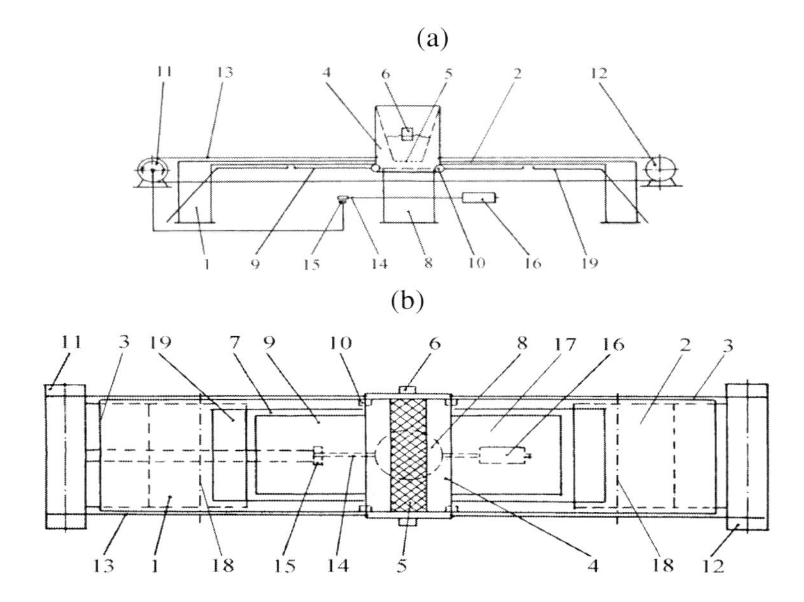

Figure 38. Schematic diagram of a device for mechanized filling of a metal-fibrous capillary structure layer on a substrate: (a) Front view; (b) Top view.
1 – case; 2 – working area; 3 – boundaries of the working area; 4 – basket; 5 – grid (sieve); 6 – vibration generator; 7 – window; 8 – scales; 9 – substrate; 10 – rollers; 11 – electric motor; 12 – idle shaft; 13 – draft; 14 – control element; 15 – switch; 16 – master element; 17 – uniform motion zone; 18 – boundaries of the uniform motion zone; 19 – plate.

The case of the heat pipe and the capillary structure were made of copper, distilled water was used as coolant. The walls of the HP case were stamped out of thin copper sheets and welded together around the perimeter by argon-arc welding thus forming a 6 × 60 × 150 mm hollow hermetic case. The inner surface of the body is covered with a metal-fiber CS sintered from 3 mm long copper fibers 50 μm in diameter. The sintering was a single technological cycle in a vacuum furnace at a temperature close to the melting point of copper. The 0.5 mm thick CS layer has a porosity of 85%. The technology of mounting integrated circuits to HPs via a special intermediate plate soldered to the HP case is also original (see Figure 37, a, b). This made it possible to place the heat-dissipating base of the IC directly on the plate securely fixing the IC on the HP with screws and nuts maintaining the case airtight. The flat leads of the IC could be connected directly to the printed circuit board conductors without any previous molding (see Figure 37, c).

The developed and implemented technological solutions provide a reliable thermal contact between the IC and the HP case. Such design is easy to manufacture, assemble and maintain.

7.2. Mechanized Filling of a Metal-Fibrous Capillary Structure Layer

A method and a device for mechanized filling of a metal-fibrous capillary structure (see Figure 38) have been developed for the small-scale fabrication of HPs (see Figure 38) [56].

Mechanizing the process of filling fibers onto a substrate allows significantly reducing the time of manufacturing fiber-optic cases, thus reducing labor costs and increasing the efficiency of capillary structure production. The use of this method made it possible to reproduce structural and geometric characteristics of the metal-fiber CS with high accuracy and to improve the quality of heat pipes with such a capillary structure.

7.3. New Technological Solution for Group Filling of Heat Pipes

In order to provide a high-efficiency group filling of HPs with a heat carrier, a new device has been developed [57]. Unlike the existing ones, this device allows filling a package of several tens of HPs with working fluid at once, distributing the working fluid evenly over all HPs in the package. This improves filling accuracy, productivity and reproducibility during batch production.

7.4. New Technological Solutions for Sealing Titanium Heat Pipes

In order produce lighter HPs for thermal management of electronic devices, the problem was solved of finding a reliable sealing technology for HPs made of titanium or other materials which after welding acquire low flexibility. The new solution consists in creating a filling tube made of bimetallic material (see Figure 39) [58]. One of the two metals is the basic material of the HP and is a low-plastic material (e.g., titanium), while the second one is a high-plastic metal (e.g., copper). The filling tube is made of a cylindrical bimetallic blank (copper–titanium). As a result, one half of the filling tube (lengthwise) is low-plastic (titanium), and the second one is high-plastic (copper) [59].

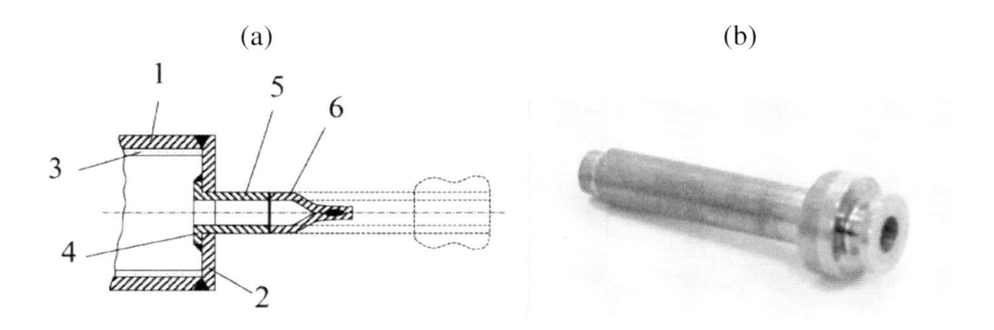

Figure 39. Sealing the heat pipe made of a low-plastic material: (a) Schematic diagram of sealing a filling unit with a bimetallic filling tube using clamping and cold welding; (b) Photograph of a bimetallic filling tube.
1 – HP case; 2 – end cap; 3 – capillary structure; 4 – welded seam; 5, 6 – low and high plastic parts of the filling tube.

In order to produce bimetallic titanium–copper blanks for filling tubes, two methods of welding titanium and copper have been developed and investigated. The first method is welding in the solid phase without intermediate layers [60]. The second one is welding under pressure and heating in vacuum using intermediate layers [61]. The most rational technological regimes have been experimentally developed for obtaining copper–titanium welded seams using the proposed methods.

The developed technological solution allowed producing reliable vacuum-tight sealing of titanium HPs in the copper part of the filling tube using cold welding by pressure and diffusion. The developed method can be used in the manufacture of both heat pipes and thermosyphons.

7.5. New Technological Solutions for Manufacturing Capillary Structures of Miniature Heat Pipes

Key aspects of MHP manufacturing technology arise from their small size. Using plug-in CSs is limited by the narrow vapor space of the MHP. The most promising approach is to make the CS directly inside the MHP wall. To this end, a new method has been proposed for manufacturing CSs for copper MHPs by making a network of microcracks on the inner surface of the case [62]. This allows for a porosity of 50 to 60%. In essence, it is necessary to create the CS on the inner surface of a copper MHP case by forming an oxidized capillary-porous layer and then reducing it to the base metal. Under special heat treatment conditions, the copper surface in an oxidizing environment is covered with a thin oxidized layer, which consists of an oxide and a cuprous oxide. These two chemical compounds have different coefficients of thermal expansion. This allows develop a significant mechanical stress, when creating a thermal gradient between the metal of the case and the oxidized layer. This mechanical stress leads to the cracking of the oxidized layer and formation of a system of interconnected capillaries and pores. The thin capillary-porous oxidized layer is then reduced in a reducing environment, and a capillary-porous structure of a certain thickness is formed, connected by an ideal thermal contact to the MHP case.

CONCLUSION

The described concepts and approaches to the design and technological implementation of two-phase heat transfer technologies in the main types of electronic devices, taking into account the main aspects of their design, make it possible to expand the functionality of leading-edge electronic systems using modern components with increased heat release.

1. The essence of these concepts lies in using a closed evaporation-condensation heat transfer cycle with a high temperature uniformity throughout the vapor space, which allows reducing the overall temperature gradient between the electronic component and the cooling water, reducing the temperature of electronic components, simplifying the main hydraulic pipeline of the basic supporting structure, thereby reducing its hydraulic resistance and energy costs for pumping of cooling water. This concept became a basis for new design and technological solutions of a functional module with a heat sink in the form of sectional heat pipes; for a ceramic switching board of a module with a built-in collector thermosyphon; for an instrument cabinets with shelves made of heat-pipe panel collectors; for a more reliable high-power electronic devices and transformer blocks with an evaporation-condensing cooling system arranged in water-cooled instrument cabinets; for control systems of infrared devices with a combined cooling system based on a contour heat pipe and thermoelectric coolers.

2. A new approach to maintaining a normal temperature of electronic components in functional modules of modernized complex electronic systems (with a significant increase in heat generation) was proposed and implemented. The approach consists in arranging parallel channels for removing heat from the most critical electronic components to the module case by using heat pipes of an unorthodox design. This made it possible to almost double (from 27 to 51.5 W) the heat output from the most heat-loaded electronic component (a powerful transistor) without increasing the overall size of the module.

3. The use of sectional heat pipes and evaporative thermosyphons in ceramic and metal plates as effective heat sinks in the basic supporting structures of the second level was proposed and experimentally justified. This allowed increasing the power of the functional module up to 204 W, of the ceramic commutation board up to 160 W, of the metal board unit up to 570 W.

4. It is shown that the most effective form of parallel-plate channels in the collector thermosyphon of a ceramic plate is a channel with a complex cross section: four shallow channels of 0.72 × 1.43 mm, connected by a common 10 × 0.78 mm channel. The studies

of evaporation and hydrodynamics in the miniscale channels of the collector TS allowed establishing that with an increase in the density of the heat flux in the evaporation zone from 0.3 to 1.0 W/cm^2, the intensity of heat exchange increases in all evaporation channels, and starting from 1.0–1.2 W/cm^2 it is practically independent of the heat flux density, and reaches values of 5000 W/(m^2°C), depending on the roughness of the channel surface. Experiments confirm that with the total thermal power of the electronic components of 160 W, the maximum surface temperature of the module's ceramic board does not exceed +60°C when the water temperature is +19°C at the inlet of the water-cooling unit.

5. A conceptually new approach to the improvement of thermal management devices based on HPs is proposed and implemented. The new method consists in the use of HP collector structures not only in second-level BSSs (cassettes, modules, blocks), but in third-level BSSs (cabinets, racks, shelves) as well. This made it possible to double the heat output (from 15 to 30 kW), which is removed from one instrument cabinet, while reducing the hydraulic resistance of the water path and energy consumption.

6. Studied were thermal management techniques, heat transfer and hydrodynamics in third-level basic supporting structures with a water cooling system using U-shaped heat pipes designed to create advanced heat-loaded electronic systems. Comparison of experimental data and computer simulation results showed that the developed models adequately describe (inconsistency within 25–30%) thermal and hydromechanical processes, and that the most effective mode of cooling water supply is feeding it from the bottom upwards. This makes it possible to use two-phase technologies to create highly efficient new generation thermal management devices for the BSSs that would remove up to 30 kW of heat from one instrument cabinet with a high uniformity of the temperature field both within one section (5 to 7°C) and within a multisection BSS (3 to 5°C), with a minimum hydraulic resistance of the main water pipeline.

7. A set of constructive and technological solutions have been developed for the use of efficient heat sinks based on miniature and conventional direct and curved heat pipes with metal-fiber and wire capillary structures to maintain a normal temperature of the modern electronics with an increased specific heat such as laser diodes, LEDs, matrices and modules based on such devices.

8. The concept of heat removal from microelectronic photodetector devices of the IR control systems to a high-temperature environment was proposed and implemented. The concept consists in cooling the photodetector devices to a specified operating temperature of 213 K using a multistage TES, heat transfer via a closed evaporation-condensation cycle to an LHP with a single-stage TEC, which serves as a heat pump, to the heat-dissipating surface with a simultaneous increase in its temperature to a level higher than the ambient environment (400 K), and further dissipation of heat into the environment.

9. A set of new technological solutions is presented for the manufacture of heat pipes for the developed highly efficient thermal management techniques for modern and advanced electronic systems.

REFERENCES

[1] *Pentium 4 computing performance dependence on temperature,* (in Russian), http://fcenter.ru/online/hardarticles/processors/5798 (Accessed 18.02.2018).

[2] V.P. Konyaev, A.A. Marmalyuk, M.A. Ladugin, T.A. Bagaev, M.V. Zverkov, V.V. Krichevsky, A.A. Padalitsa, S.M. Sapozhnikov, V.A. Simakov, Laser-diode arrays based on epitaxial integrated heterostructures with increased power and brightness of the pulse emission, *Semiconductors 48* (1) (2014) 99–103. http://dx.doi.org/10.1134/S1063782614010175.

[3] F.J. Arques-Orobon, N. Nucez, M. Vazquez, C. Segura-Antunez, V. Gonzalez-Posadas, High-power UV-LED degradation: Continuous and cycled working condition influence, *Solid-State Electronics* 111 (2015) 111–117. http://dx.doi.org/10.1016/j.sse.2015.05.039.

[4] Moon-Hwan Chang, Diganta Das, P.V. Varde, Michael Pecht, Light emitting diodes reliability review, *Microelectronics Reliability* 52 (2012) 762–782. http://dx.doi.org/10.1016/j.microrel.2011.07.063.

[5] S. Chhajed, Y.Xi, Y.-L. Li, Th. Gessmann, E.F. Schubert, Influence of junction temperature on chromaticity and color-rendering properties of trichromatic white-light sources based on light-emitting diodes, *Journal of Applied Physics* 97 (2005) 054506-1-054506-8. http://dx.doi.org/10.1063/1.1852073.

[6] S.M. Sohel Murshed, C.A. Nieto de Castro, A critical review of traditional and emerging techniques and fluids for electronics cooling, *Renewable and Sustainable Energy Reviews* 78 (2017) 821–833. http://dx.doi.org/10.1016/j.rser.2017.04.112.

[7] Valery Kosihin, *Processors overview Intel Xeon E5 v4,* 31.03.2016. https://servernews.ru/930831 (Accessed 18.02.2018).

[8] Jack Dongarra. *Report on the Sunway TaihuLight System.* June 24, 2016, http://technodocbox.com/amp/68488238-PC_Support/Report-on-the-sunway-taihulight-system-jack-dongarra-university-of-tennessee-oak-ridge-national-laboratory.html (Accessed 18.02.2018).

[9] S.W. Chi, *Heat pipe theory and practice,* Hemisphere publishing, Washington, DC, 1976.

[10] D. Reay, P. Kew, *Heat Pipes Theory, Design and Applications,* fifth ed., Butterworth-Heinemann, Oxford, UK, 2006.

[11] M.N. Ivanovskii, V.P. Sorokin, I.V. Yagodkin, *The physical properties of heat pipes,* Clarendon Press, Oxford, 1982.

[12] L.L. Vasiliev, *Low-temperature heat pipes,* A.V. Luikov Heat and Mass Transfer Institute of NAS of Belarus, Minsk, 2006, (in Russian).

[13] Yu.E. Nikolaenko, V.I. Cherny, A.P. Yakovenko, A.T. Tunik, *SU Author's certificate for invention No. 851806,* 1981, (in Russian).

[14] Yu.E. Nikolaenko, V.I. Cherny, A.P Yakovenko, *SU Author's certificate for invention No. 894891,* 1981, (in Russian).

[15] Yu.E. Nikolaenko, *UA Patent No. 36268,* 2001, (in Ukraine).

[16] Yu.F. Kiselev, *Investigation of heat-mass transfer processes in two-phase thermosyphons with low-temperature heat carriers,* PhD thesis, Kyiv Polytechnic Institute, Kiev, 1980, (in Russian).

[17] Yu.E. Nikolaenko, *UA Patent No. 109840,* 2016, (in Ukraine).

[18] D.V. Kozak, Yu.E. Nikolaenko, Thermal resistance of a gravitational heat pipe with threaded capillary structure, in: *Modern Problems of Power Engineering Science: Materials of the XIV International Scientific and Practical Postgraduate Students, Graduates and Students,* dedicated to the 85 th anniversary of the Heat and Power Faculty, Kyiv, April 18-21, 2016. In 2 Vol., Kyiv: NTUU "KPI", 2016, Vol. 1, P. 81. ISBN 978-966-622-697-9 (Zag.), ISBN 978-966-622-695- 5 (Vol. 1). http://tef.kpi.ua/rub_334.htm.

[19] Yu.E. Nikolaenko, D.V. Kozak, V.Yu. Kravets, S.M. Khairnasov. Comparison of the thermal characteristics of a thermosyphon and a gravitational heat pipe of the same size. *Proceedings of the XVII International Scientific and Practical Conference "Modern Information and Electronic Technologies",* May 23-27, 2016, Odessa, P. 164–165, (in Russian). http://ela.kpi.ua/handle/123456789/17486. ttp://tkea.com.ua/siet/archive/2016/164-165.pdf.

[20] V.M. Baturkin, Yu.E. Nikolaenko, D.M. Galyautdinov, I.T. Vladimirov, Effective cooling of powerful super high-frequency microelectronic unit, *Tekhnologiya i Konstruirovanie v Elektronnoi Apparature* (3) (2007) 46–50. http://dspace.nbuv.gov.ua/handle/123456789/52814.

[21] V.M. Baturkin, Yu.E. Nikolaenko, D.M. Galyutdinov, I.T. Vladimirov, V.N. Savina, *UA Patent No. 29528*, 2008, (in Ukraine).

[22] V.Yu. Kravets, Yu.E. Nikolaenko, *UA Patent No. 50435*, 2002, (in Ukraine).

[23] Yu.E. Nikolaenko, V.Yu. Kravets, The effect of mode parameters on Heat transfer characteristics of miniature heat pipes, *Tekhnologiya i Konstruirovanie v Elektronnoi Apparature* (6) (2001) 36-38, (in Russian). http://dspace.nbuv.gov. ua/handle/123456789/70892.

[24] V.Yu. Kravets, Yu.E. Nikolaenko, Ya.V. Nekrashevich, Experimental Studies of Heat-Transfer Characteristics of Miniaturized Heat Pipes, *Heat Transfer Research* 38 (6) (2007) 553–563. http://dx.doi.org/10.1615/HeatTransRes.v38.i6.70.

[25] M.G. Semena, B.M. Rassamakin, S.K. Zhuk, R.A. Petrenko, Yu.E. Nikolaenko, *SU Author's certificate for invention No. 1044944,* 1983, (in Russian).

[26] M.G. Semena, R. Muller, Yu.E. Nikolaenko, *SU Author's certificate for invention No. 1017900*, 1983, (in Russian).

[27] S.M. Khayrnasov, Ye.N. Pis'mennyy, Yu.E. Nikolaenko, B.M. Rassamakin, Experimental modeling of a loop heat pipe, *Teplovyye rezhimy i okhlazhdeniye*

radioelektronnoy apparatury [*Thermal works and cooling of electronic equipment*] (1) (1999) 57–61, (in Russian). http://ela.kpi.ua/ handle/123456789/17729.

[28] Yu.E. Nikolaenko, B.M. Rassamakin, S.M. Khayrnasov, Loop heat pipes with aluminum evaporator for combined cooling systems Radio electronic equipment, *Tekhnologiya i konstruirovaniye v elektronnoy apparature* [*Technology and design in electronic equipment*] (3) (2002) 22–26, (in Russian). http://www.tkea.com.ua/tkea/2002/3_2002/content.html.

[29] Yu.E. Nikolaenko, Structural and technological features of design of power supply units of REE with heat pipes, in: *Proceedings of the International Scientific and Practical Conference "Modern Information and Electronic Technologies"*, Politehperiodika, Odessa, 2000, pp. 54–55, (in Russian).

[30] Yu.E. Nikolaenko, Distribution of thermal resistance of an experimental batch of heat pipes, in: *Proceedings of the 3rd International Scientific and Practical Conference "Modern Information and Electronic Technologies"*, Politehperiodika, Odessa, 2002, p. 150, (in Russian).

[31] Yu.E. Nikolaenko, *UA Patent No. 45075,* 2002, (in Ukraine).

[32] Yu.E. Nikolaenko, Ceramic commutation boards for functional modules of a computer with increased heat release, *Upravlyayushchie Sistemy i Mashiny* [*Control systems and machines*] (5) (2006) 30–39, (in Russian). http://ela.kpi.ua/handle/123456789/16446.

[33] Yu.E. Nikolaenko, A.A. Tsygansky, Research of processes heat exchange in collector thermosyphons of commutation boards high degree of integration, *Tekhnologiya i Konstruirovanie v Elektronnoi Apparature* [*Technology and design in electronic equipment*] (6) (2007) 36–41, (in Russian). http://www.tkea.com.ua/tkea/2007/6_2007/content.html.

[34] B.I. Basok, Yu.E. Nikolaenko A.A. Tsygansky, Study heat exchange in boiling in vertical mini-scale channels complex form, in: *Theses of reports and reports, VI Minsky International Forum on Heat and Mass Transfer,* Minsk, 2, 2008, pp. 12–13, (in Russian). http://ela.kpi.ua/handle/123456789/18360.

[35] Yu.E. Nikolaenko, Schematic solutions for the organization of heat removal from functional computer modules using two-phase heat transfer elements and devices, *Upravlyayushchie Sistemy i Mashiny* [*Control systems and machines*] (2) (2005) 29–37, (in Russian). http:// ela.kpi.ua/handle/123456789/16362.

[36] Yu.E. Nikolaenko, A.A. Tsigansky, *UA Patent No. 40635,* 2009, (in Ukraine).

[37] N.M. Vakiv, A.A. Tsygansky, Yu.E. Nikolaenko, R.M. Goncharik, Thermosyphon collector cooling system of the electronic unit, in: *Proceedingsd of the 11th International Scientific and Practical Conference "Modern Information and Electronic Technologies"*, Politehperiodika, Odessa, 2, 2010, p. 29, (in Russian).

[38] Yu.E. Nikolaenko, *UA Patent No. 58839,* 2003, (in Ukraine).

[39] Yu.E. Nikolaenko, Hardware construction of high-performance computing systems with increased efficiency of heat removal, *Tekhnologiya i Konstruirovanie v Elektronnoi Apparature* [*Technology and design in electronic equipment*] (5) (2005) 31–34, (in Russian). http://www.tkea.com.ua/ tkea/2005/5_2005/content.html.

[40] A.F. Verlan, I.O. Goroshko, Yu.E. Nikolaenko, Computer modeling of heat transfer processes in promising basic bearing designs of rack-type with heat pipes, *Mathematichny machiny i systemy* [*Mathematical mechanism and systematization*] (2) (2008) 90–99, (in Russian). http://www. immsp.kiev.ua/publications/articles/ 2008/2008_2/Verlan_02_2008.pdf.pdf.

[41] Yu.E. Nikolaenko, Experimental modeling of the system cooling cabinet with heat pipes, in: *Proceedings of the seventh International scientific-practical conference "Modern Information and Electronic Technologies",* Politehperiodika, Odessa, 2, 2006, p. 36, (in Russian).

[42] Yu.E. Nikolaenko, Heat transfer characteristics of the instrument Cabinet with U-shaped heat pipes, in: *Proceedings of the eighth International scientific-practical conference "Modern Information and Electronic Technologies",* Politehperiodika, Odessa, 2007, p. 271, (in Russian).

[43] *Handbook of infrared technology*, ed. W. Wolf, G. Cisis. In 4 volumes, vol. 3 Instrument base of IR systems, Mir, Moscow, 1999, (in Russian).

[44] R. Hudson, *Infrared systems,* ed. N.V. Vasilchenko, Mir, Moscow, 1972, (in Russian).

[45] A.K. Grezin, V.S. Zinoviev, *Microcryogenic technology*, Mechanical Engineering, Moscow, 1977, (in Russian).

[46] A.V. Molodyk, N.I. Nosov, G.A. Smolyar, Yu.E. Nikolaenko, Development and research of the combined cooling system for microelectronic photodetective devices, in: *Proceedings of the Sixth International Scientific and Practical Conference "Modern Information and Electronic Technologies",* Politehperiodika, Odessa, 2005, p. 226, (in Russian).

[47] V.I. Osinsky, V.G. Verbitsky, Yu.E. Nikolaenko, S.K. Zhuk, S.V. Bobzhenko, P.A. Mierzvinski, Thermal processes in microlaser devices of information systems, *Tekhnologiya i Konstruirovanie v Elektronnoi Apparature* [*Technology and design in electronic equipment*] (2-3) (2000) 27-36, (in Russian). http://www.tkea.com.ua/ tkea/2000/2–3_2000/content.html.

[48] Yu.E. Nikolaenko, S.K. Zhuk, V.M. Baturkin, D.N. Olefirenko, Modeling and selection of systems for ensuring the thermal regime of laser Modules, *Tekhnologiya i Konstruirovanie v Elektronnoi Apparature* [*Technology and design in electronic equipment*] (2) (2001) 31–36, (in Russian). http://www.tkea.com.ua/tkea/ 2001/2_2001/content.html.

[49] Yu.E. Nikolaenko, *UA Patent No. 38931,* 2001, (in Ukraine).

[50] Yu.E. Nikolaenko, *UA Patent No. 960,* 2001, (in Ukraine).

[51] A.L. Weiner, Yu.E. Nikolaenko, *UA Patent No. 4317,* 2005, (in Ukraine).

[52] B.M. Rassamakin, S.M. Khairnasov, Yu.E. Nikolaenko, V.K. Zaripov, Yu.Yu. Rosver, Development, production and research of heat pipes for a cooling system with a large thermoelectric module, in: *Proceedings of the 11th International Scientific and Practical Conference "Modern Information and Electronic Technologies",* Politehperiodika, Odessa, 2, 2010, p. 33, (in Ukraine).

[53] L.I. Anatychuk, L, N. Vikhor, Yu.Yu. Rozver, Yu.E. Nikolaenko, B.M. Rassamakin, S.M. Hairnasov, Research and development on performances of large-sized thermoelectric module with heat pipes, in: *Proceedings of the VIII Minsk International Seminar "Heat Pipes, Heat Pumps, Refrigerators, Power Sources",* Minsk, 1, 2011, p. 239–242.

[54] Yu.E. Nikolaenko, L.N. Vikhor, Simulation of load characteristics of an optimal cascade thermoelectric cooler, *Tekhnologiya i Konstruirovanie v Elektronnoi Apparature* [*Technology and design in electronic equipment*] (6) (2001) 33–35, (in Russian). http://www.tkea.com.ua/tkea/2001/ 6_2001/content.html.

[55] Yu.E. Nikolaenko, L.N. Vikhor, Effect of heat transitions on the temperature drop of a multistage thermoelectric battery, *Tekhnologiya i Konstruirovanie v Elektronnoi Apparature* [*Technology and design in electronic equipment*] (4-5) (2002) 16–18, (in Russian). http://www.tkea.com.ua/ tkea/2002/4–5_2002/content.html.

[56] M.G. Semena, S.K. Zhuk, V.M. Baturkin, Yu.A. Khmelev, A.S. Savchenko, Yu.E. Nikolaenko, *SU Author's certificate for invention No. 1054662,* 1983, (in Russian).

[57] M.G. Semena, Yu.F. Kiselev, Yu.E. Nikolaenko, V.I. Malov, V.A. Dincin, I.L. Rosenstein, *SU Author's certificate for invention No. 1300288,* 1987, (in Russian).

[58] Yu.E. Nikolaenko, *UA Patent No. 47806,* 2002, (in Ukraine).

[59] Yu.E. Nikolaenko, Solution of technological problems in the manufacture of titanium heat pipes, in: *Proceedings of the 4th International Scientific and Practical Conference "Modern Information and Electronic Technologies"*, Politehperiodika, Odessa, 2003, p. 238, (in Russian).

[60] Yu.E. Nikolaenko, G.K. Kharchenko, Yu.V. Falchenko, Welding in the solid phase of titanium with copper in the technology of manufacturing heat pipes, *Technological Systems* (2) (2003) 24–28, (in Russian). http://technological-systems.com/ru/zhurnal/arkhiv- nomerov/2003/ts-18.

[61] Yu.E. Nikolaenko, G.K. Kharchenko, Yu.V. Falchenko, O.A. Novomlynets, Pressure welding with heating in a vacuum of titanium with copper in the technology of manufacturing heat pipes, *Technological systems* (2) (2004) 56–59, (in Russian). http://ela.kpi.ua/handle/123456789/18444.

[62] Yu.E. Nikolaenko, V.Yu. Kravets, Investigation of the modes of temperature treatment of copper shells of thermal microtubes, *Tekhnologiya I Konstruirovanie v Elektronnoi Apparature* [*Technology and design in electronic equipment*] (1) (2000) 19–22, (in Russian). hhttp://www.tkea.com.ua/ tkea/2000/1_2000/content.html.

In: Heat Pipes: Design, Applications and Technology ISBN: 978-1-53613-908-2
Editor: Yuwen Zhang

Chapter 4

PULSATING HEAT PIPES: PERFORMANCE PERKS BASED ON DIFFERENT WORKING FLUIDS

Xiaoyu Cui* and Hua Han
School of Energy and Power Engineering,
University of Shanghai for Science and Technology, Shanghai, China

ABSTRACT

The pulsating heat pipe (PHP) is a relatively new and promising addition to the family of passive two-phase energy transport devices. By charging with water, methanol, ethanol, acetone and binary mixtures at various volume mixing ratios, a vertical closed-loop PHP of ten turns with an inner/outer diameter of 2.0/4.0 mm was experimentally investigated in the bottom-heating mode. Three levels of filling ratios were categorized: small (35%, 45%), medium (55%) and large (62%, 70%) with heat inputs ranging from 10 W to 100 W. Extensive comparisons were made based on experimental results. This paper presents research on the PHP flow pattern and heat transfer mechanism as well as the effect of the working fluids' physical properties on PHP thermal resistance. The PHP mixtures' zeotropic properties and the complex molecular interactions among the components were analyzed; moreover, the PHPs charged with the mixtures were much more complex than those with pure fluids. At small or medium filling ratios, most of the binary mixtures have a better anti-dry-out performance than at least one of the pure fluids (and possibly both). At a relatively high heat input and large filling ratio, the thermal performances of the PHPs charged with the binary mixtures at various mixing ratios showed no superiority over the pure fluids, especially pure water.

Keywords: pulsating heat pipe, oscillation, binary mixture, thermo-physical property, heat transfer, thermal resistance, working fluid

---

* Corresponding Author Email: xiaoyu_cui@usst.edu.cn.

NOMENCLATURE

D	inner diameter, m
FR	filling ratio
g	gravitational acceleration, m/s^2
I	electric current, A
Q	heat load, W
R	thermal resistance,
T	temperature, °C
t	time, s
U	electric voltage, V

Subscript

e	evaporation section
c	condensation section
l	liquid
v	vapor
s	saturation state

1. INTRODUCTION

Because of the increasing density of heat discharge and the application needs in the harsh, power-hungry environment of some electronic devices, the requirement for novel and promising heat-transfer devices have never been more urgent. Several types of heat-transfer devices characterized by high thermal performance and stable operation—even in extreme conditions—are introduced as ideal alternatives to traditional thermal devices. The pulsating heat pipe (PHP), patented by Akachi [1], possesses excellent thermal performance for electronic devices of significantly compact size. PHP innovations have provided solutions to many problems in the field of electronic cooling. PHP has two types of structure, that is, the open and closed loop pulsating heat pipe. Since the working fluid can flow circularly inside PHP pipes, the closed loop pulsating heat pipe (CLPHP) usually possesses better heat-transfer performance when compared to the open loop pulsating heat pipe. It is, in fact, a gas-liquid two-phase flow heat transfer system with certain randomness, jointly driven by multiple parameters, such as the type of the working fluid, filling ratio (FR), heat input, inclination and number of turns [2]. The research on PHP can be broadly classified into two groups, theoretical and experimental. Theoretical research on PHP is still in the exploratory stage. Existing theoretical models include the mass-spring combination model, the fundamental equations models of mass, momentum and energy, the chaotic dynamic analysis model and the neural network model. Nevertheless, great deviations can still exist between theoretical models and the corresponding physical objects they represent. Experimental research includes both visualization and non-visualization experiments.

In flow visualization experiments, PHP is made of glass or transparent resin. This makes it possible to observe the growth and elimination of bubbles, the flow of liquid slugs, and the distribution and motion of working fluids, as well as two-phase flow-type variations. For example, Tong et al. [3] found that at the start-up stage, because of the noise generated by bubbly boiling, the oscillation amplitude was relatively large, the flow direction was somewhat random, and it occasionally appeared that the majority of working fluid flowed to the condensation section, which together caused a temporary dry-out in the evaporation section and a corresponding stagnation in the condensation section. Moreover, during the normal working state, circulation flow had a constant direction. As heat input increased, circulation accelerated and in the entire cycle, liquid slugs vibrated locally. This observation led Khandekar et al. [4] to determine that an elevated heating load caused the slug flow to gradually change to an annular flow. Thus, the flow pattern transformation caused the PHP operation mechanisms and heat-transfer characteristics to change. Further study [5] focused on the visualized PHP with an inner diameter of 2 mm and indicted that the gravity effect truly affected distributions of working fluids in various PHP orientations. Visualization experiments led Xu et al. [6] to conclude that from start-up to stable operation, flow modes of the working fluid would undergo changes from bubble flow, slug flow, and plug flow to mixing flow. Fu et al. [7] found that pulsating was a part of the chaotic random motion at the start-up stage and the one-way circulation at steady state, while the flow patterns of the working fluid changed from bubble flow to slug flow and plug flow. Li et al. [8] found that, under an appropriate filling ratio (FR), with an increase in heat input, the flow direction successively experienced slight oscillations, substantial oscillations, and circulation flows (which included instability, periodic stagnation or backflow) as well as steady unidirectional loop flows. Qu et al. [9] investigated the typical flow patterns of a PHP with an inner diameter of 2.5 mm charged with deionized water. With the increase of heat flux in the evaporation section, the working fluid flow altered from bubble flow to slug flow to annular flow. When the working fluid entered the condensation section, the bubble collapsed due to condensation and the flow transformed from annular flow back to slug flow and bubble flow. In general, the flow patterns of the PHP could be traced as follows: With the increase of heat input, the typical changing sequence of a flow pattern was bubble flow to slug flow and finally to annular flow; moreover, the flow direction changed from up-down oscillation and an unstable circulation flow to a steady one-way circulation flow.

Under practical applications, PHP is always made of metal, where the phenomena in PHP is invisible and its heat-transfer performance can be judged only by measuring outer wall temperatures in various sections. Non-visualization experimental research concentrated on obtaining higher heat-transfer performance, investigating how a geometric structure affects heat transfer performance, the type of working fluid and the operation condition. For example: Miyazaki et al. [10] investigated different heating methods in a copper-R142b closed-loop PHP with a 1 mm inner diameter and 30 turns. Subsequently, they found that different heating methods corresponded to different optimum FRs. Ponnappan et al. [11] carried out experimental research on a mesh wick closed-loop PHP with quadrilateral grooves. Charoensawan et al. [5] indicated that the number of turns in the PHP possessed a critical value affected by the type of working fluids, the pipe diameter, and the heat flux. The heat-transfer performance of closed-loop PHPs with different pipe diameters was also compared, charged with water, R123 and ethanol. Xu et al. [12] conducted a PHP start-up experiment associated with power spectral density (PSD) analyses and found that the start-up process presented a temperature overshoot before steady oscillation at a low heating temperature; however, a smooth oscillation period can be attained at high heat input.

In Section 5, the authors found that by combining the flow characteristics obtained in visualization experiments with the heat-transfer characteristics acquired in non-visualization experiments, a close relationship between both types of experiments was established. Moreover, the variation of temperature and thermal resistance from experimental results was also investigated in detail to extract rules pertaining to local dry-out and thermal resistance. To the best of the authors' knowledge up to this point, the relationship between local dry-out and thermal resistance has not received any kind of noteworthy attention in PHP research so far.

Among the numerous methods used to enhance PHP heat transfer, the most direct and effective way is to find out the best functional working fluid and the best liquid filling ratio. Pure fluids employed in PHP are mainly water, methanol, ethanol, acetone, n-pentane, butane, R123, FC72, FC-75, R142b and HFE-7100.

Clement et al. [13] tested a closed-loop PHP made of copper tubing with an inner diameter of 1.657 mm and 15 total turns. Working fluids include acetone, methanol, and deionized water. Methanol has outperformed other fluids tested. Clement et al.'s PHP copper tube had a filling ratio of 45% and a power input of 120 W. Two different open loop PHPs made of a single copper tube were built by Lips et al. [14]. The tubes had an inner diameter of 1.2 mm (40 total turns, PHP1) and 2.5 mm (20 total turns, PHP2) for water and acetone as working fluids, respectively. Experimental results showed that for low heat fluxes, the operating curve (overall thermal resistance vs. heat rate) is irregular, and the PHP performance is sensitive to orientation. For high heat fluxes, the curve is smooth and independent from the orientation. Charoensawan et al. [5] investigated closed loop PHPs filled with water, ethanol and R-123. The results indicate a strong influence of gravity and the numerous turns on the performance. The thermophysical properties of working fluids affect the performance which also strongly depends on the PHP operation boundary conditions. Burban et al. [15] characterized an air-cooling open looped PHP (ID = 2.5 mm) with four working fluids: acetone, methanol, water, and n-pentane; they found that acetone and n-pentane could be classified into one group that could always achieve good results for low power input and low air temperature, while water and methanol steadily improved their performance as temperature and heating power increased. Tseng et al. [16] employed working fluids including distilled water, methanol, and HFE-7100 to examine the performance of closed-loop pulsating heat pipes with a uniform tube diameter (ID = 2.4 mm) or with an alternating tube diameter. The results showed that thermal resistance in a vertical arrangement is much lower than that in a horizontal arrangement, and for a low input power, the PHP with HFE-7100 exhibited the least thermal resistance; however, when the input power was increased to over 60 W, the thermal resistance of the PHP with distilled water became the smallest.

Lin et al. [17] tested PHPs using FC-72 and FC-75 as the working fluids. They found that when charge ratios were less than 40%, the evaporator was dried out. At a charge ratio of 50%, the PHP could dissipate 2040 W of heat. They concluded that PHP performance was independent of orientation when using the two working fluids. Sarangi et al. [18] experimentally investigated the pulsating start-up heating power, maximum heating power and optimal filling ratio of a PHP with an inner diameter of 1 mm and an outer diameter of 2 mm, which was charged with pure water and ethanol. [18] found that the pulsating start-up heating power is unrelated to the filling ratios and that the maximum heating power is related to the filling ratios. Under the same working temperature, the optimal filling ratio at the maximum heating power relates to the working fluids. PHPs with water and ethanol had maximum heating powers at filling ratios of 62.5% and 50%, respectively, which were referred to as the optimal filling ratios. Shafii et al. [19] experimentally investigated the heat transfer performance of a PHP with an inner diameter of 1.8 mm with water and ethanol. The results indicated that the optimal filling ratio was 40% for water and 50% for ethanol and that the heat transfer performance decreased when the filling ratio was lower than 30% or higher than 70%.

Researchers have also investigated the influences of physical properties of working fluid on the running state and heat transfer performance of PHPs. PHP properties include boiling point, specific heat, latent heat of vaporization (LHP), dynamic viscosity, density, $(dp/dT)_{sat}$, etc. Zhang et al. [20] studied the thermodynamic oscillation characteristics of a PHP (ID = 1.18 mm) with three different working fluids, FC72, water, and ethanol, and determined that the period of oscillation was mainly related to the latent heat of vaporization (LHV), while amplitude was influenced mainly by surface tension. The looped PHP with water provides a better overall thermal performance once the heating power is greater than its minimum value. FC-72 was recommended for use in low-heat-flux situations, due to its lower minimum heating power in initiating PHP operation. The optimal filling ratio was reported as around 70% for all three working fluids. Liu et al. [2] investigated the start-up performance of a closed-loop pulsating heat pipe (ID = 2.6 mm) and determined that the optimal liquid filling ratio for start-up was about 41% for water, 52% for ethanol, and fell within the range from 35% to 41% for methanol. The working fluid with small dynamic viscosity, small specific heat, and especially a large ($(dp/dT)_{sat}$) is beneficial to the start-up performance of the PHP. Rittidech et al. [21] studied open-loop PHPs. After comparing water, ethanol and R123, they determined that the lower the latent heat of vaporization (LHV), the greater the ratio was between the maximum heat transfer and the heat transfer when horizontally placed. Lin et al. [22] used a high-definition camera to study the heat transfer performance of a PHP at a filling ratio of 60% with methanol and ethanol under different heating powers. They found the thermal resistance of methanol to be lower than the thermal resistance with ethanol within a range of heating power levels, which primarily depend on $(dp/dT)_{sat}$ of methanol being higher than that of ethanol.

After further analyzing the literature, we could not find an acceptable explanation of the effect of working fluid's physical properties on PHP flow and heat transfer performance. A closed-loop PHP (ID = 2.0 mm) was charged with a different fluid to determine each fluid's effect on heat transfer performance. The test fluids were deionized water, methanol, ethanol and acetone. The research aim was to reveal the effect of each fluid's properties (i.e., specific heat, latent heat, viscosity, etc. on the thermal performance of PHP (specified in Section 6).

The effects of various mixtures have been investigated [23–28] including nano- or micro-additive mixtures and the mixtures of pure fluids. For nano- or micro-additives fluids, the additives include Cu [29], CuNi [30], silver [31, 32], Al_2O_3 [33-35], SiO_2 [33], diamond [36] and multi-walled carbon nanotubes (MWCNT) [37]. Generally, most researchers fixed the filling ratio at 50% and the major conclusion was that PHP heat transfer and start-up performance could be enhanced by additives (although the enhancement can vary for different materials, sizes and concentrations of nano-particles) under most PHP conditions with inner diameters from 1.6 mm [30] to 3.0 mm [37]. Meanwhile, Qu et al. [33] reported that the SiO_2/water nanofluids deteriorated the thermal performance of PHP when compared to pure water; moreover, the different effects induced by the addition of different nanoparticles to pure water were attributed to the change of surface conditions at the evaporator and condenser due to different nanoparticle deposition behaviors.

Ma et al. [36] determined that higher thermal conductivity, lower viscosity of nanofluids, and stronger oscillating motion of nanoparticles were the primary factors enhancing the heat transport capability in nanofluid-charged PHPs. Besides, although many preparation methods were proposed, it is still a challenge to make a nanofluid homogeneous and to provide long-term stability with negligible agglomeration, and without adversely affecting the thermo-physical properties like lower specific heat, high viscosity, etc. [38]

For mixtures of various pure fluids, despite the studies on heat pipes where N2-Ar mixtures [23], 2-propanol-water mixtures [24], methanol-water and TEG-water mixtures [25], water-alcohol mixtures [26] and ethanol-water mixtures [39] were tested and proved to be effective in performance promotion or in enlarging functional temperature range, the studies of PHP with mixtures of various pure fluids have been limited.

Burban et al. [15] mentioned the immiscible mixture of water and n-pentane for PHP application; they also conducted a preliminary study with a mixing ratio of 2/3 and 1/3 for the immiscible water n-pentane mixture. They used an open-loop PHP of 16 turns in a single 2.5 mm inner and 3.2 mm outer diameter copper tube. The filling ratio was fixed at 60%. The results showed improved performance at low temperatures when compared with the PHP filled with only water or n-pentane; notably, the improvement was attributed to the high pressure of the n-pentane driving the flow, and the high specific heat capacity and latent heat of water, which effectively transferred the heat. It is expected that the diversity of working fluids can be multiplied by the use of mixtures and the performance be effectively improved. Mameli et al. [40] carried out experiments on a closed-loop PHP made of copper tubes at the evaporator and condenser sections with transparent glass tubes at the adiabatic section (ID = 2.0 mm) with 4 total turns. The PHP was filled with ethanol at a filling ratio of 65%. One exploratory test with azeotropic mixture used a fixed boiling point of 78.2°C at 1 atm. The mixture consisted of ethanol and water (95.5%:4.5% by weight). Equivalent thermal conductivities on the order of 10 to 15 times that of pure copper were achieved. Charoensawan et al. [41] experimentally investigated the heat transfer performance of a closed-loop PHP with a water-ethanol mixture. It was reported that the water-ethanol mixture had a filling ratio of 30% with an evaporating pipe length of 150 mm or at a filling ratio of 30% or 50% with an evaporating pipe length of 50 mm. Both pipe lengths and filling ratio combinations could effectively improve the heat transfer performance of the PHP. While this is a preliminary research finding, it will certainly be of interest to check the applicability of low volatility aqueous mixtures, which have the advantage of high latent heat of water but a large $(dp/dT)_{sat}$ of the low volatility component [42].

Determining final optimum mixtures of various PHP fluids offers promising solutions for the thermal performance promotion of PHPs. This study investigated zeotropic binary mixtures from the four pure fluids among water, methanol, ethanol, and acetone, so as to benefit from their properties for energy carrying. The various mixing ratios were charged into a closed-loop PHP at different filling ratios and extensive experiments have been carried out under various heat inputs (to the evaporation section). A detailed comparison will be made and valuable conclusions drawn, with the hope that it will greatly enrich the research on PHPs and thermal management, and accordingly, provide useful information for future study and applications (Section 7).

2. Working Fluid Physical Properties and the Interaction of Binary Mixtures

2.1. Working Fluid Physical Properties

The reason why these four working fluids were selected is that they are widely applied in PHPs, easily obtained, and are miscible liquids. The major thermo-physical properties of the four working fluids that most likely affect PHP heat transfer performance are listed in Table 2.1.

Table 2.1 Thermo-physical properties of the working fluids at standard atmospheric pressure [29]

Working fluids	Boiling point T_S	Liquid density ρl	LHV Hfg	Liquid specific heat Cpl	Dynamic viscosity $\upsilon l \times 10^6$	$(dp/dT)_{sat}$* $\times 10^3$	Thermal Conductivity λl	Surface tension $\sigma \times 10^3$
	℃	kg/m^3 (20℃)	kJ/kg	kJ/(kg℃) (20℃)	Pa·s (20℃)	Pa/℃ (80℃)	W/(m·℃) (20℃)	N/m (20℃)
Deionized water	100.0	998	2257	4.18	1.01	1.92	0.599	72.8
Methanol	64.7	791	1101	2.48	0.60	6.45	0.212	22.6
Ethanol	78.3	789	846	2.39	1.15	4.23	0.172	22.8
Acetone	56.2	792	523	2.35	0.32	6.27	0.170	23.7

*$(dp/dT)_{sat}$ was calculated from RefProp NIST (version 8.0) and based on its great temperature variations. The value at 80℃ instead of 20℃ is listed as representative.

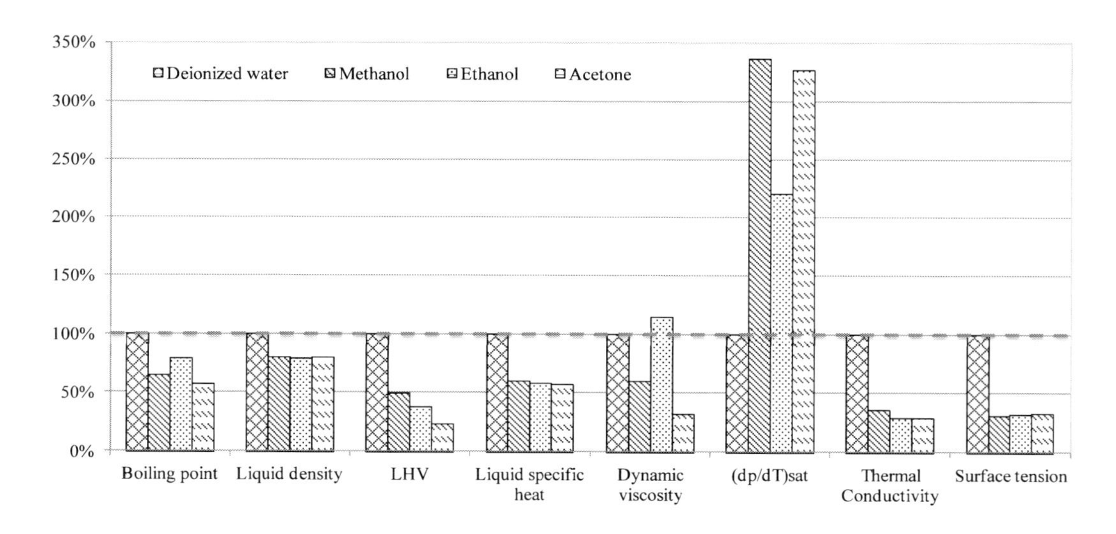

Figure 1. The relative thermo-physical properties as they relate to deionized water.

Taking the properties of the deionized water as 100%, the relative values of other fluids are plotted in Figure 1. Most of the properties do not vary that much with temperature and the relative values seldom change (larger one remains larger, and smaller one remains smaller except for $(dp/dT)_{sat}$). The value of $(dp/dT)_{sat}$ at 80^0C was given in the table to account for the great variations and the interchange of values for methanol and acetone (Figure 2). Therefore, although the fluids are not working at the values in Table 2.1, it is still reasonable to make a qualitative analysis and discussion based on their performance at different temperatures.

The best working fluid employed for PHPs have been suggested by earlier studies [30, 32–35] to have such properties as low dynamic viscosity, high value of $(dp/dT)_{sat}$, low latent heat, high specific heat and low surface tension. Higher dynamic viscosity (vl) requires higher force to drive the oscillation. Higher $(dp/dT)_{sat}$ means that a small rise in temperature could cause a large rise in pressure and thereby initiate a greater driving force for fluid motion. Higher boiling point (T_S) and higher LHV (Hfg) require more energy for the working fluid to actualize phase change and generate bubbles for the start-up and operation of oscillation [43]. However, from our perspective, it does not necessarily mean that the lower LHV is better because despite the hindering of the generation and growth of bubbles for the start-up of oscillation, a higher LHV represents a better ability to carry and accordingly transfer latent heat by phase change under the same flow condition. Another property that characterizes the energy-carrying ability of the working fluid is specific heat (Cpl). Higher Cpl may result in better sensible heat transfer because the working fluid could bring more energy within each degree centigrade of temperature difference. Higher surface tension in conjunction with dynamic contact angle hysteresis may create an additional pressure drop [33], especially for oscillation start-up [20].

Both Table 2.1 and Figure 1 show deionized water to have the highest LHV (2257 kJ/kg) and liquid specific heat (4.18 kJ/(kg°C)), which makes it the best in energy carrying and accordingly an excellent candidate for PHP operation. The other three, methanol, ethanol and acetone, have certain complementary properties compared to the deionized water. Acetone has the lowest dynamic viscosity (0.32×10^{-6} Pa·s) and almost the highest $(dp/dT)_{sat}$ (6.27 × 10-3 Pa/°C), together with the lowest boiling point (56.2°C) and the lowest LHV (523 kJ/kg), which makes it excellent in oscillation start-up and fluid motion, whereas its LHV and liquid specific heat are not favorable for energy carrying. The LHV of methanol (1101 kJ/kg) is closest to the best choice (deionized water), since the $(dp/dT)_{sat}$ is the highest up to 6.45 × 10-3 Pa/°C at 80 °C; the dynamic viscosity is just 0.6 × 10-6 Pa·s, the second lowest, and the liquid specific heat is only slightly higher than ethanol and acetone. Ethanol has the highest dynamic viscosity (up to 1.15 × 10-6 Pa·s), and has medium $(dp/dT)_{sat}$ (4.23 × 10-3 Pa/°C) and LHV (846 kJ/kg).

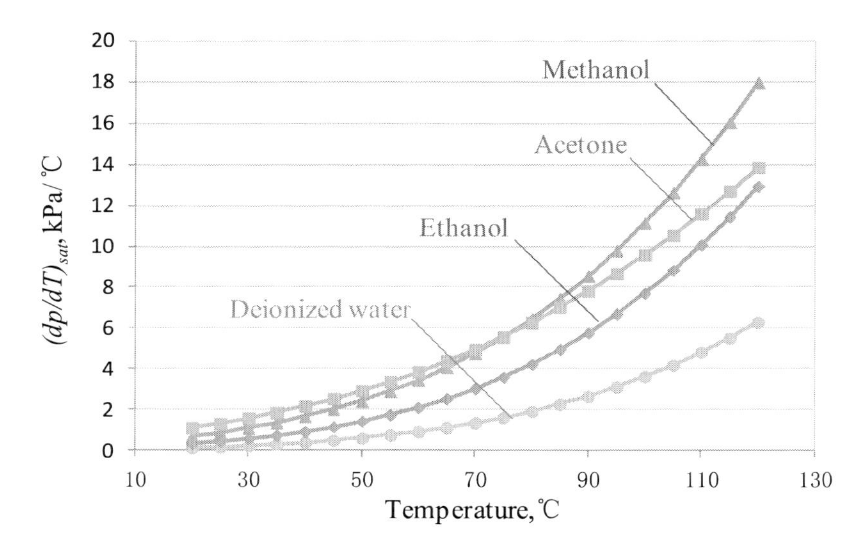

Figure 2. $(dp/dT)_{sat}$ at different temperatures for the working fluids concerned.

2.2. Binary Mixtures Phase Diagram

Unlike the single fluid (deionized water, methanol, ethanol or acetone), the phase change of the zeotropic binary mixtures shows a wide range of possible concentration shifts between the vapor and liquid phase and wide ranges of saturation temperatures and pressures, as shown in Figure 3 using the example of a water-acetone mixture. Figure 3 was plotted by Aspen Plus (version 11.1) with the pressure fixed at 1atm (Figure 3(a)) and shows the phase diagram of the temperature versus the acetone molar fraction of the water–acetone mixture, where the T_{sat} component water and T_{sat} component acetone are the saturation temperatures of water and acetone, respectively. Thus, '*b*' represents the acetone fraction (concentration) of the mixture before phase change, i.e., below the bubble line where the low boiling point component starts to evaporate; however, when the temperature is increased above the bubble line, e.g., 70°C at 1 atm, phase change happens and the acetone concentrations in the liquid phase (*b1*) and the vapor phase (*b2*) diverge from that of the original mixtures (*b*) with more fraction of acetone in the vapor phase than the liquid phase ($b2 > b > b1$) due to the lower boiling points of acetone (56.2 ℃). A concentration shift exists between the vapor and liquid phase for the component in the water–acetone mixtures, which is different from that of the pure fluid. The saturation temperatures are those on the dew line, on the bubble line, and those in-between, which are not fixed single values even under the same pressure and the same mixing ratio. Thus, saturation pressures must be measured and monitored at different temperatures. It is, therefore, possible for binary mixtures to exhibit new heat-transfer characteristics depending on the application of the PHP under concern.

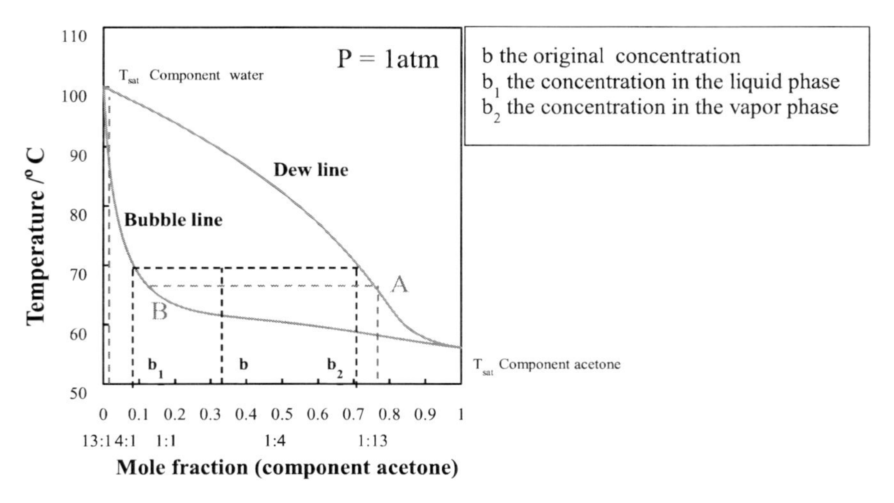

(a) Temperature-mole fraction (component acetone)

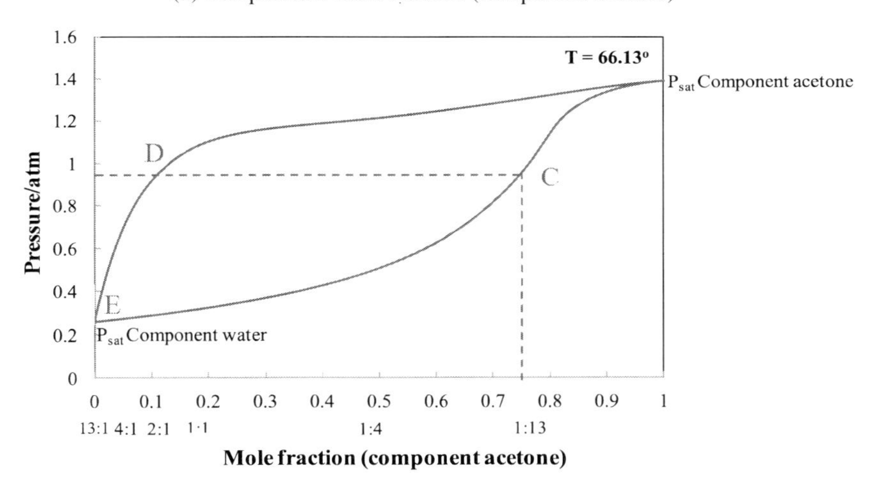

(b) Pressure-mole fraction (component acetone)

Figure 3. Phase diagram of water-acetone mixture.

The water-methanol, water-ethanol, ethanol-acetone and methanol-acetone binary mixtures are all positive-deviation solutions and have the similar features as those of the water-acetone mixtures, except that the dew line and bubble line become a little closer to each other, see Figure 4, Figure 5. Phase diagrams of the ethanol-based mixtures, obtained from Aspen Plus 8.4, are shown in Figure 4, Figure 5, and Figure 6 (the red line is the bubble line and the blue line is the dew line).

The ethanol-methanol mixture can be regarded as an ideal mixture (a mixture that obeys Raoul's law) [44]; its phase-change range is very small as Figure 6 shows.

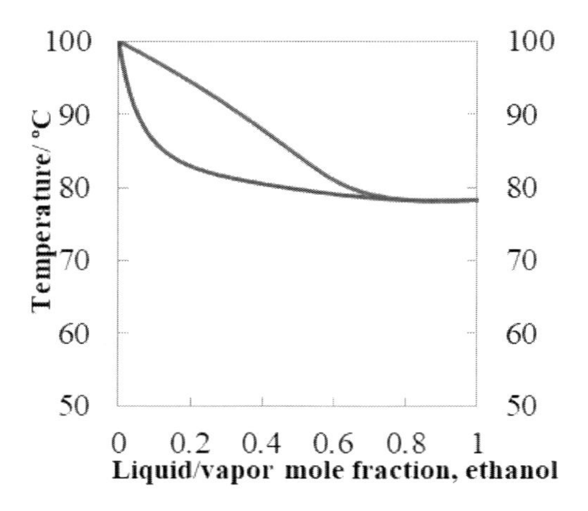

Figure 4. Temperature/mole fraction (ethanol-water).

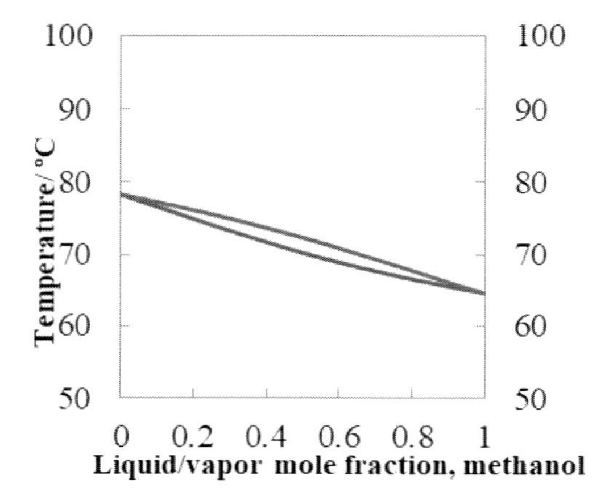

Figure 5. Temperature/mole fraction (ethanol-acetone).

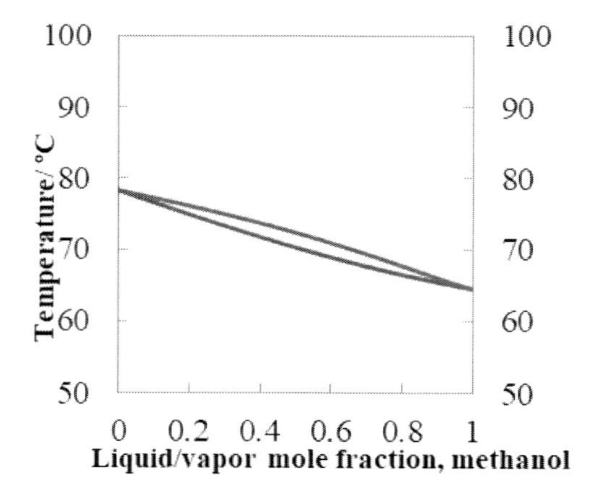

Figure 6. Temperature/mole fraction (ethanol-methanol).

3. EXPERIMENTAL SETUP

The size of the channel cannot be too large or too small, as it will influence the working of the PHP. The best range for the channel diameter [45] can be calculated as:

$$0.7\sqrt{\frac{\sigma}{(\rho_l - \rho_v)\ g}} \leq D \leq 1.8\sqrt{\frac{\sigma}{(\rho_l - \rho_v)\ g}} \tag{1}$$

where σ, g, ρ_l and ρ_v represent the surface tension, gravitational constant, density of liquid and density of vapor. For the pure fluids involved in our study, the ranges were calculated and listed in Table 3.1 below.

Based on Table 3.1, the inner diameter of the sample PHP (Figure 7(a)) was selected at 2 mm and the outer diameter is 4 mm. The PHP specimen was made of a bent copper capillary tube, connected end to end. It consisted of three parts: the evaporation section, the adiabatic section and the condensation section, with lengths of 80 mm, 20 mm and 80 mm, respectively. 10 parallel tubes were oriented vertically, and a horizontal tube was there on the upper side of the condensation section to form a closed loop. The radius of each meandering turn was 10 mm while the distance of the adjacent pipes was 20 mm. The PHP, as Figure 7(a) indicates, was in a vertical position with electric wires on the bottom, where the evaporation and the adiabatic section were put into an organic glass chamber fabricated into a dual-layer structure.

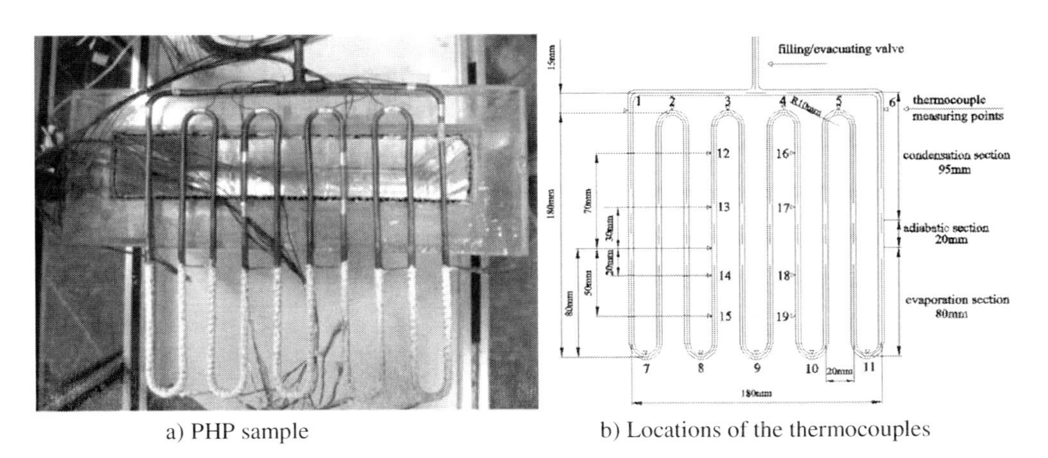

a) PHP sample b) Locations of the thermocouples

Figure 7. PHP sample and the locations of the thermocouples.

Table 3.1. The range of diameter of PHP with different working fluids

Working fluids	Ethanol	Water	Methanol	Acetone
Range of diameter(mm)	1.09 ~ 2.85	1.75 ~ 4.85	1.13 ~ 2.94	1.15 ~ 3.02

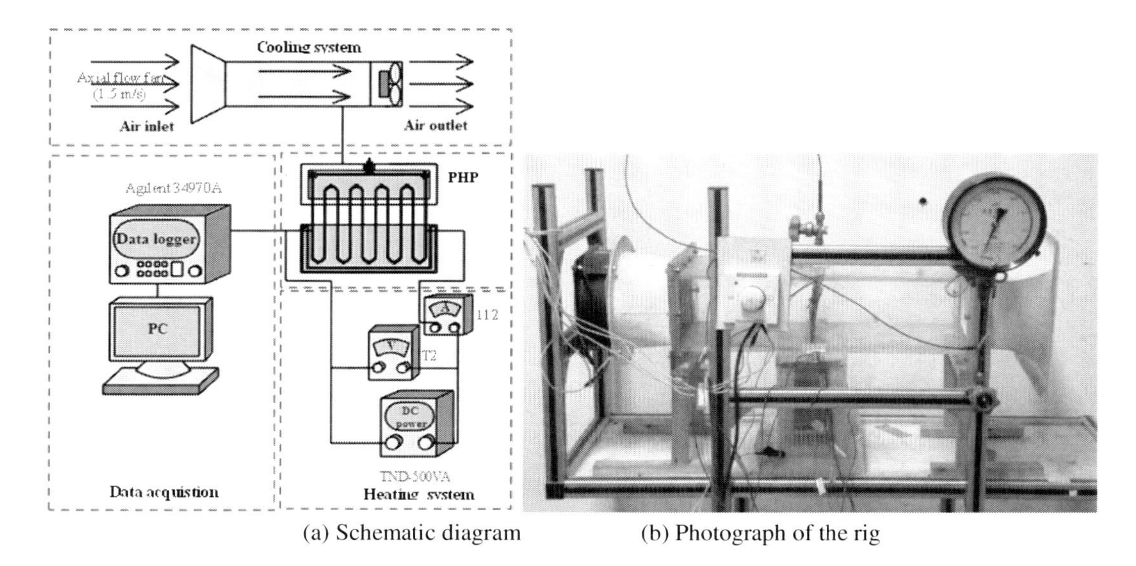

(a) Schematic diagram (b) Photograph of the rig

Figure 8. Schematic of experimental apparatus based on function and the laboratory experimental setup of the rig.

Table 3.2. Experimental apparatus and instrumentation

Experimental apparatus and instrumentation	Type	Remarks
Heating wire	Ni-Cr	Specific resistance: 1.09μΩ•m
Thermocouple	T	Accuracy after calibration:±0.1°C Response time: 0.3s
Hot-bulb anemometer	QDF-3	Range: 0.05 ~ 10m/s Measuring error: ≤5%
Data acquisition	Agilent 34970A 34901A module	Scan rate: 60 channels per second Accuracy: 0.0256°C(1μV)

Figure 7(b) shows the placement of the thermocouples. Twenty T-type thermocouples (±0.1^{0}C after calibration) were installed. Nineteen were attached on the outer pipe wall to monitor the temperatures of the PHP, and one thermocouple was fixed on the outer wall of the chamber to detect heat loss from the insulation chamber. Among the 19 thermocouples, six were fixed on the turns for the condensation section, numbered 1–6, five were fixed on turns of the evaporation section, numbered 7–11, and the remaining thermocouples were fixed at the center of the pipe lengths, numbered 12–19.

Figure 8 is a schematic of the experimental apparatus consisting of a PHP, a system for evacuation and charging, an electric heating system, an air cooling system and a multichannel data acquisition device. At the PHP condensation section, the forced air cooling constant wind speed of 1.5 m/s was adopted; the cooling duct is rectangular in the cross section, with the axial fan installed at the outlet. The duct inlet was designed in a lemniscate form to help create an evenly-distributed air flow; Agilent 34970A with a 34901A multi-channel module was utilized to collect data, where the scanning velocity approaches 60 channels per second. In consideration of the response of thermocouples (0.1s) and the average PHP period of temperature oscillation (longer than 3s [46]), the data acquisition system is reasonably satisfying.

4. Data Reduction and Error Analysis

The specific performance of a PHP is reflected by the pulsating temperature of each measuring point, while the heat transfer performance of a PHP is reflected by the mean temperature of the evaporation section and overall thermal resistance. The positions of the measuring points are shown in Figure 7(b); the number of measuring points in the evaporation section ranges from 7 to 11 and in the condensation section from 1 to 6. T_e (the mean temperature of the evaporation section) and T_c (the mean temperature of the condensation section) are determined after the smooth running of the PHP:

$$T_c = \frac{1}{6}\sum_{i=1}^{6} T_{ci} \tag{2}$$

$$T_e = \frac{1}{5}\sum_{i=7}^{11} T_{ei} \tag{3}$$

The overall resistance of the PHP is determined by:

$$\mathrm{R} = \frac{\mathrm{T}_e\text{-}\mathrm{T}_c}{Q} \tag{4}$$

Q is the heat load supplied by the heating wire. Because the heat leak of the heat preservation block is small, the heat absorbed in the evaporation section is equal to Q.

Standard uncertainties can be expressed:

$$\frac{\delta Q}{Q} = \sqrt{\left(\frac{\delta \mathrm{U}}{\mathrm{U}}\right)^2 + \left(\frac{\delta I}{I}\right)^2} \tag{5}$$

$$\frac{\delta R}{R} = \sqrt{\left(\frac{\delta \mathrm{T_e}}{\mathrm{T_e} - \mathrm{T_c}}\right)^2 + \left(\frac{\delta \mathrm{T_c}}{\mathrm{T_e} - \mathrm{T_c}}\right)^2 + \left(\frac{\delta Q}{Q}\right)^2} \tag{6}$$

Given a coverage factor (K) of 2 to cover those unincluded elements, the expected maximum uncertainty is expressed as: $U_{max} = \frac{\delta R}{R} \times \mathrm{K} < 5.0\%$.

5. Combination Study of Operation Characteristics and Heat Transfer Mechanism

5.1. Analysis of Oscillation Characteristics and Heat-Transfer Mechanisms

Four types of pure working fluids, deionized water, methanol, ethanol, acetone, and their binary mixtures were adopted under FRs ranging from 35% to 90%; and the heating capacity was 5-100 W. Data collected to calculate thermal resistances were all in the last ten minutes of the experiment to ensure the oscillation flow reached steady status.

For the oscillation characteristics and heat-transfer mechanism, the performance of the PHP charged with deionized water was similar to the other working fluids; thus, the analysis in this section will all be based on the deionized water. In this study, PHP (FR = 62%) exhibited a comparatively superior thermal performance, of which variations in temperature and thermal resistance were quite typical. Hence, temperature variation features were associated with flow patterns of the PHP (deionized water, FR = 62%), which were employed to create a representative example to illustrate the heat-transfer process under different operation conditions.

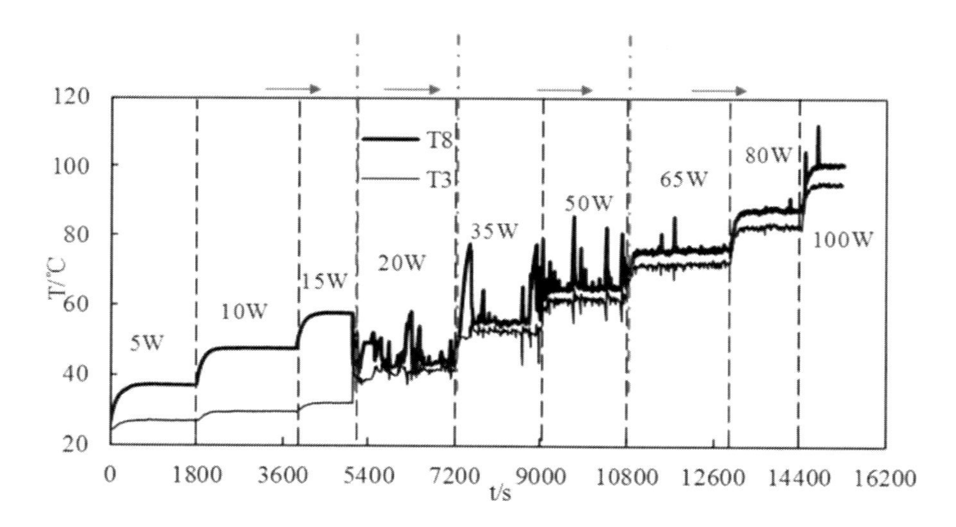

Figure 9. Operation characteristics of the PHP with deionized water under different heat inputs (FR = 62%).

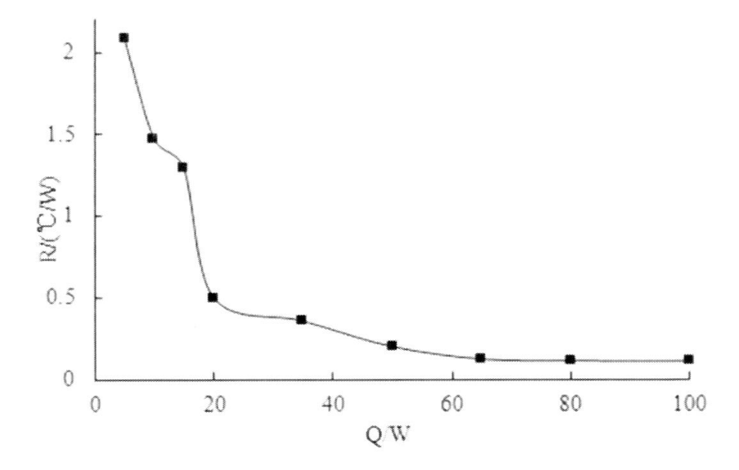

Figure 10. Thermal resistances of PHP with deionized water under different heat inputs (FR = 62%).

Figure 9 plots an experimental process of the PHP (deionized water, FR = 62%) where the heat input increased gradually after the temperature stabilized. Due to typical temperature variations and flow motion variations in the middle parallel pipes, the temperature at test point 3 (T3) and test point 8 (T8) were used to represent the temperature of the condensation section and evaporation section, respectively. As the heat input rose, the pipe wall temperature of the evaporation and condensation section successively experienced non-oscillation, large-amplitude with low-frequency oscillation and a small-amplitude coupled with high-frequency oscillation. Corresponding variations of thermal resistances are highlighted in Figure 10. Details are included in the following analysis.

Visualization experiments [47] indicate that random distribution of the vapor slug and liquid slug immediately emerged once the working fluid was charged in, changing when the heating and the forced air-cooling were imposed into the evaporation and condensation section. How did it change? What would happen to heat-transfer and why? These questions will be discussed in detail below with four heating sections divided according to corresponding flow and heat-transfer characteristics: Thus, heat inputs of 5–15 W, 20 W, 35–50 W and 65–100 W are shown in Figure 9.

(1) Within the heating section from 5 W to 15 W, no oscillation was observed, and the temperature variation was basically imperceptible. Within this heating section, when the heat input reached a certain value, the PHP entered a thermally steady state after a transient augment of temperature, and the temperature of the evaporation and condensation section exhibited a stable condition with no oscillation observed. With the increase in heat input, due to the deficient delivery of heat, the temperature increase of the evaporation section was evidently greater than that of the condensation section. A flat and smooth temperature variation at a certain heat input indicated that there was no macroscopic motion of the vapor and liquid slugs. At that time, the superheat of the vapor slugs was too small to generate enough driving force. The heat-transfer mode at this stage mainly depends on natural

convection and the thermal conductivity between the vapor and liquid slugs. Vapor slugs expanded in the evaporation section and shrunk in the condensation section. No bubbles were produced in the liquid slug or if they were produced, they grew, agglomerated and collapsed, but could not produce a large turbulence due to their isolation. In that case, the thermal resistance was relatively high as shown at the heat input of 5 W in Figure 10. As the heat input increased, the vapor slug and liquid slug continuously expanded and shrunk, respectively, and the pressure of the vapor slugs also rose. Variations in the evaporation section fostered changes in other sections of the PHP, which caused driving forces to accelerate, enhancing heat transfer and accordingly, the decline of thermal resistance, as shown in Figure 10. When the heat input attained a certain value, the current heat-transfer pattern was approaching its limit, where the elevation of temperature difference between the evaporation and condensation section equaled to or even exceeded that of the heat input, which resulted in less heat transfer from the evaporation section to the condensation section. Therefore, a slow-down declining trend appeared in the curve, as shown for the transition from 10W to 15W in Figure 10. It is possible to have even a slight rising trend, but this is not the norm.

(2) When the heat input rose to 20 W, the temperature oscillation showed large-amplitude and low-frequency; thus, the temperature in the evaporation and condensation sections presented a sharp decrease and increase, respectively and simultaneously, which led to a significant drop in temperature difference. The two temperatures presented a change much like an inverse phase featuring a large amplitude and low frequency, as indicated in Figure 11. The thermal resistance at this stage was much lower than that at 15 W; thus, Figure 10 shows a sharp drop in thermal resistance while the heat input increases from 15 W to 20 W.

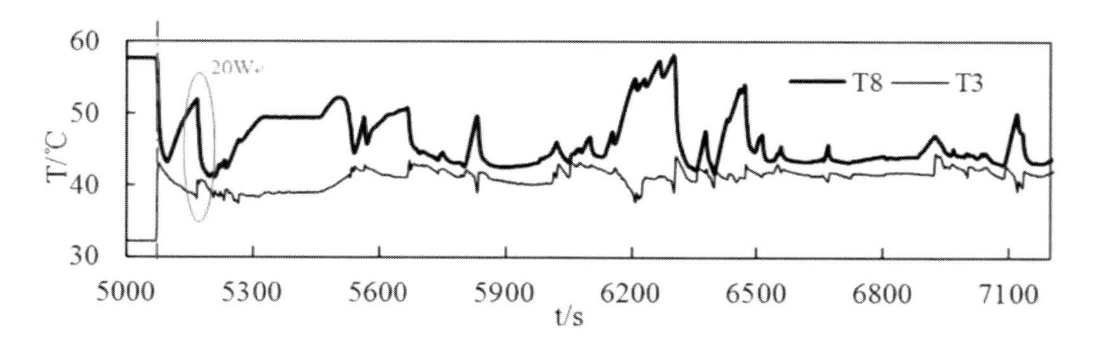

Figure 11. Temperature oscillations of PHP evaporation and condensation section with deionized water (FR = 62%, Q = 20W).

The temperature oscillation indicated that macroscopic motions or transitions between vapor and liquid slugs had occurred, which caused mass transportation. Furthermore, as the temperature of the evaporation section kept rising and that of the condensation section kept declining (like those depicted in the red ellipse in Figure 11), it showed the working fluid almost was stagnant. With the accumulation of energy and the generation and growth of bubbles in the evaporation section, the stagnation is broken up and the vapor slugs start to move upwards, consequently driving the upper liquid slugs upward. New liquid enters in and pours into the original locations, which is what led to the sharp drop of temperature shown in Figure 11. While the working fluid is transmitted into the condensation section, the bubbles agglomerate and collapse, which results in heat and mass transfer, and the temperature rise in the condensation section. Hence, the nearly inverse phase change of temperature emerged in the evaporation and condensation sections.

Referring to the results of the visualization experiments in open literature, it can be deduced that the flow modes at this stage were mainly the bubble flow and the slug flow, moving up and down in the vertical pipe (depicted in Figure 12(a)) under the action of factors like the destabilization of bubbles and the unbalance between pipes. Due to the insufficient heat input, it took a long time to accumulate enough energy to generate macroscopic motion in the PHP. Thus, the stagnation period was somewhat long, which caused low-frequency underlying the temperature oscillation. Moreover, when the fluid stagnated, there was no moving inertia to foster the entire flow; thus, a larger pressure was required to drive the working fluid, causing the amplitude of the temperature oscillation to become relatively large.

Apparently, the sensible and latent heat transfer of the working fluid were tremendously enhanced by the oscillation of the vapor and liquid slugs in the PHP. Moreover, the thermal resistance at this stage was sharply lowered compared to that at 15 W (shown in Figure 12).

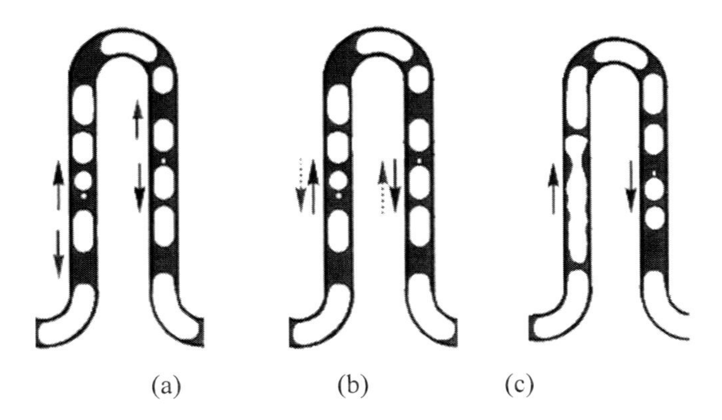

Figure 12. Oscillation flow of the working fluid in the PHP: (a) The oscillation start-up, (b) the circulatory oscillation, and (c) the steady one-way circulatory oscillation.

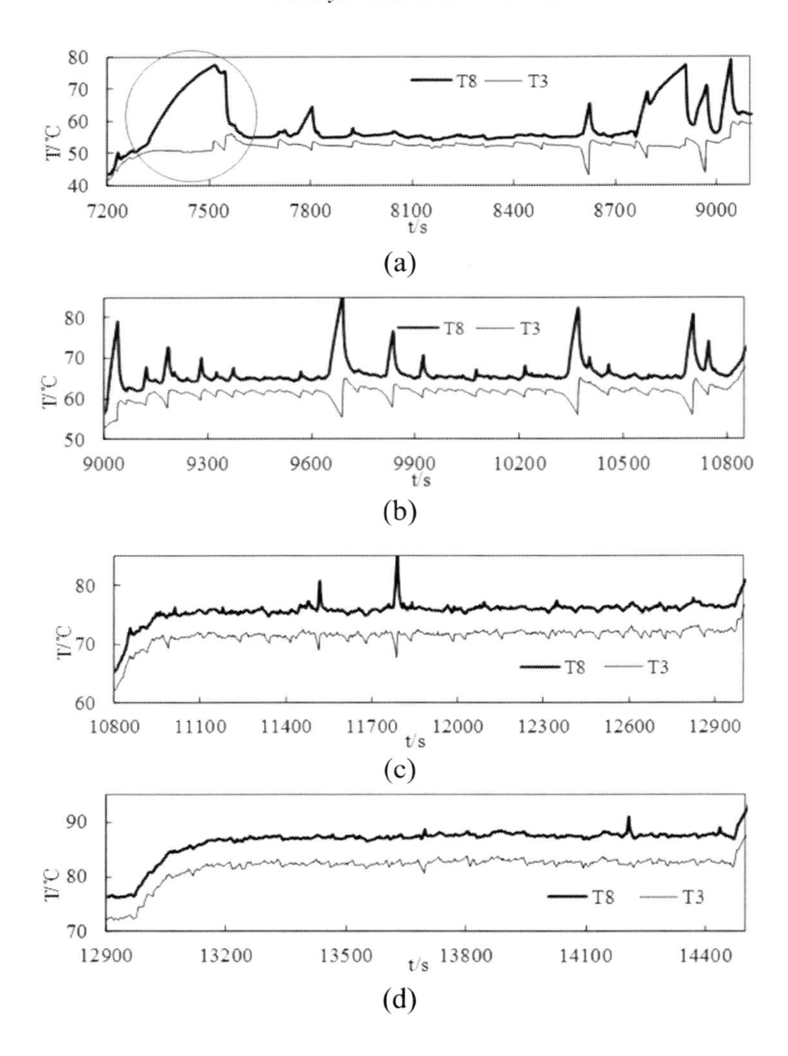

Figure 13. Temperature oscillations of the PHP evaporation and condensation sections charged with deionized water (FR = 62%): (a) Q = 35 W, (b) Q = 50 W, (c) Q = 65 W, and (d) Q = 80 W.

(3) At the interval of 35 W and 50 W (highlighted in Figure 13), temperature oscillation presented a modest amplitude associated with occasionally large-amplitudes, while the oscillation frequency increased.

As mentioned above, the oscillatory flow at 20 W was merely an up-and-down movement in the vertical pipe because the energy of the vapor slugs was not enough to drive the working fluid through the pipe turns, especially the long horizontal connection pipe above the condensation section. With the heat input continued to increase, the energy of the vapor slugs at the evaporation section rapidly accumulated and the pressure imbalance among parallel pipes was significantly boosted, which led to an entire flow of circulatory oscillation. The greater the heat input, the faster the flow and so did the generation, growth, agglomeration and collapse of bubbles carried by the flow. Therefore, the temperature oscillation at this stage was featured by small amplitude and high frequency. However, within this range of heat input, the pressure growth of bubbles can occasionally contend against the flow's inertial force, which leads to the stagnation of the working fluid and thus made the working fluid consistently heated or cooled, thereby providing it with an opportunity to change flow direction, as Figure 12(b) indicates. The arrow with the dotted line in the Figure 12(b) represents an alternative flow direction. Several times at 35 W, the temperature oscillation appeared with large amplitude and long-lasting time, as shown in Figure 13(a); when the heat input reached 50 W, greater heat input accelerated the flow, the duration of the occasional stagnation dwindled and thus, the amplitude of temperature oscillation was shortened (depicted in Figure 13(b)). Therefore, it can be concluded that the greater the heat input, the shorter the period of the large-amplitude temperature oscillation in the evaporation and condensation sections.

From 35 W to 50 W, the flow of the PHP has transformed to slug flow or plug flow in a circulatory oscillating manner with occasional stagnations and changes of direction. Compared to the state at 20 W, the sensible heat transfer was greatly enhanced, and the thermal resistance significantly dropped (exhibited by Figure 10).

(4) The temperature oscillation (heat input = 65~100 W) was characterized by small amplitude and high frequency.

Due to the augmented heat input, the oscillatory flow maintained a high speed to flush the pipe wall, and gradually entered a stage known for its steady, one-way circulation (Figure 12(c)) with an increasingly higher flow speed as well as fewer and shorter stagnations. As the heat input increased, the amplitude of the temperature oscillation continued to dwindle (shown in Figure 13(c) and Figure 13(d)), and the bubbles had no time to develop before being carried away by flow. At that time, the heat transfer of PHP was jointly influenced by both sensible and latent heat. The ability of the sensible heat transfer has been further reinforced, and the thermal resistance continued descending but in a slower manner (depicted in Figure 10).

Based on the above study and analysis on the characteristics of the operation and the thermal resistance of the PHP (deionized water, FR = 62%) under different heat inputs, it could be concluded that, as the heat input increased, before the flow oscillation start-up, there was no oscillating change in the temperatures of the evaporation and condensation section. When the flow oscillation was triggered, the temperature oscillation started and successively experienced a large-amplitude with low-frequency. The large-amplitude was accompanied by comparatively high frequency, and the small amplitude became associated with higher frequency. The working fluid in PHP shifted from no macroscopic motion to oscillations featured successively by up-and-down movements, unsteady circulatory oscillation and steady, one-way circulatory oscillation. The transformation of the flow pattern was put in a chronological order beginning with bubble flow, followed by slug flow, plug flow and annular flow. With the alteration of the operation characteristics, the heat transfer mechanisms changed, and the thermal performance was promoted as long as no dryout emerged.

5.2. Characteristics of the Local Dryout

When FR was 62%, due to the limit of the material temperature tolerance, experiments at higher heat input (>100 W) had not been carried out, and no temperature soaring had been observed in the evaporation section. However, for the FR of 45%, the temperature in the evaporation section soared in some locations at 35 W and reached even higher temperature at 50W, as shown in Figure 14.

Figure 15 showed temperature variations of four test points with the heat input at 35W. Two of them were fixed at the evaporation section (T8 and T10), and the other two at the condensation section (T3 and T4); their locations are shown in Figure 7(b). Figure 15 shows a period of time which allowed T8 to soar and keep exceeding the peak of the normal oscillation temperature before returning to normal afterwards. This behavior indicated that a local and periodical dryout existed for T8 in the evaporation section. Due to the uneven distribution of the vapor and liquid slugs as well during the uneven heat transfer in the evaporation section pipes, the liquid flowing back with T8 was small and insufficient at certain periods of time; hence, T8 was totally vaporized while being heated, which made the dryout happen. After this period, enough liquid slugs were driven by the fluid flow to cause the coexistence of vapor and liquid; thus, the temperature dropped and returned to its normal oscillation level. During the same period, T10 maintained its normal oscillation, indicating that there was no dryout at test point 10 in the evaporation section. T3 and T4 of the condensation section showed an inverse phase oscillation against that of T8 and T10, which explained that part of the coming vapor condensation, where the vapor and the liquid coexisted in the condensation section.

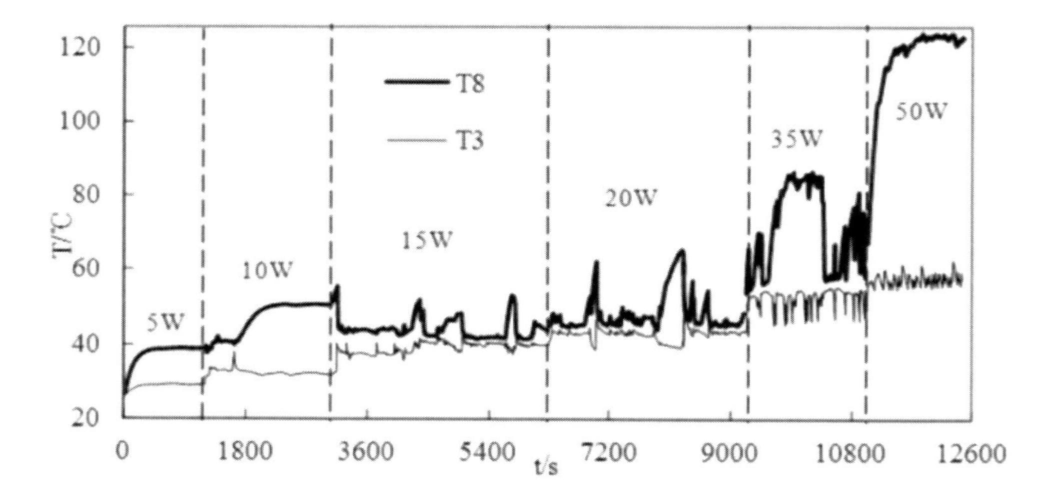

Figure 14. PHP operation characteristics with deionized water at different heat inputs (FR = 45%).

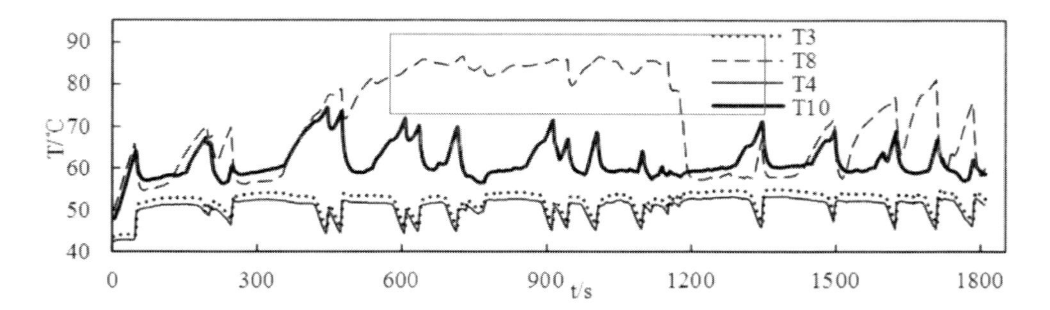

Figure 15. Temperature oscillations of the evaporation section and condensation section for the PHP charged with deionized water (FR = 45%, Q = 35W).

Owing to the existence of the local and periodical dryout, the average temperature of the evaporation section increased. The more serious the dryout, the more rapid the growth of the average temperature was. The decline of the thermal resistance slowed down and even changed into a climbing trend along with the development of dryout, as shown in Figure 16.

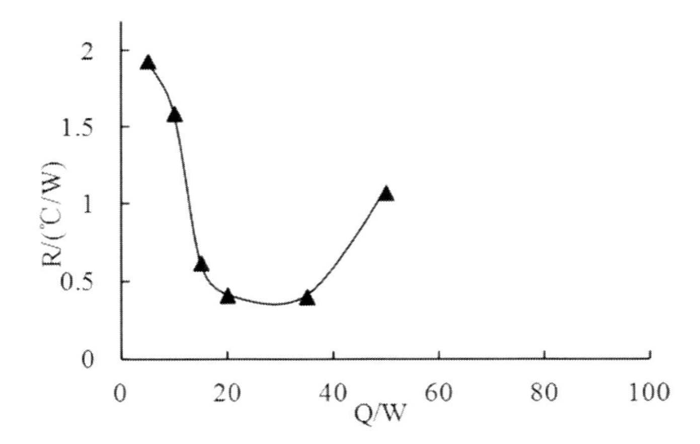

Figure 16. Thermal resistance points of the PHP charged with deionized water (FR = 45%).

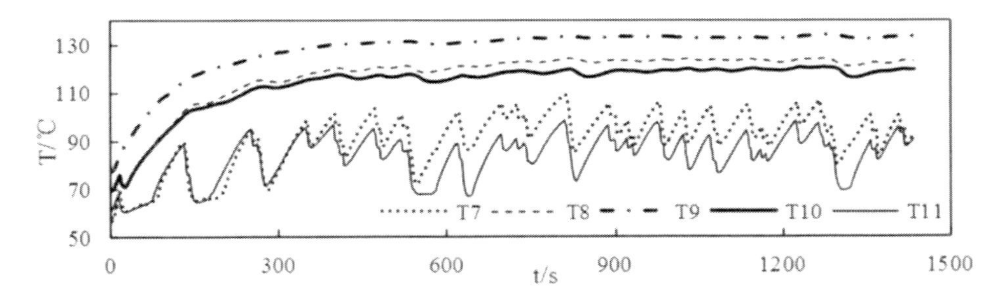

Figure 17. Temperature oscillation of the evaporation section and condensation section for the PHP charged with deionized water (FR = 45%, Q = 50 W).

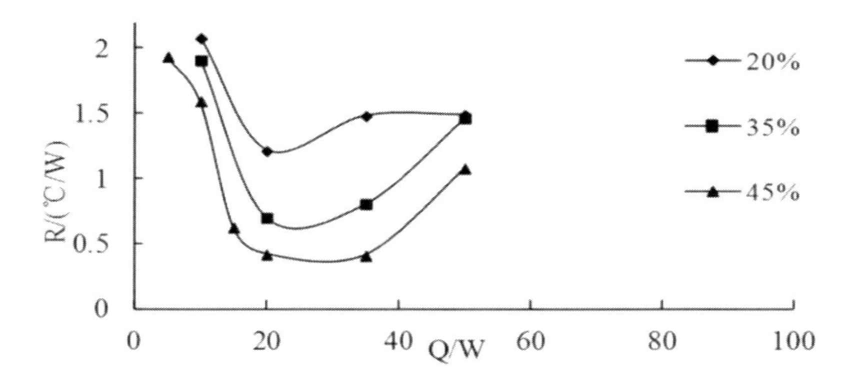

Figure 18. Experimental rules indicating dryout emerges in the pipes.

Figure 17 indicates the temperature of the evaporation section. Temperatures of each measuring point for the condensation section maintained the oscillation changes featured by a small amplitude and high frequency, which indicates that part of incoming vapor was condensed into liquid. It can be observed that this dryout happened in the middle three pipes of the evaporation section (T8, T9, and T10), but the temperatures there became stable eventually instead of continuously soaring without returning the normal level like T7 and T11, which showed that vapor can be condensed at the condensation section and supplied to the evaporation section in liquid, however for the small FR (e.g., 45%), the liquid supply was not enough to counter against the heat input, which means it totally vaporized and the evaporation section was superheated.

The temperatures of the two marginal pipes (T7, T11) were lower than those of the other three in the middle and no dryout happened because sufficient liquid could be provided, due to the long horizontal condensing pipe connecting them. Test point 8 and 10 pipes formed a neighboring connection to test point 7 and 11 pipes. Therefore, dryout happened there but in a relatively light manner compared to that at the middle T9 pipe where the supplementary liquid was the most scarce and the temperature was the highest. With dryout, there was more vapor than liquid in the pipes and because the vapor is quite inferior to the liquid in its energy carrying capability, the heat in the evaporation section could not be transferred to the condensation section promptly, which resulted in an increase of thermal resistance (as shown in Figure 18).

With regard to the PHP dryout characteristics that were distinct from the conventional heat pipe, Cai et al. [48], through a high-speed camera, recorded the fast backflow of liquid from the condensation section while dryout happened at the evaporation section, which corresponds to the variation of the temperatures described above.

Figure 18 depicts the thermal resistance variations under the FR of 20%, 35% and 45%, where dryout emerged with greater heat input. Due to dryout, the thermal resistance ascended. The smaller the FR, the more the pipes dried out and the less liquid could be supplied to the evaporation section; consequently, the greater thermal resistance corresponded to the earlier sharp climbing period of the growth heat input. However, there was no continuous temperature soaring like that found in the conventional heat-pipe; thus, the PHP pipes cannot easily be broken.

5.3. Variation Rules of PHP thermal Resistance to the Change of Heat Input and FRs

Figure 19 shows the thermal resistance of the PHP charged with deionized water at different FRs versus heat inputs. The rules of the PHP charged with methanol, ethanol and acetone are similar.

When the heat input was relatively small (e.g., ≤ PHP 20 W), thermal resistance trends significantly descended with relatively large differences existing among different FRs. Basically, the smaller the FRs, the lower the thermal resistance and the better the heat transfer performance is with FR = 20%. This seems like an obvious abnormality, probably because of the sooner vaporization for its small FR. Before the start-up of the oscillation, no obvious macroscopic motion existed in the PHP, and the initial distribution of the vapor and the liquid slugs were characterized by a certain randomness, which exerted an impact on thermal resistance. After the start-up, oscillations were always featured by large amplitudes and high frequencies with occasional stagnation, as discussed in Section 5.1. The number and lengths of the vapor and the liquid slugs were determined by the FRs, which required various driving forces. Given the same heating and cooling conditions, liquid slugs under smaller FRs were fewer, and thus, would consequently move faster and more frequently, which contributed to a relatively smaller thermal resistance. Different FRs corresponded to different oscillation-triggering heat input, while smaller FRs required smaller heat inputs for initiating oscillations. For example, an FR of 45% and 62% was reported for deionized water. Heat inputs of oscillation start-ups were 15 W (depicted by Figure 14) and 20 W (exhibited by Figure 9), respectively.

As heat input was augmented, the characteristics of thermal resistance may be divided into two groups: with dryout and without dryout. a) Below an FR of 45%, the thermal resistance first decreases and then increases after a certain heat input, which indicates dryout in the pipes. b) When FR was equal to or higher than 55%, due to the growth of the working fluid, no evident increase in thermal resistance was observed; thus, no dry-out appeared. For FRs from 55% to 70%, the overall performance of thermal resistance was superior, especially when FR was 62%. When FR increased to 90%, due to the incompressible liquid occupying the majority of the internal space, the generation and development of bubbles was restricted and a greater driving force was required to maintain its movement; thus, given the same heating and cooling conditions, the flow would be relatively slow and the thermal resistance would be comparatively great.

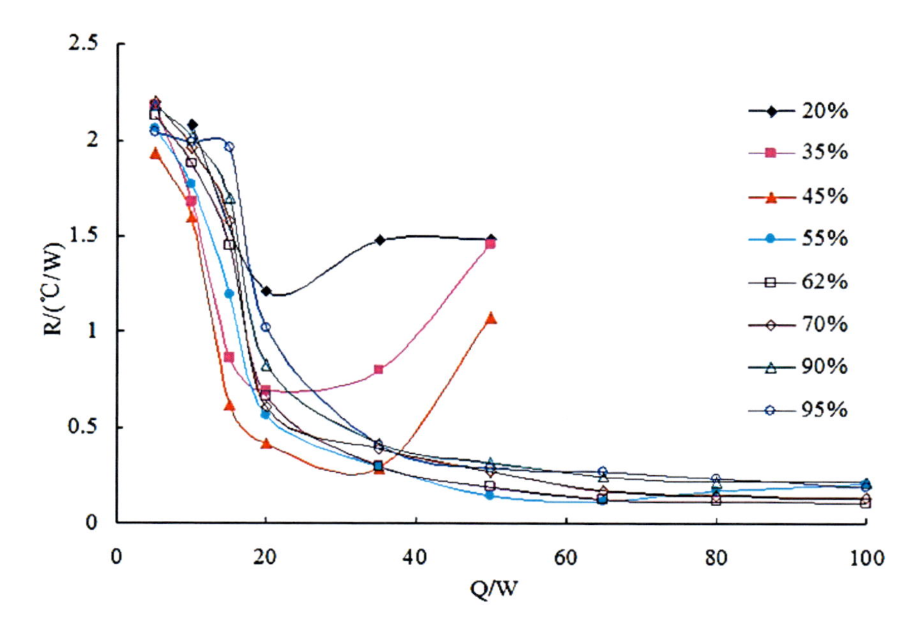

Figure 19. Thermal resistances of the PHP charged with deionized water at different FRs.

As discussed in Section 5.1, when the heat input was great enough, the oscillation was mainly characterized by small amplitude, high frequency. The entire circulatory flow was also determined. The flow went fast with the further increase of heat input and flushed the internal wall even intensely, which resulted in the enhancement of heat transfer and the reduction of thermal resistance. However, when the heat input reached 60 W, the changing of thermal resistance for each FR became slow or even stayed flat, and little difference existed between the different FRs, which indicates that there might exist a limit for the phase-changing pattern of heat-transfer under fast flow, but the limit had little relationship with FRs.

The thermal resistances for different working fluids are shown in Figure 20. It can be seen that under great heat input (e.g., $\geq$ 65W), if no dryout occurs, the thermal resistances for different working fluids and different FRs tend to come closer to one another, which illustrated that this part featured by small and relatively stable thermal resistances could be selected as the basis for the practical application of PHP.

From the above discussions, it can be concluded that, in the case of great heat input without dryout happening in the pipes, no matter what type of working fluid being used or how much working fluid is filled, the thermal resistance of PHP is relatively small and hardly varies. For the same PHP specimen, it should have little relationship with the type of working fluids and the FRs, may largely relate to the structure, the material, the size and the inclinations of the PHP. Existing mathematical models have not yet focused on this feature.

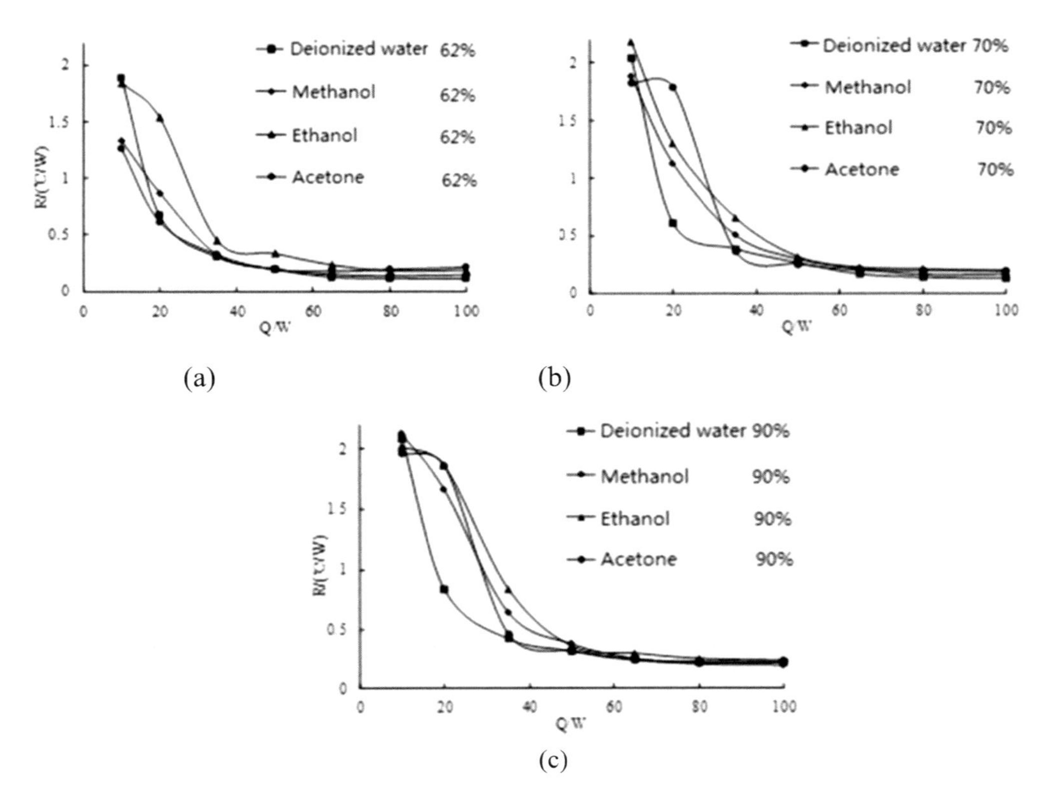

Figure 20. Thermal resistances of the PHP with different working fluids: (a) FR = 62%, (b) FR = 70%, (c) FR = 90%.

The details of Section 5 are summarized as follows:

(1) PHP flow patterns change with the augment of heat input, so do the characteristics of temperature oscillation and the heat-transfer mechanism is altered accordingly, which contributes to the enhancement of heat-transfer performance as long as no dryout emerges.
(2) Dryout emerged locally in some individual pipes at the evaporation section, is continued for a while, and then returns to the normal working condition without dryout. With the increase of heat input, dry-out would expand to more areas, but, unlike conventional heat pipes, continuous temperature soaring would not happen.
(3) It should also be noted that, given certain cooling conditions, there exists a limited heat-transfer performance, where the thermal resistance for different working fluids and FRs tend to become closer if no dryout occurs after a heat input of 65 W and eventually becomes quite similar. It may also relate to the structure, material, size and placement of PHP.

6. WORKING FLUIDS AND THEIR PROPERTIES ON THE PHP PERFORMANCE

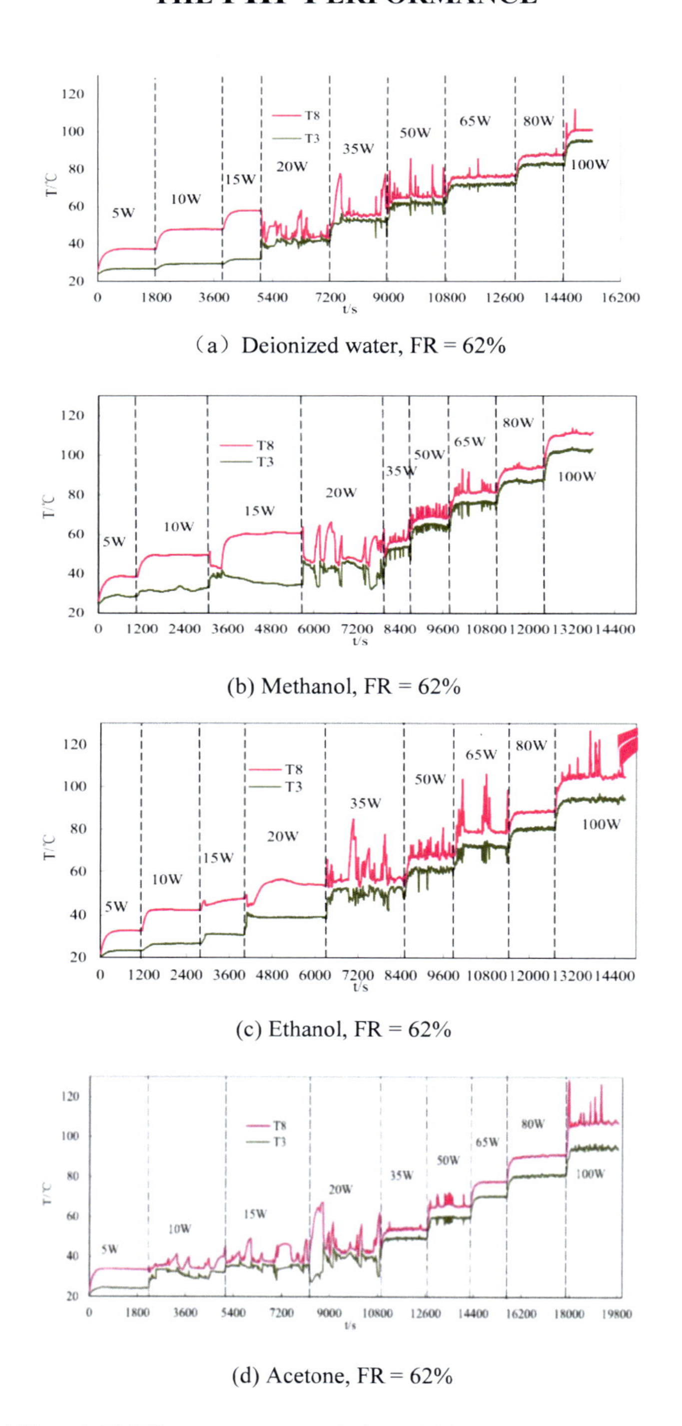

(a) Deionized water, FR = 62%

(b) Methanol, FR = 62%

(c) Ethanol, FR = 62%

(d) Acetone, FR = 62%

Figure 21. The ES(T8) and CS(T3) temperature variation with heat input for PHP.

Experiments have been carried out for a closed-loop PHP charged with deionized water, methanol, ethanol and acetone. By comparative study, the effects of the thermo-physical properties of the working fluids on the temperature, oscillation, and heat transfer have been analyzed and elaborated.

6.1. Oscillation Operation and Dynamic Viscosity (υl)

Characteristics after oscillation were triggered and can also be observed in Figure 21(a)–(d). At 35 W, compared to the temperature oscillations of deionized water (Figure 21(a)) and ethanol (Figure 21(c)), those of methanol (Figure 21(b)) and acetone (Figure 21(d)) presented a characteristic of high frequency and low amplitude. The working fluids show high-frequency and low-amplitude oscillations, which always flow relatively fast [8, 9]. Comparing their thermo-physical properties, methanol and acetone have a lower boiling point (T_s) and smaller dynamic viscosity than deionized water and ethanol. The influence of T_s and LHV could be analyzed based on Figure 21(a) and (c), where at a 35 W heating power, the working fluid with a lower boiling point and LHV (ethanol) exhibited an oscillation of lower frequency and higher amplitude than the higher one (deionized water); hence, these two properties were not the key factors that led to a high-frequency and low-amplitude oscillation, or fast flow. The dynamic viscosity (υl) of deionized water is smaller than that of ethanol and greater than that of methanol and acetone. It demonstrates a higher-frequency and lower-amplitude oscillation (faster flow) than ethanol and a lower-frequency and higher-amplitude oscillation (slower flow) than the other two. Therefore, besides its predominant role in oscillation start-up, υl is also the key factor for fluid flow during early oscillation. As the heat input further increases to 50 W, 65 W, and 80 W, the oscillation was reinforced, and with the growing effect of inertial force and the dwindling effect of viscous force (partly due to the decrease of the dynamic viscosity with the temperature augmentation), the frequency and amplitude for different working fluids came closer to one another.

6.2. Oscillation Operation and Latent Heat of Vaporization (LHV)

The temperature oscillation (Figure 21) shows that at filling ratio of 62% and heat input of 100 W, the ES temperatures of ethanol (Figure 21(c)) and acetone (Figure 21(d)) with a lower LHV exhibited high amplitude and high-frequency oscillation several times, which may indicate a sharp short-time superheat locally; whereas this phenomenon seldom happens for deionized water and methanol with higher LHV. After oscillation was triggered, the working fluids would experience different flow patterns with the increase in heat input, from up-and-down wandering, oscillating circulation of vapor/liquid slugs to vapor/liquid two-phase annular circulation [8, 49]. No matter what pattern the amount of liquid fluid coming to the ES follows, it is always unstable—sometimes more and sometimes less. At 100 W heating power, it is most likely that the working fluids flowed in the pattern of a vapor/liquid two-phase annular circulation with incoming liquid fluid in the form of a thin film clinging to the inner pipe wall. For lower LHV, the heat input in ES would occasionally exceed what was needed for the liquid film to vaporize, causing locally short-time dry-out and accordingly the sharp rise of the corresponding temperature. Therefore, it can be conjectured that at the same filling ratio, the working fluids with lower LHV may be easy to dry out, i.e., they may start to dry out at a relatively small heating power, which can be further proved by the following discussions. In fact, as the heat input increases, the fluid flow for different working fluids comes closer, the effect of the working fluids' energy carrying ability becomes prominent, and the advantage of higher specific heat and higher LHV manifests itself consequently.

6.3. Comparisons between Different Filling Ratios (FRs)

For a filling ratio of 55% (Figure 22(a1) and (a2)), while heat input increased from 35 W to 50 W (the first section of lines), a gentle rise occurred in the value of *Te* and a relatively obvious reduction in the value of *R* occurred for all working fluids. The *Te* of acetone was the lowest due to its lowest boiling point. When the heat input reached 65 W, except for deionized water, other working fluids all presented a sharp rise in both *Te* and *R* that demonstrated the existence of local dryout in the ES with acetone having the most serious dryout problem (the highest values of *Te* and *R*). The value of *R* for deionized water at this filling ratio (55%) could still be kept low after 65 W although it exhibited a gentle increase because of the gradual appearance of mild, local superheat. As pre-discussed in Section 2, low surface tension of the three fluids (methanol, ethanol and acetone) is a favorable property for PHP operation, but they were still easier to dry out. Therefore, surface tension was not dominant in the dry-out of PHP. The liquid density or the gravity has a dual-effect as a driving force for the flowing down of the fluid and a drag force for the rising up of the liquid slugs. At low filling ratio, the effect of the driving force may manifest itself due to the few liquid slugs that have come up from the evaporation section, which accordingly could assist in the return of the liquid to the ES and contribute to the prevention of dry-out, especially for the fluid with larger liquid density like deionized water. However, the liquid density may not be the dominant factor either, because the liquid density of the three fluids other than deionized water is about the same, and the dry-out conditions varied. In fact, it has been observed that the fluids of a low boiling point and low LHV tend to be easier to dry out.

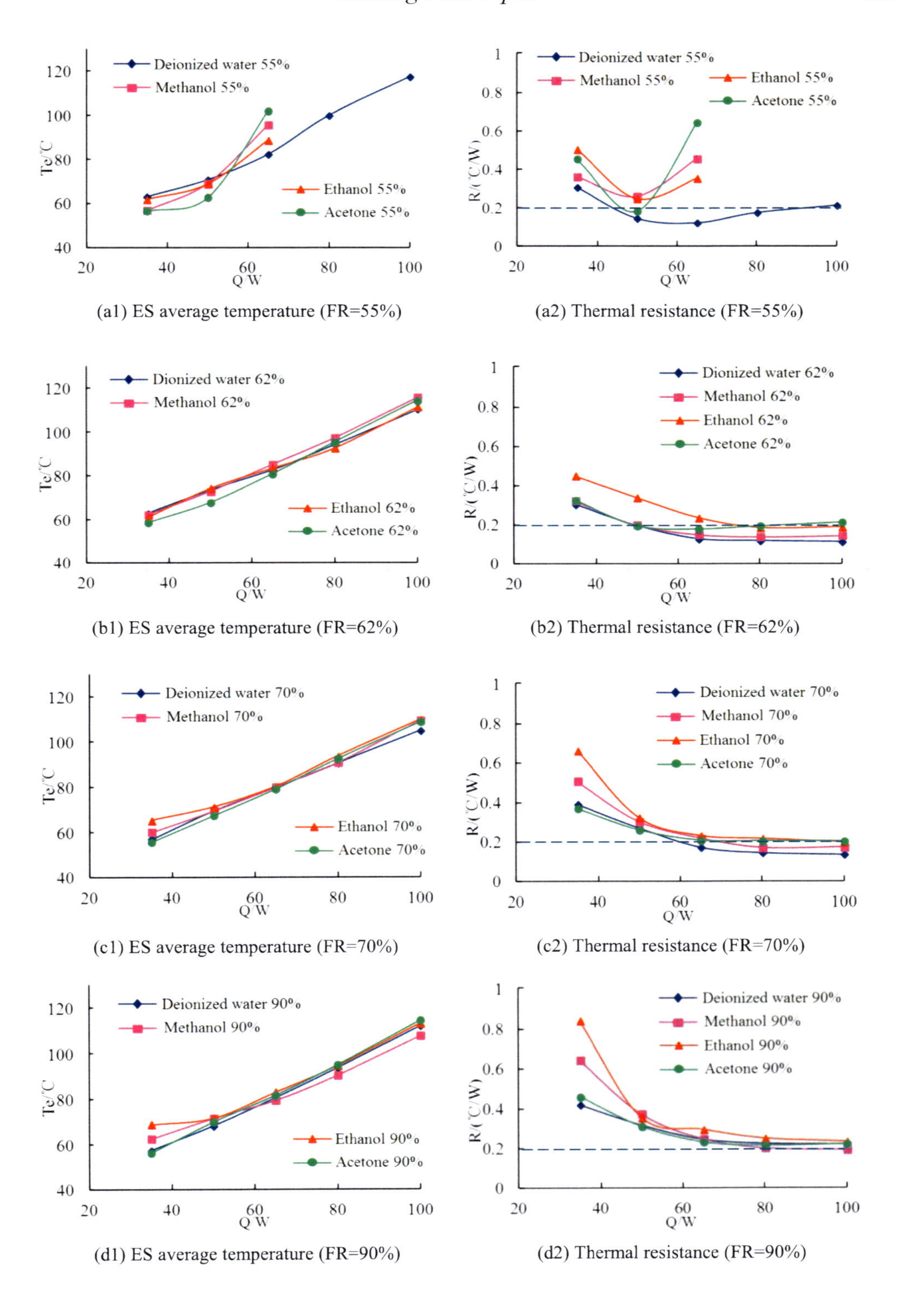

(a1) ES average temperature (FR=55%)

(a2) Thermal resistance (FR=55%)

(b1) ES average temperature (FR=62%)

(b2) Thermal resistance (FR=62%)

(c1) ES average temperature (FR=70%)

(c2) Thermal resistance (FR=70%)

(d1) ES average temperature (FR=90%)

(d2) Thermal resistance (FR=90%)

Figure 22. The variations in ES average temperatures and PHP thermal resistance with heat input for the working fluids at different filling ratios.

For filling ratios of 62 to 90% (Figure 22(b1)–(d2)), the value of T_e for each working fluid rose continuously and the value of R declined continuously with the reduction between 35 W and 50 W being the most obvious, 50 W and 65 W were the second most obvious, and after 65 W, almost became stable. The flow velocity of oscillations at 35 W was relatively low, which hindered the heat transfer, and the value of R was accordingly relatively large. As the heat input increased to 50 W, the value of T_e grew, the fluid flowed faster with the oscillation amplitude and period dwindling, and the heat transfer was reinforced consequently with the value of R reduced distinctly. Further increase in the heat input would not cause an obvious reduction in R, and the enhancement in heat transfer became smaller than that of the oscillation start-up period and the initial acceleration period. The possible reasons will be further discussed in Section 6.5.

6.4. Comparison between Different Working Fluids

If no dry-out happened, e.g., the filling ratio is greater than 55%, and the heat input is less than 65 W, the values of T_e and R for acetone were generally lower (Figure 22(b1)–(d2)) because of the lower boiling point (Ts) and the lower dynamic viscosity (υl) that might cause faster flow and better heat transfer accordingly. While the heat input reached 80 W and above, the values of T_e and R for acetone rose obviously, sometimes even higher than all others (e.g., FR = 62% and HI = 100 W in Figure 22 (b2)). It might be because the flow velocities were all great at this period due to the dwindling of dynamic viscosity (υl) and the augmentation of $(dp/dT)_{sat}$ along with the rising of temperature and heat input, that the energy carrying ability, characterized by liquid specific heat (Cpl) and LHV (Hfg), exhibited greater effect on heat transfer, and the values of Cpl and Hfg for acetone were lower than all other working fluids. Except for the dry-out situations like FR = 55% and HI = 65 W in Figure 22(a1) and (a2), the value of R for ethanol was generally somewhat higher. It can be conjectured that the heat transfer may be impeded by the great flow resistance caused by the highest dynamic viscosity (υl) and the unsatisfactory Cpl and Hfg for carrying energy.

Comparing the changing trends of R for acetone and ethanol, it could be observed that from the heat input of 50 W on, the difference of the value of R between acetone with the lowest dynamic viscosity (υl) and the difference between ethanol with the highest dynamic viscosity (υl) was dwindling. For one thing, the values of υl came closer to one another as the temperatures rose. For another, the advantage of fast flow caused by lower dynamic viscosity (υl) was fading away and a better energy carrying ability started to show its preferential effect on heat transfer.

With the highest liquid specific heat (Cpl) and LHV (Hfg), deionized water has a distinct advantage in energy carrying ability and may generally be a good working fluid for PHP except for two situations: 1) deionized water has a rather small heat input (e.g., less than 35 W), making it relatively difficult for the PHP charged with deionized water to start oscillation; moreover, the flow velocity was low even during the early oscillation, due to the high boiling point (TS), great LHV (Hfg), low $(dp/dT)_{sat}$, great dynamic viscosity (υl), and liquid density (ρl); 2). A rather large filling ratio (e.g., FR = 90% or 95%, Figure 22(d1)–(d2)) is used where there might be poor flow due to an impediment from great density and viscosity, and the subsequent poor heat transfer. For large filling ratios like 90% and 95%, methanol performs better at higher heat input with lower value of R (Figure 22(d2), (e2)), which could be attributed to the relatively low liquid density (ρl) and dynamic viscosity (υl), as well as the greatest $(dp/dT)_{sat}$ and higher LHV than ethanol and acetone. Therefore, for large filling ratios and great heat input, methanol is a better choice. Further observation shows that compared to FR = 62% and 70%, an increase in the value of R occurred for most situations requiring large filling ratios (90% and 95%), below 0.2^0C/W vs. above 0.2^0C/W for high HI, which indicated that large filling ratios over 90% were generally not favorable for PHP applications due to the deterioration of fluid motion and heat transfer.

From the above discussion, conclusions could be drawn and decisions made, such as choosing which working fluid to use and the best charging to use at which filling ratio. These decisions should be determined based on a practical situation. From the aspect of bringing the advantage of low thermal resistance into full play, 60 to 80% is a good filling ratio for acetone and the heat input or heating power should be low; the optimal filling ratio for deionized water is about 55 to 70% and the heat input can be great; methanol is a better choice for large filling ratios (e.g., 90% or above) with great heat input.

6.5. Thermal Resistance Comparison for the Working Fluids at Different Filling Ratios

With the increase in heat input, the oscillation is strengthened, the effect of viscosity can dwindle and that of liquid specific heat and especially LHV will grow, as the thermal resistances (R) for different working fluids come closer (Figure 22) and as long as no obvious dryout happens. It may be deduced, therefore, that at higher heat input, the influence of the thermo-physical properties of different working fluids on the heat transfer of PHP is fading away.

By putting different filling ratios of each working fluid into one figure (Figure 23) and by comparing the thermal resistances (R), the influence of filling ratio on the heat transfer of PHP can be observed. As long as the filling ratio is large enough to avoid dryout, (e.g., greater than 55%), the value of R for different filling ratios also came closer as the heat input increased, which means that at higher heat input, the influence of the filling ratio is also fading away.

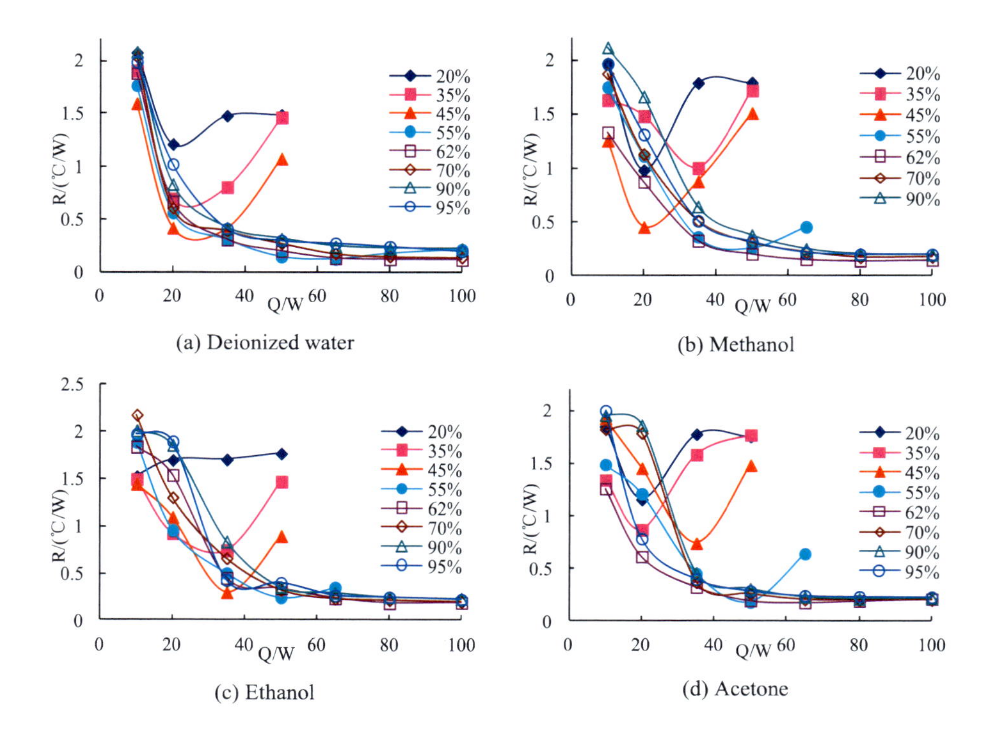

Figure 23. Thermal resistances for different filling ratios of each working fluid.

No matter what working fluid is used to fill at which ratio, the thermal resistance of a given PHP will always come to an approximate value if no continuous dryout happens. Obviously, this ultimate value of R has little relationship to the type of working fluids and its filling ratio, but largely depends on the PHP itself (the material and the structure) and the cooling condition. Given certain PHP characteristics, it may be because that with increased heating power and a corresponding growth in the flow velocity of the two-phase working fluids, the equivalent thermal resistance inside the tube wall will dwindle and become negligible and be ultimately compared to the outside thermal resistance of air convection. Thus, the limit would rise if the proper water cooling is adopted.

Major Conclusions for Section 6 are:

(1) For the same filling ratio (FR), the PHP charged with a working fluid, which had a lower boiling point and lower latent heat of vaporization (LHV) (e.g., acetone), making it easier to dry out.

(2) Dynamic viscosity (υl) outperformed all other thermo-physical properties investigated becoming the dominant property for the oscillation start-up of the closed-loop PHP. Together with $(dp/dT)_{sat}$, dynamic viscosity also played an important part in the flow velocity at early oscillation and accordingly the heat transfer of the PHP at a lower heat input (e.g., <50 W). Generally, the higher the

vl and the smaller the $(dp/dT)_{sat}$, the slower the working fluid flowed and the poorer the heat transfer was at this stage.

(3) At low heat input, the heat transfer depended heavily on whether the oscillation had been triggered, while at relatively high heat input, the effect of viscosity on heat transfer dwindled and that of the energy carrying ability characterized by liquid specific heat (*Cpl*) and latent heat of vaporization (LHV, *Hfg*) manifested, especially the latter (LHV). Greater LHV replaced the lower dynamic viscosity to become a dominant property favorable for the heat transfer of PHP at high heat input.

(4) At high heat input (e.g., >65 W), without dryout, the thermal resistance (*R*) of the PHP charged with different working fluids at different filling ratios came closer to one another, indicating that under this condition, the PHP heat transfer reached an upper limit that cannot be broken through. The limit depends largely on the PHP itself (the material, structure, and the cooling condition). In some PHPs, certain heating powers could cause the equivalent thermal resistance inside the tube wall to become negligible compared to the outside thermal resistance of air convection. Better cooling might lead to a higher heat transfer limit. Notably, the results, discussion and conclusion are all based on an air cooled PHP with an air velocity of 1.5 m/s at about 23^{0}C (Re ≈ 1.38x104).

7. Binary Mixture PHPs

Based on abundant experiments, this section presents a closed-loop PHP charged with various zeotropic binary mixtures (water-methanol, water-ethanol, water-acetone, methanol-acetone and ethanol-acetone) of various volume mixing ratios, and compared with the performances of a PHP charged with the corresponding pure fluids (deionized water, methanol, ethanol or acetone). Three levels of filling ratios were organized according to the behavior of the PHP—small filling ratios (35% and 45%), medium filling ratio (55%) and large filling ratios (62% and 70%).

The zeotropic binary mixtures described herein are all positive-deviation solutions with similar features affecting PHP behavior. The next section demonstrates the characteristics using water-based binary mixtures as an example. A discussion of the ethanol-methanol mixture follows

7.1. Water-Based Binary Zeotropes in PHP

To benefit from the complementary nature of each component, binary mixtures based on deionized water (water-methanol, water-ethanol and water-acetone) were explored as the working fluid of the CLPHP. Detailed comparisons among PHPs with the corresponding single fluid and their binary mixture are made. Various volume mixing ratios were adopted for each mixture, such as 13:1, 4:1, 2:1, 1:1, 1:4, and 1:13, with the counterpart serving more as a component than water. Thus, methanol, ethanol or acetone was added to the water. It should be noted that the mixtures at 1:4 and 1:13 ratios have less water than their counterpart. Filling ratios have been categorized into three levels according to the behavior of the PHP; thus, small filling ratios of 35% and 45%, medium filling ratios of 55%, and large filling ratios of 62% and 70% were established. Observations were made for the overall thermal resistances (R,^{0}C/W) at the heat inputs of 10 W, 20 W, 35 W, 50 W, 65 W, 80 W and 100 W, as a major representative for the PHP's thermal performance.

7.1.1. Small Filling Ratios (35%, 45%) and Medium Filling Ratio (55%)

The overall thermal resistance (R) of PHP charged with binary fluids of water-methanol, water-ethanol and water-acetone are shown in Figure 24, Figure 25 and Figure 26, respectively. Here, we see the different volume mixing ratios filled at 35% and 45%. Dashed lines with denotations (× or +) represent pure fluids like water, methanol ethanol or acetone for the convenience of comparison. The sharp peak in the line indicates a sharp increase in the value of R and accordingly the occurrence of local dryout in the evaporation section (hot end) where the liquid returned from the condensation section was not enough to supplement the evaporation, causing a sharp rise in the local temperature. Further experiments were not possible due to potential fire damage to copper tubing.

Despite finding that when pure fluids are at low filling ratio, the working fluids with lower boiling point and lower LHV were easier to dry out [50], and finding that acetone has the lowest boiling point and the smallest LHV, the water-acetone mixture (Figure 26) showed a better dryout prevention ability with most of the lines (except for the water-acetone ratio of 13:1) still keeping low at or greater than 50 W, as compared not only to the pure components (water or acetone) but to other binary fluids or pure fluids like methanol or ethanol (Figure 24 and Figure 25). This might be due to the remarkable complementary features of water and acetone in their thermo-physical properties like boiling point, LHV, $(dp/dT)_{sat}$ and dynamic viscosity. Reasons given in Zhu et al.[51] are based on the discrepancy of the evaporation of component water and component acetone: During phase change, acetone would first evaporate and abundantly enter the vapor phase; vapor slugs enriched with acetone vapor possessed relatively high saturation pressure and $(dp/dT)_{sat}$, which not only functioned as the impetus to drive the working fluid flow (favorable to replenishing liquid into the evaporation section), but also suppressed the evaporation of component water keeping the water largely in the liquid phase, thereby preventing the onset of dryout. This may also be the major reason why most of the binary mixtures, especially for water-methanol/ethanol ratio of 4:1 at filling ratio of 35%, water-ethanol ratio of 4:1 and water-methanol ratio of 13:1 at filling ratio of 45%, have better anti-dry-out performance than at least one of the pure fluids (even both of them), with relatively low value of R at relatively high heat input (50 W and above). Water–based mixtures are complex from a microscopic view. There may be hydrogen bonds at different scales for different mixing ratios and different types of mixtures, which might also affect the thermal performance of the PHP, but few references could be found in the open literature to confirm this aspect. Further study should be made.

The anti-dry-out performance of the pure ethanol is better than that of the pure methanol and close to that of deionized water (Figure 24 and 25). It can be conjectured that for the four pure fluids involved, comprehensively taking the phase change characters (mainly, boiling point and LHV) and the possible flow features (mainly, $(dp/dT)_{sat}$ and dynamic viscosity) into consideration, the boiling point played a more important role in the anti-dryout performance when the other properties were not so discrepant from each other, even more than the LHV component. This might be one of the reasons why water-ethanol mixtures generally perform better than water-methanol mixtures in dryout prevention, given the similarities in the small-scale complementary properties of methanol and ethanol to water.

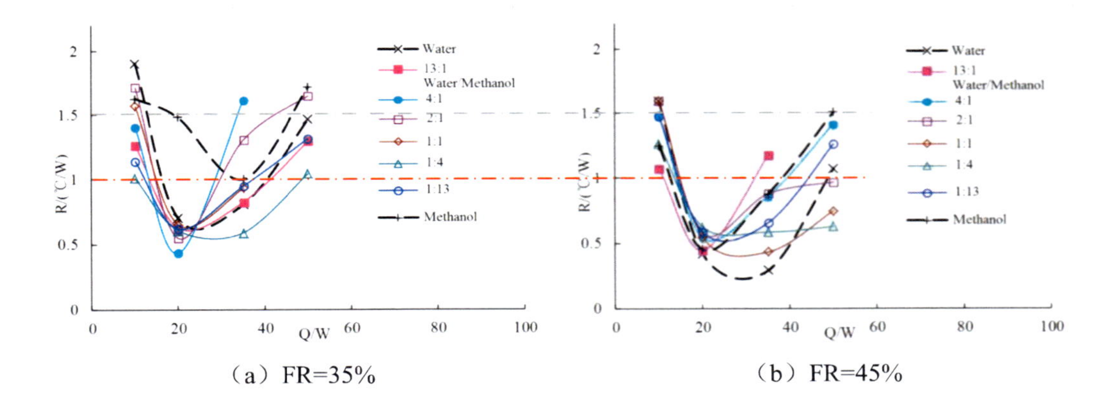

Figure 24. Overall thermal resistance of PHP with water-methanol mixtures (35%, 45%).

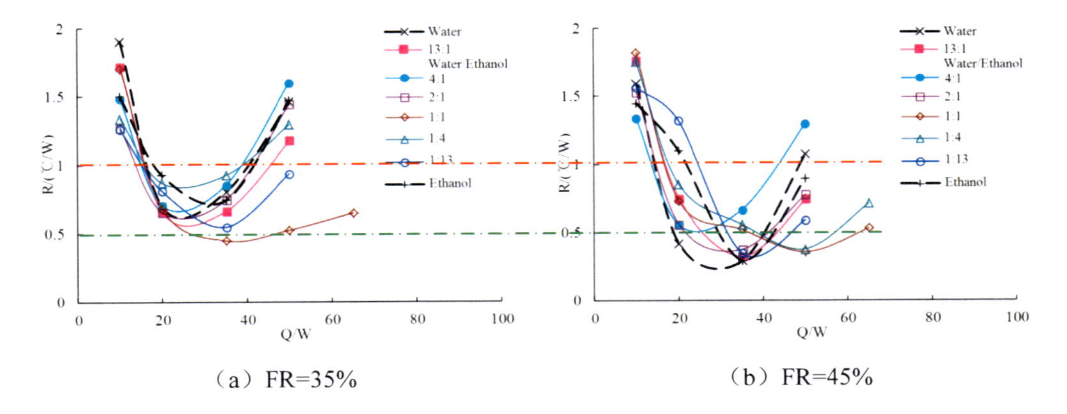

Figure 25. Thermal performance of PHP with water-ethanol mixtures (35%, 45%).

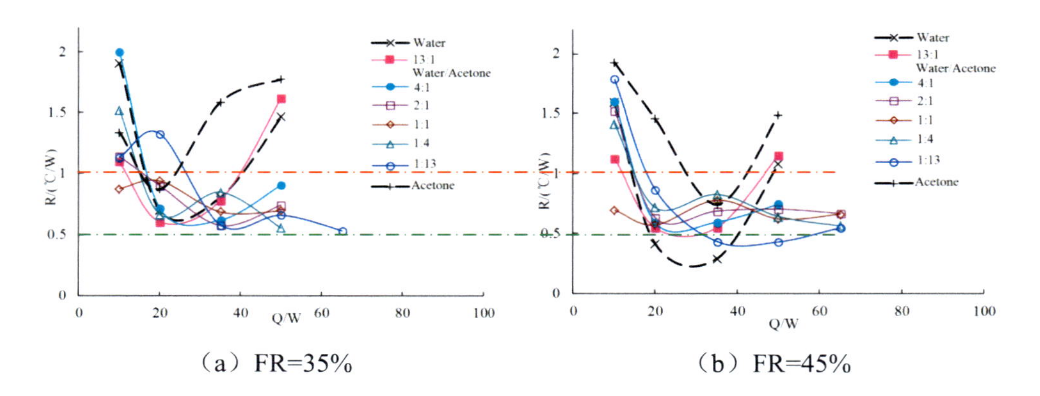

Figure 26. Thermal performance of PHP with water-acetone mixtures (35%, 45%).

Observing the dotted and dashed lines across the subfigures, it could be found that with the increase in filling ratio, most of the fluids, pure or binary, grew better in dry-out prevention with the value of *R* becoming smaller, i.e., given the working fluid, increasing the filling ratio could improve the anti-dry-out performance of the PHP, simply because of the incremented fluid quantity and accordingly, the relatively abundant replenishment of liquid to the evaporation section.

The thermal performances of the CLPHP charged with the binary mixtures and the pure fluids at filling ratio of 55% are shown in Figure 27. Pure water shows excellent anti-dryout performance with the value of *R* coming as low as 0.12^{0}C/W at 65 W with just a little rise at 100 W (0.22^{0}C/W), which has been greatly improved from those at filling ratios of 35% and 45%. In fact, the anti-dry-out conditions were generally better for all the fluids. Many of the binary mixtures came closer to pure water, like water-methanol mixtures at a ratio of 13:1 (low concentration of methanol), water-ethanol mixtures at ratios of 4:1, 1:1, and 1:13, and water-acetone mixtures at ratios of 4:1, 2:1, 1:1, and 1:13. The mixing ratios of each binary mixture with the best and worst performance of dry-out prevention at filling ratios of 35%, 45% and 55% were summarized and listed in Table 7.1 for the convenience of reference. It was observed that from the perspective of anti-dryout performance at small and medium filling ratios below or equal to 55%, the mixing ratio of 1:1 is always a good choice for water-ethanol mixtures, 1:13 is always good, and 13:1 is bad for water-acetone mixtures. Adding a small fraction of water (about 7% (volume)) into acetone (1:13) could greatly improve the anti-dryout performance of pure acetone, even better than that of pure water for small filling ratios like 35% and 45%, whereas adding the same small fraction of acetone into water made its performance similar to pure water for small filling ratios (35% and 55%) and worse than pure water for medium filling ratio (55%). Water-acetone mixing ratios of 1:13 (the acetone molar fraction is 0.76) and 13:1 (the acetone molar fraction is 0.02) were both depicted in the water-acetone phase diagram (Figure 3(a)). Point *A* is a possible point for a mixing ratio of 1:13 at high heat input where there was a great difference between the concentration of acetone in the vapor phase and the liquid phase (point *B*). Moreover, the corresponding saturation pressure (point *C* and *D* in Figure 3(b)) was higher than that of the pure water (point *E*), providing a favorable condition for water to stay in the liquid phase. However, for a water-acetone mixing ratio of 13:1 where water dominates (the water molar fraction is 0.98). The difference between the concentrations of acetone in the vapor phase and the liquid phase is quite small since that in the liquid phase is almost zero, which makes the pressure in the liquid phase almost the same as that of pure water; thus, no superiority to water exists in this aspect. Meanwhile, the adverse influence of the energy carrying ability due to the adding of acetone (small LHV and specific heat) manifests itself especially at high heat input. Due to the difficulty in measuring pressure inside the tube, the actual state of the working fluids cannot be obtained in a timely and exact manner, making the detailed quantitative analysis based on the phase diagram difficult. Nevertheless, it is a prospective path for further study of the mixtures.

Generally speaking, dryout in the evaporation section of the PHP is an undesired situation that may deteriorate the PHP's operation, cause a sharp rise of local temperature and can even melt the tube. For small and medium filling ratios, anti-dryout should be the first priority consideration for the application of PHPs, whereas for large filling ratios, dryout seldom happen; thus, the focus should be placed on the heat transfer performance of the PHP (see below).

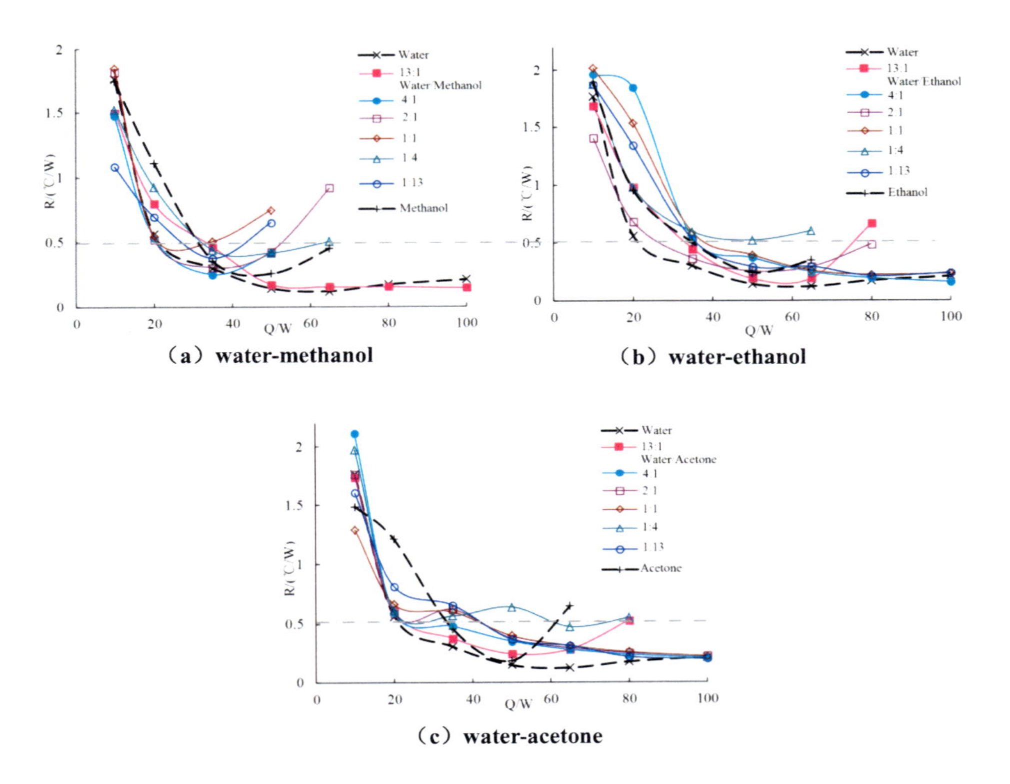

Figure 27. Thermal performance of PHP with water-based binary mixtures (55%).

Table 7.1 List of anti-dryout performances for the water-based binary mixtures

		FR = 35%	FR = 45%	FR = 55%
Water-methanol	best	1:4	1:4, 1:1	13:1
	worst	4:1	13:1, 4:1	1:1, 1:13
Water-ethanol	best	1:1, 1:13	1:1, 1:4	4:1, 1:1, 1:13
	worst	4:1	4:1	1:4
Water-acetone	best	1:13	1:13, 1:4, 1:1, 2:1	4:1, 2:1, 1:1, 1:13
	worst	13:1	13:1	1:4, 13:1

7.1.2. Large Filling Ratios (62%, 70%)

For large filling ratios of 62% and 70%, the overall thermal resistances (R) of PHP charged with binary fluids of water-methanol, water-ethanol and water-acetone with different volume mixing ratios are shown in Figure 28, Figure 29 and Figure 30, respectively. The general tendency is about the same: With the increase in heat input, the values of R tend to decrease, sharply at first (before 35 W or 50 W), to decrease slightly afterwards and become almost stable at last (after 65 W or 80 W).

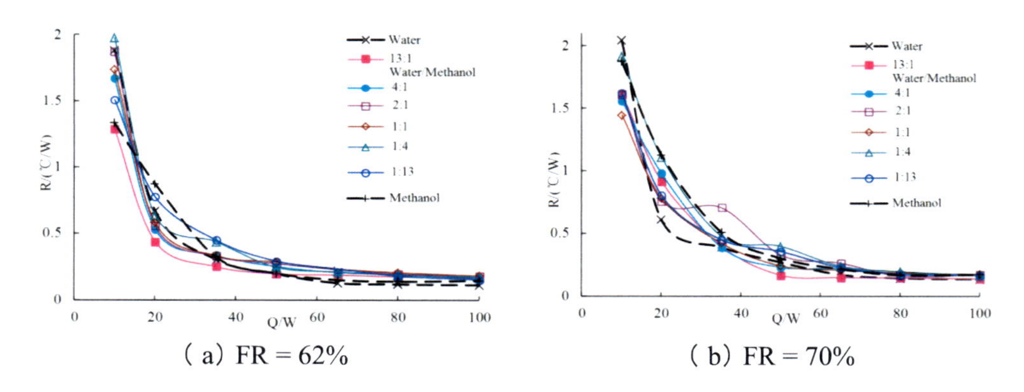

Figure 28. Thermal performance of PHP with water-methanol mixtures (62%, 70%).

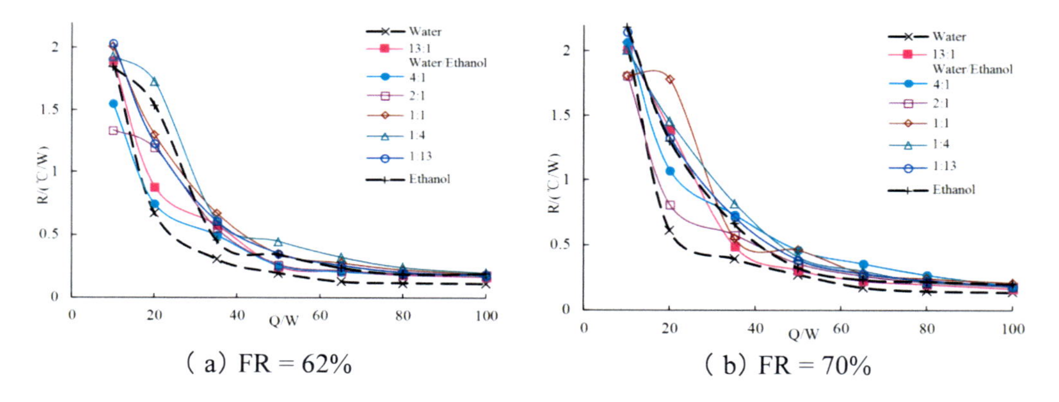

Figure 29. Thermal performance of PHP with water-ethanol mixtures (62%, 70%).

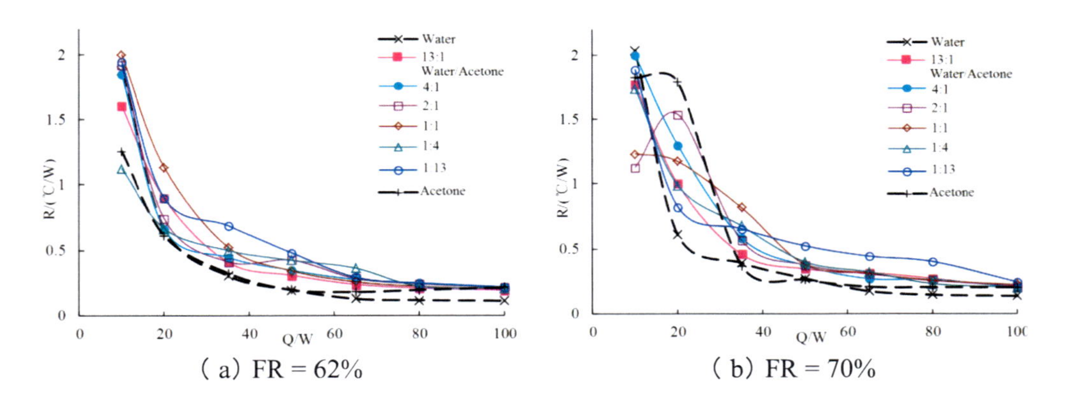

Figure 30. Thermal performance of PHP with water-acetone mixtures (62%, 70%).

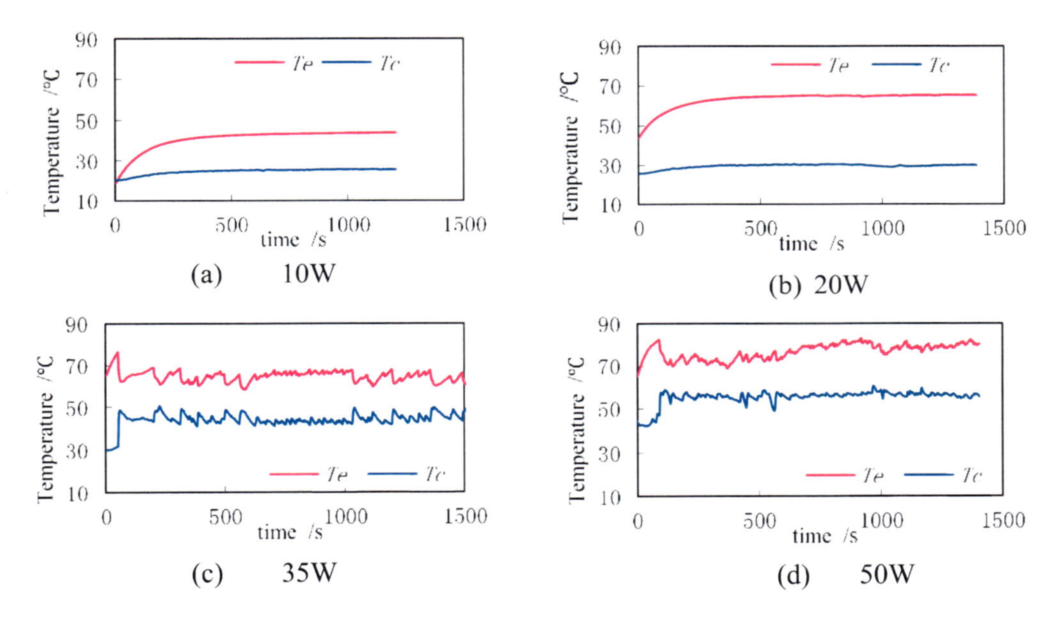

Figure 31. Temperature trends of the evaporation section (T_e) and the condensation section (T_c) for water-ethanol mixture at mixing ratio of 1:1 and a filling ratio of 70%.

The repeatability of the experiments has been previously discussed in [50] which determined that despite the basically consistent trends of descending, inconsistencies appeared between different test runs for heat input lower than 35 W, probably due to the inadequate heat force, which could not push the fluid up and across the tube bend to break the initially uneven distribution. It is therefore possible for the inconsistency to cause the seemingly abnormal rise of the lines, e.g., 10 W to 20 W for water-ethanol mixtures of 1:1 at a filling ratio of 70% (Figure 30 (b)). Temperature trends of the evaporation section (T_e) and the condensation section (T_c) at 10 W, 20 W, 35 W and 50 W are plotted in Figure 31 to observe the operation for this condition (water-ethanol 1:1 at 70%). No temperature fluctuations were found for the T_e and T_c at 10 W and 20 W, which meant that the PHP oscillation had not started yet. This confirmed the conjecture of the fluid not coming up and across the tube bend and the possible inconsistency caused. Further observations of the temperature trends for each condition showed that the drastic decrease from 10 W to 20 W like most of the lines in Figure 31 (e.g., the water methanol mixture of 13:1 (FR = 62%) or the pure water (both 62% and 70%)) demonstrated an early start-up of oscillation (at 20 W).

For large filling ratios like 62% and 70%, the PHPs charged with the pure water (deionized water), which has always had an excellent thermal performance with the *R* value keeping its low—if not the lowest value for most situations. Almost all lines of the binary mixtures on each sub-figure (Figure 28(a), (b), Figure 39(a), (b), Figure 30(a), (b)) came closer to or even coincident with one another at the higher heat input of over 50 to 80 W, and came closer to or was with one of the pure fluids (the one with relatively high thermal resistances) or both of them (pure water and pure methanol), e.g., the thermal resistances of the PHP charged with water-methanol mixtures of various mixing ratios (13:1 to 1:13) at a filling ratio of 62% (Figure 28(a)) became similar to one another after a heat input of 50 to 65 W (about 0.15-0.18^{0}C/W at 100 W). This was much closer to that of the pure methanol at 100 W (about 0.14^{0}C/W), indicating that no matter what the mixing ratio, the water-based binary mixtures investigated had no superiority in terms of thermal performance to the pure fluids at large filling ratios and relatively high heat input. From the perspective of the energy-carrying ability characterized by LHV and the specific heat, the binary mixtures should be between the pure fluids, e.g., the water-acetone mixture should be between pure water and pure acetone, but the thermal performances were not in-between. It, therefore, may be conjectured that the oscillation flow of the mixtures was not as fast as that of the pure fluids probably because of the zeotropic properties and the effect of the complex hydrogen bonds of the water-based mixtures, which will be further studied. Some of the possible reasons may be: 1) with the increase in heat input, as soon as the temperature reached the dew line (Figure 3), the concentrations of the components in the liquid phase and the vapor phase could no longer change; if the concentration shift ceased, the phase change of the mixture attained its limit and the advantage of the driving force over pure water brought about by the zeotropic properties disappeared. 2) the hydrogen bonds between the molecules of the two components made the dynamic viscosity of the mixtures greater than one or both of its components within the mixing ratios concerned, causing more flow resistance than its pure counterpart. For instance, the dynamic viscosities of the water-methanol and water-ethanol mixtures are shown in Figure 32 [52] (no such information was found for water-acetone mixtures in the literature; thus, it was determined that within the mixing ratios of this study (13:1, 4:1, 2:1, 1:1, 1:4, 1:13), the dynamic viscosity of the water-methanol mixture is always higher than that of pure methanol. The same holds true for the dynamic viscosity of the water-ethanol mixture to that of pure ethanol. Moreover, the hydrogen bonds can also affect other thermodynamic properties like boiling point, LVH, density, etc., which makes the mixtures even more complex and should be further explored. Notably, the dynamic viscosity of each mixture still differs at different mixing ratios, which reminds us that the thermal performances of the PHP were comprehensively influenced by the energy-carrying ability, the driving force, and the flow resistance of the working fluids.

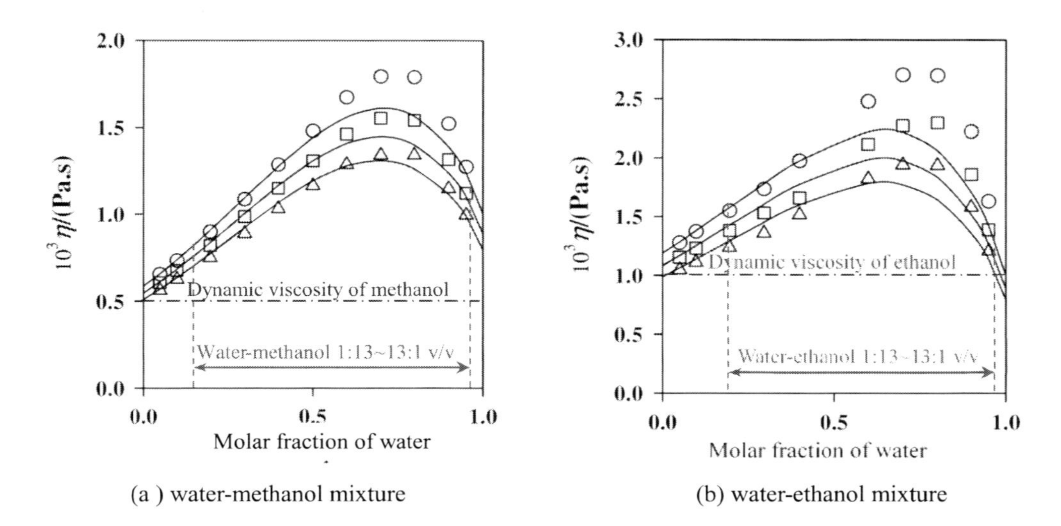

(a) water-methanol mixture (b) water-ethanol mixture

Figure 32. Dynamic viscosity of water-methanol mixture and water-ethanol mixture [52] (Lines were obtained from UNIQUAC equation and points from experiments for different temperatures (○ is 293.15 K, □ is 298.15 K, and Δ is 313.15 K)).

For water-methanol mixtures at a filling ratio of 70% (Figure 28(b)), the PHPs charged with the mixtures of different mixing ratios (especially 13:1) all have comparable heat transfer performance to the PHPs charged with pure water or pure methanol at relatively high heat input like 80W and 100W. Previous study [50] has stated that large filling ratio and great heat input were favorable to the working fluid of methanol due to 1) methanol's high $(dp/dT)_{sat}$ at high temperature, 2) the relatively large LHV and liquid specific heat compared to those of ethanol and acetone, and 3) methanol's relatively small liquid density compared to that of water, whereas water performed poorly at high filling ratios like 70% and above due to the highest liquid density. Thus, the methanol performance was boosted and the water performance was weakened. They then became closer and their thermal performances could be compared.

7.1.3. Characteristics of Different Mixtures at Certain Mixing Ratio (FR = 62%)

The thermal resistance of PHP was charged with different mixtures (water-methanol, water-ethanol and water-acetone) and prepared at certain mixing ratios. A filling ratio of 62% was demonstrated in one of the subfigures of Figure 33 to show the differences between the mixtures.

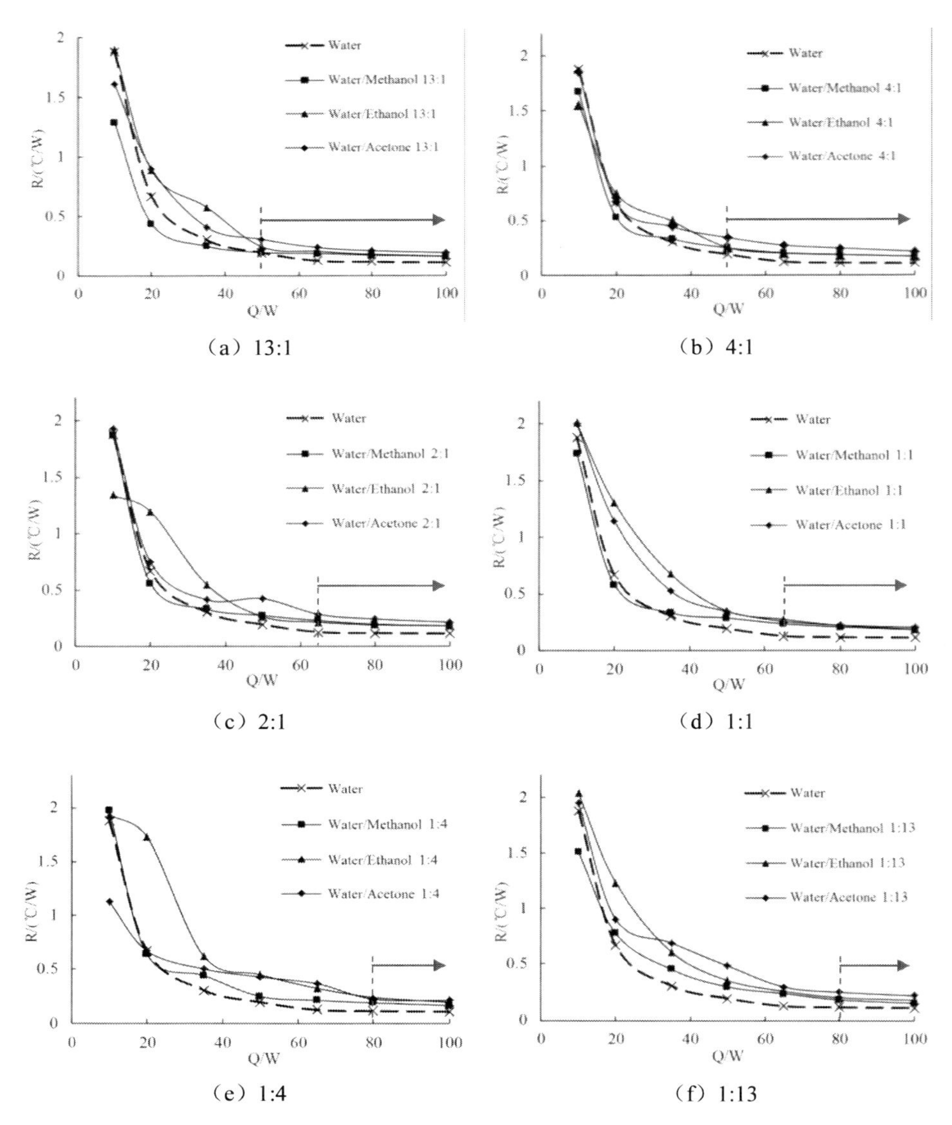

Figure 33. Thermal resistances of the PHP charged with water-based binary mixtures at different mixing ratios (FR = 62%).

For the volume mixing ratios of 13:1, 4:1, 2:1 and 1:1, the PHPs charged with the water-alcohol mixtures (the water-methanol and the water-ethanol mixture) behaved similarly at the heat input of 50 W or 65W (1:1) and above. This indicated that if the water is the majority or at least reaches half of the volume fraction in the water-alcohol mixture (≥50%/v, really water-based), the thermal performance of the PHP was insensitive to the types of the component alcohol (methanol or ethanol) at relatively high heat input. At relatively low heat input (<50 W), the water-methanol mixtures of these mixing ratios were better with *R* values drastically decreased at an early stage from 10 W to 20 W demonstrating an early and easy start-up of oscillation, even better (13:1) than the pure water. From Figure 33(a), the dynamic viscosity of the water-methanol mixture at 13:1 is higher than that of the pure water, which shows that unlike the pure working fluids [50] at relatively low heat input, the dynamic viscosity was not the only major factor that affected the start-up of the PHPs while charged with zeotropic mixtures, but possibly had a close relationship with the zeotropic properties previously mentioned and the molecular interactions between the components. The PHPs charged with the water-methanol mixtures of various mixing ratios (13:1 to 1:13 in Figure 33) at this filling ratio (62%) almost always had an excellent thermal performance throughout the heat inputs investigated (no matter how low or high), as compared to the other two mixtures (water-ethanol and water-acetone mixtures). This could probably be attributed to the greater $(dp/dT)_{sat}$ (Figure 2) of the component methanol, together with the better energy-carrying ability (liquid specific heat and LHV (Table 2.1)) compared to that of ethanol or acetone.

For the mixing ratio of 1:1(Figure 33(d)), at a relatively high heat input, 65 W and above, the three binary mixtures (water-methanol, water-ethanol and water-acetone) exhibited similar thermal performance. This is interesting because of the mixing ratio (1:1), half water and half counterpart (volume). Does it necessarily mean that at relatively high heat input, half-half (1:1/v) is always one of the good choices for the mixing of different water-based binary mixtures, at least for the filling ratio of 62%? Despite the excellent anti-dryout performance at small and medium filling ratios, the water-acetone mixture was almost always the worst one in terms of thermal resistance or thermal performance when compared to the other two water-based mixtures at this filling ratio. It did not matter what mixing ratio was used (except for 1:1) during heat input. In contrast with the excellent behavior of the PHPs charged with pure water (the dashed lines in Figure 33) or the water-methanol mixtures where both of the components had relatively high LHV, this phenomenon illustrates that for large filling ratios like 62%, the energy-carrying ability of the working fluids manifested a preferential effect on PHP thermal performance, especially at relatively high heat input where it seems that the flow ability of the working fluids was approaching its limit. It is also possible that the nearly flat part of all the lines at heat inputs higher than a certain value in the figure should be attributed partly or mainly to the air-convection cooling condition in the condensation section (cool end) instead of water cooling.

7.2. The PHP with Methanol-Ethanol Mixture

Figure 34 shows the thermal resistance of the PHP with methanol, ethanol and their mixture as the heating power changes. As shown in Figure 35(a), when the heating power exceeds 20W, the thermal resistance of PHP (FR = 45%) with methanol increasing sharply. Dry-out occurs because of its low boiling point. When the heating power exceeded 35 W, the thermal resistance of the PHP (FR = 45%) with ethanol and mixtures increased continuously, and dryout occurred. The thermal resistance of the PHP with the methanol-ethanol mixture was almost between pure methanol and ethanol. Because methanol and ethanol are of the same type, this two group-element fluid can be treated as an ideal mixture [16, 53]. As shown in Figure 6, the dew line and bubble line are close to each other. The physical properties of the mixture fall in between the lines, which is why the thermal resistance of the PHP with a methanol-ethanol mixture is almost between that with pure methanol and ethanol and why using this PHP has no advantage.

As shown in Figure 34(b), for a filling ratio of 62%, the thermal resistances of the PHP with different working fluids are similar to each other. Because methanol and ethanol are alcohols, many of their physical properties, such as density, specific heat and surface tension, are similar to each other. Moreover, methanol has a higher latent heat of vaporization than ethanol, which makes the thermal resistance of the PHP (62%) with methanol relatively low. As the filling ratio further increases (FR = 70%, 90%), the vapor space decreases, which leads to less latent heat transfer and more sensible heat transfer. Because the specific heats of pure methanol, ethanol and their mixture are similar to each other, the thermal resistances of the PHPs with different working fluids are extremely similar, as shown in Figure 34(c) and (d). Figure 35 shows the recorded temperature of the PHP with methanol-ethanol (FR = 62%, mixing ratio = 4:1). It is concluded that the temperature pulsating frequency of methanol-ethanol does not exhibit obvious changes compared with that of pure methanol as shown in Figure 36. Even when ethanol is added to methanol, the velocity of the pulsating flow does not exhibit obvious changes.

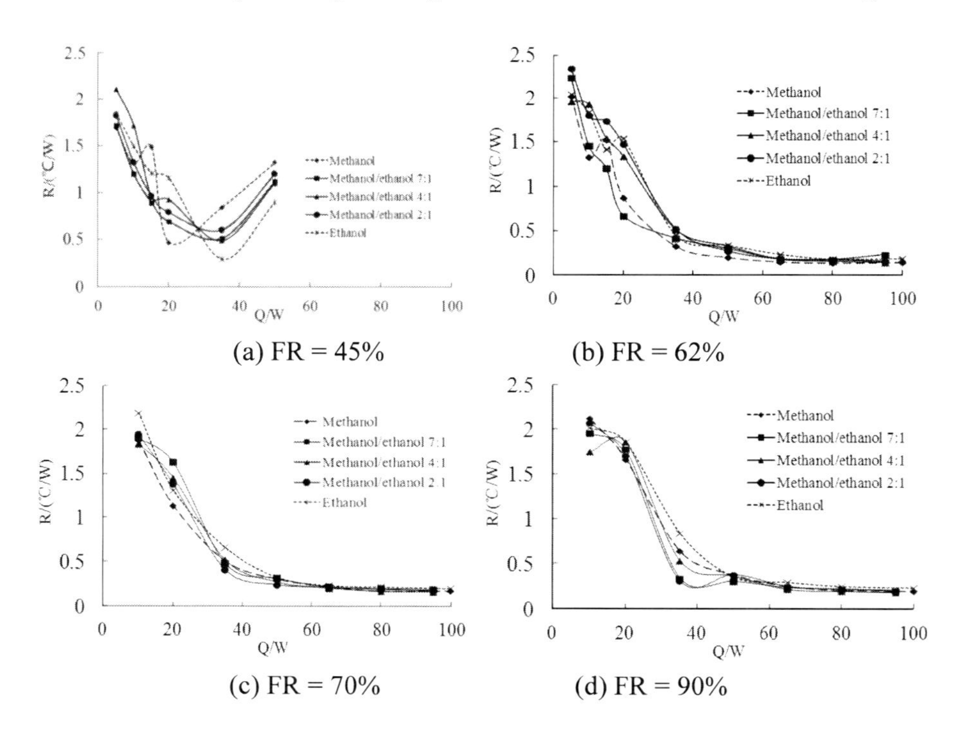

Figure 34. Thermal resistances of PHP with methanol-ethanol.

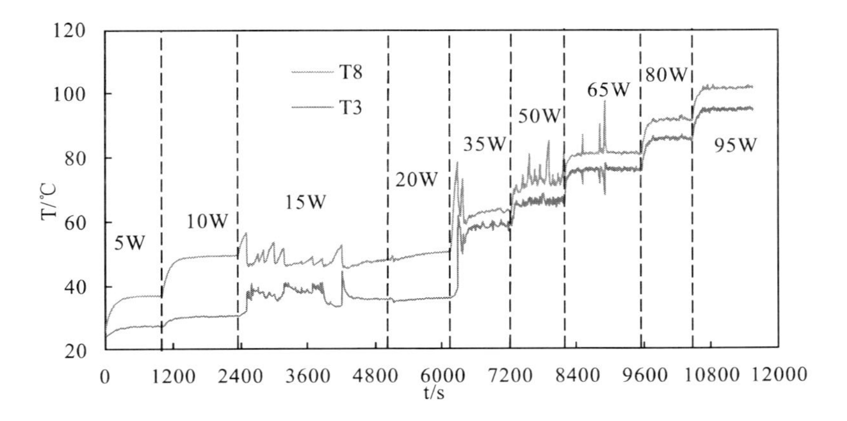

Figure 35. Real-time recorded temperatures as the heating power increases (FR = 62%, methanol-ethanol, mixing ratio = 4:1).

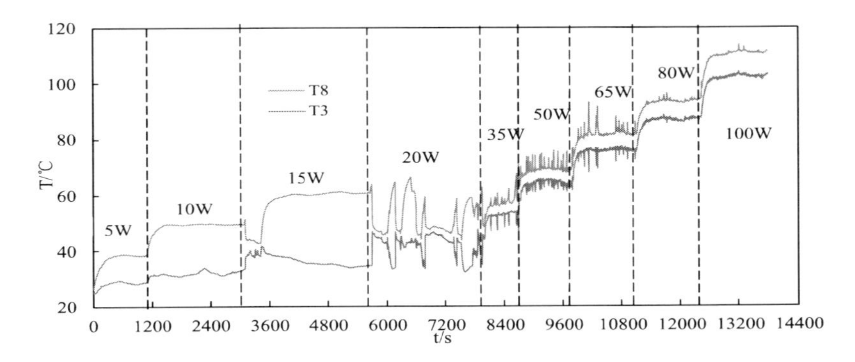

Figure 36. Real-time recorded temperatures with increasing heating power (FR = 62%, methanol).

Conclusions: PHP with Binary Mixtures

(1) At small or medium filling ratios (35%, 45%, and 55%), most of the binary mixtures of various volume mixing ratios have a better anti-dryout performance than at least one of the pure fluids (or even both of them). This could be mainly attributed to the zeotropic properties of the mixtures. The component first evaporated during the phase change was always one other than the component water due to the lower boiling point. The vapor slugs were abundant with this component of higher saturation pressure and higher $(dp/dT)_{sat}$. Not only do the vapor slugs enhance the flow driving force, they also suppress evaporation of the higher boiling point fluid.

(2) At large filling ratios (62% and 70%) where dryout seldom happens, the thermal performance of the PHPs charged with the binary mixtures at various mixing ratios had no superiority to the pure fluids especially the pure water. Most of the thermal resistances could only come closer to the pure fluid of the weaker one at a relatively

high heat input of 65 W and above. Pure water is always a good choice for PHPs at these filling ratios, which can be attributed to its excellent energy-carrying ability characterized by the largest latent heat of vaporization (LHV) and liquid specific heat. It was conjectured that the oscillation flow of the mixtures was not as fast as that of the pure fluids probably because of the zeotropic properties and the effect of the complex hydrogen bonds. This is than needs further study.

(3) Because of the remarkable complementary features for water and acetone in their thermo-physical properties like boiling point, LHV, $(dp/dT)_{sat}$ and dynamic viscosity, etc., the water-acetone mixtures have distinctive zeotropic characteristics showing a better dry-out prevention ability at small or medium filling ratios. At the large filling ratio where dryout seldom happens, the PHP with water-acetone mixtures did not show any advantage because of its weaker energy-carrying ability, which manifested a preferential effect on the thermal performance of the PHP at this stage (especially for relatively high heat input).

(4) PHP with methanol-based binary mixtures have different filling ratios, and additives will cause different effects. For a filling ratio of 45%, adding water to methanol can delay dryout, and the thermal resistance of the PHP with methanol and water is lower than that with pure methanol and water. However, adding ethanol and acetone to methanol cannot effectively improve the thermal performance of PHPs. When the filling ratio increases to 62%, 70% and 90%, adding water, ethanol and acetone cannot strengthen heat transfer.

(5) For a near-azeotropic binary mixture and a methanol-ethanol mixture, the thermal performance of PHP falls between that of the pure fluids (methanol and ethanol).

Conclusion

This study investigated the oscillation characteristics and heat transfer performance of a closed-loop PHP (inner diameter = 2 mm) charged with deionized water, methanol, ethanol, acetone and all the possible binary mixture of these single fluids with several different ratios. Section 5 discussed the flow patterns and the heat-transfer mechanism of pulsating heat pipes in detail. In Section 6, the study experimentally investigated the PHP based on its performance using four single fluids: deionized water, methanol, ethanol and acetone. A detailed comparison of the PHPs with single fluid confirmed that the higher the dynamic viscosity (vl) and the smaller the $(dp/dT)_{sat}$, the slower the working fluid flows, and the poorer the heat transfer occurs, especially during the oscillation start-up period and the early oscillation period. For the PHP with binary mixtures (Section 7), the experimental results showed that at small or medium filling ratios (35%, 45%, and 55%), most of the binary mixtures of various volume mixing ratios have a better anti-dry-out performance than at least one of the pure fluids. At large filling ratios (62% and 70%), the thermal performance of the PHPs charged with binary mixtures at various mixing ratios had no superiority over the pure fluids, especially the pure water. Detailed conclusions are at the end of each section.

Acknowledgment

The authors gratefully acknowledge the National Natural Science Foundation of China (Grant No. 51076104) for their support of the research presented herein.

References

[1] H. Akachi, *Structure of a heat pipe*. US, US4921041, 1990.
[2] X. Liu, Y. Chen, M. Shi, Dynamic performance analysis on start-up of closed-loop pulsating heat pipes (CLPHS). *International Journal of Thermal Sciences* 65 (2013) 224-233.
[3] B. Y. Tong, T. N. Wong, K. T. Ooi, Closed-loop pulsating heat pipe. *Applied Thermal Engineering* 21 (18) (2001) 1845-1862.
[4] S. Khandekar, N. Dollinger, M. Groll, Understanding operational regimes of closed loop pulsating heat pipes: an experimental study, *Applied Thermal Engineering* 23 (6) (2003) 707-719.
[5] P. Charoensawan, S. Khandekar, M. Groll, P. Terdtoon, Closed loop pulsating heat pipes: part a: parametric experimental investigations, *Applied Thermal Engineering* 23 (16) (2003) 2009-2020.
[6] X. U. Rong-Ji, R. X. Wang, W. Cong, W. U. Ye-Zheng, Design of pulsating heat pipe experiment rig and visual experiment study, *Fluid Machinery* 35 (6) (2007) 59-61.
[7] L. I. Yan, J. Li, Experimental research on heat transfer performance of pulsating heat pipe, *Refrigeration* 29 (11) (2008) 75-80.
[8] L. I. Jing-Tao, Z. X. Han, L. I. Zhi-Hong, L. Shi, Visual experiment study on operation and heat transfer characteristics of pulsating heat pipes, *Modern Chemical Industry*, 2008.
[9] Q. U. Jian, W. U. Hui-Ying, H. M. Tang, Flow visualization of micro/mini pulsating heat pipes, *Journal of Aerospace Power* 24 (4) (2009) 766-771.
[10] Y. Miyazaki, H. Akachi, Heat Transfer Characteristics of Looped Capillary Heat Pipe, in: 5th *Int. Heat Pipe Symposium*, Melbourne, Australia, 1996, pp. 208-217.
[11] R. Ponnappan, Novel groove-shaped screen-wick miniature heat pipe, *Journal of Thermophysics and Heat Transfer*, 16 (1) (2002) 17-21.

[12] J. L. Xu, X. M. Zhang, Start-up and steady thermal oscillation of a pulsating heat pipe, *Heat and Mass Transfer*, 41 (8) (2005) 685-694.

[13] J. Clement, X. Wang, Experimental investigation of pulsating heat pipe performance with regard to fuel cell cooling application, *Applied Thermal Engineering,* 50 (1) (2013) 268-274.

[14] S. Lips, A. Bensalem, Y. Bertin, V. Ayel, C. Romestant, J. Bonjour, Experimental evidences of distinct heat transfer regimes in pulsating heat pipes (php), *Applied Thermal Engineering*, 30 (8–9) (2010) 900-907.

[15] G. Burban, V. Ayel, A. Alexandre, P. Lagonotte, Y. Bertin, C. Romestant, Experimental investigation of a pulsating heat pipe for hybrid vehicle applications, *Applied Thermal Engineering*, 50 (1) (2013) 94-103.

[16] C. Y. Tseng, K. S. Yang, K. H. Chien, M. S. Jeng, C. C. Wang, Investigation of the performance of pulsating heat pipe subject to uniform/alternating tube diameters, *Experimental Thermal and Fluid Science*, 54 (4) (2014) 85-92.

[17] L. Lin, R. Ponnappan, J. Leland, Experimental investigation of oscillating heat pipes, *AIAA J. Thermophys, Heat Transfer*, 15 (2001) 395–400.

[18] R. K. Sarangi, M. V. Rane, Experimental investigations for start up and maximum heat load of closed loop pulsating heat pipe, *Procedia Engineering*, 51 (2013) 683-687.

[19] M. B. Shafii, S. Arabnejad, Y. Saboohi, H. Jamshidi, Experimental investigation of pulsating heat pipes and a proposed correlation, *Heat Transfer Engineering*, 31 (10) (2010) 854-861.

[20] X. M. Zhang. Experimental study of a pulsating heat pipe using fc-72, ethanol, and water as working fluids, *Experimental Heat Transfer,* 17 (1) (2004) 47-67.

[21] S. Rittidech, P. Terdtoon, P. Tantakom, Effect of inclination angles, evaporator section lengths and working fluid properties on heat transfer characteristics of a closed-end oscillating heat pipe, *6th International Heat Pipe Symposium*, ChiangMai, Thailand, 2000.

[22] Y. H. Lin, S. W. Kang, T. Y. Wu, Fabrication of polydimethylsiloxane (pdms) pulsating heat pipe, *Applied Thermal Engineering,* 29 (2) (2009) 573-580.

[23] Z. Q. Long, P. Zhang, Heat transfer characteristics of thermosyphon with N2 –Ar binary mixture working fluid, *International Journal of Heat and Mass Transfer* 63 (15) (2013) 204-215.

[24] K. M. Armijo, V. P. Carey, An analytical and experimental study of heat pipe performance with a working fluid exhibiting strong concentration marangoni effects, *International Journal of Heat and Mass Transfer* 64 (3) (2013) 70-78.

[25] K. Wakabayashi, A. Nagai, A. M. Sheikh, Y. Shiota, D. Narantuya, T. Watanabe, et al. Enhancement of heat transport in thermosyphon air preheater at high temperature with binary working fluid: a case study of teg–water, *Applied Thermal Engineering* 22(3) (2002) 251-266.

[26] R. Savino, N. D. Francescantonio, R. Fortezza, Y. Abe, Heat pipes with binary mixtures and inverse marangoni effects for microgravity applications, *Acta Astronautica* 61 (1–6) (2007)16-26.

[27] H. Jouhara, Z. Ajji, Y. Koudsi, H. Ezzuddin, N. Mousa, Experimental investigation of an inclined-condenser wickless heat pipe charged with water and an ethanol–water azeotropic mixture, *Energy* 61 (6) (2013) 139-147.

[28] P. Y. Wang, X. J. Chen, Z. H. Liu, Y. P. Liu, Application of nanofluid in an inclined mesh wicked heat pipes, *Thermochimica Acta* 539 (26) (2012) 100-108.

[29] F. Shang, D. Liu, H. Xian, Y. Yang, X. Du, Flow and heat transfer characteristics of different forms of nanometer particles in oscillating heat pipe, *Journal of Chemical Industry and Engineering* 58 (9) (2007) 2200-2204.

[30] K. Park, H. Ma, Nanofluid effect on the heat transport capability in a well-balanced oscillating heat pipe, *Journal of Thermophysics and Heat Transfer* 21 (2) (2007) 443-445.

[31] N. Bhuwakietkumjohn, S. Rittidech, Internal flow patterns on heat transfer characteristics of a closed-loop oscillating heat-pipe with check valves using ethanol and a silver nano-ethanol mixture, *Experimental Thermal and Fluid Science* 34 (8), (2010) 1000-1007.

[32] Y. H. Lin, S. W. Kang, H. L. Chen, Effect of silver nano-fluid on pulsating heat pipe thermal performance, *Applied Thermal Engineering* 28 (11) (2008) 1312-1317.

[33] J. Qu, H. Y. Wu, P. Cheng, Thermal performance of an oscillating heat pipe with Al_2O_3–water nanofluids, *International Communications in Heat and Mass Transfer* 37 (2) (2010) 111-115.

[34] J. Qu, H. Wu, Thermal performance comparison of oscillating heat pipes with SiO_2/water and Al_2O_3/water nanofluids, *International Journal of Thermal Sciences* 50 (10) (2011) 1954-1962.

[35] Y. Ji, C. Wilson, H. H. Chen, H. Ma, Particle shape effect on heat transfer performance in an oscillating heat pipe, *Experimental Thermal and Fluid Science* 35 (4) (2011) 724-727.

[36] H. B. Ma, C. Wilson, Q. Yu, K. Park, U. S. Choi, M. Tirumala, An experimental investigation of heat transport capability in a nanofluid oscillating heat pipe, *Journal of Heat Transfer* 128 (11) (2006) 1213-1216.

[37] M. R. Tanshen, B. Munkhbayar, M. J. Nine, H. Chung, H. Jeong, Effect of functionalized mwcnts/water nanofluids on thermal resistance and pressure fluctuation characteristics in oscillating heat pipe, *International Communications in Heat and Mass Transfer* 48 (11) (2013) 93-98.

[38] N. A. C. Sidik, H. A. Mohammed, O. A. Alawi, S. Samion, A review on preparation methods and challenges of nanofluids, *International Communications in Heat and Mass Transfer* 54 (5) (2014) 115-125.

[39] H. Jouhara, Z. Ajji, Y. Koudsi, H. Ezzuddin, N. Mousa, Experimental investigation of an inclined-condenser wickless heat pipe charged with water and an ethanol–water azeotropic mixture, *Energy* 61 (6) (2013) 139-147.

[40] M. Mameli, M. Marengo, S. Khandekar, Local heat transfer measurement and thermo-fluid characterization ofa pulsating heat pipe, *International Journal of Thermal Sciences* 75 (1) (2014) 140-152.

[41] P. Charoensawan, P. Terdtoon, Thermal performance of horizontal closed-loop oscillating heat pipe, *Applied Thermal Engineering* 28 (5–6) (2008) 460-466.

[42] M. Groll, S. Khandekar, Pulsating heat pipes: Progress and prospects, Energy and the Environment - *Proceedings of the International Conference on Energy and the Environment,* 2003, pp. 723- 730.

[43] S. Khandekar, A. P. Gautam, P. K. Sharma, Multiple quasi-steady states in a closed loop pulsating heat pipe, *International Journal of Thermal Sciences* 48 (3) (2009) 535-546.

[44] B. G. Lone, P. B. Undre, S. S. Patil, P. W. Khirade, S. C. Mehrotra, Dielectric study of methanol–ethanol mixtures using tdr method, *Journal of Molecular Liquids* 141 (1–2) (2008) 47-53.

[45] K. H. Chien, Y. T. Lin, Y. R. Chen, K. S. Yang, C. C. Wang, A novel design of pulsating heat pipe with fewer turns applicable to all orientations, *International Journal of Heat and Mass Transfer* 55 (21–22) (2012) 5722-5728.

[46] J. Qu, *Study on operational mechanism and enhanced heat transfer of micro/ miniature pulsating heat pipes and micro–grooved heat pipes*, PhD thesis, Shanghai Jiao Tong University, China, 2010.

[47] S. U. Lei, H. Zhang, J. Zhuang, Study on working principle of looped pulsating heat pipe, *Journal of Zhejiang Ocean University*, 2004.

[48] Q. Cai, R. L. Chen, C. L. Chen, An Investigation of Evaporation, Boiling, and Heat Transport Performance in Pulstating Heat Pipe, ASME 2002 *International Mechanical Engineering Congress and Exposition*, 2002, pp. 99-104.

[49] X. Cui, Y. Zhu, Z. Li, S. Shun, Combination study of operation characteristics and heat transfer mechanism for pulsating heat pipe, *Applied Thermal Engineering* 65 (1–2) (2014) 394-402.

[50] H. Han, X. Cui, Y. Zhu, S. Sun, A comparative study of the behavior of working fluids and their properties on the performance of pulsating heat pipes (PHP), *International Journal of Thermal Sciences* 82 (1) (2014) 138-147.

[51] Y. Zhu, X. Cui, H. Han, S. Sun, The study on the difference of the start-up and heat-transfer performance of the pulsating heat pipe with water−acetone mixtures, *International Journal of Heat and Mass Transfer* 77 (4) (2014) 834-842.

[52] B. González, N. Calvar, E. Gómez, Ángeles Domínguez, Density, dynamic viscosity, and derived properties of binary mixtures of methanol or ethanol with water, ethyl

acetate, and methyl acetate at t = (293.15, 298.15, and 303.15) k, *Journal of Chemical Thermodynamics* 39 (12) (2007) 1578-1588.

[53] P. R. Pachghare, A. M. Mahalle, Effect of pure and binary fluids on closed loop pulsating heat pipe thermal performance, *Procedia Engineering* 51 (2013) 624-629.

[54] X. Cui, Z. Qiu, J. Weng, et al. Heat transfer performance of closed loop pulsating heat pipes with methanol-based binary mixtures. *Experimental Thermal & Fluid Science* 76 (2016) 253-263.

[55] H. Han, X. Cui, Y. Zhu, et al. Experimental study on a closed-loop pulsating heat pipe (CLPHP) charged with water-based binary zeotropes and the corresponding pure fluids. *Energy* 109 (2016) 724-736.

[56] S. Shi, X. Cui, H. Han, J. Weng, and Z. Li, A study of the heat transfer performance of a pulsating heat pipe with ethanol-based mixtures. *Applied Thermal Engineering* 102 (2016) 1219-1227.

[57] W. Wang, X. Cui. Y. Zhu. Heat transfer performance of a pulsating heat pipe charged with acetone based mixtures. *Heat & Mass Transfer* (2016) 1-12.

In: Heat Pipes: Design, Applications and Technology
Editor: Yuwen Zhang
ISBN: 978-1-53613-908-2

Chapter 5

CHAOS IN PULSATING HEAT PIPES

***Sam M. Pouryoussefi*[1] *and Yuwen Zhang*[2,*]**
[1]Department of Mechanical Engineering, University of California, Merced, CA, US
[2]Department of Mechanical and Aerospace Engineering, University of Missouri, Columbia, MO, US

ABSTRACT

The oscillating or pulsating heat pipe (OPH or PHP) is a very promising heat transfer device. In addition to its excellent heat transfer performance, it has a simple structure. Due to essential applications of pulsating heat pipes in several areas of industry, study of its mechanism, behavior and performance is crucial. Several theoretical, numerical and experimental investigations have been carried out to get better understanding and solving of the issues in pulsating heat pipes. Despite simple structure of PHPs, it has a complicated behavior. Existence of chaos in the system is one of the significant reasons for such complicated behavior. In this chapter, numerical study of the chaotic behavior in pulsating heat pipes has been implemented using quantitative approaches such as spectral analysis of time series, correlation dimension, autocorrelation function (ACF), Lyapunov exponent, and phase space reconstruction. In addition, volume fraction of liquid and vapor has been simulated and studied under different operating conditions. Constant temperature and heat flux have been employed to investigate thermal performance such as thermal resistance and axial mean temperature for several boundary conditions. Volume of Fluid (VoF) method was used for liquid-vapor two-phase flow simulation. Water and ethanol have been used as working fluids in pulsating heat pipe. This chapter is divided into three major parts depending on the PHP's geometry and dimension. First part studies a simple two dimensional geometry. Second part studies a multi-turn two dimensional geometry and third part investigates a three dimensional geometry for the pulsating heat pipe. An experimental study from other work has been used for comparison and validation in three dimensional simulation.

* Corresponding Author Email: zhangyu@missouri.edu.

Keywords: pulsating heat pipe, chaos, two phase flow, numerical simulation

INTRODUCTION

Oscillating Heat Pipes (OHP) or pulsating heat pipes (PHP) were invented in 1990 by Akachi [1]. An OHP has heating, adiabatic, and cooling sections along the direction of heat transport. Under nominal operating conditions, the working fluid separates into liquid slugs and vapor plugs, ultimately forming a slug flow state because of the strong capillary force in the tube. Upon application of heat to the heating section, liquid slugs commence self-excited oscillation from the pressure difference along the flow path. OHPs transport heat by forced convection of liquid slugs and the phase change phenomena in liquid films surrounding vapor plugs [2]. Many experimental and theoretical studies have been conducted to understand the complicated behaviors of PHP. Shafii et al. [3, 4] presented analytical models of thermal behavior and heat transfer for both open and closed-loop PHPs with multiple liquid slugs and vapor plugs. The results showed that the gravity does not have significant effect on the performance of the unlooped PHPs with top heat mode. In addition, higher surface tension resulted in a slight increase in total heat transfer.

Zhang et al. [5] investigated heat transfer process in evaporator and condenser sections of the PHP. They developed heat transfer models in the evaporator and condenser sections of a PHP with one open end by analyzing thin film evaporation and condensation. Zhang and Faghri analyzed the oscillatory flow in PHPs with arbitrary numbers of turns [6]. The results showed that the increase in the number of turns has no effect on the amplitude and circular frequency of oscillation when the number of turns is less than or equal to five. Zhang and Faghri [7] reviewed advances and unsolved issues in PHPs. Shao and Zhang [8] studied thermally induced oscillatory flow and heat transfer in a U-shaped minichannel. The sensible heat transfer coefficient between the liquid slug and the minichannel wall was obtained by analytical solution for a laminar liquid flow and by empirical correlations for a turbulent liquid flow.

Kim et al. [9] analyzed entropy generation for a PHP. It was observed that the entropy generation is significantly affected by the initial temperature in the PHP. They also studied the effects of fluctuations of heating and cooling section temperatures on performance of a PHP [10]. They found that both amplitude and frequency of the periodic component of the temperature fluctuation affected the liquid slug displacements, temperatures, and pressures of the two vapor plugs, as well as the latent and sensible heat transfer. Ma et al. [11] used a mathematical model to predict the oscillating motion in an oscillating heat pipe. The model considered the vapor bubble as the gas spring for the oscillating motions including effects of operating temperature, nonlinear vapor bulk modulus, and temperature difference between the evaporator and the condenser. Experimental results indicated that there exists an onset power input for the excitation of oscillating motions in an oscillating heat pipe.

When the input power or the temperature difference from the evaporating section to the condensing section was higher than this onset value, the oscillating motion started, resulting in an enhancement of the heat transfer in the oscillating heat pipe.

Tong et al. [12] applied flow visualization technique for the closed loop pulsating heat pipe (PHP) using a charge coupled device (CCD). It was observed that during the start-up period, the working fluid oscillated with large amplitude; however, at steady operating state, the working fluid circulated. Borgmeyer and Ma [13] conducted an experimental investigation to evaluate the motion of vapor bubbles and liquid plugs within a flat-plate pulsating heat pipe to determine the effects of working fluids, power input, filling ratio, and angle of orientation on the pulsating fluid flow. Qu et al. [14] performed an experimental study on the thermal performance of an oscillating heat pipe (OHP) charged with water and spherical Al2O3 particles of 56 nm in diameter. The effects of filling ratios, mass fractions of alumina particles, and power inputs on the total thermal resistance of the OHP were investigated.

Qu et al. [15] conducted an experimental study on thermal performance of a silicon-based micro-PHP. The effects of gravity, filling ratio, and working fluids on the overall thermal resistance were discussed. The experimental results showed that a micro-PHP embedded in a semiconductor chip could significantly decrease the maximum localized temperature [16]. Researchers have investigated PHPs from different aspects because of their importance and useful applications [17–20]. Turkyilmazoglu investigated the effects of nanofluids on heat transfer enhancement in single and multi-phase flows [21–23]. It was shown that by increasing the diffusion parameter in the multiphase model, more enhancements in the rate of heat transfer could be obtained. In addition, a rescaling method to simplify the evaluation of flow and physical parameters, such as skin friction and heat transfer rate in single-phase nanofluids, was proposed. Xian et al. [24] experimentally investigated dynamic fluid flow in OHP under pulse heating. They employed a high-speed camera to conduct experiments and fluid flow visualization in PHP.

Jiaqiang et al. [25] studied the pressure distribution and flow characteristics of a closed OHP under different operating conditions. They analyzed the relationship of flow pattern distribution and pressure distribution. Recently there have been experimental studies that proposed the existence of chaos in PHPs under some operating conditions. The approach in these studies is to analyze the time series of fluctuation of temperature of a specified location on the PHP tube wall (adiabatic section) by power spectrum calculated through Fast Fourier Transform (FFT). The two dimensional mapping of the strange attractor and the subsequent calculation of the Lyapunov exponent have been performed to prove the existence of chaos in PHP system [26]. Dobson [27, 28] theoretically and experimentally investigated an open oscillatory heat pipe including gravity. The theoretical model used vapor bubble, liquid plug and liquid film control volumes. Experimental model was constructed and tested using water as the working fluid. By calculating Lyapunov exponents, it was shown that the theoretical model is able to reflect the characteristic chaotic behavior of experimental devices. Xiao-Ping and Cui [29] studied the dynamic properties for the micro-channel phase change heat transfer system by theoretical method combined with experiment. Liquid–vapor interface dynamic systems were obtained by introducing disjoining pressure produced by three phase molecular interactions and Lie algebra analysis. Experiments for a 0.6 × 2mm rectangular micro-channel were carried out to obtain the pressure time serials. Power spectrum density analysis for these serials showed that the system is in chaotic state if the frequency is above 7.39 Hz. Song and Xu [30] have run series of experiments to explore chaotic behavior of PHPs. Different number of turns, inclination angles, filling ratios and heating powers were tested. Their study confirmed that PHPs are deterministic chaotic systems.

Louisos et al. [31] conducted a numerical study of chaotic flow in a 2D natural convection loop with heat flux boundaries. System of governing equationswas solved using a finite volume method. Numerical simulations were performed for varying levels of heat flux and varying strengths of gravity to yield Rayleigh numbers ranging from 1.5 × 102 to 2.8 × 107. Ridouane et al. [32] computationally explored the chaotic flow in a 2D natural convection loop. They set constant temperatures for the boundary conditions in heating and cooling sections. Numerical simulations were applied for water corresponding to Prandtl number of 5.83 and Rayleigh number varying from 1000 to 150,000. Results in terms of streamlines, isotherms, and local heat flux distributions along the walls were presented for each flow regime. Louisos et al. [33] numerically investigated chaotic natural convection in a toroidal thermosiphon with heat flux boundaries. Nonlinear dynamics of unstable convection in a 3D toroidal shaped thermal convection loop was studied. The lower half of the thermosyphon was subjected to a positive heat flux into the system while the upper half was cooled by an equal-but-opposite heat flux out of the system. Delineation of multiple convective flow regimes was achieved through evolution of the bulk-mass-flow time-series and the trajectory of the mass flow attractor. Qu et al. [34] employed ethanol as the working fluid to investigate the chaotic behavior of wall temperature oscillations in a closed-loop PHP. The experimental results were analyzed by using nonlinear analyses of the pseudo-phase-plane trajectories, the correlation dimension, the largest Lyapunov exponent, and the recurrence plots. Chaotic states were observed in the recurrence plots of the temperature oscillations.

Pouryoussefi and Zhang [35-37] recently conducted a numerical simulation of the chaotic flow in the closed-loop PHP. They applied simple two dimensional, multi-turn two dimensional and three dimensional structure for the heat pipes geometry. Constant temperature and heat flux have been employed to investigate thermal performance such as thermal resistance and axial mean temperature for several boundary conditions. Volume of Fluid (VoF) method was used for liquid-vapor two-phase flow simulation. Water and ethanol were used as working fluids in pulsating heat pipe. Quantitative approaches such as spectral analysis of time series, correlation dimension, autocorrelation function (ACF),

Lyapunov exponent, and phase space reconstruction were employed to investigate chaos in PHPs.

1. Chaos

Most students of science or engineering have seen examples of dynamical behavior whish san be fully analyzed mathematically and in which the system eventually (after some transient period) settles either into periodic motion (a limit cycle) or into a steady state (i.e., a situation in which the system ceases its motion). When one relies on being able to specify an orbit analytically, these two cases will typically (and falsely) appear to be the only important motions. The point is that chaotic orbits are also very common but cannot be represented using standard analytical functions [38]. Chaotic motions are neither steady nor periodic. Indeed, they appear to be very complex, and, when viewing such motions, adjectives like wild, turbulent, and random some to mind. Despite the complexity of these motions, they commonly occur in systems whish themselves are not complex and are even surprisingly simple. There are some major parameters and approaches to investigate chaos in dynamical systems.

1.1. Power Spectral Density

Most of the signals encountered in applications are such that their variation in the future cannot be known exactly. It is only possible to make probabilistic statements about that variation. The mathematical device to describe such a signal is that of a random sequence which consists of an ensemble of possible realizations, each of which has some associated probability of occurrence. Of course, from the whole ensemble of realizations, the experimenter can usually observe only one realization of the signal, and then it might be thought that the deterministic definitions of the previous section could be carried over unchanged to the present case. However, this is not possible because the realizations of a random signal, viewed as discrete–time sequences, do not have finite energy, and hence do not possess discrete-time Fourier transforms (DTFTs). A random signal usually has finite average power and, therefore, can be characterized by an average power spectral density [39]. The power spectrum of a time series describes the distribution of power into frequency components composing that signal. According to Fourier analysis any physical signal can be decomposed into a number of discrete frequencies or a spectrum of frequencies over a continuous range. The statistical average of a certain signal or sort of signal (including noise) as analyzed in terms of its frequency content is called its spectrum. PSD can be defined as [35]:

$$S_x(f) = \lim_{T\to\infty} E\left\{\frac{1}{2T}\left|\int_{-T}^{T} x(t)e^{-j2\pi ft}dt\right|^2\right\} \quad (1)$$

where $x(t)$ is the random time signal and E is the energy of the signal. The PSD is particularly useful for studying the oscillations of a system. There will be sharper or broader peaks at the dominant frequencies and at their integer multiples, the harmonics. Periodic or quasi-periodic signals show sharp spectral lines. Measurement noise adds a continuous floor to the spectrum. Thus in the spectrum, signal and noise are readily distinguished. Deterministic chaotic signals may also have sharp spectral lines but even in the absence of noise there will be a continuous part of the spectrum [40].

1.2. Correlation Dimension

In chaos theory, the correlation dimension is a measure of the dimensionality of the space occupied by a set of data points, often referred to as a type of fractal dimension. The correlation dimension is well known in fractal geometry and it is used to calculate a fractal dimension from a time series. The time series is embedded in n-dimensional space which is done by forming vectors of length n. The embedding dimension of an attractor dataset is the dimension of its address space. In other words, it is the number of attributes of the attractor dataset. The attractor dataset can represent a spatial object that has a dimension lower than the space where it is embedded [41]. For example, a line has an intrinsic dimensionality one, regardless if it is in a higher dimensional space. The intrinsic dimension of an attractor dataset is the dimension of the spatial object represented by the attractor dataset, regardless of the space where it is embedded. The above definition of correlation dimension involves phase space vectors as the location of points on attractor. Thus, given a scalar time series, it is important to reconstruct an auxiliary phase space by an embedding procedure. Finding the correlation dimension for a chaotic process from the time series data set is usually applied for getting information about the nature of the dynamical system. Correlation dimension is popular because of its simplicity and flexibility data storage requirements. It is an alternate definition for true dimension or fractal dimension. As discussed in the previous section, the time series data set can reconstruct phase space formed by the trajectories of the system, the attractor. Suppose that our attractor consists of N data points obtained numerically or experimentally, the correlation sum is defined [40]:

$$C(r) = \lim_{N\to\infty}\left(\frac{number\ of\ pairs\ of\ points\ whitin\ a\ sphere\ of\ radius\ r}{N^2}\right) \quad (2)$$

In practice, we increase N until $C(r)$ is independent of N. The correlation sum typically scales as

$$C(r) \approx \alpha\, r^{D_c} \tag{3}$$

where D_c is the correlation dimension and r is chosen to be smaller than the size of attractor and larger than the smallest spacing between the points. In practice, we plot ln $C(r)$ as a function of $ln\ r$ and obtain the dimension D_c from the slope of the curve. The above definition of the correlation sum and dimension involve phase space vectors as the locations of points on an attractor. Thus given a scalar time series, we first have to reconstruct a phase space by an embedding procedure.

1.3. Autocorrelation Function

Autocorrelation is the cross-correlation of a signal with itself at different points in time (that is what the cross stands for). Informally, it is the similarity between the observations as a function of the time lag between them. It is a mathematical tool for finding repeating patterns, such as the presence of a periodic signal obscured by noise, or identifying the missing fundamental frequency in a signal implied by its harmonic frequencies. It is often used in signal processing for analyzing functions or series of values, such as time domain signals. The autocorrelation of a random process describes the correlation between values of the process at different times, as a function of the two times or of the time lag. Often in time series it is required to compare the value observed at one time point to a value observed one or more time points earlier. Such prior values are known as lagged values. If there is some pattern in how the values of time series change from observations to observation, it would be so useful to analyze the time series. The correlation between the original time series values and the corresponding τ -lagged values is called autocorrelation of order τ [37]. Time domain analysis is the approach based on the autocorrelation functions to make inference from an observed time series to estimate the model. The theoretical autocorrelation function is an important tool to describe the properties of a stochastic process. An important guide to the persistence in a time series is given by the series of quantities called the sample autocorrelation coefficients, which measure the correlation between observations at different times. The set of autocorrelation coefficients arranged as a function of separation in time is the sample autocorrelation function, or the *ACF*. *ACF* helps to determine how quickly signals or process change with respect to time and whether a process has a periodic component [30]. The autocorrelation function will have its largest value of $ACF=1$ at $\tau=0$. Autocorrelation function of the time series are computed as [42]:

$$ACF(\tau) = \frac{\sum_{i=1}^{N}(T_{i+\tau}-\bar{T})(T_i-\bar{T})}{\sum_{i=1}^{N}(T_i-\bar{T})^2} \tag{4}$$

where T_i and $T_{i+\tau}$ are the temperature observations at the time domain, $\bar{T}$ is the overall mean temperature and τ is the lag.

1.4. Lyapunov Exponent

Lyapunov exponents are a fundamental concept of nonlinear dynamics. They quantify local stability features of attractors and other invariant sets in state space. Positive Lyapunov exponents indicate exponential divergence of neighboring trajectories and are the most important attribute of chaotic attractors. While the computation of Lyapunov exponents for given dynamical equations is straight forward, their estimation from time series remains a delicate task [43]. Consider two points in a space, X_0 and $X_0+\Delta x_0$, each of which will generate an orbit in that space using some equation or system of equations. These orbits can be thought of as parametric functions of a variable that is something like time. If we use one of the orbits as reference orbit, then, the separation between the two orbits will also be a function of time. Because sensitive dependence can arise only in some portions of a system (like the logistic equation), this separation is also a function of the location of the initial value and has the form $\Delta x(X_0, t)$. In a system with attracting fixed points or attracting periodic points, $\Delta x(X_0, t)$ diminishes asymptotically with time. If a system is unstable, like pins balanced on their points, then, the orbits diverge exponentially for a while, but eventually settle down. For chaotic points, the function $\Delta x(X_0, t)$ will behave erratically. It is thus useful to study the mean exponential rate of divergence of two initially close orbits using the formula. Let $d(0)$ be the small separation between two arbitrary trajectories at time 0, and let $d(t)$ be the separation between them at time t. Then, for a chaotic system [41]:

$$d(t) \approx d(0)\, e^{\lambda t} \tag{5}$$

where λ is the largest positive Lyapunov exponent. In the pulsating heat pipe, let X_p be the position of the liquid plug relative to its initial position. Given some initial condition X_0, consider a nearby point $X_0+\delta$, where the initial separation δ_0 is extremely small. Let δ_n be the separation after n steps. If $|\delta_n| \approx |\delta_0|\, e^{n\lambda}$, λ will be the Lyapunov exponent. A positive λ means an exponential divergence of nearby trajectories which is a signature of chaos. A negative λ corresponds to stable fixed point and if $\lambda=0$, this means stable limit cycle. A more precise and computationally useful formula for λ with the logistic map of $f(x)$ is derived by Strogatz [41]:

$$\lambda = \lim_{n\to\infty}\left[\frac{1}{n}\sum_{i=0}^{n-1} ln|f'(x_i)|\right] \tag{6}$$

1.5. Phase Space Reconstruction

To reconstruct a useful version of the internal dynamics for a given time series from a sensor on a single state variable $x_i(t)$ in a n-dimensional dynamical system, delay-coordinate embedding can be used. If the embedding is performed correctly, the involved theorems guarantee that the reconstructed dynamics is topologically identical to the true dynamics of the system, and therefore that the dynamical invariants are also identical. This is an extremely powerful correspondence; it implies that conclusions drawn from the embedded or reconstruction-space dynamics are also true of the real unmeasured dynamics. This implies, for example, that one can reconstruct the dynamics of the earth's weather simply by setting a thermometer on a windowsill [37]. Considering a data set comprised of samples $x_i(t)$ of a signal state variable x_i in a n-dimensional system, measured once every Δt seconds. To embed such a data set, we construct r-dimensional reconstruction-space vectors $\boldsymbol{T}(t)$ from r time-delayed samples of the $x_i(t)$, such that

$$\boldsymbol{T}(t)=[x_i(t),\ x_i(t+\tau),\ x_i(t+2\tau),\ldots,\ x_i(t+(m-1)\tau)] \quad (7)$$

In a pulsating heat pipe, the time series is usually a sequence of temperature values which depends on the current phase space, taken at different points in time. It is important that given enough dimensions (r) and the right delay (τ), the reconstruction-space dynamics and the true, unobserved state-space dynamics are topologically identical. More, formally, the reconstruction-space and state-space trajectories are guaranteed to be equivalent if $r \geq 2n+1$.

2. Volume of Fluid (VOF) Method and Governing Equations

In this chapter, Volume of Fluid (VOF) method has been applied for two phase flow simulation. The VOF model is a surface-tracking technique applied to a fixed Eulerian mesh. It is designed for two or more immiscible fluids where the position of the interface between the fluids is of interest. In the VOF model, a single set of momentum equations is shared by the fluids, and the volume fraction of each of the fluids in each computational cell is tracked throughout the computational domain. Applications of the VOF model include stratified flows, free-surface flows, filling, sloshing, the motion of large bubbles in a liquid, the motion of liquid after a dam break, the prediction of jet breakup (surface tension), and the steady or transient tracking of any liquid-gas interface [35]. The VOF formulation relies on the fact that two or more fluids (or phases) are not interpenetrating. For each additional phase added to the model, the volume fraction of the phase in the computational cell is introduced. In each control volume, the volume fractions of all phases sum to unity. The fields for all variables and properties are shared by the phases and represent volume-averaged values, as long as the volume fraction of each of the phases is known at each location. Thus the variables and properties in any given cell are either purely representative of one of the phases, or representative of a mixture of the phases, depending upon the volume fraction values. Then the qth fluid's volume fraction in the cell is denoted as αq. Based on the local value of α_q, the appropriate properties and variables will be assigned to each control volume within the domain.

Tracking of the interfaces between the phases is accomplished by the solution of a continuity equation for the volume fraction of one (or more) of the phases. For the qth phase, this equation has the following form [37]:

$$\frac{1}{\rho_q}\left[\frac{\partial}{\partial t}(\alpha_q\rho_q) + \nabla.(\alpha_q\rho_q\boldsymbol{v}) = \sum_{p=1}^{n}(\dot{m}_{pq} - \dot{m}_{qp})\right] \tag{8}$$

where $\dot{m}_{qp}$ is the mass transfer from phase q to phase p, and $\dot{m}_{pq}$ is the mass transfer from phase p to phase q due to phase change. The primary-phase volume fraction will be computed based on the following constraint:

$$\sum_{q=1}^{n}\alpha_q = 1 \tag{9}$$

The volume fraction equation was solved using explicit time discretization. In the explicit approach, finite-difference interpolation schemes are applied to the volume fractions that were computed at the previous time step [37]:

$$\frac{\alpha_q^{n+1}\rho_q^{n+1} - \alpha_q^n\rho_q^n}{\Delta t}V + \sum_f(\rho_q U_f^n \alpha_{q,f}^n) = \left[\sum_{p=1}^{n}(\dot{m}_{pq} - \dot{m}_{qp}) + S_{\alpha_q}\right]V \tag{10}$$

where $n+1$ is the index for new (current) time step, n is the index for previous time step, $\alpha_{q,f}$ is the face value of the qth volume fraction, V is the volume of cell and U_f is the volume flux through the face, based on normal velocity.

The properties appearing in the transport equations are determined by the presence of the component phases in each control volume. In the vapor-liquid two-phase system, the density and viscosity in each cell are given by [37]:

$$\rho = \alpha_v\rho_v + (1 - \alpha_v)\rho_l \tag{11}$$

$$\mu = \alpha_v\mu_v + (1 - \alpha_v)\mu_l \tag{12}$$

One momentum equation is solved throughout the domain, and the resulting velocity field is shared among all phases. The momentum equation, shown below, is dependent on the volume fractions of all phases through the properties ρ and μ [37]:

$$\frac{\partial}{\partial t}(\rho\vec{v}) + \nabla \cdot (\rho \boldsymbol{v}\boldsymbol{v}) = -\nabla p + \nabla \cdot [\mu(\nabla \boldsymbol{v} + \nabla \boldsymbol{v}^T)] + \rho \boldsymbol{g} + \boldsymbol{F} \tag{13}$$

One limitation of the shared-fields approximation is that in cases where large velocity differences exist between the phases, the accuracy of the velocities computed near the interface can be adversely affected.

The energy equation, also shared among the phases, is [37]:

$$\frac{\partial}{\partial t}(\rho E) + \nabla \cdot \big(\boldsymbol{v}(\rho E + p)\big) = \nabla \cdot \big(k_{eff} \nabla T\big) + S_h \tag{14}$$

The VOF model treats energy, E, and temperature, T, as mass-averaged variables [37]:

$$E = \frac{\sum_{q=1}^{n} \alpha_q \rho_q E_q}{\sum_{q=1}^{n} \alpha_q \rho_q} \tag{15}$$

where E_q for each phase is based on the specific heat of that phase and the shared temperature. The properties ρ and k_{eff} (effective thermal conductivity) are shared by the phases and the source term, S_h is equal to zero.

The surface curvature is computed from local gradients in the surface normal at the interface. The surface normal $\boldsymbol{n}$, defined as the gradient of α_q, the volume fraction of the qth phase is:

$$\boldsymbol{n} = \nabla \alpha_q \tag{16}$$

The curvature, K, is defined in terms of the divergence of the unit normal, $\hat{\boldsymbol{n}}$:

$$\boldsymbol{K} = \nabla \cdot \hat{\boldsymbol{n}} \tag{17}$$

where

$$\hat{\boldsymbol{n}} = \frac{\boldsymbol{n}}{|\boldsymbol{n}|} \tag{18}$$

is the unit surface normal vector. The surface tension can be written in terms of the pressure jump across the surface. The force at the surface can be expressed as a volume force using the divergence theorem. It is this volume force that is the source term which is added to the momentum equation. It has the following form [37]:

$$F_{vol} = \sigma_{12} \frac{\rho k_1 \nabla \alpha_2}{\frac{1}{2}(\rho_1 + \rho_2)} \quad (19)$$

where ρ is the volume-averaged density computed using Eq. (11). Equation (19) shows that the surface tension source term for a cell is proportional to the average density in the cell.

To model the wall adhesion angle in conjunction with the surface tension model, the contact angle that the fluid is assumed to make with the wall is used to adjust the surface normal in cells near the wall. This so-called dynamic boundary condition results in the adjustment of the curvature of the surface near the wall. If θ_w is the contact angle at the wall, then the surface normal at the live cell next to the wall is [37]:

$$\hat{\boldsymbol{n}} = \hat{\boldsymbol{n}}_w cos\theta_w + \hat{\boldsymbol{t}}_w sin\theta_w \quad (20)$$

where $\hat{n}_w$ and $\hat{t}_w$ are the unit vectors normal and tangential to the wall, respectively. The combination of this contact angle with the normally calculated surface normal one cell away from the wall determine the local curvature of the surface, and this curvature is used to adjust the body force term in the surface tension calculation.

3. Simple Two-Dimensional Pulsating Heat Pipe

The pulsating heat pipe structure consists of three sections: heating section (evaporator), cooling section (condenser), and adiabatic section. Figure 1 illustrates a schematic configuration of the PHP which has been used in this study. The three different sections have been distinguished by two consistent lines. The width of the tube is 3 mm. The structure and dimension of the PHP were the same for all different operating conditions. But different evaporator and condenser temperatures and filling ratios were tested for the numerical simulation. Water was the only working fluid. The time step for the numerical simulation was 10^{-5} seconds. Figure 2 shows meshing configuration used in this study. Only condenser section has been depicted to show the quadrilaterals mesh which were employed for simulation. The quadrilaterals mesh was used for the entire pulsating heat pipe. In this specific test case, surface tension plays a key role in the performance of the pulsating heat pipe. It should be noted that the calculation of surface tension effects on triangular and tetrahedral meshes is not as accurate as on quadrilateral and hexahedral meshes. The region where surface tension effects are most important should therefore be meshed with quadrilaterals or hexahedra.

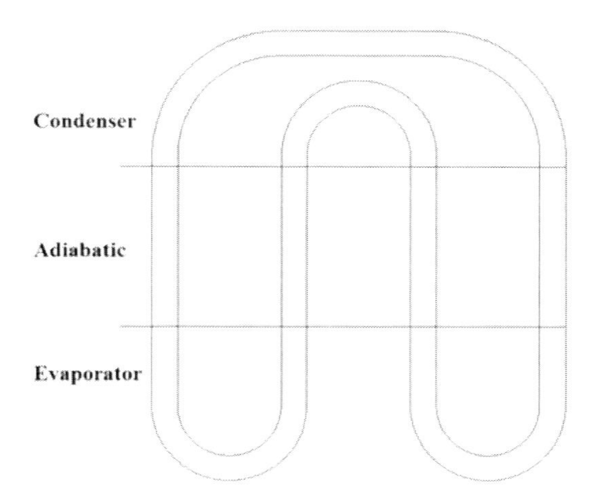

Figure 1. Pulsating heat pipe structure [35].

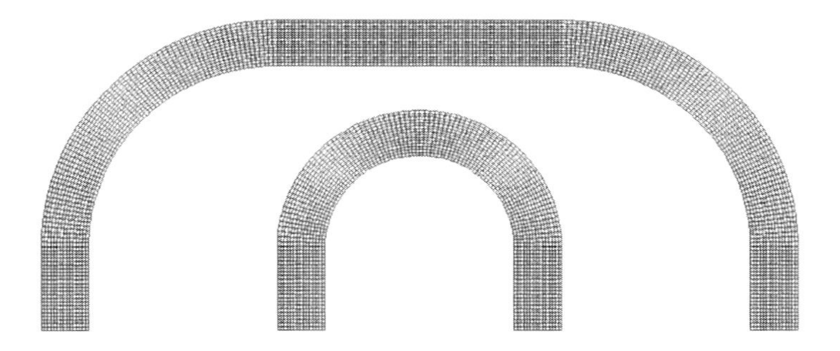

Figure 2. Meshing configuration (not in scale) [35].

3.1. Volume Fractions and Time Series

Volume fraction contours at different points in time have been brought for two different operating conditions in this section (Red color represent the vapor and Blue color represent the liquid). The first four figures (a to d) in Figure 3 show the evolution process for the flow in PHP. Figure 3 (e) to (h) show the volume fraction contours after the fluid flow has been formed. Different structures, filling ratios, evaporator temperatures and condenser temperatures were investigated to form flow with perfect liquid and vapor plugs in PHP. It is obvious in Figures 3 and 4 that liquid and vapor plugs are formed perfectly instead of stratified flow in PHP. The results showed that the fluid flow finally circulates in one direction (clockwise or counterclockwise) in the pulsating heat pipe. This direction is based on a random process and could be different even under the same operating and boundary conditions [26]. The liquid and vapor plug transport is due to the pressure pulsation in the system. Since the pressure pulsations are fully thermally driven, there is no external mechanical power source required for the fluid transport. Difference in temperatures for evaporator and condenser and fluid flow in the system cause temperature gradients inside the PHP. The net effect of all these temperature gradients within the system is to cause non equilibrium pressure condition which, as mentioned earlier is the primary driven force for thermo-fluidic transport. One of the most important advantages of current numerical simulation comparing to the other experimental and theoretical studies for PHPs is that it is easier to set and test many different operating conditions. But for such an experimental investigation it may need some equipment that could be expensive and difficult to prepare and set up in a proper way which may cause many difficulties and limitations.

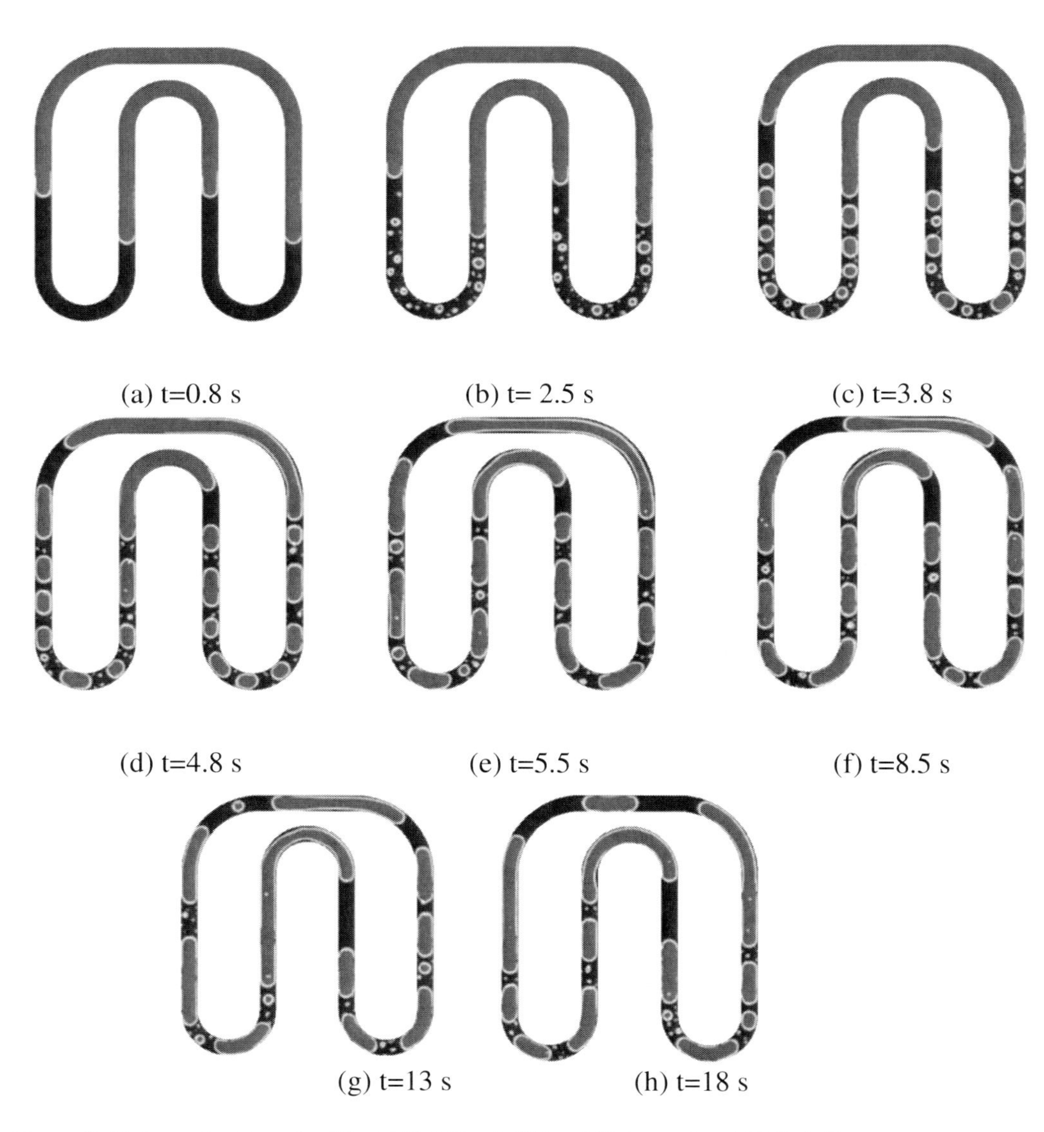

Figure 3. Volume fractions for Th=145°C, Tc=35°C and filling ratio of 30% [35].

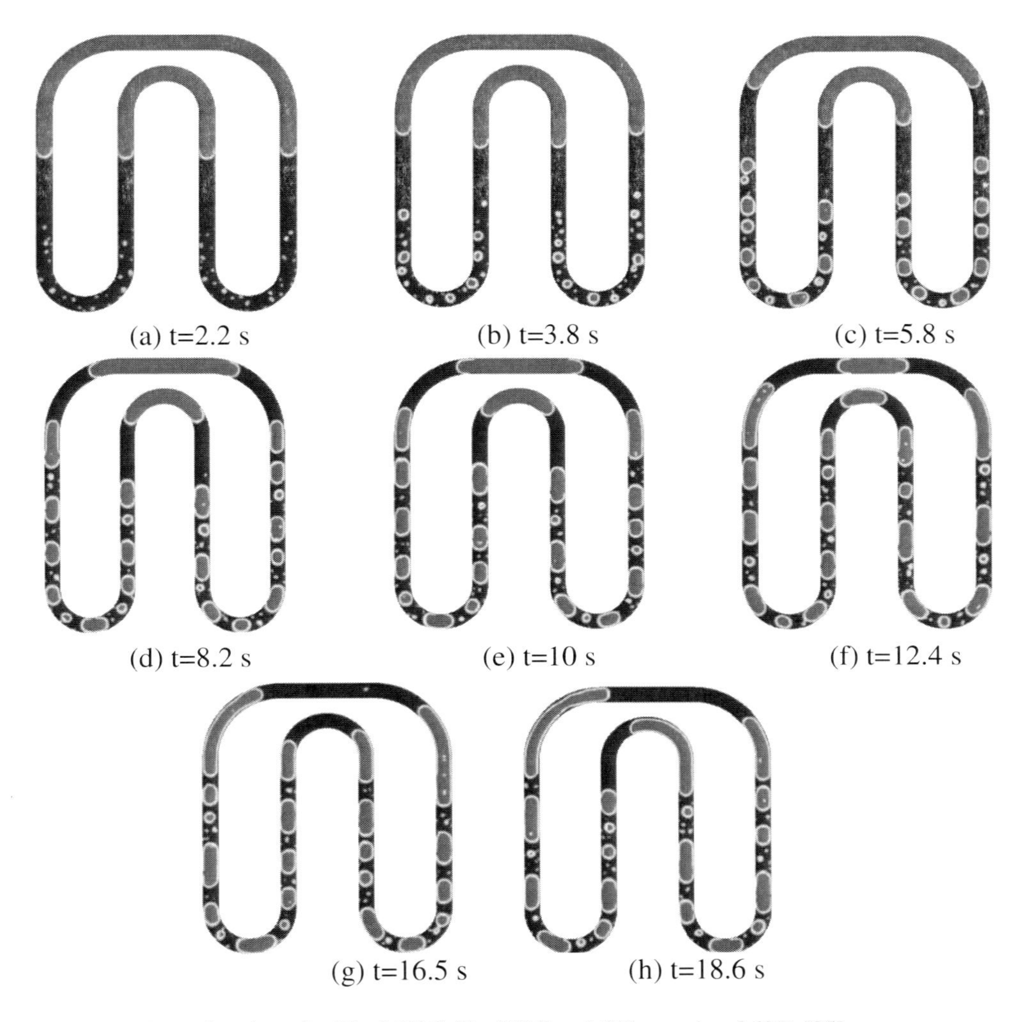

Figure 4. Volume fractions for Th=150°C, Tc=35°C and filling ratio of 60% [35].

Initially the liquid length was at the same level for all four tubes of the PHP. Figure 3(a) shows that after 0.8 seconds there is level difference for the liquid lengths in the tubes before boiling starts. This is because of the pressure distribution in the PHP. Figure 3(b) illustrates that boiling has started and small vapor bubbles are visible in the PHP. Formation of perfect vapor plugs is evident in Figure 3(c). In Figure 3(d) most of the vapor bubbles have been grown to the perfect vapor plugs. Formation of liquid films on the condenser surface as result of vapor condensation can be seen in Figure 3(d) but the fluid flow or circulation has not started yet. A large vapor plug with a thick liquid film is shown in Figure 3(e) in the condenser. In fact, at this time of 5.5 seconds both left and right hand sides of the flow in PHP have been reached together and fluid flow has been completed. Figures 3(f) – (h) show volume fractions in the PHP during the flow circulation. Finally, there is a counterclockwise flow circulation at the PHP. Mostly a similar process occurs in the PHP with filling ratio of 60% (Figure 4). The effect of change in filling ratio can be investigated qualitatively and quantitatively. The qualitative investigation can be carried out by observing and comparing the volume fraction figures in different times. One of the most significant effects due to the increasing the filling ratio is pressure increment in the PHP which lead to a longer time duration for boiling start (because of increasing the saturated temperature) and slower flow motion in the PHP. In addition, it was observed in Figure 4 that the flow circulation is clockwise in contrast with the filling ratio of 30%. As mentioned earlier the flow circulation direction is a random process. A detailed quantitative investigation on the effect of filling ratio will be carried out in this study later.

Figure 5 shows a snapshot of liquid-vapor plugs as result of simulation in vertical part of the system. The flow pattern in the PHP is categorized as capillary slug flow. Because of the capillary structure of the PHP, liquid plugs having menisci on the plug edge are formed due to surface tension forces.

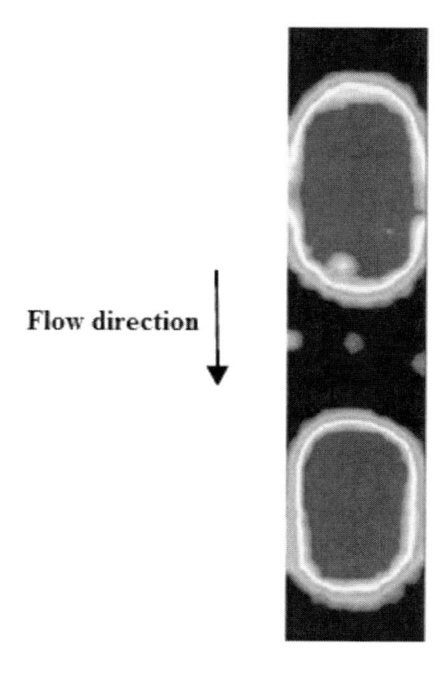

Figure 5. Liquid and vapor plugs in the vertical tube of the PHP [35].

Liquid thin film exists around the vapor plug, and the thickness of the liquid film may vary depend on some boundary conditions. One of the most important parameters affecting the liquid film is the wall temperature in the PHP. Since evaporator, condenser and adiabatic sections have different working temperatures liquid film with different thicknesses may form in these regions. Figure 6 shows three different liquid films surrounding the vapor plugs with different thicknesses as result of simulation. Figure 6 illustrates vapor plugs in the evaporator, condenser and adiabatic sections respectively. It can be seen that by increasing the wall temperature, the liquid film thickness decreases.

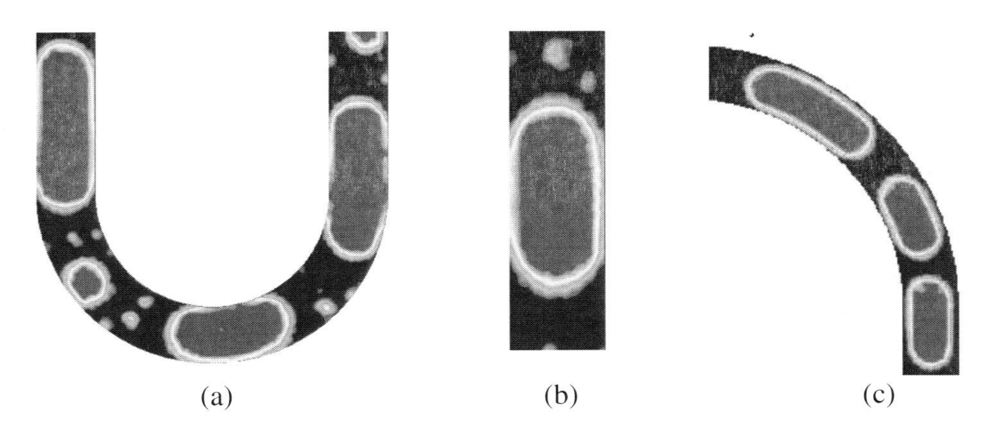

Figure 6. Liquid film around the vapor plugs at (a) evaporator, (b) adiabatic, and (c) condenser sections [35].

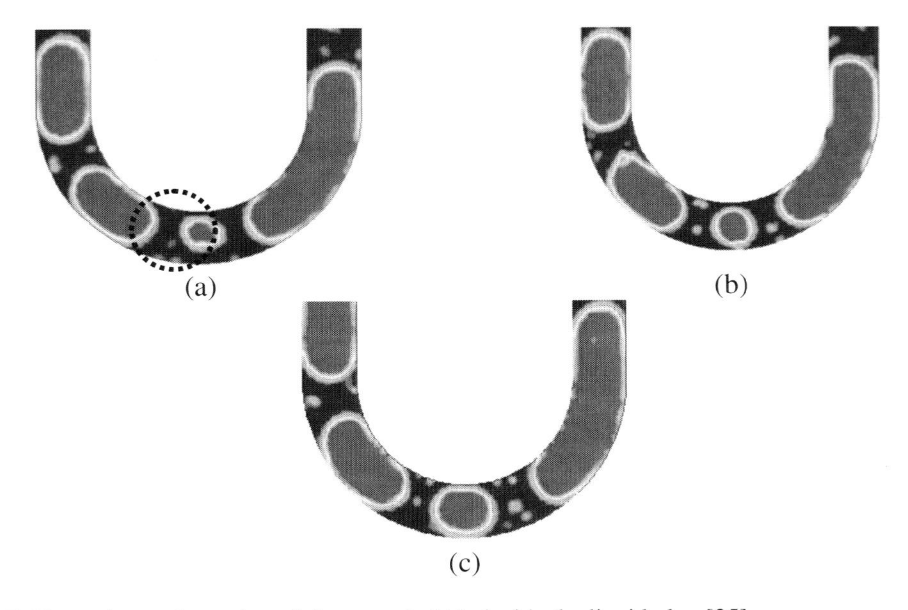

Figure 7. Formation and growing of the vapor bubble inside the liquid plug [35].

In evaporator section the temperature of the liquid plugs increases which is followed by evaporation mass transfer to the adjoining vapor plug or splitting the liquid plug by formation of new bubbles inside due to the nucleate boiling in the slug flow regime. Such a process is shown in Figure 7. It is obvious that by initiation and growing the vapor bubble (at the bottom of the evaporator), the liquid plug is breaking up. Sometimes two vapor bubbles combine together to form a larger vapor slug. Such a combination process in evaporator is illustrated in Figure 8 as a result of simulation; this phenomenon mostly was observed in the adiabatic section. Figure 9 shows vapor bubbles combination in adiabatic section. This type of behavior was observed rarely in the condenser.

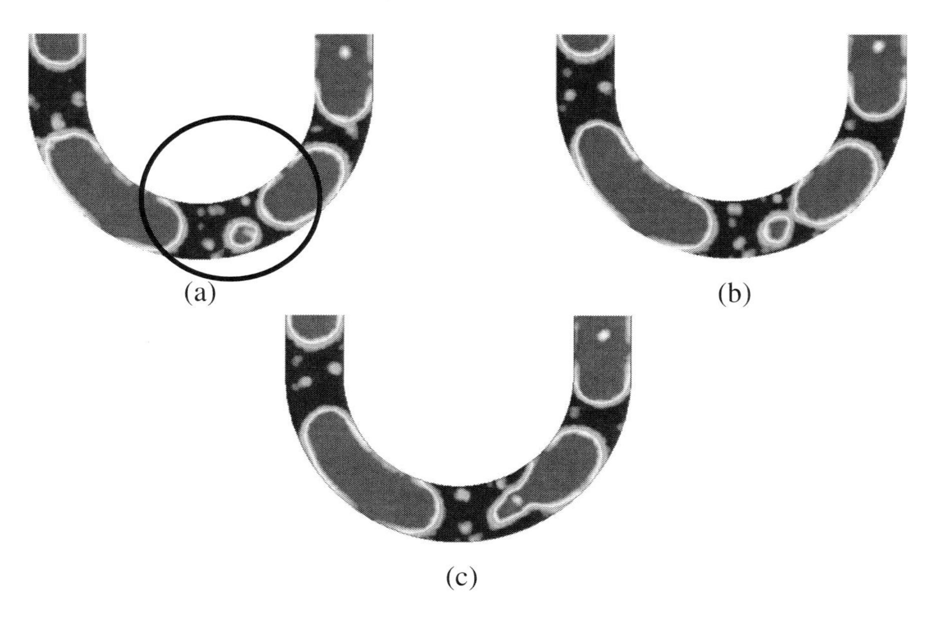

Figure 8. Vapor bubbles combination in evaporator [35].

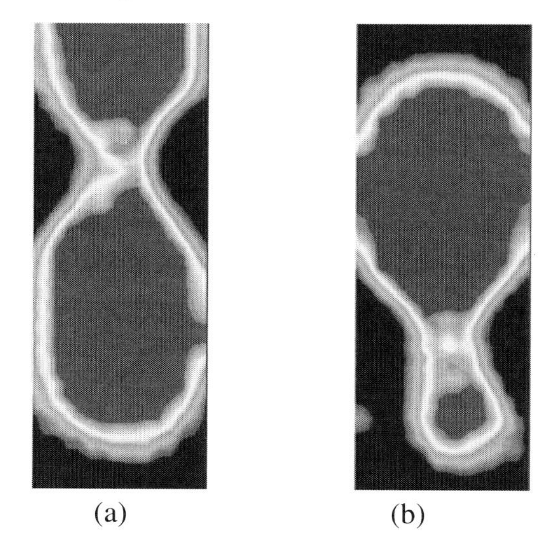

Figure 9. Vapor bubbles combination in adiabatic section [35].

Liquid accumulation as a result of vapor condensation occurred on the condenser surface. This condensation process was more visible on the bending parts due to the surface tension. Figure 10 shows liquid accumulation on the condenser surfaces. Sometimes the condenser surfaces can be distinguished from the adiabatic surfaces by observing and tracking the thick liquid films which form on the condenser surfaces. Sometimes the gravity acceleration causes the liquid film sliding down and forming small droplets at the bottom of the condenser surface.

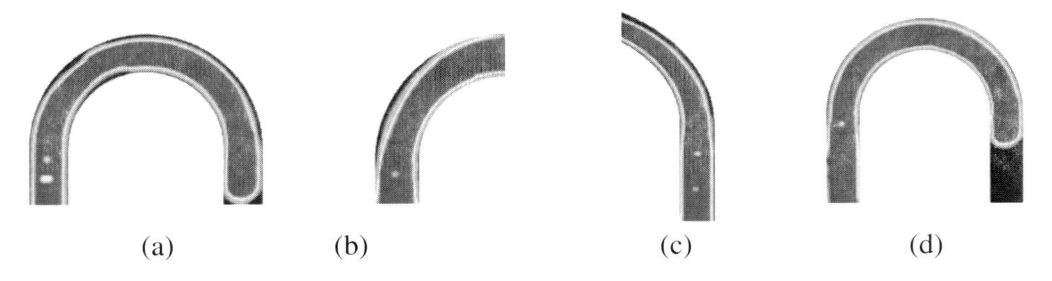

Figure 10. Vapor condensation and liquid accumulation on the condenser surfaces [35].

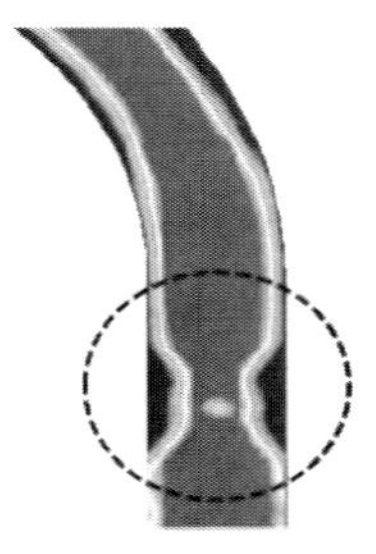

Figure 11. Liquid droplet formation at the bottom of the condenser [35].

Figure 10(a) shows vapor condensation and liquid film formation on the condenser surface. Figure 10(b) shows liquid accumulation on the bending part. Figures 10 (c) and (d) show how the condenser section could be distinguished from the adiabatic section by tracking the liquid films thickness.

Figure 11 shows liquid droplet formation at the bottom of the condenser because of the liquid sliding due to the gravity acceleration.

The investigated temperature range for the evaporator and condenser in this study were 100 – 180°C and 20 – 50°C, respectively. The range of filling ratio was from 30 to 80%. It was observed that many of the simulation results failed to get a successful performance for the pulsating heat pipe because of the operating condition. The volume fraction contours of vapor and liquid plugs in a time series were the most important and visible evident to recognize the flow behavior in this numerical simulation. The primary step to investigate the chaotic flow was to observe this vapor and liquid plugs in a time interval. Temperature of the adiabatic section was one of the other criteria that have been investigated to identify the chaos in PHP. Time series of temperature for different operating conditions were tested. Aperiodic, irregular, nonlinear and complex time series of temperature were obtained for those simulations that had successful performance on volume fraction contours for PHP. After observing results of simulations for time series and volume fraction contours, many of them were discarded and those that seem to have proper behavior used for further investigation of chaotic flow.

The analysis of data set that has been observed at different points in time leads to specific issues in statistical modeling and its interpretation. There are several objectives in analyzing a time series. The objectives may be classified as description, explanation, prediction and control. The nonlinear time series methods are based on the theory of dynamical systems which says the time evolution is defined in some phase space. Considering the importance of phase space for the study of systems with deterministic properties, it is notable that to analyze a data set, usually there is not a phase space object but a time series, most likely a sequence of scalar values. So the time series observation should be converted to the phase space. This issue of phase space reconstruction is mostly solved by the delay method [40]. In a pulsating heat pipe, the time series is usually a sequence of temperature values which depends on the current phase space, taken at different points in time:

$$R_n = R(T(n\Delta t)) \tag{21}$$

The delay reconstruction in m dimensions will be formed by the vectors $\boldsymbol{R}_n$:

$$\boldsymbol{R}_n = (R_{n-(m-1)\tau}, R_{n-(m-2)\tau}, \dots, R_{n-\tau}, R_n) \tag{22}$$

where τ is the lag or delay time which refers to time difference in number of samples between adjacent components of the delay vectors (in times units, $\tau\Delta t$).

The vector time series could be defined as

$$\boldsymbol{X}(t)=[T(t),T(t+\tau), T(t+2\tau),...,T(t+(m-1)\tau)] \tag{23}$$

It is important for the dimension m of the system to be large enough. Then there is a defined number which is called true dimension of the system or fractal dimension of the system D. The reconstruction space and sate space trajectories are guaranteed to be equivalent if $m \geq 2D+1$.

During the simulation, different operating conditions were tested including different evaporator and condenser temperatures and different filling ratios. Water was the only working fluid that was employed for simulation. Figure 12 shows the time series of fluctuation of temperature on the walls of the adiabatic section. The numbers of 1 to 8 are assigned to the walls of the adiabatic section from left to the right. Locations for measuring the temperatures are in the middle of the adiabatic walls of the PHP. Evaporation temperature of 135°C, condenser temperature of 30°C and filling ratio of 70% are the operating conditions. Figures 12(a)-(c) illustrate the temperatures for the walls number 1, 4 and 8, respectively. It is obvious in Figure 12 that the temperature behavior for the adiabatic wall under mentioned operating conditions is totally irregular and aperiodic. As mentioned earlier, the chaotic behavior for PHPs appears only under some operating conditions. In many of the simulation results, irregularity and aperiodic behavior were not observed due to its operating condition. For example, in some cases the temperature of the adiabatic section was almost constant and close to one of the evaporator or condenser sections. In some cases, the fluid was stuck in the condenser section due to the low filling ratio and low temperature of the condenser section which leads to higher surface tension for that part.

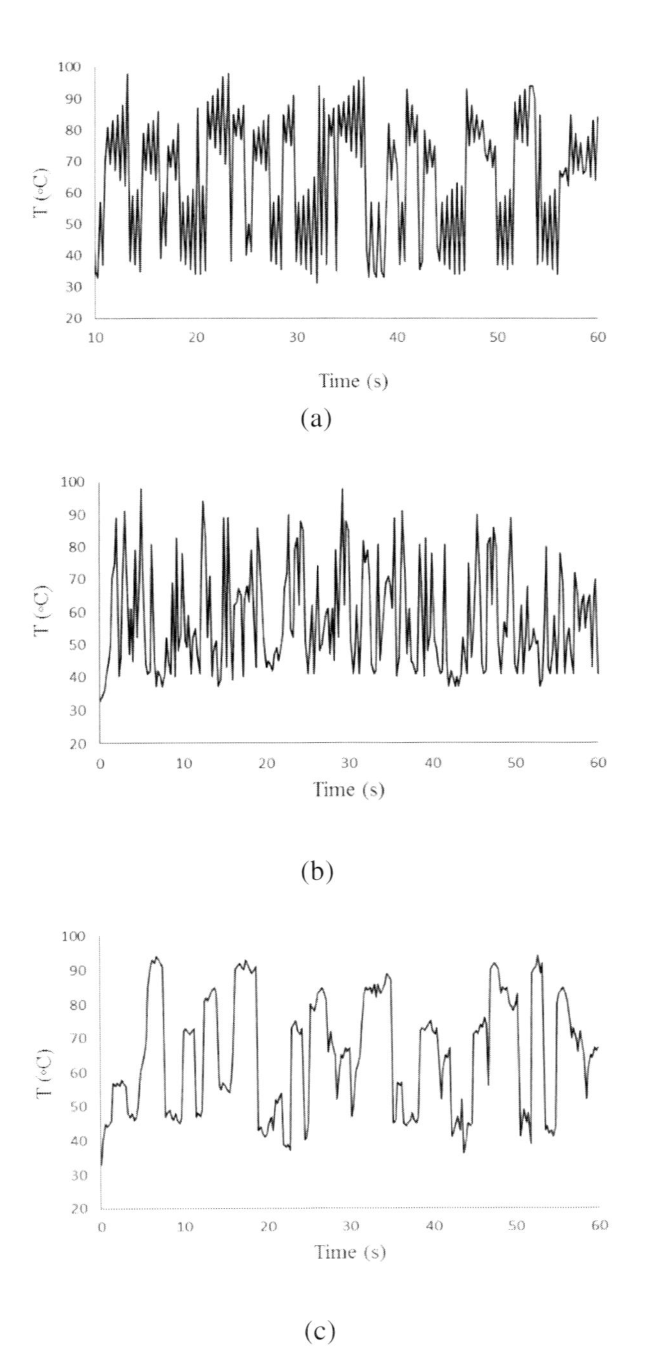

Figure 12. Time series of temperature for the adiabatic walls; (a): wall #1, (b): wall #4, (c): wall #8 [35].

In some cases, high temperature of the evaporator surface led to form abnormal vapor plugs in this section. Figure 13 illustrates forming abnormal vapor plugs due to the high temperature of the evaporator. It is evident this type of abnormal vapor plugs do not exist in the condenser section. It was found that in case of high temperatures for evaporator section, due to the intensive nucleate boiling and formation of superheated vapor, abnormal vapor plugs form in the PHP. This phenomenon mostly occurred when the evaporator temperature was higher than the saturated temperature related to the local pressure at the PHP. In addition, formation of abnormal vapor plugs had adverse effect on the fluid flow in the PHP which decreases performance of the PHP. To avoid this phenomenon, the water properties table was a good reference to find the conditions that abnormal vapor plugs form.

The first step to obtain and confirm chaos in a PHP was to observe irregular and aperiodic behavior in PHP. Since in this study numerical simulations were employed to investigate the flow in PHP, numerical results were used to observe this behavior. Some important approaches such correlation dimension, power spectrum density, Lyapunov exponents and autocorrelation function were employed to investigate the chaotic flow afterwards.

Figure 13. Formation of abnormal vapor plugs in evaporator [35].

3.2. Correlation Dimension

The definition and method to calculate correlation dimension are explained in Section 2.2.

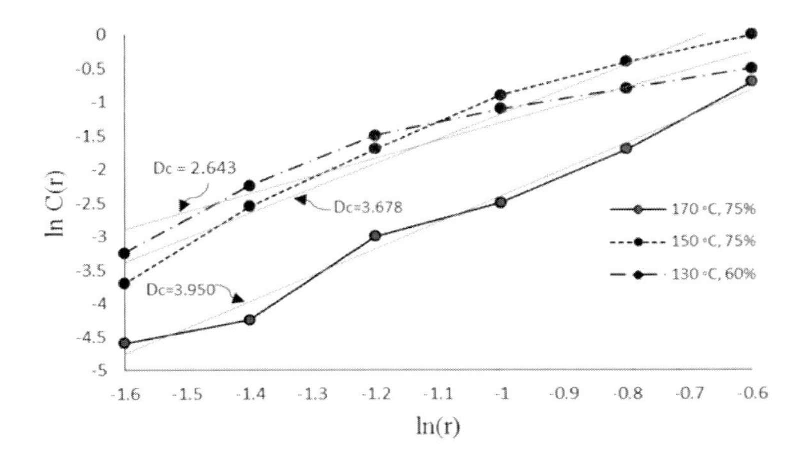

Figure 14. Correlation dimension values (D_c) [35].

Figure 14 shows the relationship between the correlation sum as a function of radius of hyper sphere or the scale in the phase space based on the time series of temperature. Figure 14 illustrates the correlation dimension for three evaporation temperatures of 130, 150, 170°C and correspondents filling ratio of 60%, 75% and 75%, respectively. It is notable that the correlation dimension value is the slop of the curves. It can be seen that by increasing the evaporator temperature and filling ration, the slope of the curves and correlation dimension increase. High values of correlation dimension refers to high frequency, small scale temperature oscillations, caused by miniature bubbles or short vapor plugs dynamically flowing in PHP tubes [30]. The lower correlation dimension corresponds to low frequency of temperature oscillations and large amplitude caused by large bubbles in the PHP. By measuring the slope of the tangent lines, correlation dimension values are 2.643, 3.678 and 3.95 for the evaporation temperatures of 130, 150 and 170°C, respectively.

3.3. Power Spectral Density

The definition and method to calculate power spectral density are explained in section 2.1.

Figure 15 shows the power spectrum density of the time series for the temperature on the adiabatic wall by Matlab analysis. The time step for the numerical simulation was 10^{-5} seconds. It does not mean that the data acquisition frequency was capable to achieve the value of 10^5 Hz. Regarding the operating condition and the capability of the numerical simulation for catching the temperature variation on the adiabatic section, it was reasonable to record the temperature values every 0.5 seconds. Then the highest value dedicated to the data acquisition frequency was 2 Hz. So it was possible to plot the power spectrum density diagram up to 1 Hz. Since the proper frequency interval for investigation of the power spectrum density is equal to the maximum data acquisition frequency divided by the number of data in the related time interval (time interval considered 60 seconds, then the number of the recorded temperature values was 120 for the frequency of 2 Hz), the power spectrum density was plotted with frequency interval of 1/60 Hz. The absence of dominating peaks in the power spectrum density suggests that the phenomenon is neither periodic nor quasi-periodic. Besides, the power spectrum decays by increasing the frequency. These types of behavior for power spectrum density demonstrate the chaotic state in such operating condition.

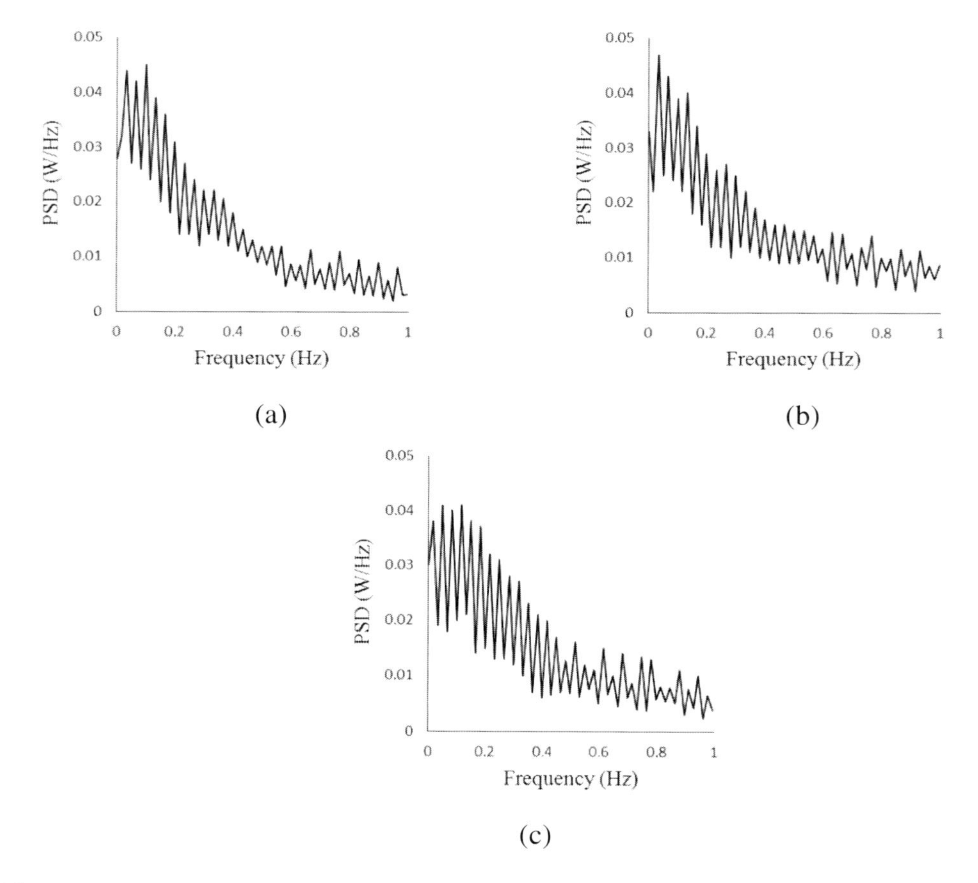

Figure 15. Power spectrum density of the times series of temperature for the adiabatic walls; (a): wall #1, (b): wall #4, (c): wall #8 [35].

3.4. Lyapunov Exponent

The definition and method to calculate Lyapunov exponent are explained in section 2.4.

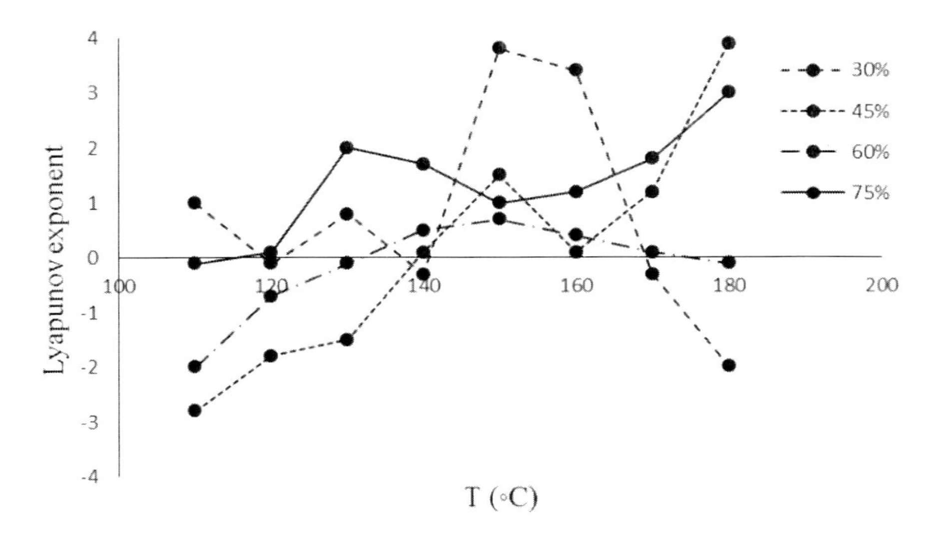

Figure 16. Lyapunov exponents as a function of evaporator temperature at different filling ratios [35].

Figure 16 shows the Lyapunov exponents versus evaporator temperature at four filling ratio of 30%, 45%, 60% and 75%, respectively. It can be seen that for filling ratio of 30% the Lyapunov exponent mostly increases by increasing the temperature except near the temperature of 150°C. Positive values of Lyapunov exponent can be observed for temperatures between 140 to 180°C. At filling ratio of 45% the Lyapunov exponent increases firstly then decreases by increasing the temperature. Lyapunov exponents are positive for temperatures between 130 to 175°C. Lyapunov exponent values are mostly positive for filling ratios of 60% and 75%. Only one negative value of Lyapunov exponent is visible for filling ratio of 75%. This signifies the chaotic state in this filling ratio for temperature values higher than 110°C and less than 180°C.

3.5. Autocorrelation Function

The definition and method to calculate autocorrelation function are explained in section 2.3. Figure 17 illustrates the autocorrelation function (ACF) of time series at filling ratio of 75% and three temperature differences between evaporator and condenser.

It is evident in Figure 17 that the autocorrelation function decreases with time. Behavior of ACF indicates the prediction ability of the system. Decreasing of the ACF shows that the prediction ability is finite. This signifies the chaotic state in the PHP. As shown in Figure 17 by increasing the temperature difference between the evaporator and condenser, ACF decreases. In addition, the ACF values decrease faster in higher values of temperature difference.

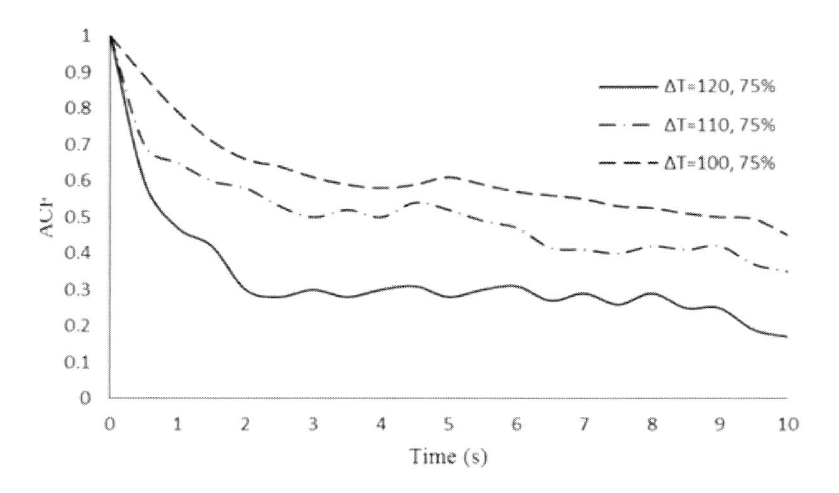

Figure 17. Autocorrelation function versus time [35].

4. Multi-Turn Two-Dimensional Pulsating Heat Pipe

The two dimensional structure of the PHP were the same for all different operating conditions in this work and water was the only working fluid. Different evaporator heating powers, condenser temperatures and filling ratios were tested for the numerical simulation. The pulsating heat pipe structure consists of three sections: heating section (evaporator), cooling section (condenser), and adiabatic section. Figure 18 illustrates a schematic configuration of the PHP which has been used in this study. The three different sections have been distinguished by two horizontal lines. The length and width of the tube are 870 mm and 3 mm respectively. The span of the PHP was considered 1 m as a 2D simulation. Figure 19 shows meshing configuration used in this study. Only a part of evaporator section has been depicted to show the quadrilaterals mesh which were employed for simulation. The quadrilaterals mesh was used for the entire pulsating heat pipe.

4.1. Volume Fractions

Figure 20 illustrates volume fractions of liquid and vapor at different times (Red color represents the vapor and Blue color represents the liquid). Figure 20(a) shows almost the initial condition of the PHP. Figure 20(b) explains formation of vapor bubbles and fluid flow development in the PHP. Figure 20(c) depicts the volume fractions of liquid and vapor in the PHP after the fluid flow has been established. Mostly a similar process occurs in the PHP with filling ratio of 60% (Figure 21). One of the most significant effects due to the increasing the filling ratio is pressure increment in the PHP which lead to a longer time duration for boiling start (because of increasing the saturated temperature) and slower flow motion in the PHP. The effect of change in filling ratio on the chaotic and thermal behavior of the PHP will be investigated in the next sections afterwards. It is evident liquid plugs having menisci on the plug edges are formed due to surface tension forces. A liquid thin film also exists surrounding the vapor plug. The angle of contact of the menisci, the liquid thin film stability and its thickness depends on the fluid-solid combination and the operating parameters which are selected. If a liquid plug is moving or tends to move in a specific direction then the leading contact angle (advancing) and the lagging contact angle (receding) may be different [26]. This happens because the leading edge of the plug moves on a dry surface (depending on the liquid thin film stability and existence) while the lagging edge moves on the just wetted surface. These characteristics can be observed in Figure 22 as result of numerical simulation.

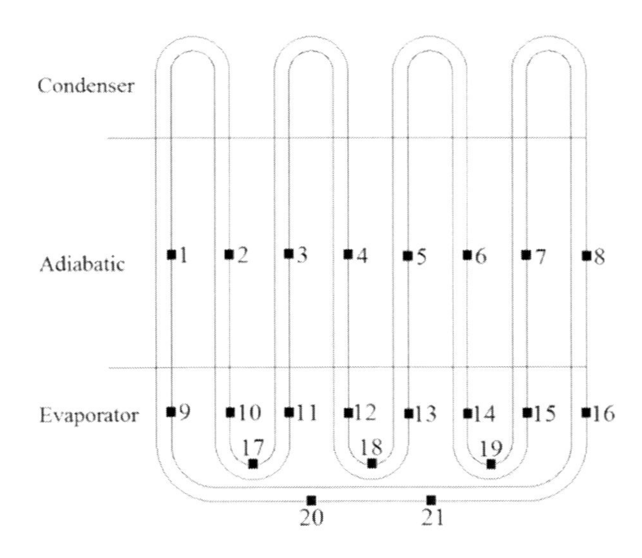

Figure 18. Pulsating heat pipe structure [36].

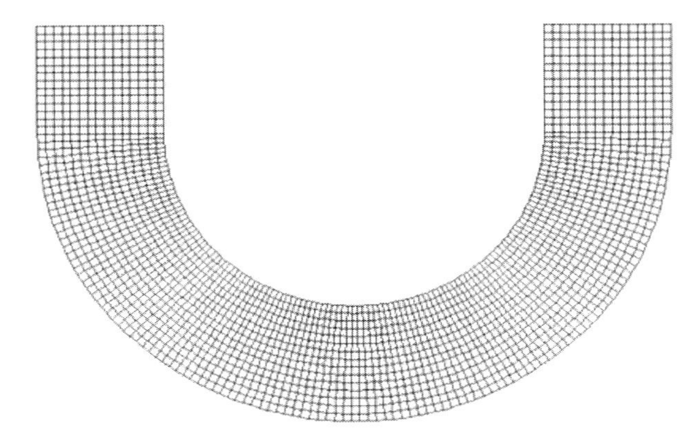

Figure 19. Meshing configuration [36].

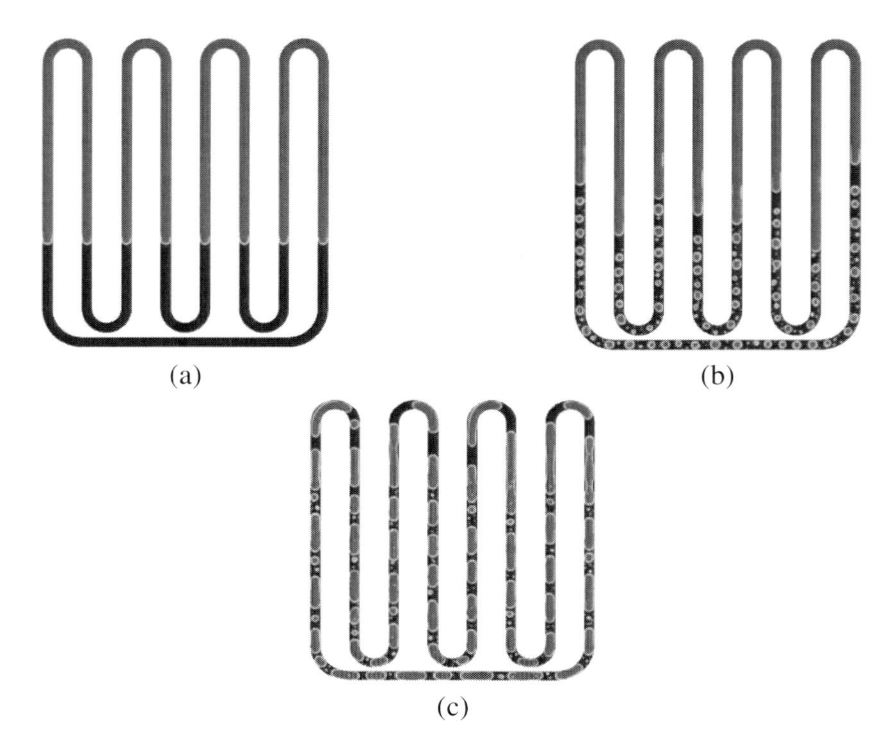

Figure 20. Volume fractions of liquid and vapor at different times with Q=40 W, T_c=30°C and FR= 35%; (a): t=0.1 s, (b): t=1.7 s, (c): t=20 s [36].

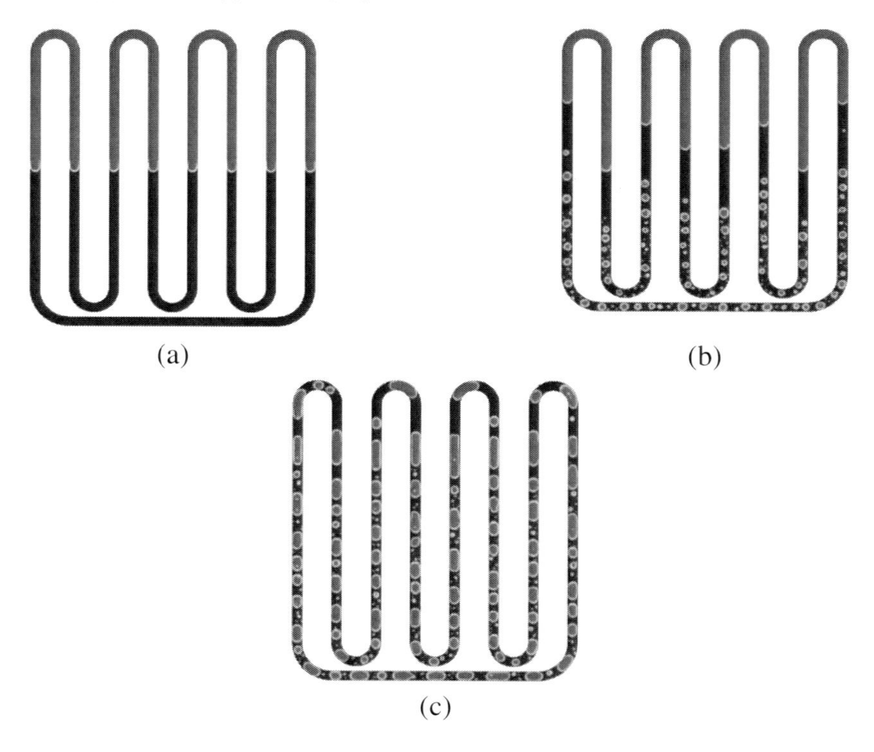

Figure 21. Volume fractions of liquid and vapor at different times with Q=55 W, T_c=25°C and FR= 60%; (a): t=0.1 s, (b): t=2.6 s, (c): t=20 s [36].

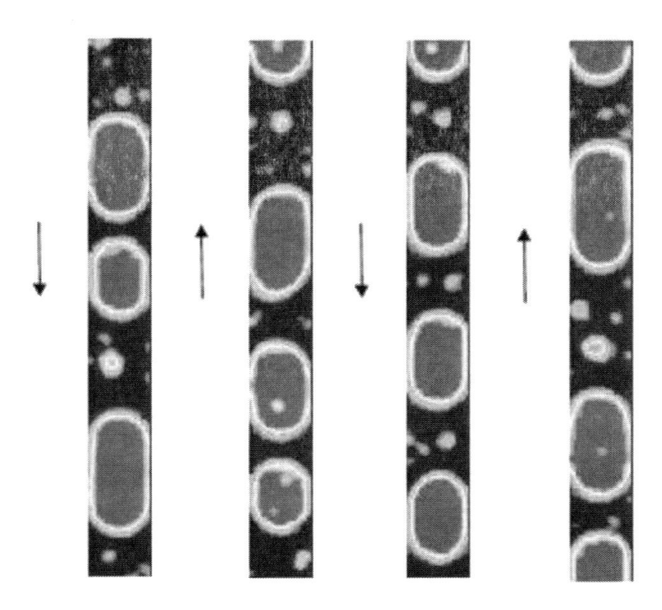

Figure 22. Vapor and liquid plugs in the PHP [36].

The results showed that the fluid flow finally circulates in one direction (clockwise or counterclockwise) in the pulsating heat pipe. This direction is based on a random process and could be different even under the same operating and boundary conditions [36]. It is seen that by increasing the wall temperature, the liquid film thickness around the vapor plugs decreases. In evaporator section the temperature of the liquid plugs increases which is followed by evaporation mass transfer to the adjoining vapor plug or splitting the liquid plug by formation of new bubbles inside due to the nucleate boiling in the slug flow regime. Sometimes two vapor bubbles combine together to form a larger vapor slug. This phenomenon was mostly observed in the adiabatic section and rarely was observed in the condenser. In addition, liquid accumulation as a result of vapor condensation occurred on the condenser surface. This condensation process was more visible on the bending parts due to the surface tension. It was observed that many of the simulation results failed to get a successful performance for the pulsating heat pipe because of the operating condition. In some cases, the fluid was stuck in the condenser section due to the low filling ratio and low temperature of the condenser section which leads to higher surface tension for that part. In some cases, high temperature of the evaporator surface led to form abnormal vapor plugs in this section due to the intensive nucleate boiling and formation of superheated vapor.

Figure 23 compares formation of such abnormal vapor plugs with normal plugs. Formation of abnormal vapor plugs had adverse effect on the fluid flow in the PHP which decreases performance of the PHP. After observing results of simulations for volume fraction contours and temperature time series, many of them were discarded and those that seem to have proper behavior used for further investigation of chaotic flow.

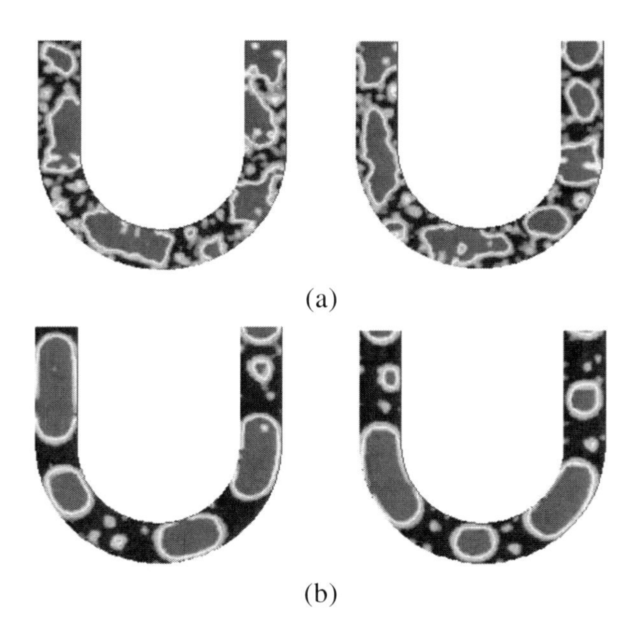

Figure 23. Comparing formation of abnormal vapor plugs (a) with normal vapor plugs (b) at evaporator [36].

4.2. Non-Linear Temperature Oscillations

The non-linear time series analysis is a popular method for investigation of complicated dynamical systems. Different non-linear analytical approaches, including power spectrum density, correlation dimensions, and autocorrelation function have been employed to analyze the temperature time series at adiabatic wall. Temperature behavior of adiabatic wall in a PHP is one of the most important factors to investigate chaos in the PHP. It can be seen in Figure 18 that there are 21 points on the wall of the PHP assigned for temperature measurements. Eight points are located at the adiabatic section and thirteen points are located at evaporator. Then, there are eight choices to observe temperature behavior of adiabatic wall. Figure 24 illustrates the temperature oscillations of point #4 and #18 on the adiabatic wall and evaporator respectively on the PHP under operating condition mentioned in Figure 21. Point selection at adiabatic section and evaporator was done randomly.

It is obvious that aperiodic, irregular, non-linear and complex time series of temperature are obtained for both evaporator and adiabatic wall. Evaporator temperature oscillations have higher amplitude and higher mean temperature comparing with adiabatic wall temperature oscillations. As mentioned earlier only adiabatic wall temperature behavior has been considered for chaos investigation purpose. To reach this aim, power spectrum density analysis has been applied to the obtained time series of adiabatic wall temperature.

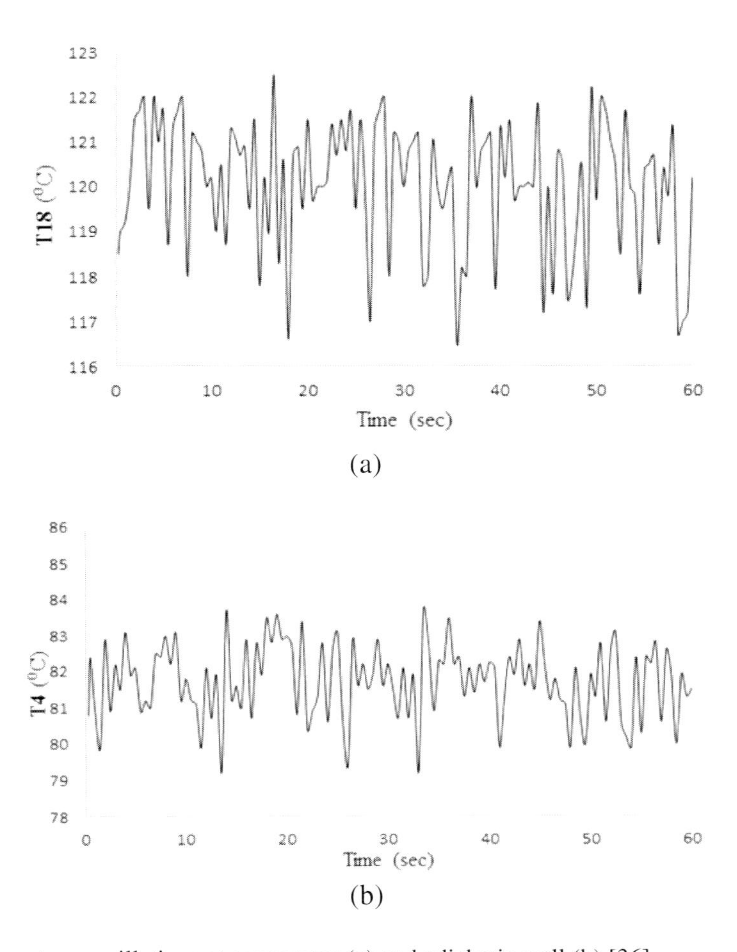

Figure 24. Temperature oscillations at evaporator (a) and adiabatic wall (b) [36].

4.3. Power Spectral Density

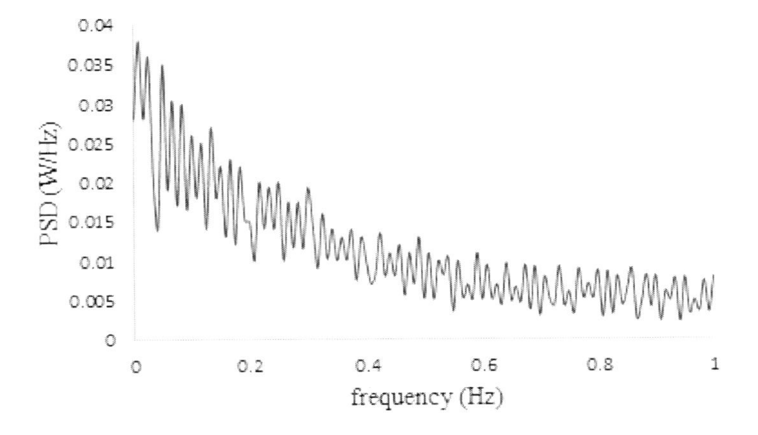

Figure 25. Power spectrum density of the adiabatic wall temperature time series (point #4) [36].

Figure 25 shows the power spectrum density of the time series for the temperature on the adiabatic wall by Matlab analysis. Absence of dominating peaks in power spectrum density behavior and its decay by increasing the frequency were signatures of chaotic state in the PHP under such operating conditions. Important approaches such correlation dimension and autocorrelation function were employed to investigate the chaotic state afterwards.

4.4. Correlation Dimension and Autocorrelation Function

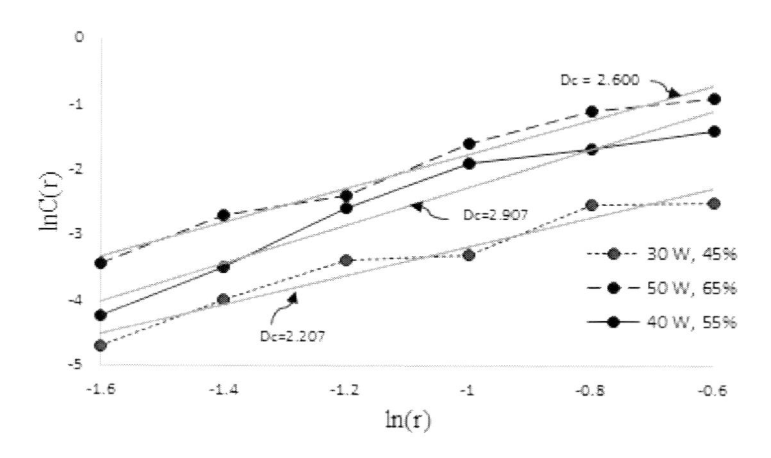

Figure 26. Correlation dimension values (D_c) [36].

Figure 26 illustrates the correlation dimension for evaporator heating powers of 30 W, 40 W and 50 W with filling ratios of 45%, 55% and 65%, respectively at condenser temperature of 35°C. Values of 2.207, 2.600 and 2.907 were obtained for correlation dimension. It is evident that by increasing the evaporator heating power and filling ratio, the slope of the curves and correlation dimension increase. High values of correlation dimension refers to high frequency, small scale temperature oscillations, caused by miniature bubbles or short vapor plugs dynamically flowing in PHP tubes [36]. The lower correlation dimension corresponds to low frequency of temperature oscillations and large amplitude caused by large bubbles in the PHP.

Figure 27 illustrates the autocorrelation function (AFC) of time series at filling ratio of 65% and three evaporator heating powers of 30 W, 45 W and 60 W at condenser temperature of 35°C.

Behavior of ACF indicates the prediction ability of the system [36]. The autocorrelation function will have its largest value of AFC=1 at τ=0. Figure 27 shows that the autocorrelation function decreases with time. Decreasing of the ACF shows that the prediction ability is finite as a signature of chaos. It is obvious in Figure 27 by increasing the evaporator heating power, ACF decreases.

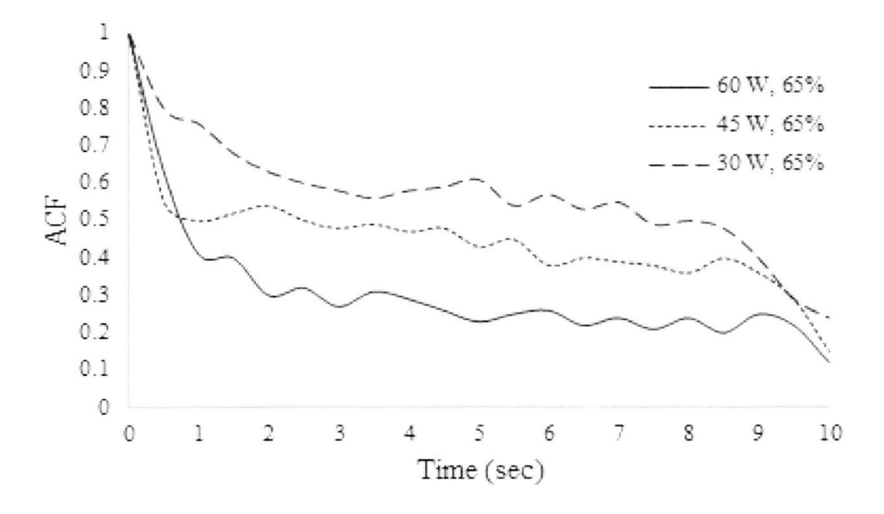

Figure 27. Autocorrelation function versus time [36].

4.5. Thermal behavior

For the traditional heat pipes, the filling ratio range of working fluid is lower comparing to the PHPs. Large amount of working fluid inside the channels causes liquid blockage and reduces pressure difference between heating and cooling sections which can stop the oscillation movement. But for the PHPs a higher filling ratio is required than traditional heat pipes. To investigate the effect of filling ration on the performance of PHP, several simulations were carried out under different heating powers, condenser temperatures and filling ratios. Figures 28(a) and 28(b) explain the axial temperature of the wall through the PHP. It should be noted that the relative distance for the point #1 is considered zero as the origin and the distance increases by moving downward from point #1 to #9. In addition, the temperature distributions in these figures do not include the condenser temperature since its value is known and constant. Temperature profiles have been illustrated at three different filling ratios of 40%, 60% and 80% under heating powers of 40 W and 55 W for evaporator and condenser temperature of 25°C as part of numerical simulations. Mean temperature for each distance is defined as the average of data points from 0 to 60 seconds on that distance.

Generally these temperature distributions were symmetric from point #2 to #3, #4 to #5, #6 to #7 and #20 to #21 for vertical centerlines at #17, #18, #19 and #18 respectively. It was found that for all the heating powers, the mean temperatures in the evaporator had its lowest values at the filling ratio of 60%. It is obvious that at higher filling ratio of 80% and lower filling ratio of 40%, the evaporator mean temperatures have higher values comparing to that of 60%. Then based on the numerical simulation results, filling ratio of 60% was found as optimal filling ratio for the thermal performance of the PHP.

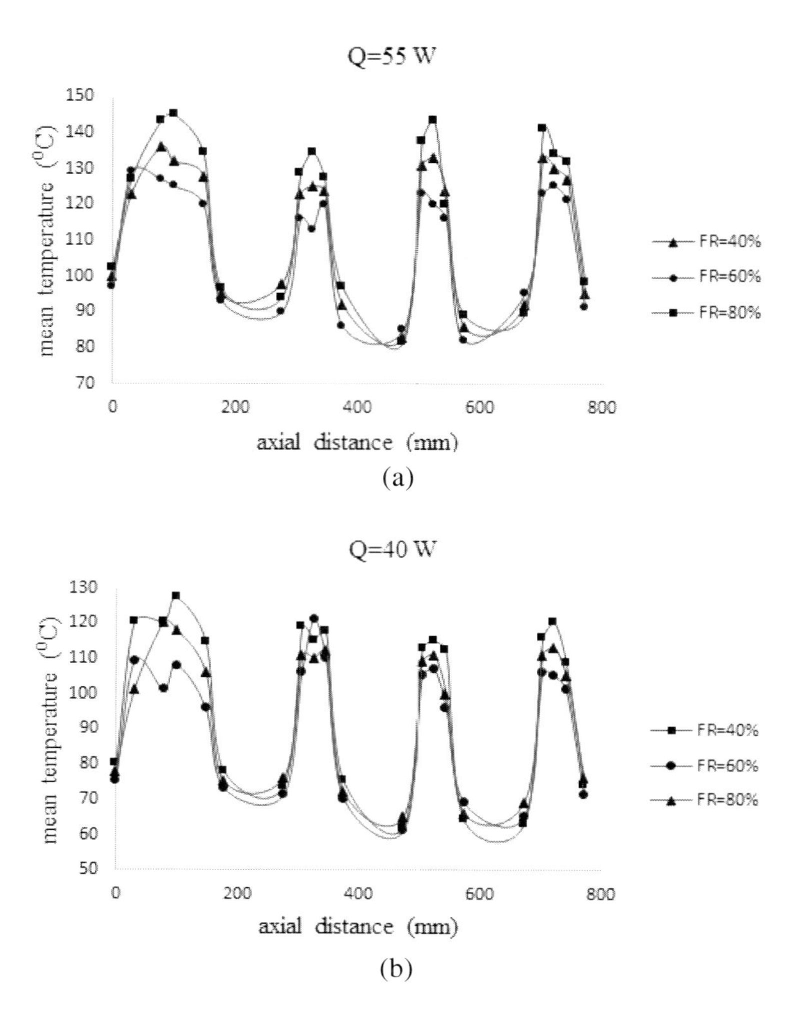

Figure 28. Axial mean temperature distribution for heating powers of 55 W (a) and 40 W (b) [36].

The overall thermal resistance of a PHP is defined as the difference average temperatures between evaporator and condenser divided by the heating power. For evaporator average temperature, thirteen points from #9 to #21 were used to calculate the average temperature. Since condenser has constant temperature, calculation is not necessary to get the average temperature which is equal to that constant value. Equation (24) defines the thermal resistance of a PHP.

$$R = \frac{Te - Tc}{Q} \tag{24}$$

where T_e is the evaporator average temperature, T_c is the condenser average temperature and Q is the evaporator heating power.

Figure 29 describes behavior of thermal resistance with respect to evaporator heating power at recommended optimal value of 60% for filling ratio and condenser temperature of 25°C. By increasing the heating power from 10 to 40 W, the thermal resistance decreases sharply. From 40 to 45 W, thermal resistance has slight decrease and remains almost constant about 1.62°C/W at heating power of 45 W. Then thermal resistance starts to increase by increasing the heating power afterwards. It was seen that the PHP has its lowest thermal resistance of 1.62°C/W at evaporator heating power of 45 W. It should be noted that the thermal resistance depends on many parameters such as PHP structure, working fluid and operating conditions. Thus, obtained values in this paper are suitable for the test case in current study. But it can be concluded that there are optimal filling ratios and minimum thermal resistance at any test case which can be obtained by numerical or experimental investigations.

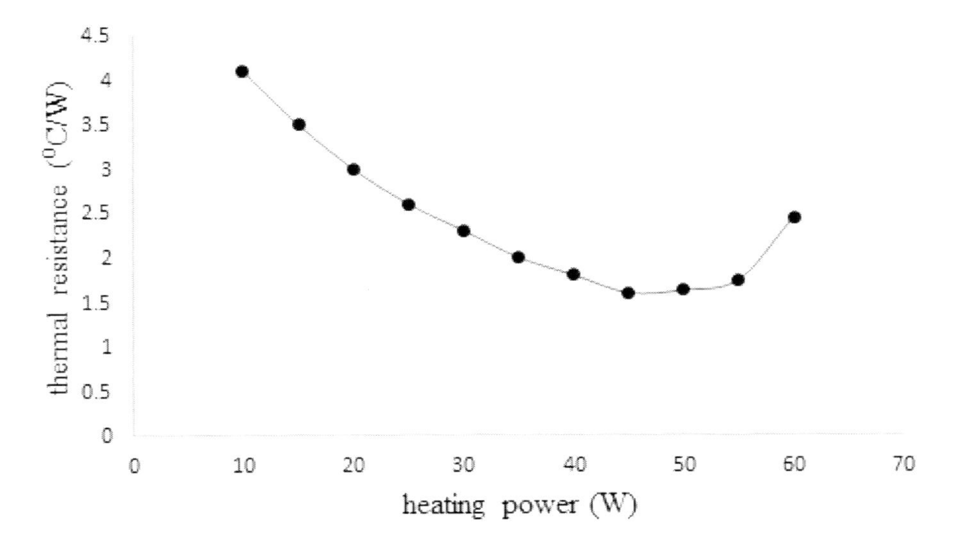

Figure 29. Thermal resistance versus heating power [36].

5. Three-Dimensional Pulsating Heat Pipe

A schematic cross sectional view of the 3D PHP is shown in Figure 30. In order to validate the simulation results of current study with an experimental test case, the structure and dimensions of the pulsating heat pipe in simulation were considered the same as experimental investigation in [20]. The diameter of the tube is 1.8 mm. The lengths of the evaporator, adiabatic and condenser sections are 60 mm, 150 mm and 60 mm, respectively. Water and ethanol were used as working fluids. Different evaporator heating powers and filling ratios were tested for the numerical simulation. There are 25 points in Figure 30 highlighted with bold rectangles and these points are assigned for temperature measurement during numerical simulation.

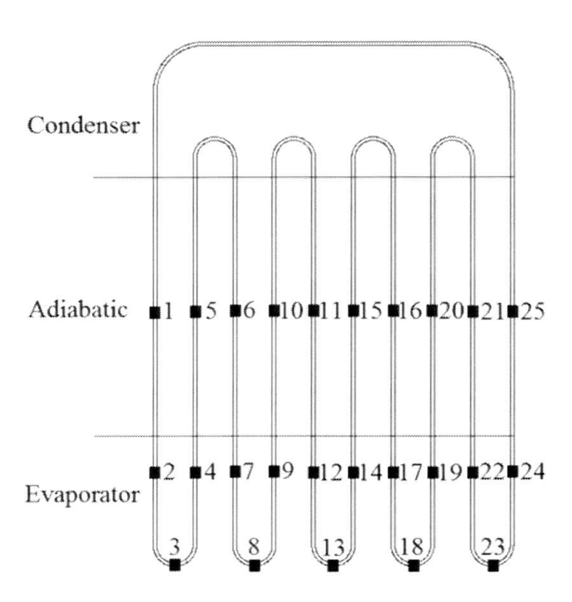

Figure 30. Schematic cross sectional view of the 3D PHP [37].

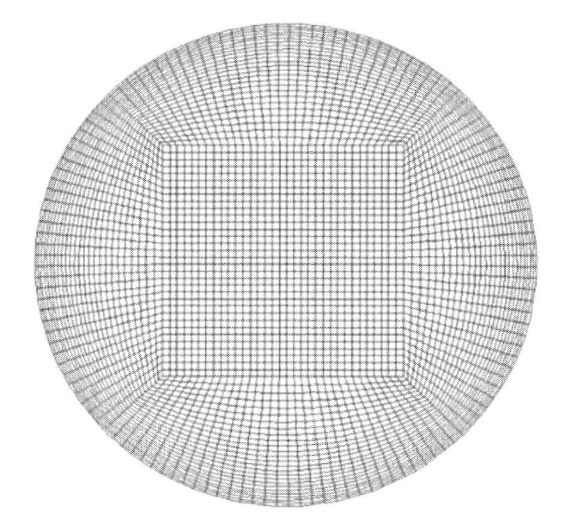

Figure 31. Meshing configuration [37].

Figure 31 shows cross sectional profile of meshing configuration used in the numerical simulation. This profile has been swept through the entire centerline of the PHP led to create hexahedral meshes. Surface tension is an important factor in the performance of the pulsating heat pipe. Employing quadrilateral or hexahedral meshes for numerical simulation leads to more accurate results in case of surface tension than those of triangular and tetrahedral. The hexahedral mesh was used for the entire pulsating heat pipe. It is evident in Figure 31 that number of layers is increased near the wall of the PHP to properly capture the flow gradients.

5.1. Validation

As mentioned earlier, in order to validate the simulation results of current study with an experimental test case, the structure and dimensions of the pulsating heat pipe in simulation were considered the same as experimental investigation in [20]. In addition, the same working fluids of water and ethanol were employed. It should be noted that due to the experiment limitations as the mentioned investigation, the maximum heating power of 70 W and 60 W were applied for water and ethanol as working fluids respectively in [20]. Besides, constant temperature of 20°C was used for the condenser boundary condition in the experiment. Figure 3 shows comparison of thermal resistance with respect to heating power between simulation results of current study and experimental results of Shafii et al. [20].

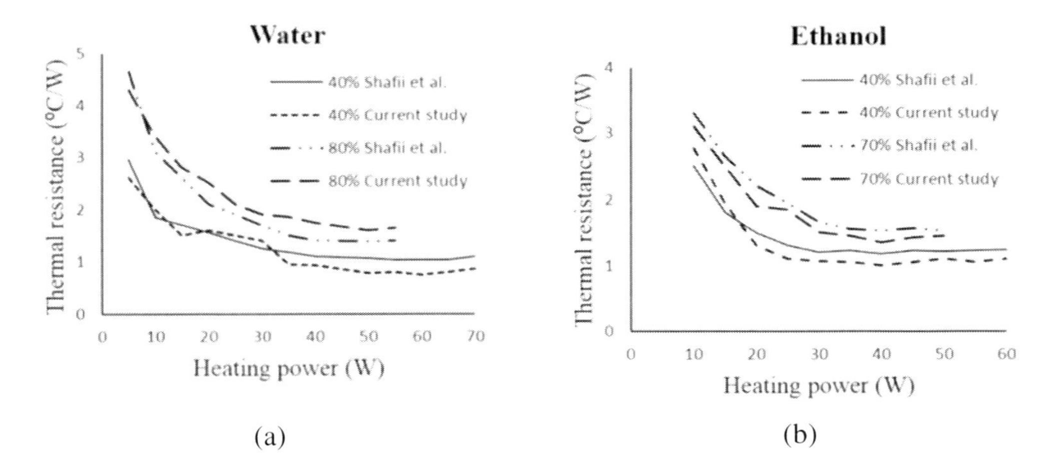

Figure 32. Comparison of thermal resistance versus heating power between simulation results of current study and experimental results [37].

Thermal resistance behavior has been illustrated at filling ratios of 40% and 80% for water as working fluid and 40% and 70% for ethanol as working fluid. As seen in Figure 32(a) the thermal resistance obtained from simulation has higher values after heating power of 10 W comparing with experimental results for water as working fluid and filling ratio of 80%. The maximum error occurs at heating power of 40 W which is almost 13%. The average error is 8.7% under these operating conditions. At filling ratio of 40%, the maximum error occurs at heating power of 60 W which is almost 15%. The average error is 9.7% under these operating conditions. Figure 32(b) shows that the thermal resistance obtained from simulation has lower values for all heating powers comparing with experimental results for ethanol as working fluid and filling ratio of 70%. The maximum error occurs at heating power of 20 W which is almost 10%. The average error is 6.3%

under these operating conditions. At filling ratio of 40%, the maximum error occurs at heating power of 25 W which is almost 13%. The average error is 7.5% under these operating conditions. Figure 3 confirms there is a good agreement between simulation and experimental results. The range of errors percentage is normal for such comparison between simulation and experimental results. There are several source of error in any experiment such as instrument and measurement inaccuracy and uncertainty. One of the most important issues in this particular test case is insulation of evaporator and adiabatic sections which cannot be ideal experimentally. Thus, ideal insulation in numerical simulation is one of the significant reasons which cause differences between simulation and experimental results.

5.2. Volume Fractions

Figures 33 to 37 illustrate volume fractions of liquid and vapor at different times (red color represents the vapor and blue color represents the liquid) under different operating conditions. Figure 33(a) shows almost the initial condition of the PHP with ethanol as working fluid and filling ratio of 40%. Figure 34 demonstrates formation of vapor bubbles and fluid flow development in the PHP. Figure 35 depicts the volume fractions of liquid and vapor in the PHP after the fluid flow has been established. Mostly a similar process occurs in the PHP with water as working fluid and filling ratio of 65% (Figures 33(b), 36 and 37). One of the most significant effects due to increasing filling ratio is pressure increase in the PHP which leads to a longer time duration for boiling to start (because of increasing the saturated temperature) and slower flow motion in the PHP. In addition to higher filling ratio, due to higher specific heat and higher saturation temperature for water comparing to ethanol, boiling process starts later for the PHP with water as working fluid than PHP with ethanol as working fluid. The effect of change in filling ratio on the chaotic and thermal behavior of the PHP will be investigated in the next subsections afterwards. It is evident that liquid plugs having menisci on the plug edges are formed due to surface tension forces. A liquid thin film also exists surrounding the vapor plug. The angle of contact of the menisci, the liquid thin film stability and its thickness depends on the fluid-solid combination and the operating parameters which are selected. If a liquid plug is moving or tends to move in a specific direction then the leading contact angle (advancing) and the lagging contact angle (receding) will be different [26]. This happens because the leading edge of the plug moves on a dry surface (depending on the liquid thin film stability and existence) while the lagging edge moves on the just wetted surface.

These characteristics can be observed in Figure 38, which shows a snapshot of liquid-vapor plugs as result of simulation in the vertical part of the system.

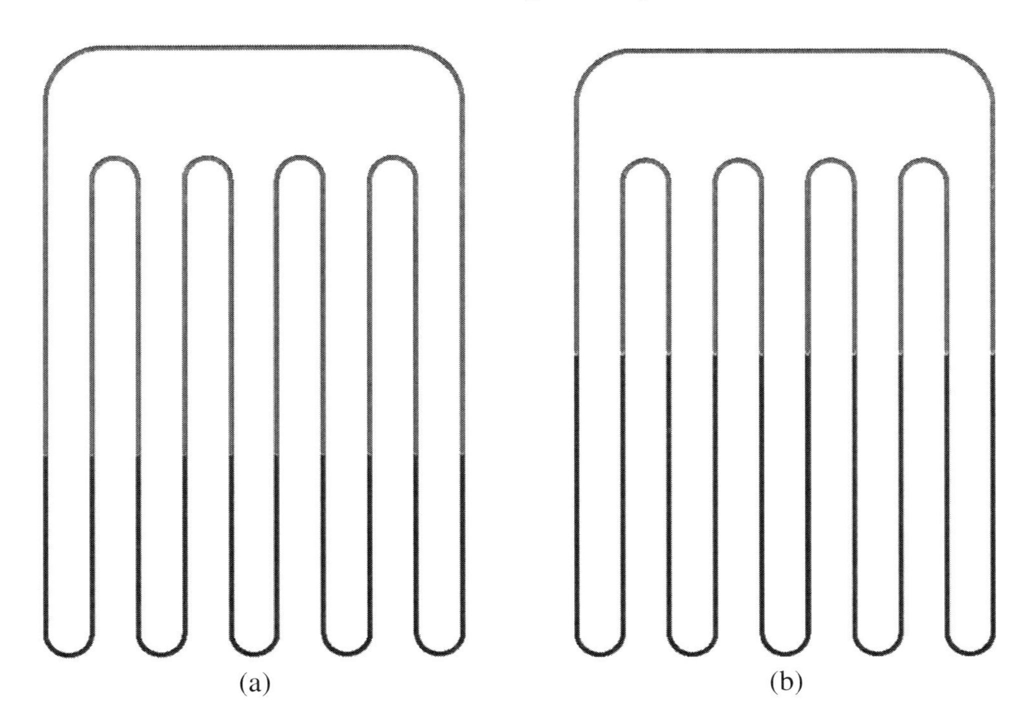

Figure 33. Volume fractions of liquid and vapor at t=0.1 s under heating power of 70 W and condenser temperature of 20°C for ethanol, FR of 40% (a) and water, FR of 65% (b) [37].

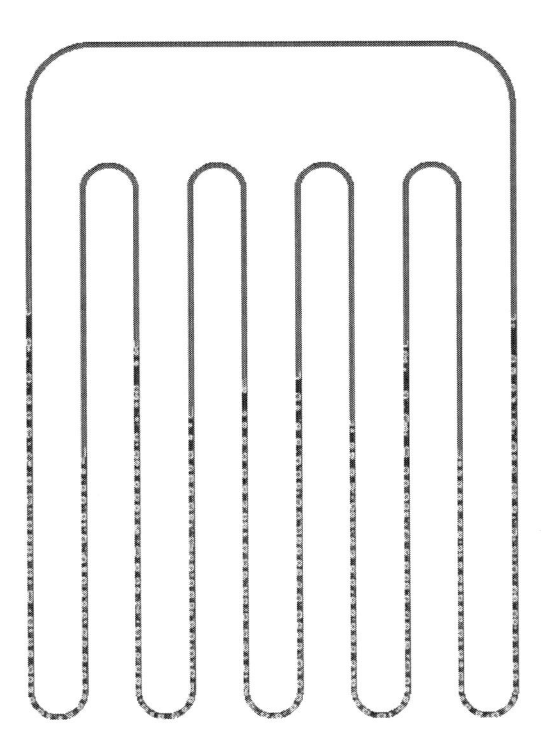

Figure 34. Volume fractions of liquid and vapor under heating power of 70 W and condenser temperature of 20°C for ethanol as working fluid with FR of 40% at t=0.8 s [37].

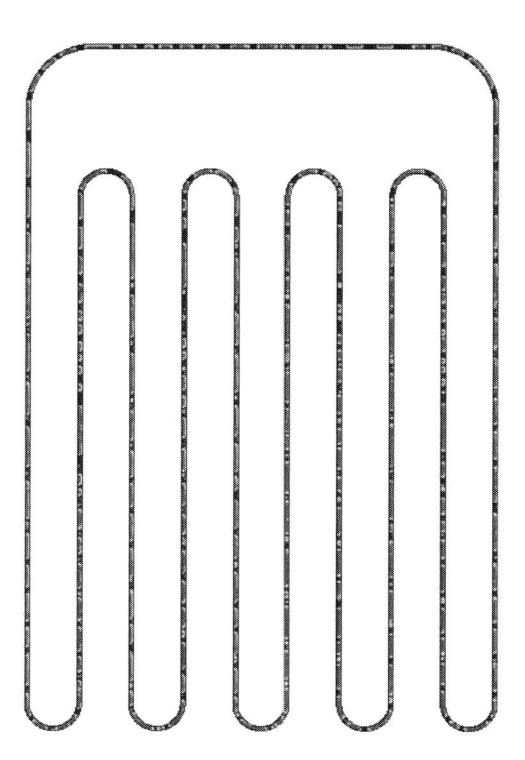

Figure 35. Volume fractions of liquid and vapor under heating power of 70 W and condenser temperature of 20°C for ethanol as working fluid with FR of 40% at t=20 s [37].

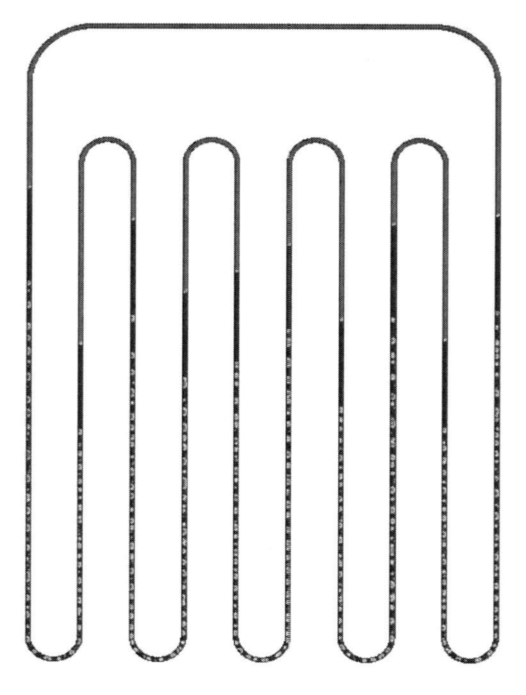

Figure 36. Volume fractions of liquid and vapor under heating power of 70 W and condenser temperature of 20°C for water as working fluid with FR of 65% at t=0.8 s [37].

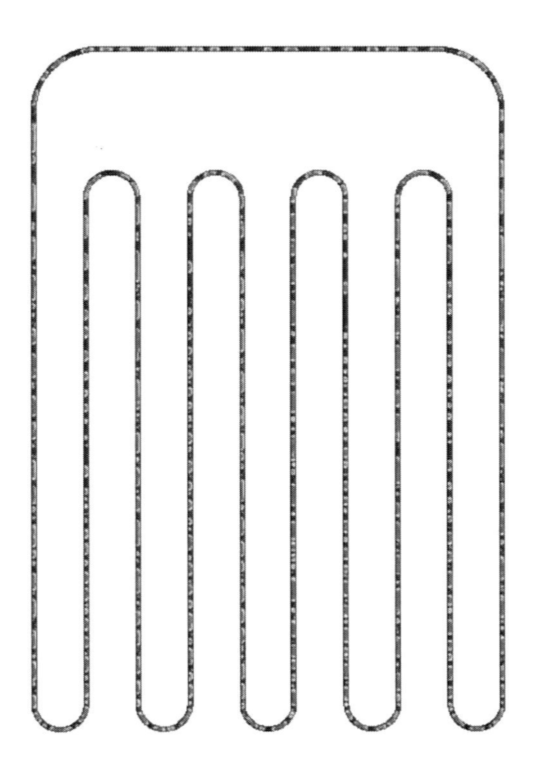

Figure 37. Volume fractions of liquid and vapor under heating power of 70 W and condenser temperature of 20°C for water as working fluid with FR of 65% at t=20 s [37].

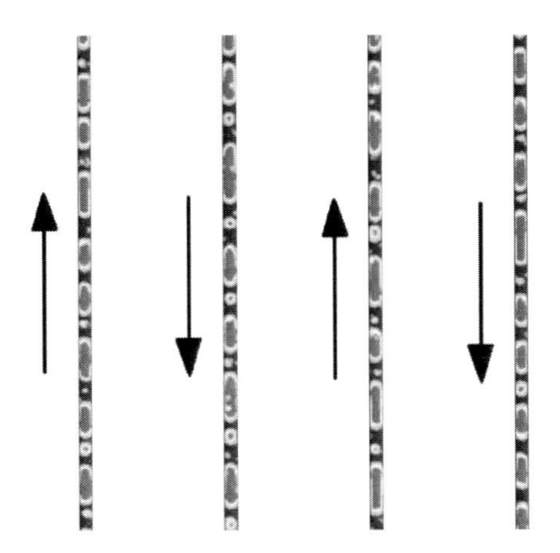

Figure 38. Vapor and liquid plugs in the PHP [37].

The results showed that the fluid flow finally circulates in one direction (clockwise or counterclockwise) in the pulsating heat pipe. This direction is based on a random process and could be different even under the same operating and boundary conditions. The liquid and vapor plugs are subjected to pressure forces from the adjoining plugs. Forces acting on the liquid plugs are due to the capillary pressures created by the menisci curvatures of the adjacent vapor bubbles. The liquid plugs and vapor bubbles experience internal viscous dissipation as well as wall shear stress as they move in the PHP tube [26]. The relative magnitude of these forces will decide the predominant force to be considered. The liquid and vapor plugs may receive heat, reject heat or move without any external heat transfer, depending on their location in the evaporator, condenser or the adiabatic section. Probability of events frequently places vapor bubbles in direct contact with the evaporator tube surface. In this case saturated vapor bubbles receive heat which is simultaneously followed up by evaporation mass transfer from the adjoining liquid plugs thereby increase the instantaneous local saturation pressure and temperature.

Liquid thin film exists around the vapor plug, and the thickness of the liquid film may vary depending on some boundary conditions. One of the most important parameters affecting the liquid film is the wall temperature in the PHP. Since evaporator, condenser and adiabatic sections have different working temperatures, liquid film with different thicknesses may form in these regions [35]. Figure 10 shows different liquid films surrounding the vapor plugs with different thicknesses as result of simulation. Figure 39 illustrates vapor plugs in the evaporator (a), adiabatic section (b) and condenser (c) respectively. It can be seen that by decreasing the wall temperature from evaporator to condenser, the liquid film thickness increases.

Figure 40 shows the comparison of the averaged thickness of liquid film for the vapor plugs at evaporator, adiabatic section and condenser under different heating power with condenser temperature of 20°C and water as working fluid with filling ratio of 65%. It is evident that condenser has thicker liquid film than the adiabatic section and evaporator at each heating power and evaporator has thinner liquid films. The thickness decreases linearly from 10 W to 40 W in evaporator. By increasing the heating power higher than 40 W, a decrease in slope can be observed from 40 W to 100 W. A similar behavior for the thickness in adiabatic section is visible. But the change in slope occurs at a higher heating power of 60 W. The thickness in condenser decreases from 0.24 mm to 0.12 mm linearly with a constant slop by increasing the heating power.

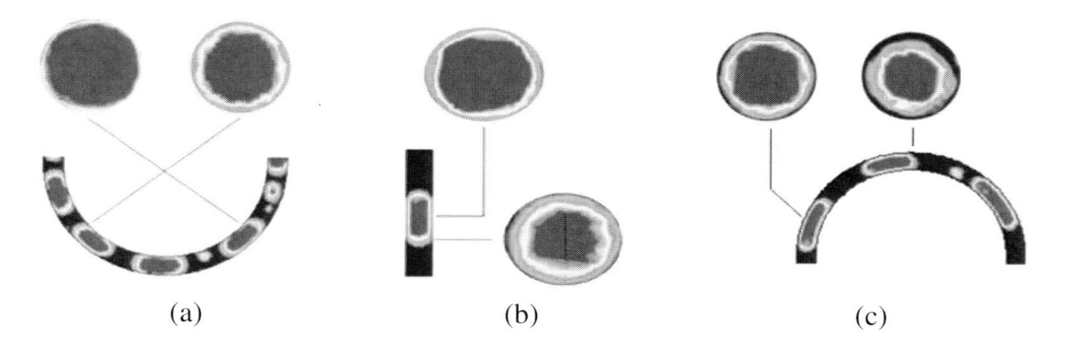

Figure 39. Liquid film around the vapor plugs at evaporator (a), adiabatic section (b) and condenser (c) [37].

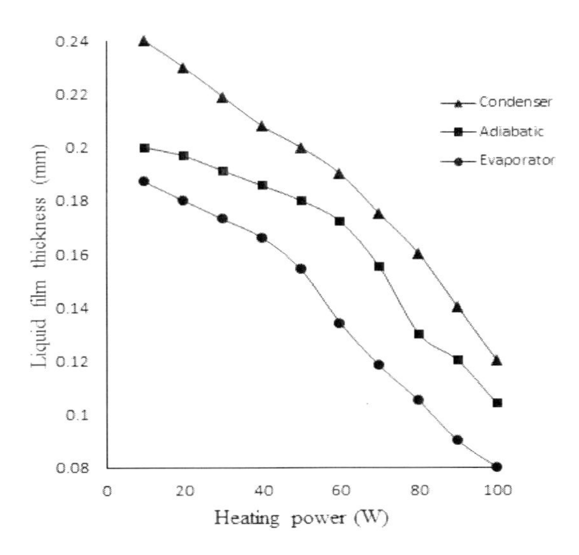

Figure 40. Liquid film thickness versus heating power [37].

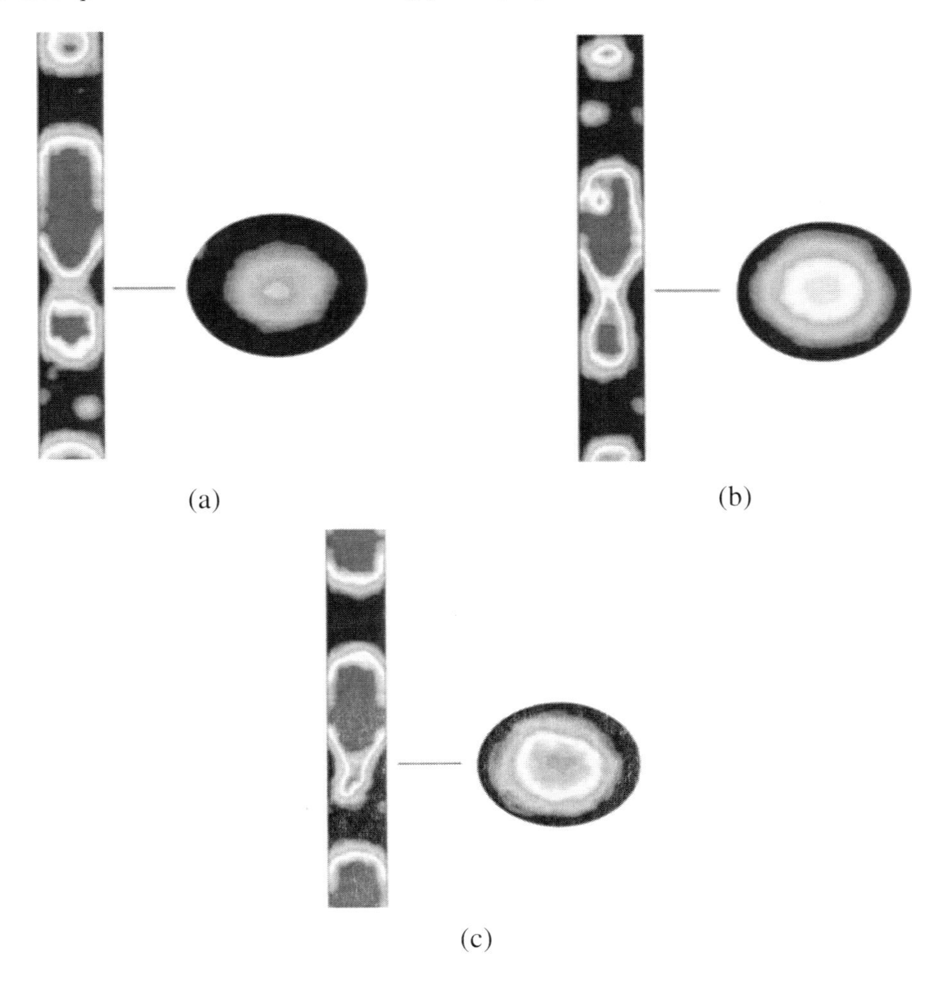

Figure 41. Vapor bubbles combination [37].

In adiabatic section, while passing from the evaporator to the condenser, the train of vapor-liquid plugs is subjected to a series of complex heat and mass transfer process. Essentially non equilibrium conditions exist whereby the high pressure, high temperature saturated liquid-vapor plugs is brought down to low pressure, low temperature saturated conditions existing in the condenser. Internal enthalpy balancing in the form of latent heat takes place by evaporation mass transfer from the liquid to the vapor plugs whereby saturation conditions are always imposed on the system during the bulk transient in the adiabatic section [26]. Sometimes two vapor bubbles combine together to form a larger vapor slug [35]. Such a combination process in evaporator and adiabatic section is illustrated in Figure 41 as a result of simulation; this phenomenon mostly was observed in the adiabatic section and was observed rarely in the condenser.

5.3. Spectral Analysis of Time Series

In this work, 10 points on the PHP are dedicated to measure the temperature of the adiabatic walls as shown in Figure 30. To investigate and analyze the temperature time series, Power Spectrum Density (PSD) approach has been employed. Figure 42 shows the time series of the adiabatic wall temperature at point #16 under heating power of 90 W, condenser temperature of 20°C, filling ratio of 45% and water as working fluid and its power spectrum density diagram. The spectral analysis of the time series in Figure 42(b) indicates a dominant peak around frequency of 0.2 Hz. This signifies an intense periodic or quasi-periodic oscillation of temperature at this dominant frequency with PSD of 0.05 (W/Hz). PSD of oscillations at other frequencies are in an order to be neglected comparing with this dominant peak. Thus, the temperature behavior in Figure 42(a) can be classified as periodic or quasi-periodic with frequency of 0.2 and PSD of 0.05 under mentioned operating conditions.

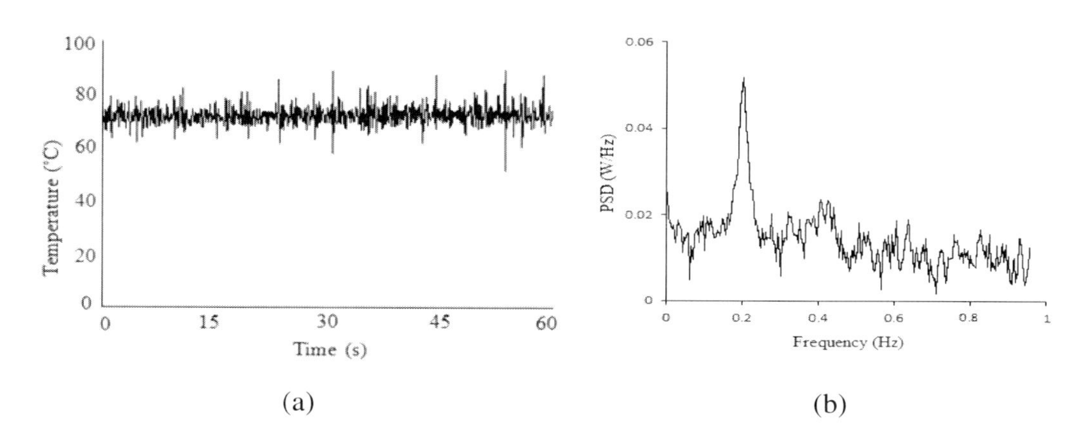

Figure 42. Time series of temperature (a) and PSD diagram (b) for point #16 under heating power of 90 W, condenser temperature of 20°C, filling ratio of 45% and water as working fluid [37].

Figure 43 shows the time series of the adiabatic wall temperature at point #11 under heating power of 90 W, condenser temperature of 20°C, filling ratio of 55% and ethanol as working fluid and its power spectrum density diagram. It can be seen in Figure 43(b) that there are two dominant peaks in PSD diagram. Oscillations at frequencies of 0.12 Hz and 0.23 Hz have higher PSD of 0.05 and 0.03 (W/Hz) comparing oscillations at other frequencies. Existence of these two dominant peaks signifies periodic or quasi-periodic oscillations of temperature at these dominant frequencies. Thus, the temperature behavior in Figure 43(a) cannot be classified as chaos under the mentioned operating conditions.

Figure 44 shows the time series of the adiabatic wall temperature at point #20 under heating power of 90 W, condenser temperature of 20°C, filling ratio of 70% and water as working fluid and its power spectrum density diagram. As shown in Figure 44(b), there is not any significant and visible dominant peak in PSD diagram. At some frequencies, the PSD has higher value comparing other frequencies. But those higher values of PSD are not in an order to be considered as a dominant peak. Thus, the temperature oscillations are neither periodic nor quasi-periodic. In addition, by increasing the frequency, the power spectrum density decays in Figure 44(b) as signature of chaotic behavior. Then, the temperature behavior in Figure 44(a) can be classified as chaos under mentioned operating conditions.

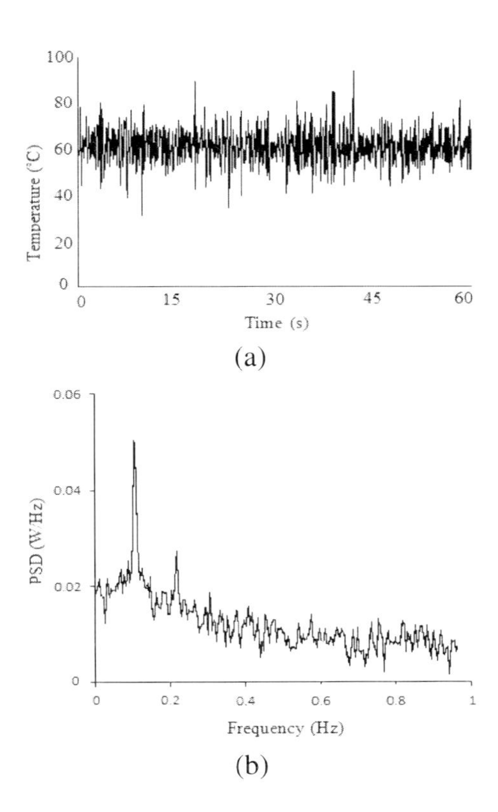

Figure 43. Time series of temperature (a) and PSD diagram (b) for point #11 under heating power of 90 W, condenser temperature of 20°C, filling ratio of 55% and ethanol as working fluid [37].

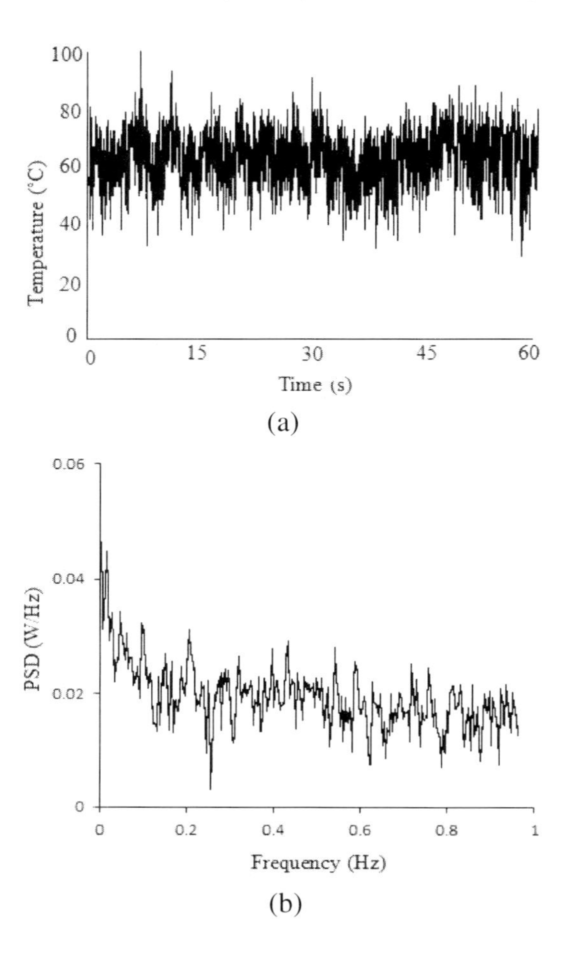

Figure 44. Time series of temperature (a) and PSD diagram (b) for point #20 under heating power of 90 W, condenser temperature of 20°C, filling ratio of 70% and water as working fluid [37].

Figure 45 shows the time series of the adiabatic wall temperature at point #25 under heating power of 90 W, condenser temperature of 20°C, filling ratio of 65% and ethanol as working fluid and its power spectrum density diagram. It is evident there is not any dominant peak in PSD diagram at all. Besides, the power spectrum density decays by frequency increment. Absence of dominant peak in PSD diagram and its decay with respect to frequency indicate the chaotic state in the system. So, the temperature time series in Figure 45(a) can strongly be classified as chaos under mentioned operating conditions.

As mentioned earlier, there are 10 points assigned for adiabatic wall temperature measurement (#1, #5, #6, #11, #15, #16, #20, #21 and #25). Numerical simulations and analytical investigations concluded that if any of these points had periodic or quasi-periodic behavior, rest of the points (in adiabatic section) had periodic or quasi-periodic behavior in terms of temperature time series. This conclusion was also applicable for chaotic behavior of the temperature time series. Then, in order to investigate the chaotic behavior in the entire PHP, only selecting one point (in adiabatic section) was sufficient.

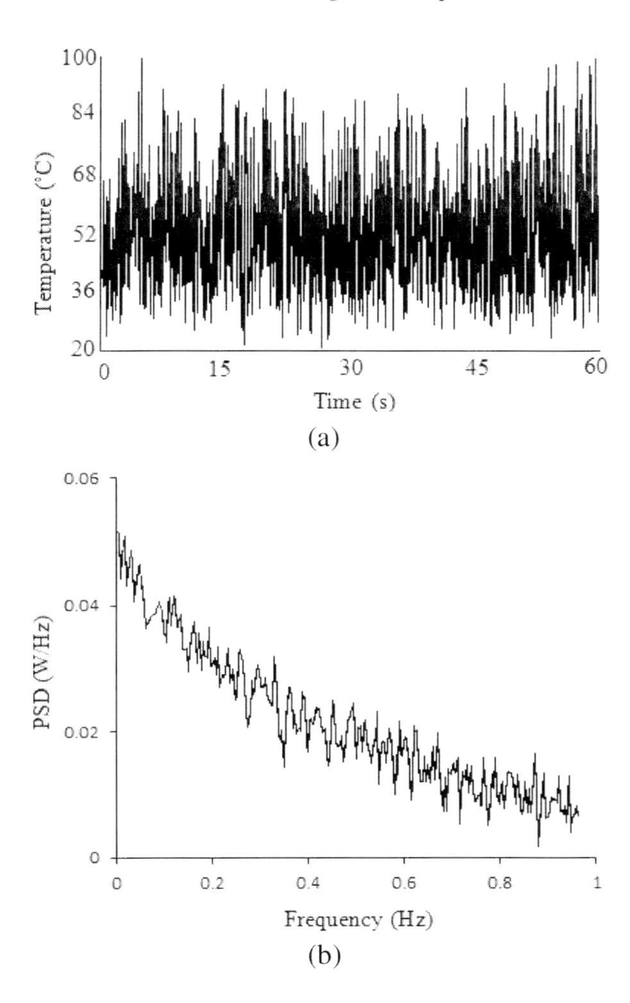

Figure 45. Time series of temperature (a) and PSD diagram (b) for point #25 under heating power of 90 W, condenser temperature of 20°C, filling ratio of 65% and ethanol as working fluid [37].

5.4. Correlation Dimension

Figure 46 illustrates the correlation dimensions under evaporator heating powers of 60 W, 80 W and 100 W with filling ratios of 45%, 55% and 75%, respectively at condenser temperature of 20°C for water and ethanol. Values of 2.674, 2.757 and 2.921 were obtained for water under mentioned operating conditions. In addition, values of 2.806, 2.922 and 3.263 were obtained for ethanol under the same operating conditions. It is evident for both water and ethanol that by increasing the evaporator heating power and filling ratio, the slope of the curves and correlation dimension increase. Besides, correlation dimension values for ethanol are higher than water under the same operating conditions.

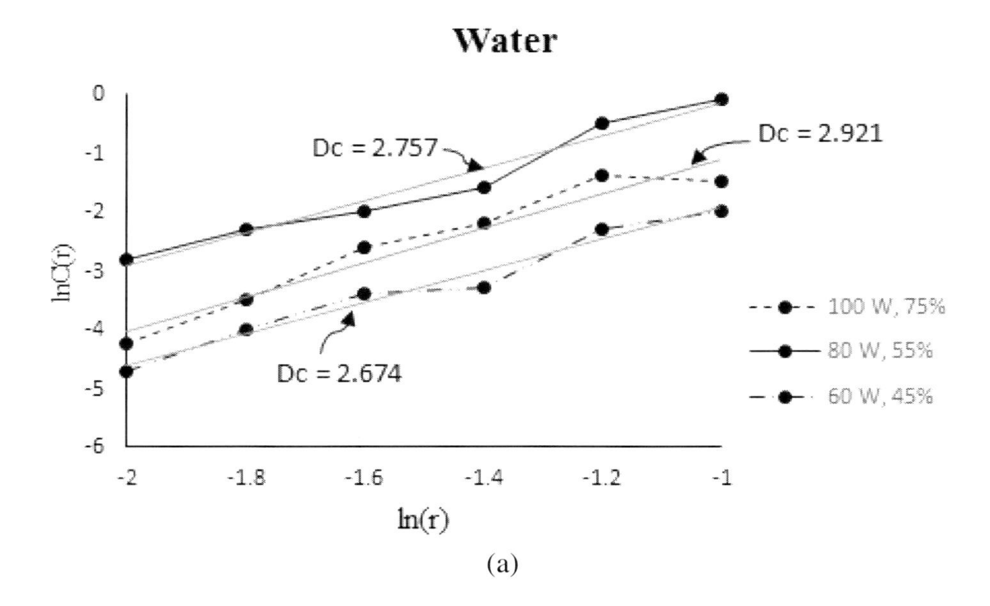

(a)

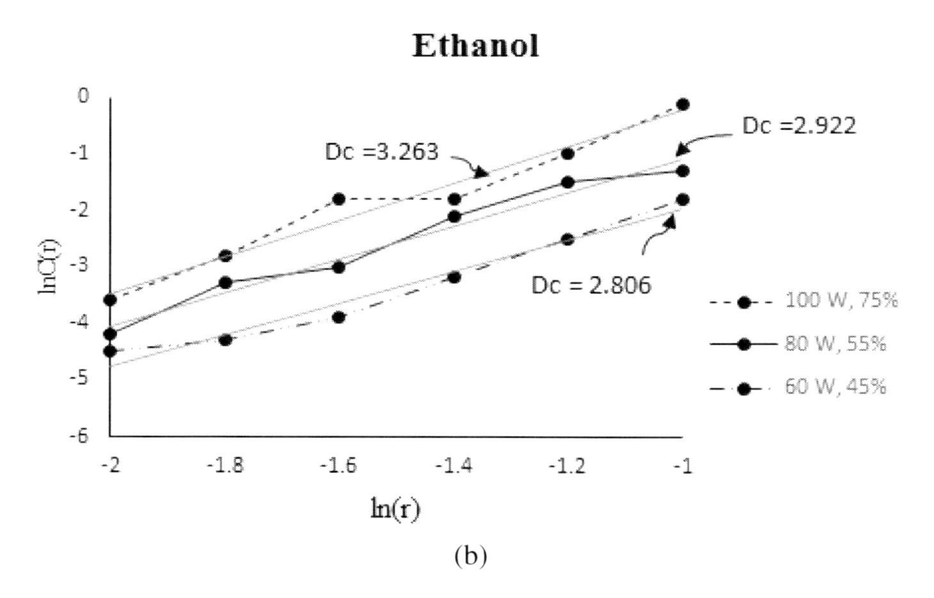

(b)

Figure 46. Correlation dimension values (Dc) with water (a) and ethanol (b) as working fluids [37].

5.5. Autocorrelation Function

Figure 47 shows that by increasing the evaporator heating power, ACF decreases for water and ethanol. In addition, linear behavior of the ACF changes gradually to exponential behavior by increasing the heating power. Although the same operating conditions were applied for the PHP with water and ethanol as working fluids, change in working fluid did not lead to any particular conclusion for ACF behavior.

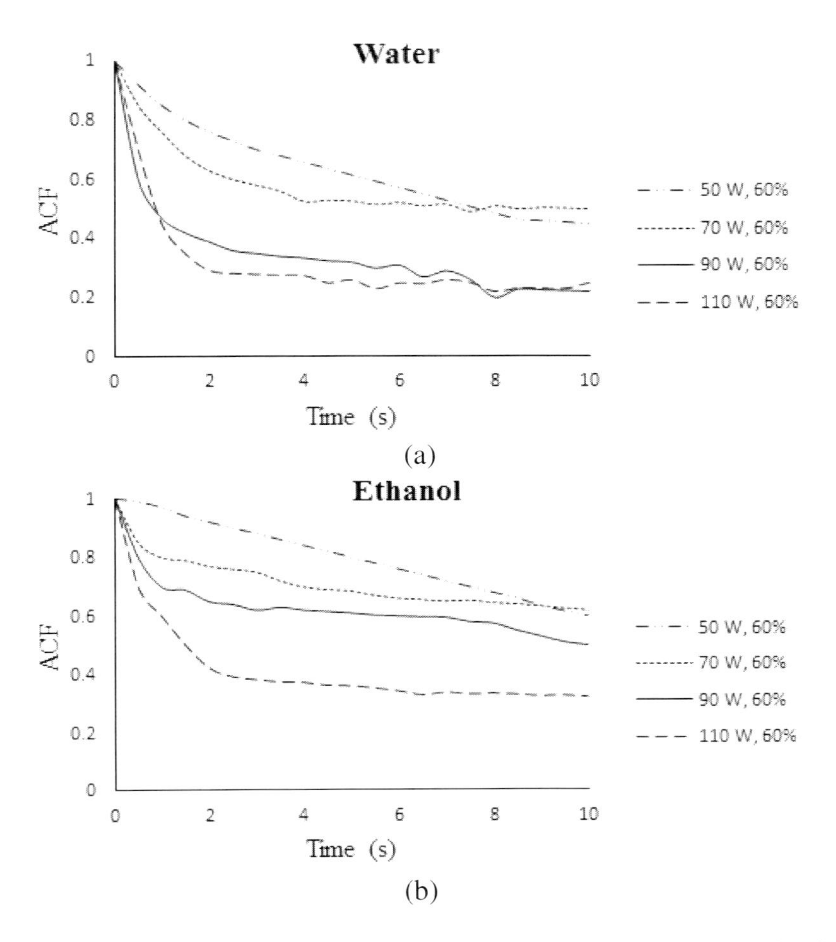

Figure 47. Autocorrelation function versus time with water (a) and ethanol (b) as working fluids [37].

5.6. Lyapunov Exponent

Figure 48 illustrates the Lyapunov exponents versus evaporator heating power at three filling ratio of 50%, 60% and 70% at condenser temperature of 20°C with water and ethanol as working fluids. Mostly, Lyapunov exponent increases by increasing heating power in Figure 48(a) with water as working fluid. Three exceptional points can be seen at heating powers of 20 W, 50 W and 60 W for filling ratios of 60%, 50% and 70%. Negative Lyapunov exponents appeared at heating power of 10 W with FR of 70%, heating power of 10 W with FR of 50% and heating power of 20 W with FR of 70% indicating PHP is not in chaotic state under these operating conditions. Lyapunov exponent increases by increasing the heating power as well for the PHP with ethanol as working fluid as shown in Figure 48(b). Exceptional points from this behavior exist under some operating conditions. Negative Lyapunov exponents appeared at heating power of 10 W with FR of 70%, heating power of 10 W with FR of 50%, heating power of 20 W with FR of 60% and heating power of 40 W with FR of 70% indicating PHP is not in chaotic state under these operating conditions. It is evident in Figure 48 that range of Lyapunov exponent value for the PHP with water and ethanol as working fluids are similar. This indicates strong dependency of Lyapunov exponent to the structure and dimensions of the PHP which is the same for all operating conditions in current study.

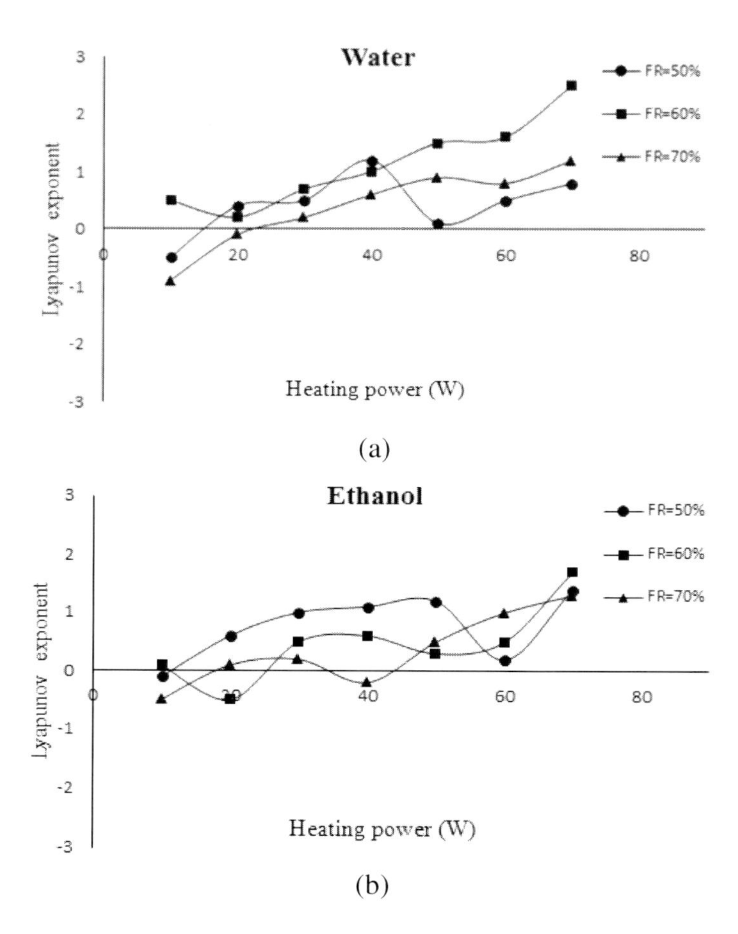

Figure 48. Lyapunov exponents versus evaporator heating power at different filling ratios with water (a) and ethanol (b) as working fluids [37].

5.7. Phase Space Reconstruction

Regarding explanations in section 2.5, the attractor reconstructions were carried out. It is evident in Figure 49(a) the attractor pattern is like an O-ring structure which is mostly for periodic or quasi-periodic systems. Then, the PHP behavior under heating power of 75 W, condenser temperature of 20°C, filling ratio of 55% and ethanol as working fluid is not in a chaotic state. Figure 49(b) illustrates the attractor pattern under heating power of 40 W, condenser temperature of 20°C, filling ratio of 60% and water as working fluid. It can be seen that the attractor pattern at this low heating power is narrow, slim and elongated, corresponding to low dimension characteristic of the temperature oscillation and chaotic properties of PHP. Figure 49(c) illustrates the attractor pattern under heating power of 80 W, condenser temperature of 20°C, filling ratio of 65% and ethanol as working fluid. At this higher heating power, the attractor pattern is more complex. In addition, it has slight wider and more scattered distribution than the attractor pattern in Figure 49(b), corresponding to a higher dimension characteristic of the temperature oscillation and chaotic properties of PHP. Figures 49(b) and 49(c) are strong evidence of deterministic chaotic systems under mentioned operating conditions.

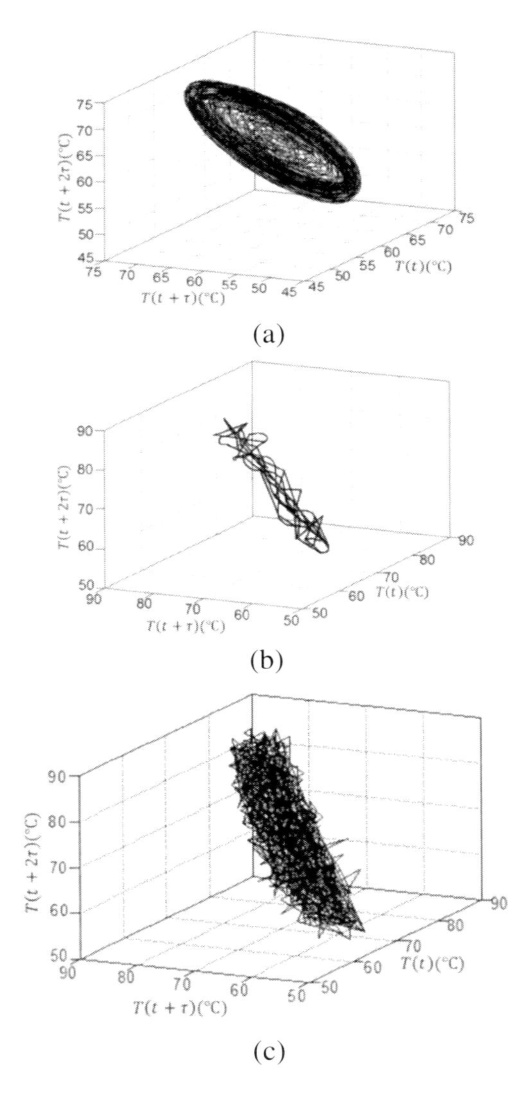

Figure 49. Reconstructed 3D attractor patterns under (a): heating power of 75 W, condenser temperature of 20°C, filling ratio of 55% and ethanol as working fluid (b): heating power of 40 W, condenser temperature of 20°C, filling ratio of 60% and water as working fluid (c): heating power of 80 W, condenser temperature of 20°C, filling ratio of 65% and ethanol as working fluid [37].

5.8. Thermal Performance

In current study, effects of filling ratio and working fluid on the thermal performance of the PHP have been investigated. Filling ratios of 35%, 45%, 55%, 65% and 75% for water and ethanol have been tested under different heating powers and condenser temperature of 20°C. For the traditional heat pipes, the filling ratio range of working fluid is lower comparing to the PHPs. Large amount of working fluid inside the channels causes liquid blockage and reduces pressure difference between heating and cooling sections which can stop the oscillation movement. But for the PHPs a filling ratio that is higher than that of traditional heat pipes is required. The overall thermal resistance of a PHP is defined as the difference average temperatures between evaporator and condenser divided by the heating power. For evaporator average temperature, 15 points (#2, #3, #4, #7, #8, #9, #12, #13, #14, #17, #18, #18, #22, #23 and #24) were used to calculate the average temperature. Since condenser has constant temperature, calculation is not necessary to get the average temperature which is equal to that constant value.

Figure 50 illustrates the thermal resistance variation with respect to evaporator heating power at different filling ratios and water as working fluid. A general behavior of thermal resistance can be seen in Figure 50. At all filling ratios, by increasing the heating power, the thermal resistance decreases initially to reach a minimum value as an optimal point. Thermal resistance increases by increasing the heating power afterwards. This optimal point is important since at related operating conditions, the pulsating heat pipe transfers the maximum heat flux with a minimum temperature difference between the evaporator and condenser leads to a better thermal performance of the PHP. The minimum thermal resistance occurs at different heating powers for different filling ratios. It is obvious in Figure 50 that thermal resistance has lower values at filling ratio of 45% comparing with other filling ratios. The minimum thermal resistance of 0.85°C/W happens at heating power of 90 W for this filling ratio.

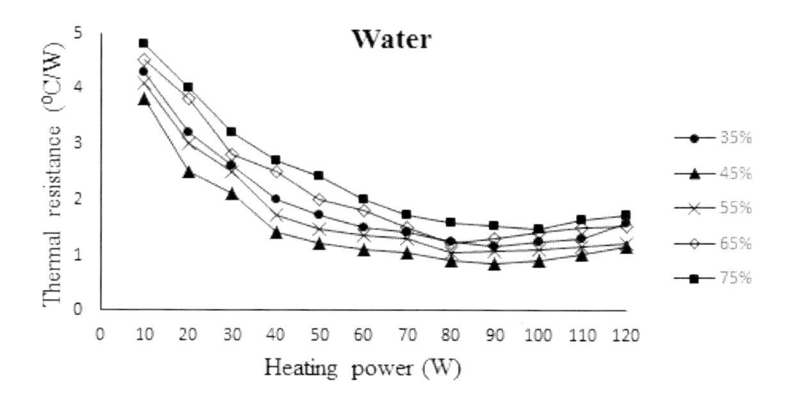

Figure 50. Thermal resistance versus heating power at different filling ratios and water as working fluid [37].

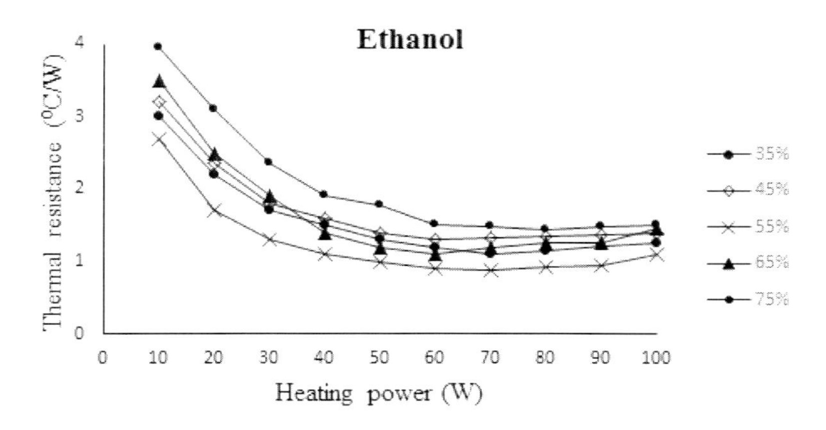

Figure 51. Thermal resistance versus heating power at different filling ratios and ethanol as working fluid [37].

As shown in Figure 51, the same conclusion of general behavior and optimal point of thermal resistance is applicable for ethanol as well. Thermal resistance has lower values at filling ratio of 55%. The minimum thermal resistance of 0.88°C/W happens at heating power of 70 W for this filling ratio.

CONCLUSION

Numerical simulations have been carried out to investigate the chaotic and thermal behaviors of simple two-dimensional, multi-turn and three-dimensional closed-loop pulsating heat pipes. Heat flux and constant temperature boundary conditions were applied for evaporator and condenser respectively. Water and ethanol were used as working fluids. Volume fractions of liquid and vapor for the working fluid were obtained under different operating conditions. Combination of two vapor plugs rarely occurred in the condenser. Vapor condensation and liquid accumulation were more visible on the bending parts of the condenser due to the surface tension. Time series of the adiabatic section temperature was found as proper criteria to study the chaotic state. It was found that the time series has complicated, irregular and aperiodic behavior in several operating conditions. Decreasing the power spectrum density of the time series with respect to frequency and absence of dominating peaks in the related curves denoted the existence of chaos in the pulsating heat pipe. Spectral analysis of temperature time series using Power Spectrum Density showed existence of dominant peak in PSD diagram indicates periodic or quasi-periodic behavior in temperature oscillations at particular frequencies. It was found for both water and ethanol as working fluids by increasing the evaporator heating power and filling ratio, correlation dimension increases. Similar range of Lyapunov exponent value for the PHP with water and ethanol as working fluids indicated strong dependency of Lyapunov exponent to the structure and dimensions of the PHP. An optimal filling ratio and minimum thermal resistance were found for better thermal performance of the pulsating heat pipe. Comparison of thermal resistance behavior versus heating power between simulation results of current study and experimental results of Shafii et al. [20] confirmed a good agreement between simulation and experimental results. An O-ring structure pattern was obtained for reconstructed 3D attractor at periodic or quasi-periodic behavior of temperature oscillations.

REFERENCES

[1] H. Akachi, *U.S. Patent No. 4921041*, 1990.

[2] T. Daimaru, S. Yoshida, H. Nagai, Study on thermal cycle in oscillating heat pipes by numerical analysis, *Applied Thermal Engineering* 113 (2017) 1219-1227.

[3] M. B. Shafii, A. Faghri, Y. Zhang, Thermal modeling of unlooped and looped pulsating heat pipes, *Journal of heat transfer* 123 (6) (2001) 1159-1172.

[4] M. B. Shafii, A. Faghri, Y. Zhang, Analysis of heat transfer in unlooped and looped pulsating heat pipes, *International Journal of Numerical Methods for Heat & Fluid Flow* 12 (5) (2002) 585-609.

[5] Y. Zhang, A. Faghri, Heat transfer in a pulsating heat pipe with open end, *International Journal of Heat and Mass Transfer* 45 (4) (2002) 755-764.

[6] Y. Zhang, A. Faghri, Oscillatory flow in pulsating heat pipes with arbitrary numbers of turns, *Journal of Thermophysics and Heat Transfer* 17 (3) (2003) 340-347.

[7] Y. Zhang, A. Faghri, Advances and unsolved issues in pulsating heat pipes, *Heat Transfer Engineering* 29 (1) (2008) 20-44.

[8] W. Shao, Y. Zhang, Thermally-induced oscillatory flow and heat transfer in an oscillating heat pipe, *Journal of Enhanced Heat Transfer* 18 (3) (2011).

[9] S. Kim, Y. Zhang, J. Choi, Entropy Generation Analysis for a Pulsating Heat Pipe, *Heat Transfer Research* 44 (1) (2013).

[10] S. Kim, Y. Zhang, J. Choi, Effects of Fluctuations of Heating and Cooling Section Temperatures on Performance of a Pulsating Heat Pipe, *Applied Thermal Engineering* 58 (1) (2013) 42-51.

[11] H. B. Ma, B. Borgmeyer, P. Cheng, Y. Zhang, Heat transport capability in an oscillating heat pipe, *Journal of Heat Transfer* 130 (8) (2008) 081501.

[12] B. Y. Tong, T. N. Wong, K. T. Ooi, Closed-loop pulsating heat pipe, *Applied Thermal Engineering* 21 (18) (2001) 1845-1862.

[13] B. Borgmeyer, H. Ma, Experimental investigation of oscillating motions in a flat plate pulsating heat pipe, *Journal of Thermophysics and Heat Tran*sfer 21 (2) (2007) 405-409.

[14] J. Qu, H.Y. Wu, P. Cheng, Thermal performance of an oscillating heat pipe with Al2O3–water nanofluids, *International Communications in Heat and Mass Transfer* 37 (2) (2010) 111-115.

[15] J. Qu, H. Wu, P. Cheng, Experimental study on thermal performance of a silicon-based Micro pulsating heat pipe, *ASME 2009 Second International Conference on Micro/Nanoscale Heat and Mass Transfer* (2009) 629-634.

[16] [16] J. Qu, H. Wu, Q. Wang, Experimental Investigation of Silicon- Based Micro-Pulsating Heat Pipe for Cooling electronics, *Nanoscale and Microscale Thermophysical Engineering* 16 (1) (2012) 37-49.

[17] S. M. Thompson, H. B. Ma, R. A. Winholtz, C. Wilson, Experimental Investigation of Miniature Three-Dimensional Flat-Plate Oscillating Heat Pipe, *Journal of Heat Transfer* 131 (4) (2009) 043210.

[18] S. M. Thompson, P. Cheng, H. B. Ma, An Experimental Investigation of a Three-Dimensional Flat-Plate Oscillating Heat Pipe With Staggered Microchannels, *International Journal of Heat and Mass Transfer* 54 (17) (2011) 3951-3959.

[19] H. B. Ma, *Oscillating heat pipes,* New York, Springer, 2015.

[20] M. B. Shafii, S. Arabnejad, Y. Saboohi, H. Jamshidi, Experimental Investigation of Pulsating Heat Pipes and a Proposed Correlation, *Heat Transfer Engineering* 31 (10) (2010) 854-861.

[21] M. Turkyilmazoglu, Anomalous Heat Transfer Enhancement by Slip Due to Nanofluids in Circular Concentric Pipes, *International Journal of Heat and Mass Transfer* 85 (2015) 609-614.

[22] M. Turkyilmazoglu, Analytical Solutions of Single and Multi-Phase Models for the Condensation of Nanofluid Film Flow and Heat Transfer, *European Journal of Mechanics-B/Fluids* 53 (2015) 272-277.

[23] M. Turkyilmazoglu, A Note on the Correspondence Between Certain Nanofluid Flows and Standard Fluid Flows, *Journal of Heat Transfer* 137 (2) (2015) 024501.

[24] H. Xian, W. Xu, Y. Zhang, X. Du, Y. Yang, Experimental Investigations of Dynamic Fluid Flow in Oscillating Heat Pipe Under Pulse Heating, *Applied Thermal Engineering* 88 (2015) 376-383.

[25] E. Jiaqiang, X. Zhao, Y. Deng, H. Zhu, Pressure Distribution and Flow Characteristics of Closed Oscillating Heat Pipe During the Starting Process at Different Vacuum Degrees, *Applied Thermal Engineering* 93 (2016) 166-173.

[26] S. Khandekar, M. Schneider, M. Groll, Mathematical modeling of pulsating heat pipes: state of the art and future challenges, in: S. K. Saha, S. P. Venkateshen, B. V. S. S. S. Prasad, S. S. Sadhal (Eds.), *Heat and Mass Transfer,* Tata McGraw-Hill Publishing Company, New Delhi, India, 2002, pp. 856–862.

[27] R. T. Dobson, Theoretical and experimentalmodelling of an open oscillatory heat pipe including gravity, *International Journal of Thermal Sciences* 43 (2) (2004) 113-119.

[28] R. T. Dobson, An open oscillatory heat pipe water pump, *Applied Thermal Engineering* 25 (4) (2005) 603-621.

[29] L. Xiao-Ping, F. Z. Cui, Modelling of phase change heat transfer system for micro-channel and chaos simulation, *Chinese Physics Letters* 25 (6) (2008) 2111.

[30] Y. Song, J. Xu, Chaotic behavior of pulsating heat pipes, *International Journal of Heat and Mass Transfer* 52 (13) (2009) 2932-2941.

[31] W. F. Louisos, D. L. Hitt, C. M. Danforth, Chaotic flow in a 2D natural convection loop with heat flux boundaries, *International Journal of Heat and Mass Transfer* 61 (2013) 565-576.

[32] E. H. Ridouane, C. M. Danforth, D. L. Hitt, A 2-D numerical study of chaotic flow in a natural convection loop, *International Journal of Heat and Mass Transfer* 53 (1) (2010) 76-84.

[33] W. F. Louisos, D. L. Hitt, C. L. Danforth, Chaotic natural convection in a toroidal thermosyphon with heat flux boundaries, *International Journal of Heat and Mass Transfer* 88 (2015) 492-507.

[34] J. Qu, H. Wu, P. Cheng, X. Wang, Non-Linear Analyses of Temperature Oscillations in a Closed-Loop Pulsating Heat Pipe, *International Journal of Heat and Mass Transfer* 52 (15) (2009) 3481-3489.

[35] S. M. Pouryoussefi, Y. Zhang, Numerical investigation of chaotic flow in a 2D closed-loop pulsating heat pipe, *Applied Thermal Engineering* 98 (2016) 617-627.

[36] S. M. Pouryoussefi, Y. Zhang, Analysis of chaotic flow in a 2D multi-turn closed-loop pulsating heat pipe, *Applied Thermal Engineering* 126 (2017) 1069-1076.

[37] S. M. Pouryoussefi, Y. Zhang, Nonlinear analysis of chaotic flow in a three-dimensional closed-loop pulsating heat pipe, *ASME Journal of Heat Transfer* 138 (12) (2016) 122003.

[38] O. Edward, *Chaos in dynamical systems,* Cambridge university press, 2002.

[39] P. Stoica, R. L. Moses, *Spectral analysis of signals,* 452, Upper Saddle River, NJ, Pearson Prentice Hall, 2005.

[40] H. Kantz, T. Schreiber, *Nonlinear time series analysis,* 7, Cambridge university press, 2004.

[41] S. H. Strogatz, *Nonlinear dynamics and chaos: with applications to physics, biology, chemistry, and engineering*, Westview Press, 2014.

[42] J. Gao, Y. Cao, W. W. Tung, J. Hu, *Multiscale analysis of complex time series: integration of chaos and random fractal theory, and beyond,* John Wiley & Sons, 2007.

[43] C. H. Skokos, G. A. Gottwald, J. Laskar, *Chaos Detection and Predictability,* 915, Springer, 2016.

In: Heat Pipes: Design, Applications and Technology ISBN: 978-1-53613-908-2
Editor: Yuwen Zhang

Chapter 6

VARIOUS SHAPED HEAT PIPE APPLICATIONS AND HSHPTM SOFTWARE

Jung-Chang Wang*
Department of Marine Engineering (DME),
National Taiwan Ocean University (NTOU),
Keelung, Taiwan, ROC

ABSTRACT

The aim of this chapter is to depict the various shaped heat pipes that work based on two-phase heat transfer devices and the HSHPTM (heat sink-heat pipe thermal module) software. There are several applications for heat pipes and HSHPTM software in structural and thermal developments as well as challenges when working with the electronic systems, desalination systems, renewable energy systems, and devices. Some of these challenges are high heat flux thermal management in smaller areas and higher power, multi-physics and electro thermal co-design of the electronics in central processing units (CPUs) and graphic processing units (GPUs). This chapter will also discuss thermal design of next-generation data centers of personal computers (PCs), Notebooks (NBs), and servers. Other applications for heat pipes and HSHPTM software are automotive light emitting diodes (LEDs) such as found in headlights and harsh environment electronic systems. Other considerations are the progress and challenges in software tools and advances in measurement and characterization. Programs and methods for some thermal module designs are introduced in this chapter. The developed complex method presented herein provides high quality and a shorter development time owing to fewer iterations of the design procedures. Based on theoretical models with empirical formulas, the author developed a computer program coded by Virtual Basic version 6.0 to develop a Windows program known as HSHPTM software for industries and researchers that specialize in

* Corresponding Author Email: jcwang@ntou.edu.tw.

thermal performance for electronic devices. These results show that the predictions made using HSHPTM software agreed with the experimental data.

Keywords: heat pipe, heat sink, thermal module, software, VB, HSHPTM, VCTM

INTRODUCTION

An extended surface such as a fin is usually added to increase the rate of heat removal for traditional cooling techniques (e.g., heat sink). Over the past few decades, significant advances in the design theory of heat sinks have been made. The heat capacity from the heat source is conducted and transferred through the heat sink to its surroundings by heat convection. Therefore, adding fins (surface areas) made from several materials including metals, plastics or special substances can help dissipate heat from the heat source. Choosing the proper software tools to be employed in structural dynamics and thermal performance designs are important. The thermal design of a heat sink-heat pipe system uses a composite approach ensuring the thermal conditions needed to provide high-quality development. Several theoretical models of thermal modules have been developed to predict their thermal performances and challenges respectively with software and algorithms. The two-phase flow heat transfer device assembly to thermal modules has become mature, and the heat pipe-based two-phase flow heat transfer module is one of the best innovations in decades [1-7]. A heat sink with embedded heat pipes transfers the total heat capacity from the heat source to the base plate with embedded heat pipes and fins sequentially, and then dissipates the heat flow into the surrounding air. A micro-channel heat pipe consists of a wick with micro capillary channels, which are partially filled with a working fluid [8-10]. The main parts of a heat pipe are evaporator and condenser. An arbitrary adiabatic section can also be present in cases where there is a distance between the heat sink and heat source. The working fluid inside a heat pipe evaporates by receiving heat in the evaporator section and is converted into liquid by heat dissipation in the condenser section; afterwards, the liquid returns to the evaporator, and this process continues to transfer heat from the evaporator to the condenser section [11-18]. The mechanism of liquid return from the condenser to evaporator depends on the wick and its capillary structure inside the heat pipe. Conventional heat pipes utilize capillary force to return liquid from the condenser to the evaporator. Capillary channels within the wick structure can be provided by sintering powder or grooved on the inner surface of heat pipes. This type of heat pipe is less sensitive to orientation and can work efficiently in the absence of gravity force. Another type of two-phase heat transfer device, which is more compact in size and geometry, is the thermosyphon or pulsating heat pipe (PHP), and vapor chamber (VC) [19-29]. PHPs are made of a capillary tube and have several bends. Pressure instabilities inside the tube, which are due to heat input to various turns, are the main reason for the fluid motion. In thermosyphons, condensed liquid in the condenser returns to the evaporator by gravity assistance. Thermosyphons are very efficient in heat transfer; however, they are very sensitive to gravity and orientation. The vapor chamber (VC) with its internal vapor-liquid interaction has a better thermal performance than metallic material in a large footprint heat sink. The overall operating principle of the VC is defined as follows: At the very beginning, the interior of the VC is in the vacuum, in front of the wall face of the cavity absorbing the heat flow from its source. The working fluid in this interior will be rapidly transformed into vapor as a result of boiling, which causes the water to evaporate and fill up the whole interior of the cavity. The resultant vapor will be condensed into liquid by the cooling action caused by the convection between the fins and fan on the outer wall of the cavity. This new liquid will then reflow to the the heat source cavity along the capillary channels in the wick structure. The effectiveness and better thermal performance of VC has already been confirmed according to the up-to-date research and mass production application in the server system and VGA thermal module [30-34].

J.C. Wang and R.T. Wang [35] derived a novel formula for effective thermal conductivity through the VC by use of an intelligent dimensional analysis in combination with the thermal-performance experimental method. It discussed the values of one, two and three-dimensional effective thermal conductivity and compared them with a metallic heat spreader. When metallic materials serve as the heat spreaders, their thermal conductivities have constant values when the operating temperature variances are not large.At operating temperatures of 27°C, the thermal conductivities of pure copper and aluminum as heat spreaders are 401 W/m°C and 237 W/m°C, respectively. When the operating temperature is 127°C, the thermal conductivities of pure copper and aluminum are 393 W/m°C and 240 W/m°C, respectively. Results show that the two and three-dimensional effective thermal conductivities of VC are more than double that of the copper and aluminum heat spreaders, proving that it can effectively reduce the temperature of heat sources. The maximum heat flux of the VC is over 800,000 W/m^2, and its effective thermal conductivity will increase with input power increasing. It is deduced from the novel formula that the maximum effective thermal conductivity is above 800 W/m°C; thus, after comparing the maximum thermal conductivity with the experimental value, the calculating error is no more than ±5%. Wang et al. [36-39] introduced a thermal-performance experiment with the illumination-analysis method to explore green illumination techniques for use on light emitting diodes (LEDs) as a solid-state luminescence source application in relative light lamps with VC. The thermal performance of the LED vapor chamber-based plate is much greater than that of the LED copper- and aluminum- based plate. The results show that the experimental thermal resistance values of LED copper- and vapor chamber-based plates are 0.41°C/W and 0.38°C/W at 6 W, respectively. Moreover, the illumination of a 6-Watt LED vapor chamber-based plate is 5% larger than 6 W. Thus, in addition to having the best thermal performance above 5 Watts, the luminance of the LED vapor chamber-based plate is the highest. Wang et al. [40] utilized experimental analysis with the

Windows program, VCTM V1.0, to investigate the thermal performance of the vapor chamber and apply it to 30-Watt high-power LEDs. Results show that the maximum effective thermal conductivity was 870 W/m°C. Comparing this thermal conductivity with the experimental value, the calculating error is no more than ±5%. Furthermore, the LED vapor chamber-based plate solves the hot-spot problem of 30-W high-power LEDs (HI-LEDs), successfully. This Microsoft Windows program, VCTM V1.0, was used to conduct a heat flow-lighting performance test to evaluate and analyze the results of exceeding 100 watts in a HI-LED (higher-power LED) projector (with a 230 W LED projection lamp). In addition, a Windows-based optimization program was developed to optimize the heat sink of the HI-LEDs project in [41] and [42]. The results indicated that the maximal LED temperature of the projector was 108°C at a steady temperature state (after approximately 1 h of operation). When the metal-core printed circuit board (MCPCB) of the projector was replaced with a vapor-chamber PCB (VCPCB), the LED temperature decreased to 87.6°C. Iterative calculations were subsequently performed using the self-developed Windows-based optimization program for determining the optimal fin interval (8.94 mm) and thickness (2.56 mm). These optimal parameters were simulated using CFD (computational fluid dynamics), and the results were compared with those of the original 150-W HI-LED projector. A Rayleigh number of 6.1304×10^5 was obtained, indicating a laminar flow model. When the optimized heat sink was used, the LED temperatures for the MCPCB and VCPCB HI-LEDs projector samples decreased by 14 and 9.74°C, respectively. The high value of the optimized HI-LEDs projector (8.094 W/m^{2}k) was higher than that of the original HI-LEDs projector (7.154 W/m^{2}k), and its weight was reduced by approximately 2%. And for the 230 W LED projection lamp cooling module in the natural convection, modification was made to the fin parameters including fin spacing, height and thickness and LED base plates materials to achieve optimal heat dissipation and performance through experimental and simulation analysis. Results show that the best values of the cooling module are a 1.1 mm aluminum fin thickness, 7.5 mm pitch, and 31 mm height, which can effectively reduce the LED junction temperature under 75°C at a 90° incline angle. The impact of the incline angle of the LED vapor chamber-based plate to the thermal performance of the present cooling module had to be assessed before the VC cooling system could be used to cool a HI-LED system. Notably, the experimental results were in good agreement with the theoretical results, with a calculating error of not more than ±10%. A novel comparative process of a 230 W LED projection light was developed. Wang [43] analyzed the performance of a novel LED-MGVC (light emitting diodes-micro-generator vapor chamber) device using experimental and illumination-analysis methods with VCTM V1.0. Energy-efficient, small and lightweight HI-LEDs were combined with a thermo-generation module (TGM) to transform the heat power generated by the LED into electric energy. Variation in the dielectric copper and solder layer thickness in the printed circuit board (PCB) composite was found to affect the thermal performance of the HI-LEDs lighting system, and a VC was shown to provide excellent heat dissipation performance when used with HI-LEDs. Therefore, VC and PCB (VCPCB) were combined for integration with the HI-LEDs package system, i.e., the micro-generator with LED vapor chamber-based plate (LED-MGVC) for performance and illumination comparison. This study results depict that the LED-MGVC system can provide significant improvement for thermal performance and illumination and thermoelectric properties. Thermal performance of the thermal module with the VC can be determined within several seconds by using VCTM V1.0.

Various parameters affect thermal performance and heat transfer capability of the VC and heat pipes. Orientation, structure of VC and heat pipes, filling ratio, material of the tube, heat input and working fluid are among the most influential factors in the thermal performance of VC and heat pipes [44-51]. Thermophysical properties of working fluids significantly affect thermal performance of various types of heat pipes. Working fluid selection depends on the application and working conditions. For instance, for cryogenic heat transfer purposes, it is necessary to use working fluids with a very low boiling temperature; while for high-temperature applications, fluids with a high boiling point are more appropriate since the possibility of dry-out exists for fluids which do not have a high boiling temperature. Fluids with a higher thermal conductivity are more appropriate for enhancement of heat transfer in heat pipes. Wang et al. [52-54] have experimentally investigated the thermal resistance of a aluminium heat sink with two and four horizontally embedded U-shaped heat pipes of 6 mm diameter; they showed that two heat pipes embedded in the base plate can carry 36 percent of the total dissipated heat capacity from the central process unit (CPU), while the remaining 64% of the heat is delivered from the base plate to the fins. Furthermore, when the CPU power was 140 W, the total thermal resistance was at its minimum of 0.27°C/W. Moreover, four embedded heat pipes can carry 48 percent of the total dissipated heat capacity from the CPU; the total thermal resistance is under 0.24°C/W. The total thermal resistance of the heat sink with embedded heat pipes is only affected by changes in the base-to-heat pipes' thermal resistance and the heat pipes' thermal resistance over the heat flow path. In other words, the total thermal resistance varies according to the functionality of the heat pipes. If the temperature of the heat source is not allowed to exceed 70°C, the total heating powers of the heat sink with two and four embedded heat pipes will not exceed 131 W and 164 W respectively. Notably, the thermal performance of a heat sink with embedded heat pipes was codified in a Windows program for rapid calculation, i.e., the Microsoft® Visual Basic™ 6.0 commercial software. This software was used to create the author's HSHPTM (heat sink-heat pipe thermal module) software. The superposition principal analytical method for thermal performance of a heat sink with embedded heat pipes was completely established [55]. Moreover, one set of risers of the L-shape heat pipes were functioning as the evaporating section while the other set acted as the condensing section. This study utilizes a versatile superposition method with thermal resistance network analysis to design and experiment on a thermal module with six embedded L-shaped or two U-shaped heat pipes and plate fins under different fan speeds and heat source areas. This type of heat pipes-heat sink module successively transfers heat capacity from a heat source to the heat pipes and to the heat sink and their surroundings.The module's embedded heat pipes and plate fins are suitable for cooling electronic systems via a forced convection mechanism. The thermal resistances contain all major components from the thermal interface through the heat pipes and fins. Thermal performance testing shows that the lowest thermal resistances of the representative L- and U-shaped heat pipes-heat sink thermal modules are respectively 0.25 and 0.17°C/W under twin fans of 3,000 RPM and 30 × 30 mm^2 heat sources. The results of this work is a useful thermal management method to facilitate rapid analysis. Wang et al. [56, 57] arranged vertically the six L-type heat pipes in such a way that the bottom section acts as the evaporating section and the risers act as the condensing section. [56] and [57] describe the design, modeling, and test of a heat sink with embedded L-shaped heat pipes and plate fins. This type of heat sink is particularly well suited for cooling electronic components such as microprocessors using forced convection. The mathematical model includes all major components from the thermal interface through the heat pipes and fins. It is augmented with measured values for the heat pipe thermal resistance. The Microsoft® Visual Basic™ 6.0 also uses an iterative superposition method to predict the thermal performance. The sum of the bypass heating power ratios is 14.4% for Q1 and Q2, 20.8% for Q3 and Q4, and 52% for Q5 and Q6, obtained using both experimental results and the HSHPTM software program based on MS VB6.0. Thermal performance testing shows that a representative heat sink with six heat pipes haws carried 160 W and has reached a minimum thermal resistance of 0.22°C/W. HSHPTM software predicted a thermal resistance of 0.21°C/W, which was within 5% of the measured value. Moreover, the total thermal resistance of the heat sink with six embedded L-type heat pipes was only affected by changes in the base to heat pipes thermal resistance and heat pipes thermal resistance over the heat flow path. That is, the total thermal resistance varied according to the functionality of the L-type heat pipes. Heat pipes-heat sink modules transfer heat from a heat source to the heat pipes, and then to the heat sink and out into the surrounding ambient air, which makes them suitable for cooling electronic components through a forced convection mechanism. The configuration and thermal performance of the heat sinks with inserted heat pipes are presented in this chapter. This chapter uses experimental procedures to investigate the thermal performance of two embedded U-shaped heat pipe thermal modules and six embedded L-shaped heat pipe thermal modules with different fan speeds and heat source areas. Results for a single U-shaped heat pipe are 0.04°C/W at 78.85 W, while sequential results for L-shaped heat pipes are 1.04°C/W, 2.07°C/W, 2.76°C/W, 2.19°C/W and 1.7°C/W between 34 W and 40 W. The geometry and heat transfer effects on heat pipes embedded in the heat sinks-cooling system were investigated and the results are in this chapter. In the forced convection system, configurations of both L-and U-shaped heat pipes were considered. This study adopts a versatile superposition method and least-square estimators with a thermal resistance network analysis to design and experiment with their geometry and heat transfer effects under different fan speeds and heat source areas. The result of this work is a useful thermal management method to facilitate rapid analysis and has provided a useful insight into the design of heat pipe cooling systems. The heat sink-heat pipe thermal module (HSHPTM) applied U-type or L-type heat pipes to transfer the total heat capacity from the heat source to the based plate and fins successfully, and then dissipated the heat flow into the surrounding air. The MS VB6.0-based HSHPTM V1.0 was used for the proper design of the heat sink-heat pipe thermal module. The computing core of the HSHPTM program employs the theoretical thermal resistance analytical approach with the iterative convergence in this study to obtain a numerical solution. The results show that this calculated error comparison with experimental results is within ±5%. The embedded U-type heat pipes carried 46% and 63% of the total dissipated heat capacity for one- and two-pair embedded U-type heat sink-heat pipe thermal modules, and dissipated an 87.2% heat flow for six embedded L-type heat pipes,. The HSHPTM V1.0 rapidly and capably calculated the thermal performance of a heat sink-heat pipe thermal module installed with a processor by inputting simple and lumped parameters.

HEAT PIPES OF VARIOUS SHAPES

There are some various types of heat pipes including PHP, VC, and thermosyphone as shown in Figure 1. A conventional heat pipe has three sections containing an evaporation section, adiabatic section, and condensation section.A heat pipe absorbs large amounts of latent heat through the phase change of working fluid and transfers rapid heat flow to the other side in vapor form without any fluid machinery. Thus, the heat pipe may be regarded as the passive component of a self-sufficient vacuum closed system, which includes the wick/capillary structure and working fluid. The working fluid is pure water or nanofluid with low oxygen content. At the beginning, the heat pipe is in a vacuum. After the evaporation section's wall surface absorbs the heat from the heat source, the working fluid in the interior is rapidly transformed into a high-pressure vapor, which fills the entire interior of the heat pipe and flows through the adiabatic section to the condensation section. The high temperature and pressure vapor is produced at the condensation section after the adiabatic section. During the cooling phase, vapor releases latent heat; condenses it into liquid by the cooling action, and flows back to the location of the heat source in the evaporation section because of the capillary force along the micro-capillary wick structure. The capillary force depends on the wick structure, viscosity, surface tension, and wetting ability of the working fluid. Therefore, the capillary force and the high vapor pressure are the main driving force for circulation in a conventional heat pipe. A two-phase vapor and liquid process occurs in a heat pipe, which maintains the saturated state. It is usually assumed that the temperatures of the vapor and liquid are equal at their interface.

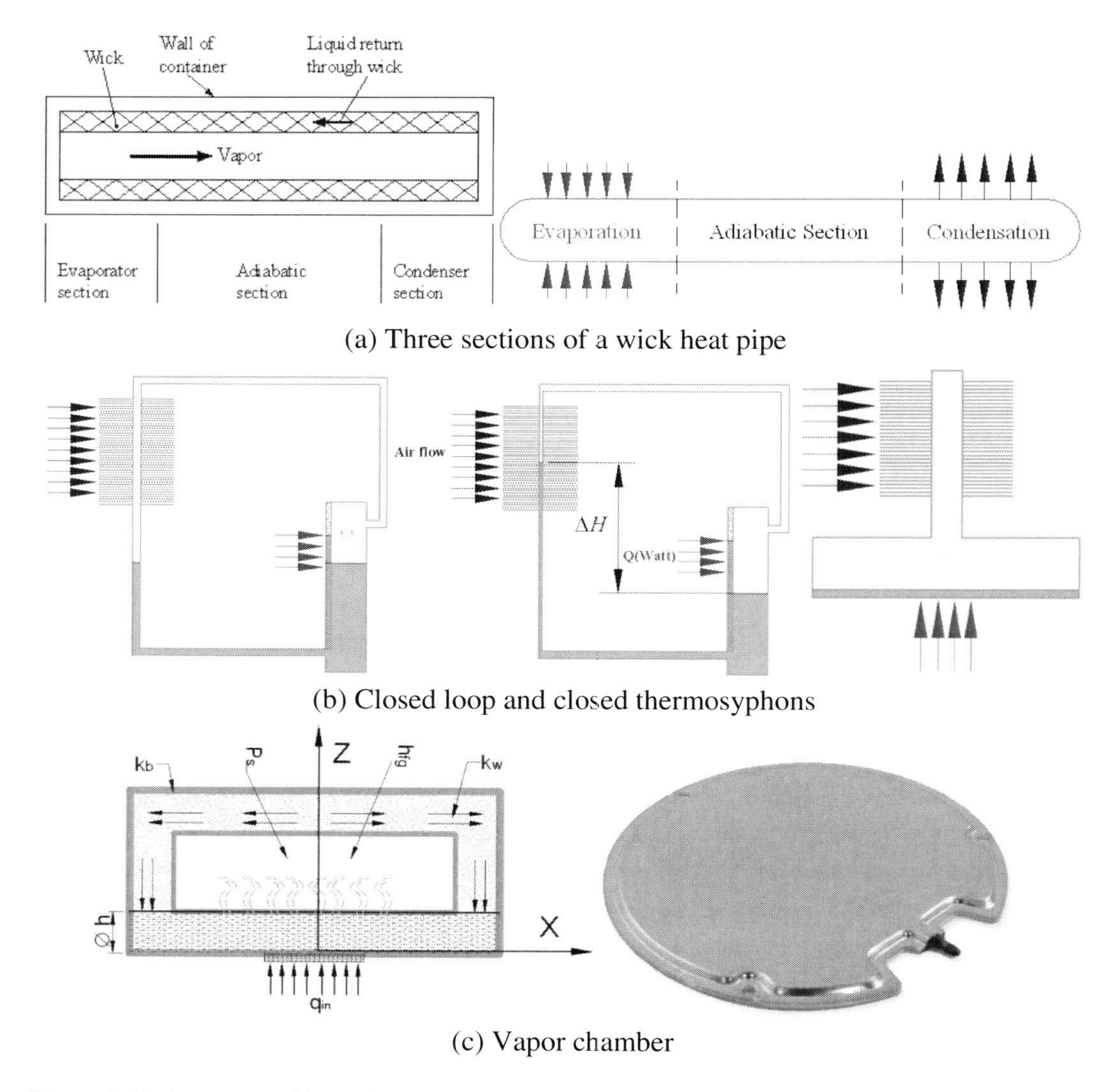

(a) Three sections of a wick heat pipe

(b) Closed loop and closed thermosyphons

(c) Vapor chamber

Figure 1. Various types of heat pipes.

The heat pipes consist of a vacuum-tight container lined with a wick with micro-capillary channels on the inner surface. These are filled with a working fluid, which exists in both liquid and vapor states within the container. Similar to wick heat pipes, various nano structure shapes can be used in screen mesh wick heat pipes. For instance, graphene sheets were dispersed in water-based fluid to obtain graphene and water nanofluid (working fluid) to be used in a miniature loop heat pipe (with a screen mesh wick). The thickness of graphene nano sheets is in the range of 1-5 nm and the concentrations are 0.003%, 0.006% and 0.009% volume fraction. Among the tested concentrations, 0.006% vol. showed the best heat transfer improvement. The working fluid absorbs heat capacity, evaporates from the heat source, and then condenses into a liquid through the condensation section to complete a thermodynamic cycle under a slight temperature difference. Much valuable research has focused on the application of nanofluids in various types of heat pipes, and the subsequent literature has clearly shown that nanofluids (working fluid) are able to significantly enhance the thermal performance of heat pipe. The improvement in thermal performance depends on several parameters such as type of nano particles, the base fluid, concentration and shape of nano particles, and working conditions. However, there are various benefits for their application in a heat pipe. A heat pipe is a uniform temperature device due to latent heat. Its thermal response time is faster than solid materials. Thus, it is quiet, reliable and is known for its longevity. Finally, a heat pipe has smaller volume, lighter weight, and higher usability. Although the heat pipe has good thermal performance when lowering the temperature of the heat source, its operating limitation is the key thermal design issue, known as the critical heat flux or greatest heat capacity quantity. This is because the temperature drop depends on the pressure drop that the vapor experiences as it flows through the vapor core. A heat pipe may bump into four kinds of heat transport limits: sonic limit, entrainment limit, capillary limit, and boiling limit. The speed of the vapor flow increases as the heat capacity of the heat source increases. At the same time, the vapor speed and pressure achieves their maximum rate at the interface between the vapor and liquid. When the internal pressure is high, the high interface pressure blocks the liquid, keeping it from reflowing to the evaporation section. The vapor temperature inside the evaporation section will increase while the liquid is trapped in the condensation section. However, when the driving forces, gravity and capillary force of the liquid are large enough to overcome the shear stress in the interface, the liquid is able to reflow to the evaporation section. Thus, the entrainment limit occurs when the heat capacity is increased and the vapor's flow speed is higher than the gateway value, forcing it to bear the shear stress in the liquid, since the driving force of vapor is larger than the surface tension of the liquid in the wick's microcapillary channel structure and interface. This phenomenon will lead to the entrainment of the liquid, affecting the flow back to the evaporation section. This is also called flooding phenomenon.

When flooding happens, the vapor temperature and pressure oscillates with time. The occurrence of the flooding phenomenon limits the performance of a heat pipe. In this case, we are considering the closed thermosyphon. As soon as the reflowing of liquid recovers, the vapor temperature drops. This phenomenon is similar to the flux of the constant mass flow rate during shrinking and expanding in the nozzle neck. Therefore, the speed of flow in this area is unable to arrive above the speed of sound. This is called the sonic limit. When the sonic phenomenon happens, flow choking phenomena may occur and cause the axial temperature to drop, thereby decreasing the thermal performance of the heat pipe. The boiling limit often takes place for the traditional metal in the heat pipe's wick structured capillaries, which means the initiation of bubble generation begins inside the capillaries, and it may result in a localized burn out if the bubbles are trapped inside the wick. If the vapor flow rate increases in the evaporation section, the working fluid between the wick and the contact surface of wall will achieve the saturated temperature of vapor in order to produce boiling bubbles. This kind of wick structured capillaries will prevent the vapor bubbles from leaving thereby encapsulating the vapor layer of the film. The boiling limit phenomenon causes a large thermal resistance resulting in a high temperatures in the heat pipe. The capillary limit is also called the water power limit, which is used in the heat pipe of the low temperature operation. Specific wick structured capillaries provide certain limitations for working fluid in circulation; this monitoring helps maintain a pressure balance between the wick and vapor core within the heat pipe including protection of the maximum capillary pumping pressure $\Delta P_{cap,max}$ (Pa), the vapor pressure drop along the heat pipe ΔP_{vl} (Pa), the pressure drop due to the evaporation and condensation sections at the liquid-vapor interface ΔP_{ll} (Pa), the pressure drop of the liquid flow in a wick structure due to the frictional drag ΔP_{wick} (Pa), and the pressure drop in the liquid die to the effect of the gravitational force in the direction of the heat pipe axis ΔP_g (Pa) defined as equation (1).

$$\Delta P_{cap,\max} = \frac{2\sigma}{r_c} \geq \Delta P_{vl} + \Delta P_{ll} + \Delta P_{wick} + \Delta P_g \quad (1)$$

where σ is the surface tension (N/m) and r_c is effective capillary radius (m).

The capillary limit means that the loading heat capacity is larger than the capability of capillary pressure rise to drive the working fluid from the condensation section to evaporation section. The maximum heat transfer attainable in a heat pipe is achieved under conditions where the capillary pressure head is greater than or equal to the sum of pressure losses along the vapor-liquid path as shown in Eq. (1). Once the heat overload exceeds the capability of the capillary pressure rise, the supply of working fluid will not be enough, and the evaporation section will dry out. The failure normally happens at the far end of the evaporation section first, drying out the wick from the evaporation section towards the condensation section. The index is the dry-out point for failure causing a reduction in the heat transfer of the evaporation section in order to stop functioning. A failure mechanism in which the capillary pressure is no longer sufficient to feed the evaporation section, or in other words, the mass flow rate of the vapor leaving the evaporation section exceeds the mass flow rate of liquid feeding the evaporation section from the condensation section, causing the evaporation section to dry out. The dry out point leads to a large and rapid reduction in the ability to transfer heat capacity into the evaporation section. The high thermal resistance for lower heat loads results from the lower merit number of water at temperatures lower than its boiling point, whereas the high thermal resistance at higher powers come from compromising the capillary limit. The selection of a heat pipe also has some important considerations besides the above four operating limits.

Table 1. Operating temperature of working fluid

Types	Melting Point (°C; 1atm)	Boling Point (°C; 1atm)	Range (°C)	Compatible Materials
Water	0	100	30~200	Cu、SUS、Ni、Si
NH_3	-77.6	-33.2	-60~100	Al、SUS、Ni、Ti
Methanol	-98	64.7	10~130	Cu、SUS、Ni、Fe、Si
Acetone	-93.1	56.3	0~120	Al、SUS、Cu、Si

Table 2. Operating pressure of working fluid

Operating temperature	0°C	100°C	Types
dP/dT	0	2.5	Water
	18	180	NH_3
	0	10	Methanol
	0	5	Ethanol
	1	16	Acetone

Table 3. Merit number of working fluid

Operating temperature	0°C	100°C	Types
$M\times10^3$	0	5.2	Water
	0.8	0.2	NH_3
	0.4	0.6	Methanol
$M = \frac{\rho_l \sigma_l h_{fg}}{\mu_l}\left(\frac{KW}{m^2}\right)$	0.2	0.5	Ethanol
	0.3	0.4	Acetone

Table 4. Capillary parameter of working fluid

Operating temperature	0°C	100°C	Types
	7.6	5.8	Water
$C_H\times10^{-5}$	4.0	1.0	NH_3
$C_H = \sigma/\rho$	3.0	2.5	Methanol
(m^3/sec^2)	3.0	2.5	Ethanol
	3.5	2.5	Acetone

Tables 1 to 4 present the major choice points of a heat pipe for the best working substance selection. First, the working environment will usually have either high or low temperature conditions which will require a compatible high or low temperature operating heat pipe. After confirming the operating environment, the material, the condition of the internal sintered body, and the type of working fluid for a heat pipe are determined. Therefore, in order to prevent the heat pipe's expiration, the consideration of the selection of a heat pipe is very important. Table 1 reveals the operating temperature of the working fluid for compatible materials. Table 2 shows the operating pressure drop of working fluid in a heat pipe. Table 3 exhibits the merit number of the working fluid. Table 4 demonstrates the capillary parameter of the working fluid.

Nanofulids are not one of the working fluids mentioned in Tables 1-4; however, nanofluids are often used as working fluids in heat pipes. Thus, the most important nanofluid requirement is that there must be an optimum concentration for nano particles. High concentrations cause unfavorable effect on thermal performance. The size and shape of nano particles are influential factors, and their effect on the thermal performance of heat pipes is based on their thermophysical properties of. Improvement in thermal performance of heat pipes by applying nanofluids are mainly attributed to their higher thermal conductivity compared with pure fluids and their increase in nucleation sites. Using a nanofluid hybrid can lead to higher improvement in thermal performance. The enhancement in thermal performance of heat pipes by using nanofluids depends on working condition (such as filling ratio and inclination angle) and the structure of the heat pipes.

Heat Sink-Heat Pipe Thermal Module (HSHPTM) Software

A heat sink-heat pipe thermal module can transfer the total heat capacity from the heat source to the based plate with embedded heat pipes and fins sequentially, and then dissipate heat flow into the surrounding air. This HSHPTM been successfully applied in PCs, servers, CPUs, GPUs, LEDs and specific consumer-electronic products. For a traditional aluminum heat sink fan, its primary function is to draw the heat load away from the heat sources that affect the performance and safety of electronic devices. By increasing the fan speed, it can quickly take away more heat capacity from the heat source, drawing it into its larger surface area and thereby transfer cooler air into its fin-like aluminum structure. This improves the heat transfer by pulling the hot air away from the electrical heat generated by computers, and increases the thermal efficiency of the thermal module. However, a noise problem will occur and the performance can be unfavorable. The difference between heat sink-heat pipe thermal module and traditional heat sink thermal modules is that the largest heat capacity of the heat source first transfers to the evaporation section of the heat pipe, so that the working fluid within the vapor of the heat pipe produces steam. Steam releases heat in the condensation process and recondenses the steam into liquid back into the evaporator using the capillary force of the wick's capillary structure to complete this thermodynamic cycle. The rest of the heat capacity is taken away from the heat sink through the forced convection of the fan. Therefore, the heat-sink heat pipes thermal module has better thermal performance at the same rotational speed of the fan with lower noise. This chapter utilizes Microsoft® Visual Basic™ 6.0 (MSVB6.0) commercial software with theoretical thermal resistances and the iterant method to develop the MSVB6.0 based HSHPTM version 1.0 for rapid calculating to accurately measure the thermal performance of the heat sink-heat pipe thermal module through inputting simple parameters. The longitudinally-finned heat sink-heat pipe thermal modules studied in this article are shown in the Figure 2. The embedded heat pipes can have an either L- or U-shaped geometry. One end of the heat pipes are inserted into the base plate as the evaporation section and the other end protrudes through the upper portion of the fins as the condensation section. The proportion of the total heat capacity carried away via the embedded heat pipes has to first be determined by iteration and theoretical thermal resistances. These input parameters have dimensions of a heat sink, heat pipes, a heat source and fan. The computing core of this MS Windows program employs the theoretical thermal resistance analytical approach with iterative convergence. This method's purpose (as stated in this article) is to obtain numerical solutions. And these individual thermal resistances of contact, base plate, and base to heat pipes, heat pipes, fins, and the total thermal resistance can be obtained through this MSVB6.0-based HSHPTM V1.0 presented in this chapter.

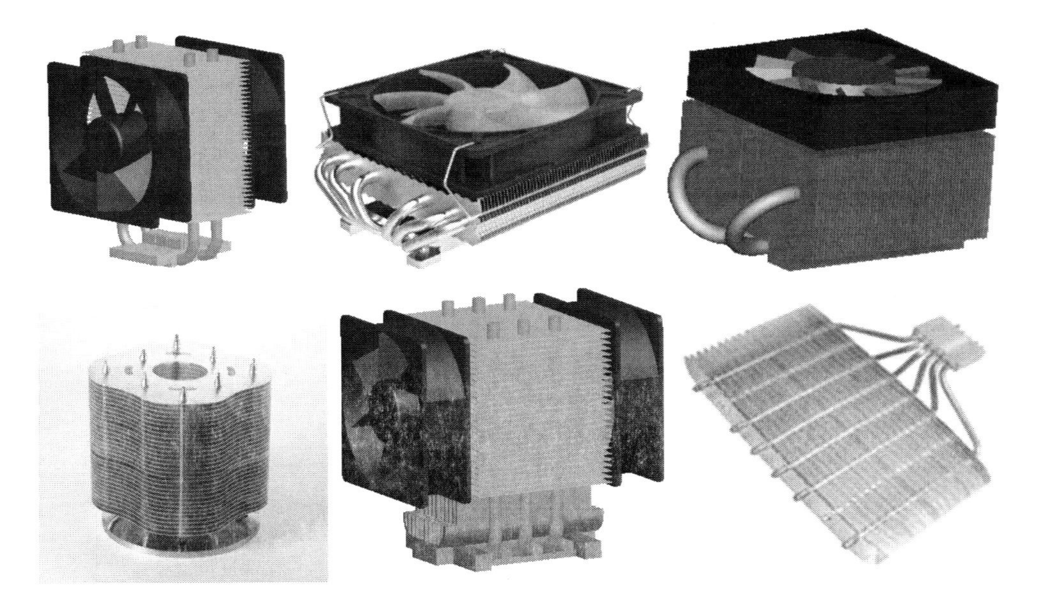

Figure 2. Various heat sink-heat pipe thermal modules.

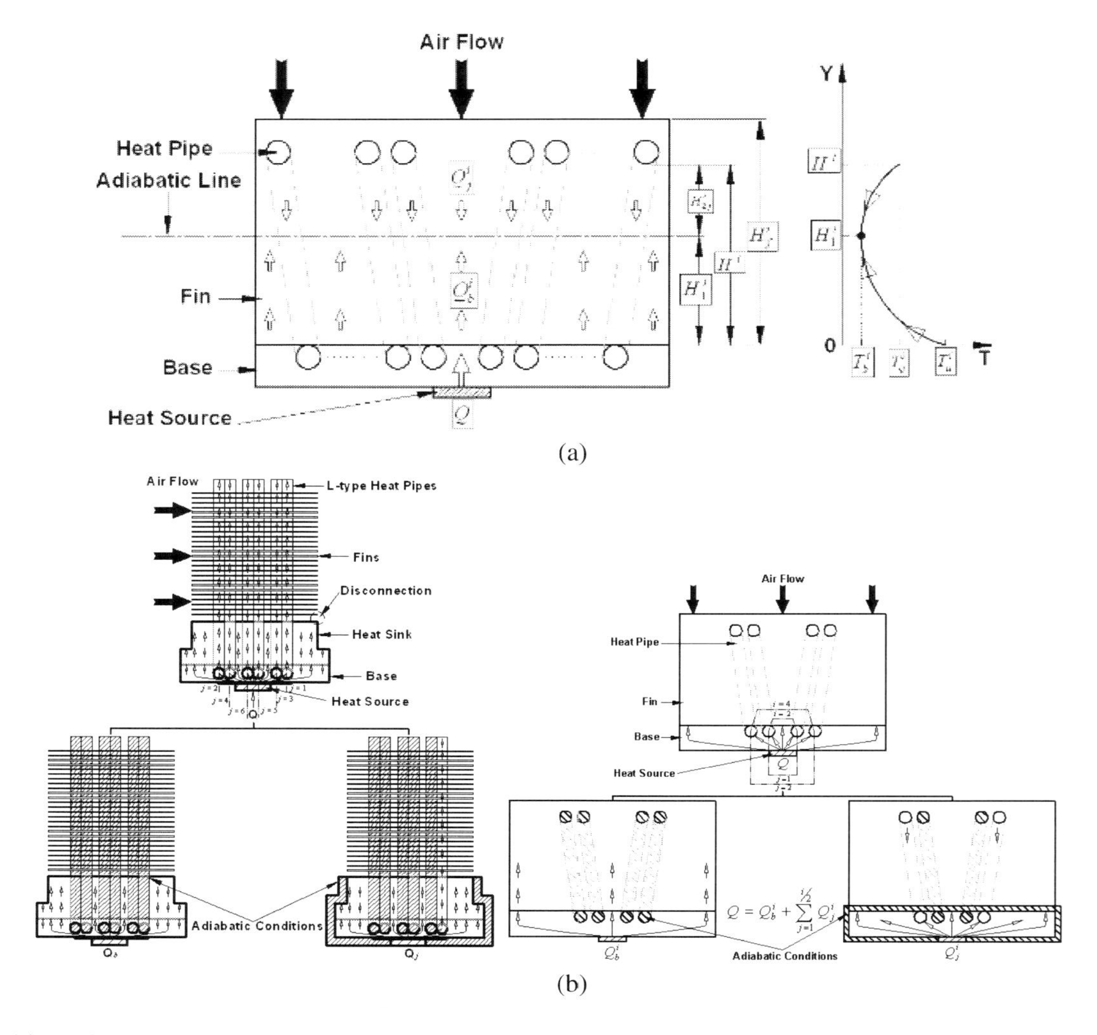

Figure 3. Heat sink-heat pipe thermal module heat flow and superposition method.

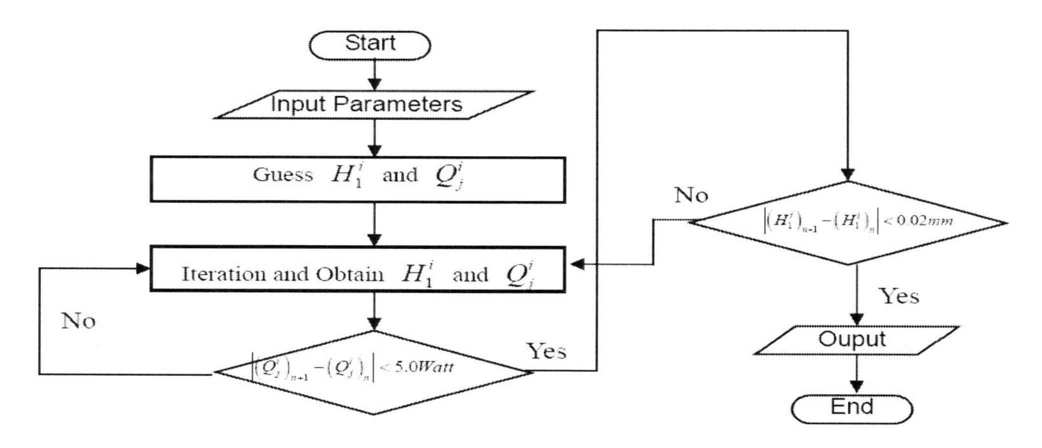

Figure 4. Iterative flowchart of program.

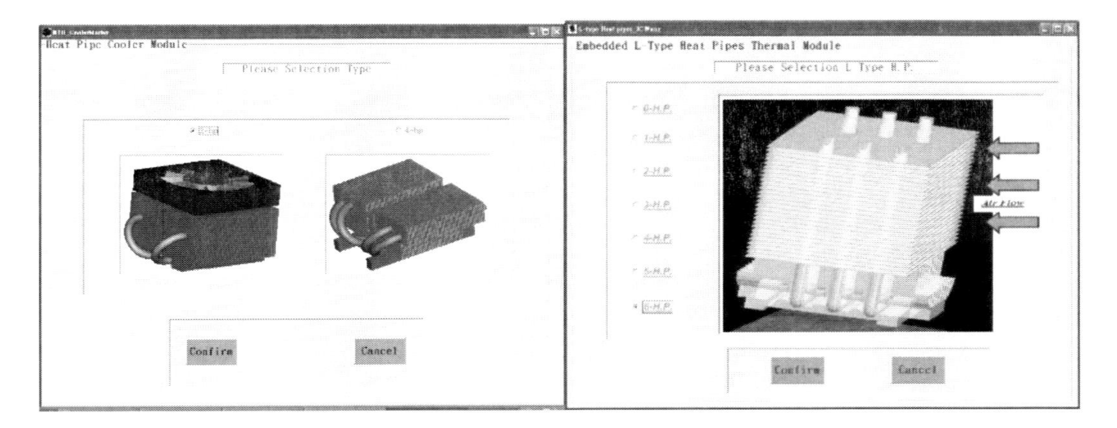

Figure 5. Selection heat sink type window.

This Windows program was coded with Microsoft® Visual Basic™ 6.0 and was used to calculate the thermal performance of a heat sink-heat pipe thermal module in the present chapter. From the energy conservation theorem, the total heating power Q equals the sum of the base plate heat capacity Q_b^i and embedded heat pipes heat capacity Q_j^i. The ratio of Q_b^i and Q_j^i depends on theses thermal resistances. The symbols i and j respectively denote the number and position of the embedded heat pipes. Figure 3 shows that the total heating power equals the sum of the base plate and embedded heat pipes heat capacities from the superposition method. The region where temperature gradient equals to zero appearing between these two heat transfer pathways is called the adiabatic line. The fins in this adiabatic position will not have heat transfer. The height of the fin is H_f^i. The distance from the upper surface of the base plate to the condensation section of the heat pipes is H^i , and those from the adiabatic position to the upper surface of the base plate and to the condensation section of the heat pipes are H_1^i and H_{2j}^i respectively. T_b^i is adiabatic temperature, T_{cj}^i is the temperature of the condensation section of the heat pipes, and T_u^i is the temperature of the upper surface of the base plate. Figure 4 shows the flowchart of the Windows MSVB6.0-based HSHPTM V1.0. Initially, these H_1^i and Q_j^i were the guessing values (estimated—not firm) and gave one percent of the H_f^i and Q. This program estimates numerical values including individual thermal resistances via the guessing values, inputs simple parameters and the ratio of Q_b^i and Q_j^i. Secondary, according to these data, inversely calculating next step values of the H_1^i and Q_j^i. Repeating these steps until satisfying conditions where the absolutions of the $\left[(Q_j^i)_{n+1}-(Q_j^i)_n\right]$ and $\left[(H_1^i)_{n+1}-(H_1^i)_n\right]$ will yield values smaller than 5.0 W and 0.02 mm respectively. Finally, these numerical values including the individual thermal resistance and temperature of each heat sink-heat pipe thermal module and heat capacity of the base plate and embedded heat pipes are obtained through the HSHPTM v1.0 program.

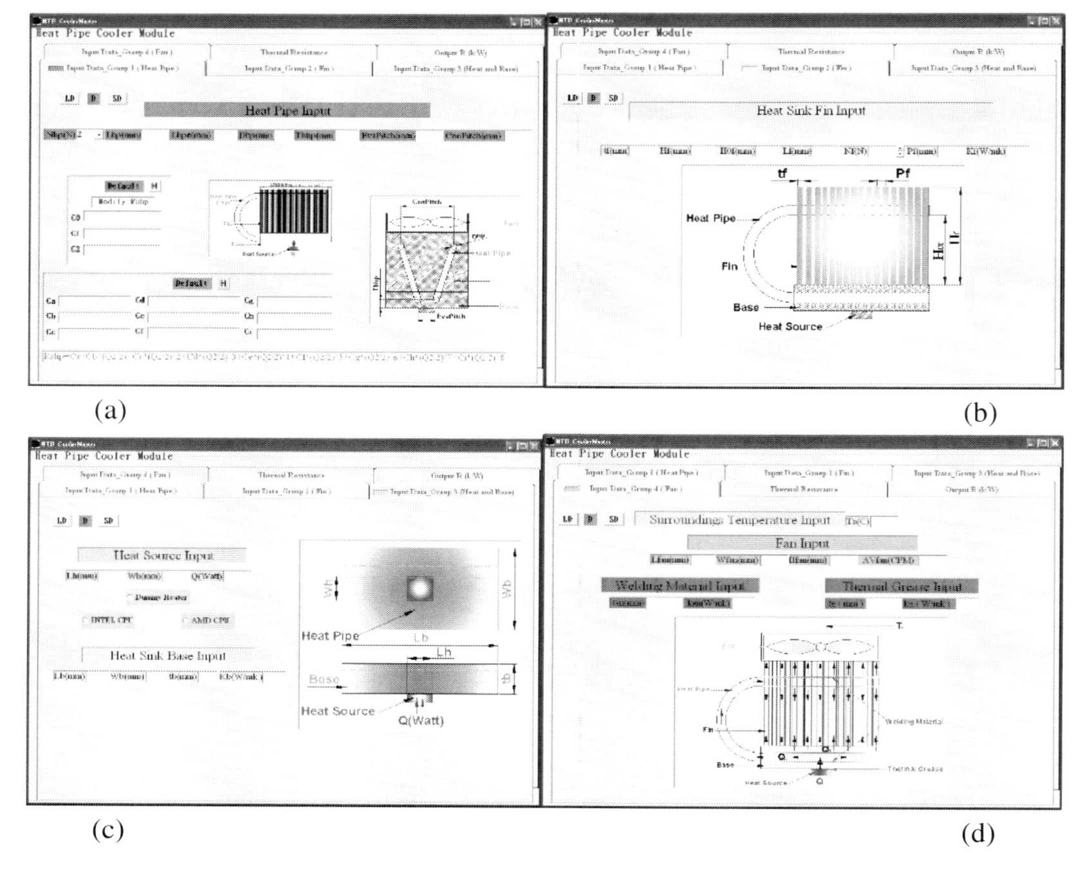

Figure 6. Input parameters before operation (a) Dimensions, position and thermal performance of embedded heat pipes, (b) dimensions and materials of heat sink, (c) dimensions, input power, and materials of heat source and base plate, and (d) performance and materials of fan and welding grease.

The preceding calculations given to determine the thermal performance of a heat sink-heat pipe thermal module must also have the correct parameters, since they affect its thermal performance including the dimensions, thermal performance and position of embedded heat pipes. Thus, it is very important for the optimum parameters to be selected to receive the best thermal performance of the heat sink-heat pipe thermal module. The program contains two main windows. The first is the selection window adjusted in the program to serve as the main menu as shown in Figure 5. In this window, the type of heat sink-heat pipe thermal module can be chosen separately. The second window has five main sub-windows. There are four sub-windows of the input parameters for the thermal module as shown in Figure 6. The first sub-window is the simple parameters of heat pipes including dimensions and thermal performance as shown in Figure 6(a). Figure 6(b) shows the second sub-window involving detailed dimensions of a heat sink. The third and fourth sub-windows are the simple parameters containing the input power of the heat source, materials of thermal grease, and the solder and performance curve of the fan as shown in Figures 6(c) and 6(d). All the input parameters required for this study of the Windows program were given and the Windows MSVB6.0-based HSHPTM V1.0 starts calculating once they are entered. Later, the program examines the situation by pressing a calculated icon. The fifth sub-window is the window showing the simulation results. In this sub-window, when the calculate icon is pressed to analyze the thermal performance of a heat sink with embedded heat pipes, we can see a figure as it is shown in Figure 7. According to the necessary iterative calculation and analog method, the heat capacity Q_j^{i} transferred by embedded heat pipes is found. According to these calculations of the heat sink-heat pipe thermal module, the simulation operation can be done through the HSHPTM V1.0.

The original program codes of HSHPTM V1.0 software exhibited with a description in theHSHPTM Software Applications section.

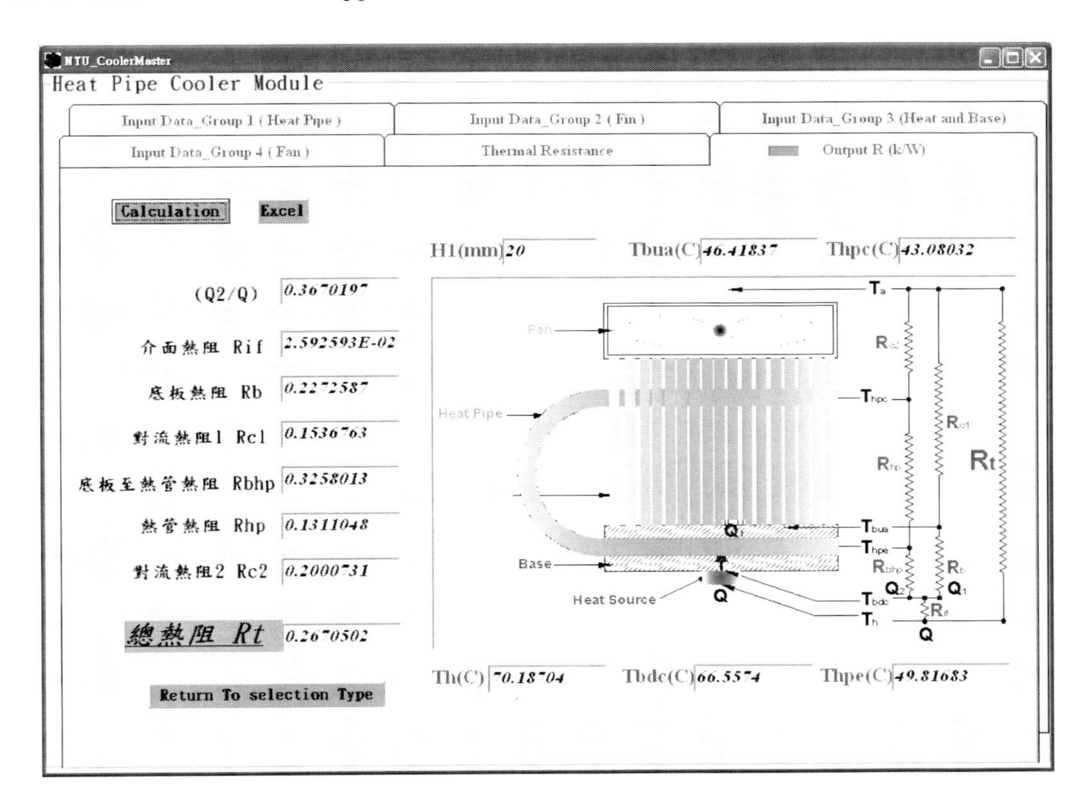

Figure 7. Main menu of program after operation.

HSHPTM Software Applications

This section compares the thermal performance between heat pipe (HP) and vapor chamber (VC). The goal of this simulation is to compare the thermal performance of the embedded heat pipe-heat sink and the vapor chamber heat sink utilizing the Windows programs of the VCEK_ML V1.exe and BaseResistance_ML.exe coding by Prof. Jung-Chang Wang to estimate the thermal performance for MicroLoops Company VC and its thermal module. The equal thermal conductivity was about 730 W/mk. Figure 8 shows the CFD models of HP and VC based on commercial designs. ANSYS Icepak commercial electronic heat transfer analysis software developed by American Fluent Inc. was adopted in this study. The dimensions of the heat sink (aluminum) are about 160 x 140 x 63 mm^3. The input power of the heat source is 100 W and 10 x10 mm^2. Free convection, laminar flow and +Y gravity (9.865m/s^2) are simulation conditions. The VC is 56 x 56 x 3 mm^3. Finally, from the results, the VC is better than HP as shown in Figure 9. The total thermal resistance may be under 0.4°C/W at an ambient temperature of 40°C. Suggesting 70 x 70 x 3 mm^3, VC has a better thermal performance than that of 56 x 56 x 3 mm^3 VC, and the temperature can be under 80°C. The performances of 70 x 70 x 3 mm^3 VC can improve above 5%.

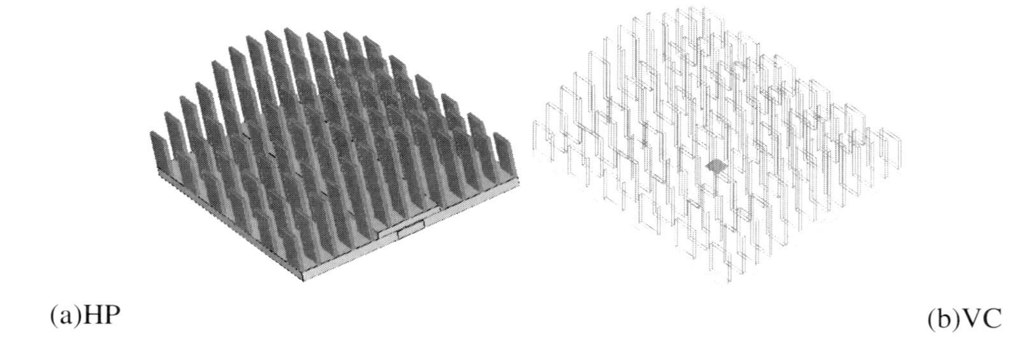

(a)HP (b)VC

Figure 8. CFD models of HP and VC Heat Sinks.

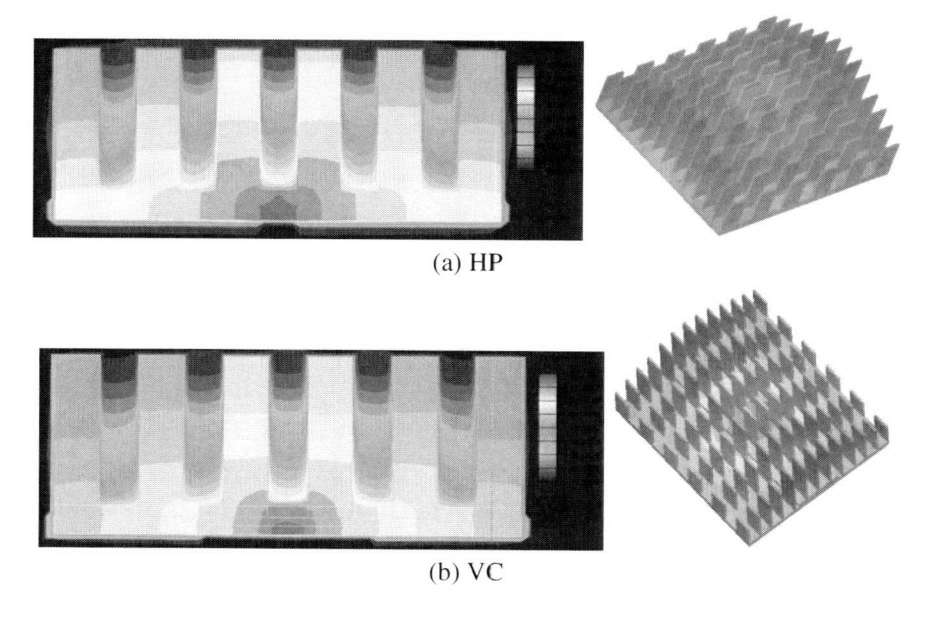

(a) HP

(b) VC

Figure 9. CFD results of temperatures.

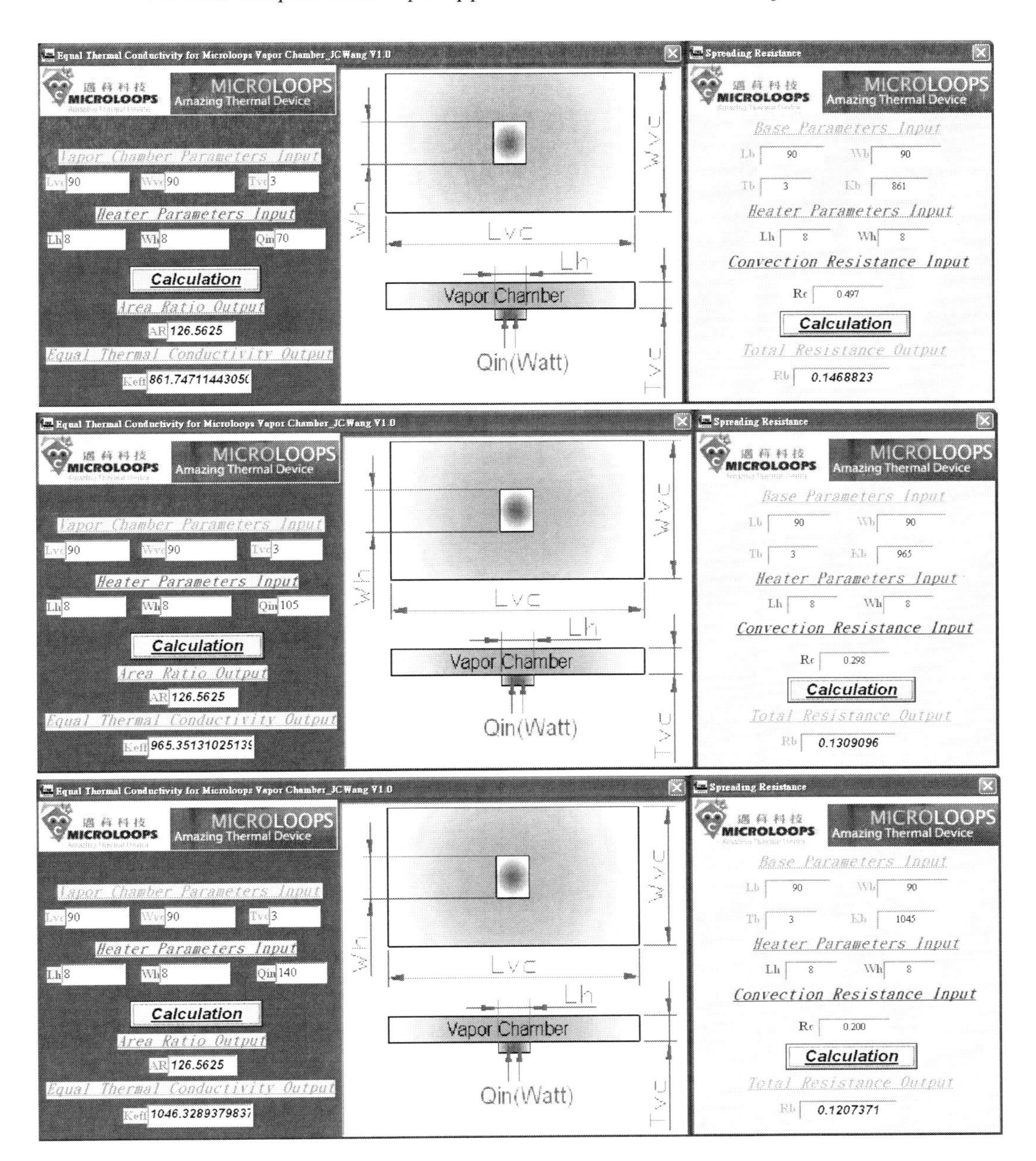

Figure 10. VCEK_ML V1.exe and BaseResistance_ML.exe coded by Prof. Wang (NTOU).

The Windows programs of the "VCEK_ML V1.exe" and "BaseResistance_ML.exe" coded by Prof. Jung-Chang Wang (NTOU) are shown in Figure 10. The equal thermal conductivity (K_{vc}) of the VC for 16_LEDs (4 x 4 die heaters is 70 W; the die area is 1 mm x 1 mm, so the total area is 8 mm x 8 mm) or about 861 W/mk. For the 105 W 4 x 4 die heaters, the area of a die is 1 mm x 1 mm, so the total area is 8 mm x 8 mm. K_{vc} is about 965 W/mk. For the 140 W 4 x 4 die heaters, the area of the die is 1 mm x 1 mm, so the total area is 8 mm x 8 mm; thus, the K_{vc} is about 1046 W/mk. A hi-end VGA card coordinates the dimensions of the dual-slot protocol control information (PCI) card. Thus, a dual-slot side-cooling thermal module for the VGA card was chose for a better thermal performance. The three dimensions of the heat sink thermal module are all 118 x 83 x 32 mm^3. The thermal design parameters are the bare-die GPU type of 18 x 16 mm^2 with a thermal input power of 165 W, and a total thermal resistance of under 0.28°C/W. Ambient inlet temperature T_a to fan assumes 45°C, and the performance curve of the blower was the maximum air volume flow of 30 cubic feet per minute (CFM) and a static pressure of 360 Pa. The size of the centrifugal fan is 90 x 25 mm^3, and the rotational speed is 3000 RPM. The reference flow rate will be 15 CFM to assist the optimum fin efficiency. The mean air velocity is about 3.5 m/s. The Reynolds number is about 10^4. The mean convection coefficient is about 35 W/m^2°C. The pressure drop can be considered as Eq. (2), where f_{app} is the fanning-friction factor, K_c and K_e are contraction and expansion pressure loss coefficients, V_a is the air velocity in the channel, and L_f and D_h are the fin length and hydraulic diameter of the fin, respectively. Equations (3) and (4) reveal the estimation of fin thickness and optimum fin gap for the rectangle fin at force convection. Studies have shown that the error involved in the one-dimensional fin analysis is negligible (less than about 1%). For optimum fin efficiency, C_s obtains 0.01. The fin material is aluminum with a required thickness of under 0.295 mm. This study selected 0.3 mm as the fin thickness. C_o is 0.03 and the optimum fin gap is about 1.63 mm. Therefore, the fin gap range may be between 1.3 and 2.1 mm with a step of 0.2 mm.

$$\Delta P = \left[\left(4 f_{app} \cdot \frac{L_f}{D_h} \right) + K_c + K_e \right] \cdot \left(\frac{1}{2} \cdot \rho_a \cdot \overline{V}_a^2 \right) \tag{2}$$

$$t_f \le C_s \cdot \left(\overline{h} / 5 \cdot K_f \right) \tag{3}$$

$$S_o \cong C_o \cdot \left(t_f \cdot H_f \right)^{0.25} \tag{4}$$

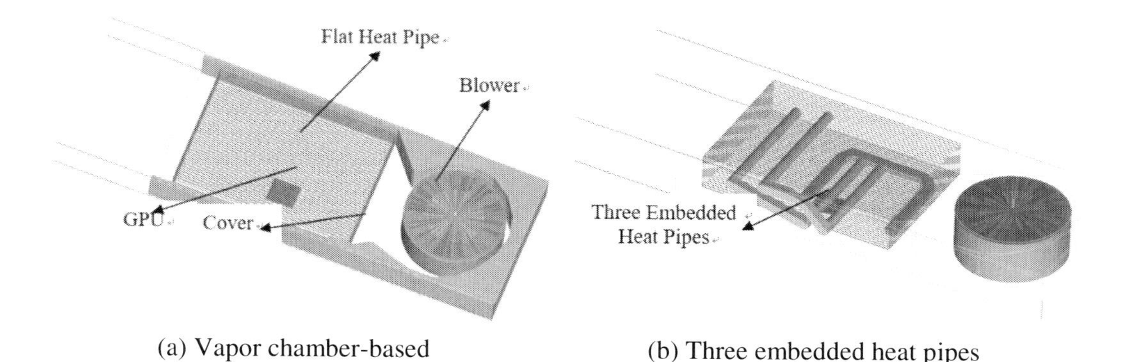

(a) Vapor chamber-based (b) Three embedded heat pipes

Figure 11. VGA Thermal module CFD models.

The entire analytical model was established by utilizing file conversion skill between CAD and CFD. There are two types of thermal modules. The overall size of a vapor chamber is 70 x 70 x 3 mm^3. A comparison of the vapor chamber-based thermal module is made with the embedded heat pipes thermal module as shown in Figure 11. The 3-D embedded heat pipes thermal module is composed of a copper plate that is 50 x 50 x 3 mm^3 with an embedded 6-mm diameter U-shaped heat pipe and two 6-mm diameter bending heat pipes cross-connecting to fins. The interior of the plastic housing for these thermal modules has a structure with a throat. A throat functions to force the air flow toward the other side side under a pressurized condition. The conditions of all CFD models containing fins, heat sink, cover and blower are the same except for the base plate. The numerical simulation analysis can be divided into preprocessing, numerical solving and post-processing. With regard to preprocessing, first of all, a geometrical model is established for the 3D VGA card thermal module. Generally, in order to reduce the computation grid elements and time taken for simulation and solving, some minor characteristics without influence or with a little influence will be ignored when establishing the 3-D geometrical model. As for the input boundary conditions and thermo-physical properties, the ambient temperature is set to 45°C, the turbulent model is the k-ε two-equations, the grid pattern is a nonstructural one, and the entire simulation analysis type is steady state. For the entire module, about 2,200 grid elements were used, with about 1000 iterations and an approximate duration of 36 hours to simulate every scenario.

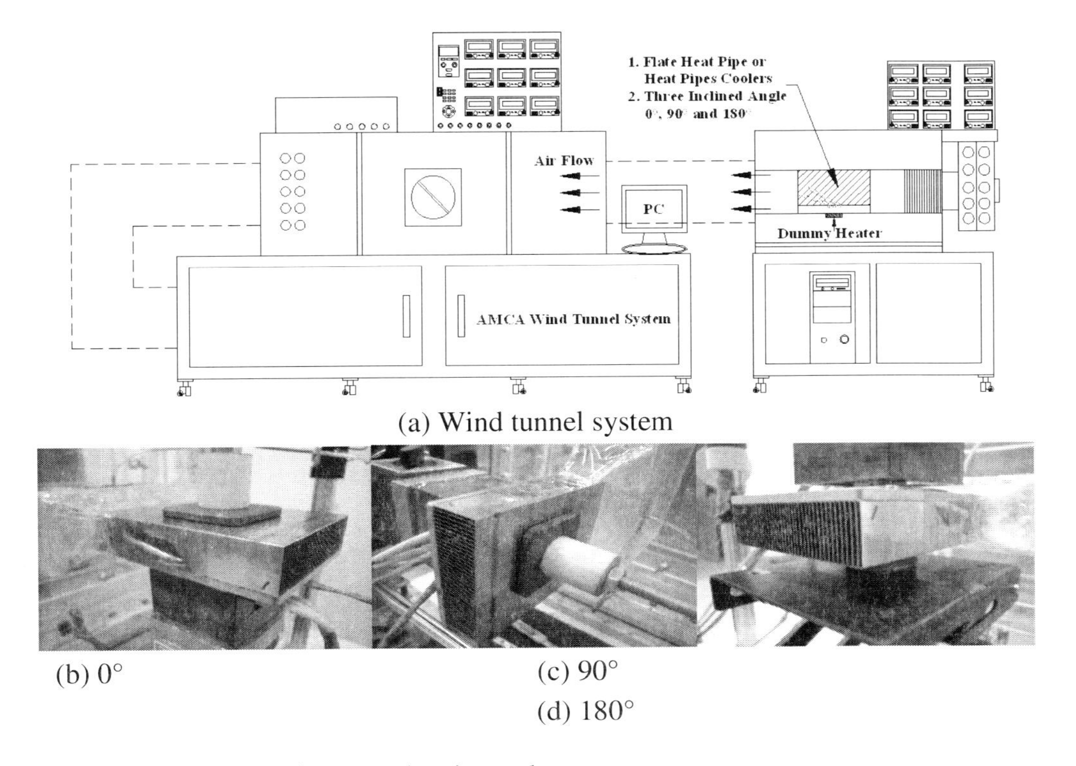

(a) Wind tunnel system

(b) 0°

(c) 90°

(d) 180°

Figure 12. Experimental equipment and testing angles.

Figure 12 is the apparatus used to test the thermal performance experiment in this study. The experimental values are obtained from the AMCA wind tunnel testing system of actual mass-production goods. The apparatus had a flat heat pipe with three embedded heat pipe thermal modules, which applied to a hi-end VGA card as shown in Figure 12(a). The AMCA wind tunnel system follows AMCA 210 standard manufactured by Taiwan LonGwin Corp., which its overall dimensions are 850 × 3600 × 1800 mm^3. The maximum operating flow rate and static pressure are 250 CFM and 200 mm-H_2O, respectively. The accuracy is within ±2% and the repeatability error is ±1%. Three types of inclined angles were used set at 0° (horizontality), 90° (verticality) and 180° (anti-gravity) as shown in Figures 12(b) to 12(d). The thermal design parameters are the bare-die 18 × 16 mm^2 GPU type. Thus, the dummy heat source for simulation GPU is a heated 36.5 mm × 25.5 mm × 11 mm copper block, and the area of heat source is 18 mm × 16 mm. Two electrical heating tubes are provided a copper block, and a T-type thermocouple is embedded to measure the temperature of the dummy heater (T_h). Another thermocouple measures the ambient temperature (T_a). When the temperature was in the range from −30 to 180°C, the measurement error of the T-type thermocouple was about ±0.2°C. A pressure of about 9 to 11 Kg/cm^2 is applied upon the fixture fixing thermal module with a simulated heat source during the experiment to control and maintain the thermal contact resistance of the system. Three kinds of input power are 150, 165 and 180 Watts in this experiment. The thermal performance experiments were conducted based on the results obtained from a vapor chamber-based thermal module to demonstrate the difference between the thermal performance of that and one of embedded three heat pipes thermal module. The procedures of the experiment introduced in the study are as follows: Step 1. Set up and calibrate the equipment of the AMCA wind tunnel testing system. Step 2. Apply thermal grease on the surface of the dummy heater uniformly. Step 3. Fix the thermal module on the fixtures. Step 4. Turn on the wind tunnel system. Step 5. Adjust the pressure applied upon the fixture to stabilize thermal contact resistance. The thermal contact resistance can be minimized by applying thermal grease, i.e., silicon oil, and soft metallic foil, i.e., tin, silver, copper, nickel, or aluminum. Step 6. Adjust the voltage and current of the power supply unit until the heating power of the simulated heat source is 150 W. Step 7. Record the steady-state temperature value through the wind tunnel system. Step 8. Change the inclination angles to 90° and 180° in order. Step 9. Repeat Step 6 to adjust the heating power to 165 W and 180 W in that order.

Figure 13 shows the simulated results of the curve fitting, and the fitting equation is shown as Eq. (5). The symbol S means space. The error has about 2% between the numerical data of VCTM V1.0 and yjr simulated data of Icepak. Also, the 1.9 mm fin gap had a better thermal performance than the 1.7 mm fin gap based on the numerical data from VCTM V1.0. Therefore, for a vapor chamber-based thermal module solution, considering the cost and performance, a 0.3 mm Al fin with a 1.9 mm fin gap is recommended. In this condition, there are 37-pcs fin counts, which saves around 12% of the fin cost. Therefore, the experiment chose the optimum fin design with a 1.9 mm gap and 37 counts to manufacture the prototype of the vapor chamber-based thermal module. The vapor chamber-based and embedded three heat pipe thermal modules were analyzed and designed to optimization in a hi-end VGA card cooling system without changing the fan performance through CFD simulation with a thermal resistance analysis and experiment in this study. And the VC-based thermal module can achieve the optimum heat dissipation and the maximum heat flux may exceed 60 W/cm^2. From the experimental results, when Al fins are 29 mm in height, 0.3 mm in thickness, 37 in number and spaced out 1.9 mm apart, the optimum total thermal resistance of a VC-based thermal module is 0.23°C/W in this study. Through numerical and experimental results, the thermal performance of a vapor chamber-based thermal module has above 8% improvement compared to that of an embedded three-heat pipe thermal module. The optimization process is done and checked by VCTM V1.0 and HSHPTM V1.0 Windows software featured in this chapter. Finally, the simulation analytical results and experimental data show that the error of the total thermal resistance value is within 5%.

$$T_h = 227 - 219 \times S + 114 \times S^2 - 20 \times S^3 \quad (5)$$

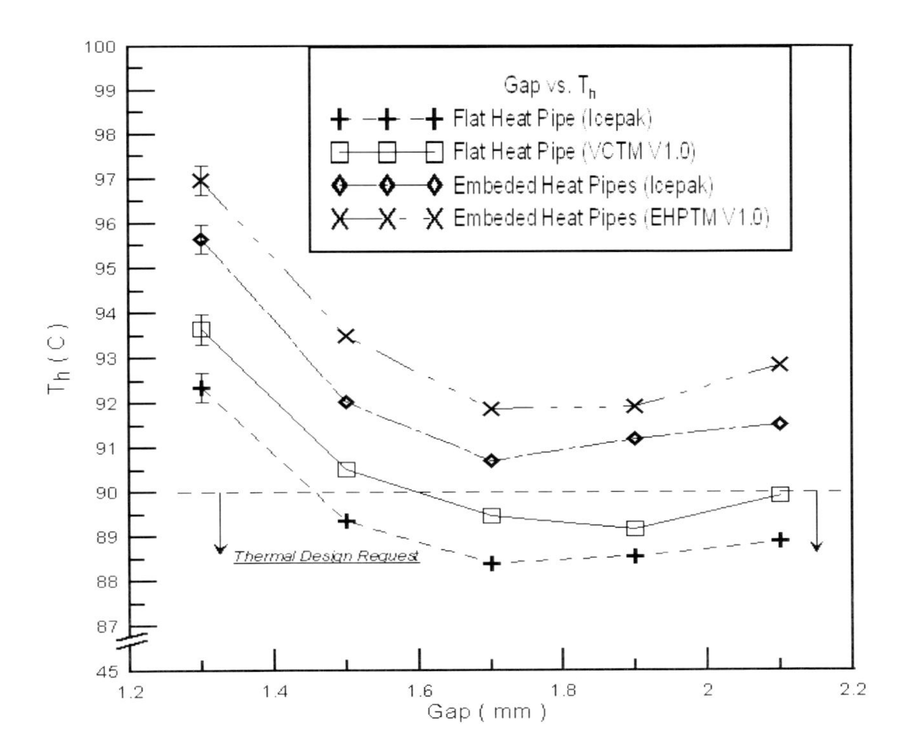

Figure 13. Relationships of T_h and space for VC-based and embedded HP thermal modules.

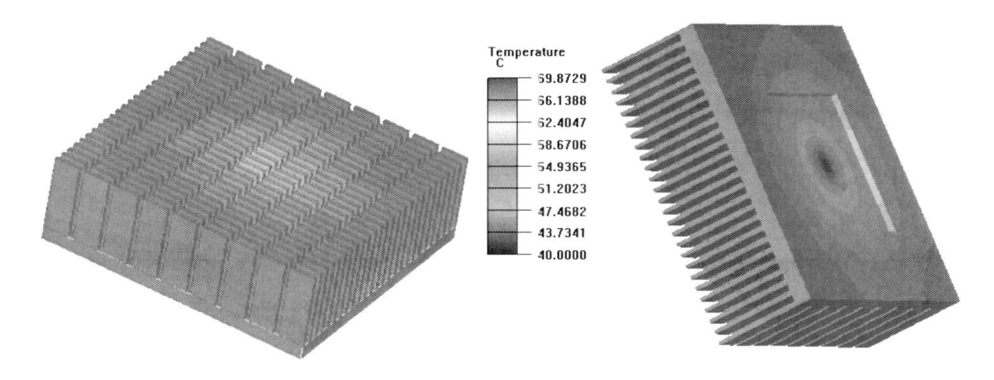

Figure 14. The better heat sink with VC thermal module design.

Another application is the utilization of the "VCEK_ML V1.exe" and "BaseResistance_ML.exe" coding by Prof. Wang to estimate the thermal performance for MicroLoops Corp. VC and its thermal module to obtain a better designed heat sink simulation of 140 x 150 x 50 mm^3. The input power of the heat source is 100 W and 10 x 10 mm^2. Free convection, laminar flow and +Y gravity (9.865 m/s^2) are simulation conditions. The vapor chamber is 70 x 7 0 x 3 mm^3. The equal thermal conductivity is about 970 W/mk. We selected a cross-cut heat sink (aluminum) as a better mechanical manufacturing extrusion process. First, the base plate thickness of 8 mm was selected as the better thickness. Second, five scenarios of different fin thicknesses were compared including 1.2 mm, 2.4 mm, 3.6 mm, 4.8 mm and 5.6 mm fins at the same 21 counts. Third, five scenarios of different fin counts were compared including 21, 30, 42, 51 and 72 counts at the same 1.8 mm fin thickness. Finally, the better heat sink was determined as having an 8-mm base plate, a 1.8-mm fin thickness and a 26 fin count as shown in Figure 14. The total thermal resistance may be under 0.4°C/W at an ambient temperature of 40°C.

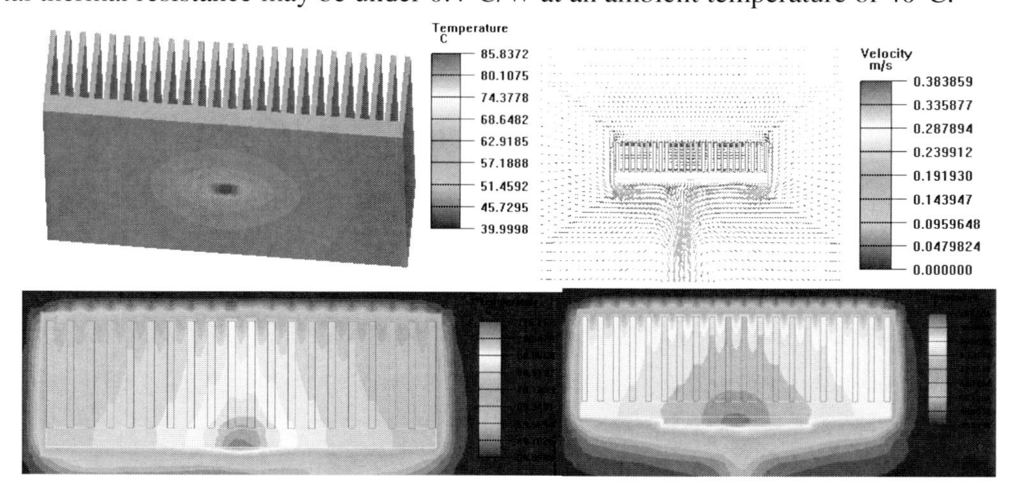

Figure 15. Temperatures and velocity of three cases.

Table 5. CFD results of three cases

	T_a (°C)	T_h (°C)	R_t (°C/W)	R_c (°C/W)
Case 1	40	85.84	0.458	0.191
Case 2	40	118.26	0.783	0.473
Case 3	40	104.51	0.645	0.473

Finally, there are three cases including 200 x 200 x 50 mm^3, 120 x 120 x 40 mm^3 with and without a vapor chamber of 56 x 56 x 3 mm^3 to estimate the thermal performance for MicroLoops VC and its thermal module. The input power of the heat source is 100 W and 10 x 10 mm^2. Free convection, laminar flow and +Y gravity (9.865m/s^2) are simulation conditions. You can utilize these window programs of the "VCEK_ML V1.exe" and "BaseResistance_ML.exe" coding by Prof. Wang to assist the simulation. Results reveals that the difference between large and small heat sinks of convection thermal resistances (R_c) is about 60%. The vapor chamber has better thermal performance of 18% from Case2 and Case3 as shown in Figure 15 and Table 5.

CONCLUSION

These thermal-module methodologies of LED lighting lamps, personal computers (PCs), Notebooks (NBs), and servers including central processing units (CPUs) and graphic processing unit (GPUs) of smaller areas and higher power in the electronics industry were investigated in this chapter. This approach presents a method of dealing with the tremendous amount of number crunching requirements posed by the thermal design of two-phase flow thermal modules. The HSHPTM V1.0 (heat pipe software) and VCTM V1.0 (vapor chamber software) are also expected to help operators make decisions related to the thermal modules' lifetime and reliability in a proper, reasonable and systematic way. The preliminary results presented here are very encouraging. Among the many thermal analysis and simulation approach techniques, HSHPTM V1.0 and VCTM V1.0 seem to be the only ones which will make the real-time thermal design of electronic devices possible. The author is looking for contributing to the LED industry, government and academia for the green energy-saving lamps through embedded heat pipe- and vapor chamber-based thermal modules design. The author's future efforts shall be dedicated to developing a LED green energy-saving lighting systems. Finally, the author would like to mention a few points concerning the contributions to the present study. This chapter is a major revised version of an earlier paper. Some of the materials presented in this chapter go back to the beginning of this line of research, and some of the authors's seminal works are included in the literature review; thus, they are in the following references.

Acknowledgments

The author gratefully acknowledges the support of the National Science Council (NSC) and Ministry of Science and Technology (MOST) for my research. He is grateful for the National Taiwan Ocean University (NTOU) for their support through the years, especially as it relates to the Thermo-Illuminanace Lab. He also thanks the Nova Science Publishers for for their guidance in gaining permission to reprint some of his material here.Lastly, the author would like to thank all his colleagues and students who contributed to this chapter.

References

[1] J.-C. Wang, 2008, Novel Thermal Resistance Network Analysis of Heat Sink with Embedded Heat Pipes, *Jordan Journal of Mechanical and Industrial Engineering,* Vol. 2, No. 1, Mar., pp. 23-30.

[2] J.-C. Wang, H.-S. Huang and S.-L. Chen, 2007, Experimental Investigations of Thermal Resistance of a Heat Sink with Horizontal Embedded Heat Pipes, *International Communications in Heat and Mass Transfer,* Vol.34, Issue 8, October, pp. 958-970.

[3] Lofland, S.J. and Chesser, J.B., Intel Corporation, 2003. *Heat sink with heat pipes and fan.* U.S. Patent 6,625,021.

[4] J.-C. Wang and S.-L. Chen, 2007, Thermal Performance of Two-Phase Closed Thermosyphon Cooling System, *Proceedings of the Third International Conference on Thermal Engineering: Theory and Applications,* May 21-23, Amman, Jordan.

[5] R.-T. Wang and J.-C. Wang, 2012, Analyzing the Pressure-Difference Phenomenon between the Condensing and Boiling Sections in a Heat Pipe Cooling System, *International Communications in Heat and Mass Transfer,* Vol. 39, Issue 3, March, pp.390-398.

[6] Saaski, E., Hannemann, R.J. and Fox, L.R., *Digital Equipment Corporation, 1989. Integral heat pipe module.* U.S. Patent 4,833,567.

[7] H.-S. Huang, J.-C. Wang and S.-L. Chen, 2007, Experimental Investigation on Thermal Performance of Heat Sink with Two Pairs of Embedded Heat Pipes, *Proceedings of IPACK2007*, ASME InterPACK`07.

[8] Eggers, P., Battelle Memorial Institute, 1974. *Heat pipes.* U.S. Patent 3,786,861.

[9] Mohammadi, M., Mohammadi, M., Ghahremani, A.R., Shafii, M.B. and Mohammadi, N., 2014. Experimental investigation of thermal resistance of a ferrofluidic closed-loop pulsating heat pipe. *Heat Transfer Engineering,* 35(1), pp.25-33.

[10] Kroebig, H.L. and Riha III, F.J., The United States Of America As Represented By The Secretary Of The Air Force, 1977. *Heat pipe system.* U.S. Patent 4,000,776.

[11] Faghri, A., 1995. *Heat pipe science and technology.* Global Digital Press.

[12] Khalili, M. and Shafii, M.B., 2016. Investigaing thermal performance of a partly sintered-wick heat pipe filled with different working fluids. *Scientia Iranica,* 23(6), pp.2616-2625.

[13] Nussbaumer, T., Wakili, K.G. and Tanner, C., 2006. Experimental and numerical investigation of the thermal performance of a protected vacuum-insulation system applied to a concrete wall. *Applied Energy,* 83(8), pp.841-855.

[14] Ma, H.B., Maschmann, M.R. and Liang, S.B., 2002, June. Heat transport capability in a pulsating heat pipe. In *8th AIAA/ASME Joint Thermophysics and Heat Transfer Conference,* Paper No. AIAA (Vol. 2765).

[15] Yang, X.F. and Liu, Z.H., 2012. Flow boiling heat transfer in the evaporator of a loop thermosyphon operating with CuO based aqueous nanofluid. *International Journal of Heat and Mass Transfer,* 55(25), pp.7375-7384.

[16] Maidanik, Y.F., 1999. State-of-the-art of CPL and LHP technology. *Heat Pipe Science and Technology,* pp.19-30.

[17] Aboutalebi, M., Moghaddam, A.N., Mohammadi, N. and Shafii, M.B., 2013. Experimental investigation on performance of a rotating closed loop pulsating heat pipe. *International Communications in Heat and Mass Transfer,* 45, pp.137-145.

[18] Reay, D., McGlen, R. and Kew, P., 2013. *Heat pipes: theory, design and applications.* Butterworth-Heinemann.

[19] Faghri, A., 2012. Review and advances in heat pipe science and technology. *Journal of heat transfer,* 134(12), p.123001.

[20] Tecchio, C., Oliveira, J.L.G., Paiva, K.V., Mantelli, M.B.H., Gandolfi, R. and Ribeiro, L.G.S., 2017. Thermal performance of thermosyphons in series connected by thermal plugs. *Experimental Thermal and Fluid Science,* 88, pp.409-422.

[21] Yang, K.S., Cheng, Y.C., Jeng, M.S., Chien, K.H. and Shyu, J.C., 2014. An experimental investigation of micro pulsating heat pipes. *Micromachines,* 5(2), pp.385-395.

[22] Lindemuth, J. and Rosenfeld, J., Thermal Corp, 2003. *Vapor chamber with sintered grooved wick.* U.S. Patent 20040069455A1.

[23] Lataoui Z, Jemni A. Experimental investigation of a stainless steel two-phase closed thermosyphon. *Applied Thermal Engineering.* 2017 Jul 5;121:721-7.

[24] Gedik, E., 2016. Experimental investigation of the thermal performance of a two-phase closed thermosyphon at different operating conditions. *Energy and Buildings,* 127, pp.1096-1107.

[25] Koubek, K.J., Kosson, R.L. and Quadrini, J.A., Grumman Aerospace Corporation, 1993. *Vapor chamber cooled electronic circuit card.* U.S. Patent 5,179,500.

[26] Zhang, Y. and Faghri, A., 2003. Oscillatory flow in pulsating heat pipes with arbitrary numbers of turns. *Journal of Thermophysics and Heat Transfer,* 17(3), pp.340-347.

[27] Han, X., Wang, X., Zheng, H., Xu, X. and Chen, G., 2016. Review of the development of pulsating heat pipe for heat dissipation. *Renewable and Sustainable Energy Reviews,* 59, pp.692-709.

[28] Daimaru, T., Yoshida, S. and Nagai, H., 2017. Study on thermal cycle in oscillating heat pipes by numerical analysis. *Applied Thermal Engineering,* 113, pp.1219-1227.

[29] Tsai, T.E., Wu, H.H., Chang, C.C. and Chen, S.L., 2010. Two-phase closed thermosyphon vapor-chamber system for electronic cooling. *International Communications in Heat and Mass Transfer,* 37(5), pp.484-489.

[30] J.-C. Wang, 2010, Development of Vapour Chamber-based VGA Thermal Module, *International Journal of Numerical Methods for Heat & Fluid Flow,* Vol.20, Issue 4, May, pp.416-428.

[31] J.-C. Wang, 2012, 3-D Numerical and Experimental Models for Flat and Embedded Heat Pipes Applied in High-end VGA Card Cooling System. *International Communications in Heat and Mass Transfer,* Vol. 39, Issue 9, November, pp.1360-1366.

[32] J.-C. Wang and W.-J. Chen, 2011, Vapor Chamber in High-end VGA Card, *Microsystems, Packaging, Assembly and Circuits Technology Conference*, 2011. IMPACT 2011. 6th International, pp.393-396.

[33] J.-C. Wang, 2014, Analyzing Thermal Module Developments and Trends in High-Power LED, *International Journal of Photoenergy, Vol. 2014*, Article ID 120452, May, 11 pages, doi:10.1155/2014/120452.

[34] J.-C. Wang and T.-C. Chen, 2009, Vapor Chamber in High Performance Server, *Microsystems, Packaging, Assembly and Circuits Technology Conference,* 2009. IMPACT 2009. 4th International, pp.364-367.

[35] J.-C. Wang and R.-T. Wang, 2011, A Novel Formula for Effective Thermal Conductivity of Vapor Chamber, *Experimental Techniques, Vol. 35,* Issue 5, September/October, pp.35-40.

[36] R.-T. Wang and J.-C. Wang, 2016, Analyzing the Structural Designs and Thermal Performance of Nonmetal LED Lighting Devices, *International Journal of Heat and Mass Transfer,* Vol. 99, August, pp.750-761.

[37] J.-C. Wang and C.-L. Huang, 2010, Vapor Chamber In High Power LEDs, *Microsystems, Packaging, Assembly and Circuits Technology Conference,* 2010. IMPACT 2010. 5th International, pp.1-4.

[38] J.-C. Wang, 2011, Thermal Investigations on LEDs Vapor Chamber-Based Plates, *International Communications in Heat and Mass Transfer,* Vol.38, Issue 9, November, pp. 1206-1212.

[39] J.-C. Wang, T.-S. Sung and W.-P. Chen, 2011, Hyper-Generation LEDs VCPCB, *Microsystems, Packaging, Assembly and Circuits Technology Conference,* 2011. IMPACT 2011. 6th International, pp.332-335.

[40] J.-C. Wang, R.-T. Wang, T.-L. Chang, D.-S. Hwang, 2010, Development of 30 Watt High-Power LEDs Vapor Chamber-Based Plate, *International Journal of Heat and Mass Transfer,* Vol. 53, Issue 19/20, September, pp.3900-4001.

[41] R.-T. Wang and J.-C. Wang, 2015, Optimization of Heat Flow Analysis for Exceeding Hundred Watts in HI-LEDs Projectors, *International Communications in Heat and Mass Transfer,* Vol. 67, October, pp.153-162.

[42] J.-C. Wang, 2014, Thermal Module Design and Analysis of a 230 Watt LED Illumination Lamp under Three Incline Angles, *Microelectronics Journal,* Vol. 45, Issue 4, April, pp.416-423.

[43] J.-C. Wang, 2013, Thermoelectric Transformation and Illuminative Performance Analysis of a Novel LED-MGVC Device, *International Communications in Heat and Mass Transfer,* Vol. 48, November, pp.80-85.

[44] Ebrahimi, M., Shafii, M.B. and Bijarchi, M.A., 2015. Experimental investigation of the thermal management of flat-plate closed-loop pulsating heat pipes with interconnecting channels. *Applied Thermal Engineering,* 90, pp.838-847.

[45] Tang, H., Tang, Y., Yuan, W., Peng, R., Lu, L. and Wan, Z., 2018. Fabrication and capillary characterization of axially micro-grooved wicks for aluminium flat-plate heat pipes. *Applied Thermal Engineering,* 129, pp.907-915.

[46] Ong, K.S., Haw, P.L., Lai, K.C. and Tan, K.H., 2017, April. Vapor chamber with hollow condenser tube heat sink. *In AIP Conference Proceedings* (Vol. 1828, No. 1, p. 020018). AIP Publishing.

[47] Wang, Y. and Vafai, K., 2000. An experimental investigation of the thermal performance of an asymmetrical flat plate heat pipe. *International journal of heat and mass transfer,* 43(15), pp.2657-2668.

[48] Cardin, N., Brik, M., Lips, S., Siedel, S., Bonjour, J. and Davoust, L., 2017. *Effect of a dc electric field on the liquid-vapor interface in a grooved flat heat pipe.*

[49] Patankar, G., Weibel, J.A. and Garimella, S.V., 2017. Working-fluid selection for minimized thermal resistance in ultra-thin vapor chambers. *International Journal of Heat and Mass Transfer,* 106, pp.648-654.

[50] Lataoui, Z. and Jemni, A., 2017. Experimental investigation of a stainless steel two-phase closed thermosyphon. *Applied Thermal Engineering,* 121, pp.721-727.

[51] Velardo, J., Singh, R., Date, A. and Date, A., 2017. An Investigation into the Effective Thermal Conductivity of Vapour Chamber Heat Spreaders. *Energy Procedia,* 110, pp.256-261.

[52] J.-C. Wang, 2014, L- and U-shaped Heat Pipes Thermal Modules with Twin Fans for Cooling of Electronic System under Variable Heat Source Areas, *Heat and Mass Transfer,* Vol. 50, issue 11, pp.1487-1498.

[53] J.-C. Wang, R.-T. Wang, C.-C. Chang and C.-L. Huang, 2010, Program for Rapid Computation of the Thermal Performance of a Heat Sink with Embedded Heat Pipes, *Journal of the Chinese Society of Mechanical Engineers,* Vol. 31, Issue 1, February, pp.21-28.

[54] J.-C. Wang, 2014, U- and L-shaped Heat Pipes Heat Sinks for Cooling Electronic Components Employed a Least Square Smoothing Method, *Microelectronics Reliability*, Vol. 54, Issues 6/7, June, pp.1344-1354.

[55] J.-C. Wang, 2009, Superposition Method to Investigate the Thermal Performance of Heat Sink with Embedded Heat Pipes, *International Communication in Heat and Mass Transfer,* Vol.36, Issue 7, August, pp. 686-692.

[56] J.-C. Wang, 2011, L-type Heat Pipes Application in Electronic Cooling System, *International Journal of Thermal Sciences,* Vol.50, Issue 1, January, pp.97-105.

[57] Wang RT. A fitting, simple and versatile window program (HSHPTM) design using lumped parameters and one-dimensional thermal resistance models. *Heat and Mass Transfer.* 2013 Feb 1; 49(2):291-7.

[58] R.-T. Wang, Y.-W. Lee, S.-L. Chen, and J.-C. Wang, 2014, Performance Effects of Heat Transfer and Geometry on Heat Pipes Thermal Modules under Forced Convection, *International Communications in Heat and Mass Transfer,* Vol. 57, October, pp.140-149.

In: Heat Pipes: Design, Applications and Technology
Editor: Yuwen Zhang
ISBN: 978-1-53613-908-2

Chapter 7

Intermediate Temperature Heat Pipe Working Fluids

William G. Anderson[1,*], Calin Tarau[1] and David L. Ellis[2]
[1]Advanced Cooling Technologies, Inc., Lancaster, PA, US
[2]NASA Glenn Research Center, Cleveland, OH, US

Abstract

There are a number of applications that could use heat pipes or loop heat pipes (LHPs) in the intermediate temperature range of 450 to 750 K, including space nuclear power system radiators, fuel cells, geothermal power, waste heat recovery systems, and high temperature electronics cooling. Potential working fluids include organic fluids, elements, and halides. This chapter reviews previous life tests conducted with 30 different intermediate temperature working fluids, including elements, organic working fluids, and halides and over 60 different working fluid/envelope combinations. During more recent investigation, life tests have been run with three elemental working fluids: sulfur, sulfur-iodine mixtures, and mercury. Other fluids offer benefits over these three liquids in this temperature range. Three sets of organic fluids stand out as good intermediate temperature fluids: (1) diphenyl, diphenyl oxide, and eutectic diphenyl/diphenyl oxide, (2) naphthalene, and (3) toluene. In addition, life tests have been conducted at temperatures up to 550 K with water and titanium and Monel alloys and at temperatures up to 673 K with titanium and superalloy envelopes paired with halides. Recently, roughly half of these heat pipes were selected for destructive evaluation. The working fluids were analyzed, and sections of the heat pipes were examined to determine the type and amount of corrosion in the wicks and heat pipes. The results showed that Titanium/water and Monel/water heat pipes are suitable for temperatures up to 550 K. The long-term life tests also established that Titanium/$TiBr_4$ at 653 K, and Hastelloy B-3, C-22 and C-2000/$AlBr_3$ at 673 K were compatible. The results indicate that the tested envelope materials and working fluids can form viable material/working fluid combinations.

* Corresponding Author Email: Bill.Anderson@1-act.com.

Keywords: heat pipe working fluid, loop heat pipe working fluid, life tests, intermediate temperature working fluids, fluid/envelope/wick compatibility

HEAT PIPE WORKING FLUID/ENVELOPE/WICK COMPATIBILITY – LIFE TESTS

Heat pipes are designed to operate without maintenance for long periods of time, e.g., for more than twenty years in Spacecraft Thermal Control Systems. One requirement is that the envelope, wick, and working fluid must be compatible. Note that compatibility requirements for heat pipes are more stringent than for many other applications. For example, aluminum and water are not compatible, since an aluminum/water heat pipe will generate large amounts of non-condensable gas in hours or days, stopping the heat pipe from operating.

In the past, life tests have been conducted on hundreds of different working fluid/envelope/wick combinations. Over time a long list of compatible envelope/wick/ material systems has been built up, including copper/water for electronics cooling, alkali metal/superalloys for high temperature, and aluminum/ammonia for spacecraft thermal management.

Most heat pipe life tests today are conducted as a Quality Control measure to confirm that the heat pipe fabrication processes are under control. For example, all extrusions for aluminum/ammonia Constant Conductance Heat Pipes (CCHPs) are on life test at elevated temperature, to demonstrate the long-term life required in heat pipes for satellites. The heat pipes are operated at elevated temperatures, 24 hours a day for 365 days a year. Periodically, the life test heat pipes are placed back into performance test fixtures, fully insulated and operated at very low temperatures to look for signs of non-condensable gas (NCG).

The one area where active research is still ongoing is the Intermediate Temperature Range, between 150°C (~425K) and 400°C (~675K). Below 150°C, water is used most commonly in copper envelopes. At higher temperatures, up to 275°C, water is still a good working fluid, however, copper is no longer suitable as an envelope material, due to the low yield strength of copper, so other envelope materials such as titanium and Monel alloys must be used. Above 400°C, cesium is used with steel and superalloy envelopes. At lower temperatures, the power that can be carried by cesium is limited by the sonic limit, so a different working fluid must be used. The halide $AlBr_3$ is compatible with superalloys at temperatures up to 400°C.

When conducting life tests, the problems with incompatible fluid/envelope pairs include:

- Fluid decomposition
- Corrosion, blocking the wick or developing leaks in the heat pipe envelope
- Non-condensable gas generation, caused by either of the problems above
- Materials transport

Non-Condensable Gas Generation

The most common symptom of incompatibility is Non-Condensable Gas (NCG) generation in the heat pipe. During operation, the NCG accumulates at the end of the condenser, reducing its effective length. In turn, this reduces the power that the heat pipe can carry at a given temperature. For compatible systems, NCG generation is mainly caused by impurities in the working fluid or contaminants on the wick or wall.

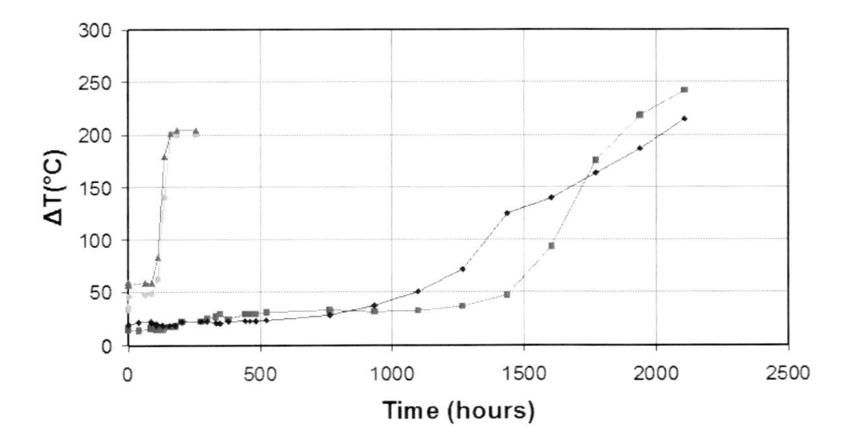

Figure 1. Therminol life tests at 400 and 450°C. The gas generation rate is dependent on the operating temperature. Gas is generated quickly at 450°C, and more slowly at 400°C. No gas was observed after 1000 hours at 350°C.

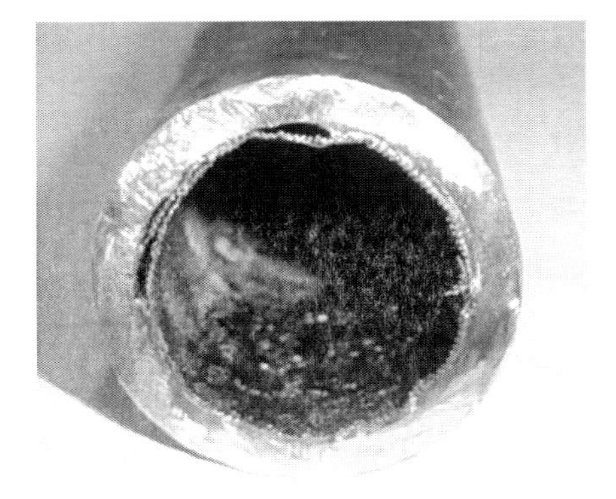

Figure 2. Charring of Therminol during life tests at 450°C.

Gas generation can also be caused in organic working fluids when the temperature is too high. Eutectic diphenyl/diphenyl oxide is an intermediate temperature fluid that is sold under the trade names of Dowtherm A and Therminol VP-1. Heat pipes with Therminol VP-1 as working fluid were life tested at three different operating temperatures, 350, 400, and 450°C, and monitored for gas generation (Anderson et al., 2007b). As shown in Figure 1, the life test pipes operating at 450°C showed a significant increase in ΔT after only 90 hours, most likely caused by excess of NCG. These pipes were taken off life test and purged, then put back on life test. NCG generation continued at a high rate. The 450°C heat pipes were taken off life testing after 300 hours, sectioned, and analyzed. As shown in Figure 2, a portion of the Therminol charred during the life test.

Corrosion

Corrosion occurs when the working fluid is not chemically compatible with the envelope or wick material (NCG can also be generated, as discussed above). Corrosion products can block a portion of the wick, reducing the heat pipe maximum power. In more extreme cases, a leak can develop, with the heat pipe ceasing operation. As discussed below in more detail, the authors have conducted a series of life tests in the intermediate temperature range with superalloy envelopes and halide working fluids. Gallium trichloride was not compatible, due to rapid corrosion. The $GaCl_3$/superalloy heat pipes all leaked at the pinch-off weld after roughly one week of operation at 360°C (633K); see Figure 3.

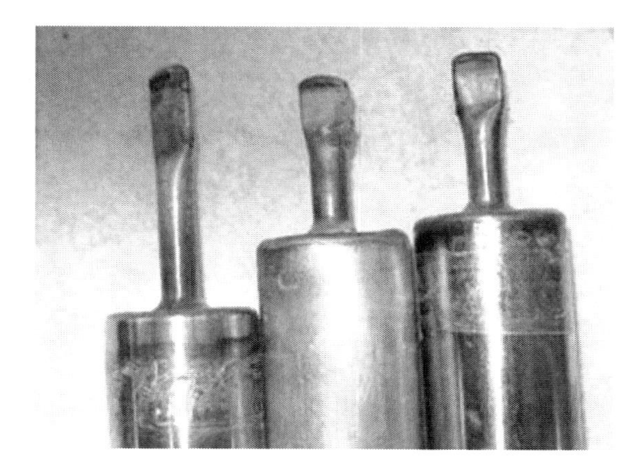

Figure 3. $GaCl_3$ is incompatible with superalloys. A leak developed at the pinch-off tubes within one week after the life test was started.

Materials Transport

Material transport of the wick/envelope can occur when the working fluid has a high solubility for one of the wick/envelope components. During heat pipe operation, working fluid is vaporized in the evaporator, and then condensed in the condenser, transferring heat. The working fluid vapor contains no impurities, so the working fluid in the condenser is also very pure. If the working fluid has a high solubility for one or more of the components in the wick or envelope, these components will dissolve in the working fluid, and be carried along to the evaporator. When the working fluid vaporizes, the dissolved components are left behind.

A beautiful example of material transport with an incompatible envelope/fluid pair is shown in Figure 4. The copper grains in the evaporator show that copper was transported in the cesium from the condenser to the evaporator, hence Monel and copper/nickel is not compatible with cesium.

REVIEW OF PREVIOUS LIFE TESTS

At present, there is no commonly accepted working fluid over the entire intermediate temperature range. Potential working fluids include elemental working fluids (such as sulfur), organic compounds, and halides. A short review of previous life tests follows. Note that Anderson (2007a) provides more detailed information on these tests.

Figure 4. Material transport after a 1000-hour life test with cesium at 475°C. Monel and Copper/Nickel 70/30 (a) Monel. (b) Copper/Nickel 70/30. Note the copper particles that were transported in solution from the condenser and deposited in the evaporator.

Elements

Pure sulfur is not suitable in the intermediate temperature range because of its high liquid viscosity, although it may be useful at higher temperatures. Sulfur has a unique temperature dependent polymerization property at 470 K, which increases its liquid viscosity peak to approximately 100 Pa-s. This is about three orders of magnitude higher than the maximum level for effective heat pipe operation. The addition of a small percentage of iodine reduces the viscosity to a level that may be acceptable for reasonable heat pipe operation (Polasek and Stulc, 1976, Timrot et al., 1981). A potential problem with both sulfur, and sulfur/iodine mixes is that they react strongly with many envelope materials.

There are several problems with mercury as a working fluid including:

- Toxicity
- Difficulty in achieving good wetting of the wick and wall material without extensive corrosion
- High density, which translates into increased mass.
- Aggressive attack or solutioning of many metals, e.g., copper

Additives have been used successfully with mercury to wet coarse wicks, but it appears to be very difficult to achieve wetting in finer pore wicks. Heat pipe tests with mercury in a sintered stainless-steel wick failed because the mercury did not wet the stainless steel (Anderson, Rosenfeld, Angirasa, and Mi, 2004).

Table 1. Summary of Sulfur, Sulfur-Iodine, and Mercury Life Tests

	Compatible	**Incompatible**
Sulfur/Al	Short Term 873 K (600°C)/~200 hrs./3003 Al (Ernst, 2006)	
Sulfur/SS		773K (Polasek, 1989)
Sulfur – 10% I/SS	Short Term 623 K 304 SS (Anderson, 2004)	833 K. (Polasek, 1989)
Sulfur – 10% I /5052 Al		623K (Anderson, 2004)
Sulfur – 10% I /Ti		623K/Ti-6Al-4V (Anderson, 2004) 523K/CP-Ti (Anderson, 2004)
Sulfur – 10% I /Nb1% Zr		623K/(Anderson, 2004)
Mercury/SS	603K/347 SS (Deverall, 1971, Reid 1991)	

Sulfur, Sulfur-Iodine, and Mercury life tests are summarized in Table 1. Mercury is compatible with 347 SS based upon long term life tests. Sulfur is compatible with pure aluminum based on a short-term life test, as is Sulfur-10% Iodine with 304 SS.

Organic Working Fluids

Life tests have been conducted with 19 different organic working fluids. Most of the suitable organic fluids are ring compounds. The reason for this was discussed by Saaski and Owzarski (1977) who pointed out that these types of compounds are more stable than the long chain hydrocarbons. Saaski and Owzarski also pointed out that replacing some (or all) of the hydrogen atoms with fluorine may make the compound more stable.

Potential problems with the organic working fluids include the possibility of polymerization and/or dissociation. Polymerized fluids generally undergo an increase in liquid viscosity, which will decrease the circulation of the working fluid in a heat pipe and therefore its heat transport capacity. Dissociation normally generates non-condensable gases (NCG), which over time will build up in the heat pipe condenser. The presence of NCG reduces the effective length of the heat pipe condenser and hence the area available for heat transfer. This will either cause the temperature to rise at a given power level or the power level to be decreased at a given temperature.

Typically, organic fluids develop problems more quickly as the temperature is increased. The maximum operating temperature for an organic fluid depends both on the operating temperature, and how long the heat pipe needs to operate.

The most commonly tested organic fluids have been diphenyl, diphenyl oxide, and a eutectic mixture of diphenyl/diphenyl oxide (Trade Names Dowtherm A, Therminol, and Diphyl). Eutectic diphenyl/diphenyl oxide is nearly an azeotrope (Basilius and Prager, 1975), so the liquid and vapor have almost the same composition. This avoids the problems encountered with other mixtures such as NaK, where fractional distillation can occur (Anderson, 1993). Life test results for these three fluids are summarized in Table 2.

When using diphenyl, diphenyl oxide, or diphenyl/diphenyl oxide at temperatures over 673 K (400°C), non-condensable gas is generated in a relatively short time periods as shown in Table 2. The exact period depends upon the fluid and material and decreases as the temperature increases. For example, Kenney and Feldman found that their diphenyl pipes took less than 72 hours to gas up at 748 K (475°C), and 366 hours to gas up at 695 K (422°C). Between 573 and 673 K (300 and 400°C), these fluids are generally suitable, for short duration tests near 673 K, and long duration tests near 573 K. For example, Groll et al. found that 321 SS was compatible for ~40,000 hours at 573 K (300°C), but not at 623 K (350°C).

Table 2. Summary of Diphenyl, Diphenyl Oxide and Eutectic Diphenyl/Diphenyl Oxide Life Tests[1]

	Compatible	Incompatible
Diphenyl/Al	514K/6061 Al (Saaski)	
Diphenyl/Mild Steel	523 K/(Groll) 498 K (Kenney, 1978)	673K (Groll) 598K (Kenney 1978) 526K (Saaski)
Diphenyl/Black Iron	523 K (250°C)/7,158 hrs (Kenney, 1978)	
Diphenyl/Stainless	623 K/316 SS (Grzyll) 548 K/304 SS (Kenney, 1978) 543 K/316L SS (Groll) ***Short Term*** 673K/304 SS (Kenney 1978	695K/304 SS (Kenney, 1978) 673K/316L SS (Groll)
Diphenyl Oxide/ Stainless Steel	***Short Term*** 573 K/347 SS & 304 SS (LASL, 1968a, 1968b, 1970)	
Dowtherm A/Mild Steel	543K/ST 35 (Groll) 523K (250°C) (Kenney, 1978)	573 K/ST 35 (Groll)
Dowtherm A/ Stainless Steel	573K/321 SS (Groll) 541K/304 SS (Kenney, 1978) ***Short Term*** 673K/304SS (Kenney, 1978)	623K/304 (Anderson) 623 K/321 SS (Groll 473K (Basilius 1975)
Dowtherm A/Cu		473K (Basilius 1975)
Dowtherm A/ Cu-Ni		523K/CuNi10Fe (Groll)
Dowtherm A/Ti	543K (Groll)	406°C (680K)

Life tests results for organic fluids other than diphenyl and diphenyl oxide are summarized in Table 3. Fluids have been ranked by the highest temperature for a compatible life test with any envelope material.

Table 3. Summary of Organic Fluid Life Tests other than Diphenyl and Diphenyl Oxide

	Compatible	Incompatible
Naphthalene/Al	488 K/6061 Al (Saaski)	
Naphthalene/Mild Steel	543K/ST 35 & 13CrMo44[2]. (Groll))	
Naphthalene/Stainless Steel	623/K/316 SS (Grzyll) 593K/316L SS (Groll) 593K/Alloy 20 Vasil'ev 1988) ***Short Term*** 653K/Alloy 20 (Vasil'ev 1988)	
Naphthalene/Cu-Ni	593K hrs/CuNi10Fe[3] (Groll)	
Naphthalene/Ti	593K (Groll) 593K (Vasil'ev, 1988)	

[1] For all tables Groll (1982, 1987, 1989), Grzyll (1991, 1994, 1995), Saaski (1977, 1980), Anderson (2007b)
[2] 13CrMo44 is a 1% Cr-1/2% Molybdenum Steel
[3] Copper Nickel Alloy, resistant to corrosion in seawater

	Compatible	**Incompatible**
O-Terphenyl/Al		580K/6061 Al (Saaski)
O-Terphenyl/Mild Steel	545K/Al-178 (Saaski)	
O-Terphenyl/SS	623K/316 SS (Grzyll)	
Decafluorobiphenyl/SS	623K/316 SS (Grzyll)	
Toluene/Al	410K/6061 Al (Saaski)	
Toluene/Mild Steel	523K/ ST 35 & 13CrMo44 (Groll)	
Toluene/SS	523K/316 SS (Groll)	
Toluene/Copper Nickel	553K/CuNi10Fe (Groll)	
Toluene/Ti	523K (Groll)	
1-Fluoronaphthalene/ Al	493K/6061 Al (Saaski)	
1-Fluoronaphthalene/ Mild Steel	530K/A178 (Saaski)	
1-Fluoronaphthalene/ Stainless Steel		530K/304 SS (Saaski)
N-Octane/Mild Steel	503K/ST 35 (Groll)	
1-Fluoronaphthalene/ Stainless Steel		530 K/304 SS (Saaski)
N-Octane/Stainless	523K/321 SS (Groll)	
Dowtherm E	493K/ST 35 (Groll)	
Octafluoronaphthalene/ Aluminum	482K/6061 Al (Saaski)	
Octafluoronaphthalene/ Mild Steel		488K/A178 (Saaski)
Quinoline/SS		623 K/316 SS (Grzyll)
Monochloronaphthalene /Stainless Steel		560K/A178 (Saaski)
Formyl-piperidine/SS		553K/304 SS/ (Kenney1978)
P-Terphenyl/SS		723K/304 SS (Kenney, 1978)
ortho- and meta-terphenyl/Mild Steel		593K/13CrMo44 (Groll)
ortho- and meta-terphenyl/SS		623K/316L SS (Groll)
diphenyl, ortho- and meta-terphenyl/Mild Steel		623K/13CrMo44 (Groll)
diphenyl, ortho- and meta-terphenyl/SS		623 K/316L SS (Groll)
Perfluoro-1,3-5-triphenylbenzene		573K/316 SS (Grzyll)

Since all of their life tests to date showed compatibility, two fluids stand out in Table 3: toluene and naphthalene. Toluene was compatible with a copper-nickel alloy, CuNi10Fe, at 553 K (280°C), as well as with aluminum, mild steel, stainless steel, and titanium at lower temperatures. This is probably close to the maximum useful range of toluene, since the critical point of toluene is 592 K (319°C).

Water is generally a better working fluid, since it can also be used in this temperature range, and has a Merit number that is roughly 50 times higher than toluene. However, toluene has three advantages over water, which may make it a suitable choice for certain conditions. These advantages are:

- Compatibility with a larger number of envelope/wick materials
- Melting temperature of 178 K (-95°C) versus 273 K (0°C)
- Lower saturation pressure (e.g., 23.4 atm. at 550 K versus 60.4 atm. for water)

Naphthalene is compatible with stainless steel, copper-nickel, and titanium, based on long term life tests at 593 K (320°C) and above. It has also been shown to be compatible at lower temperatures with aluminum and mild steel. It was compatible with Alloy 20 stainless steel for short term tests at 380°C.

While fluorinated compounds have been theorized to be more stable than the same compound without fluorine, this has not been verified in life test data. Gryzll, Back, Ramos, and Samad, (1994) found that Decafluorobyphenyl ($C_{12}F_{10}$) was less stable than diphenyl ($C_{12}H_{10}$) under the same test conditions. Perfluoro-1,3,5-triphenylbenzene underwent severe thermal decomposition. Naphthalene was compatible with mild steel at 623 K (350°C) for 5,520 hours, while Monochloronaphthalene was found to be unsuitable after 642 hours at 560 K (287°C), and Octafluoronaphthalene had NCG gas generation at 488 K (215°C). Other stable, fluorinated life tests have been conducted at temperatures of 530 K (257°C) and below.

Table 4. Halide Life Test Summary

	Compatible	**Incompatible**
$TiBr_4$/CP Ti		653K/Ti (Anderson 2013)
$TiCl_4$/Steel	432K/A-178 Steel (Saaski)	
$TiCl_4$/Superalloy	573K/Hastelloy (Anderson 2013)	
$TiCl_4$/Ti	500K/CP-Ti (Locci, 2005)	
$TiCl_4$/Al		438K/Al-6061 (Saaski)
$SnCl_4$/Al		432K/Al-6061 (Saaski)
$SnCl_4$/Steel	429K/A-178 Steel (Saaski)	
$SnCl_4$/Superalloy		553K/Hastelloy (Anderson 2013)
$AlBr_3$/Superalloy	673K/Hastelloy (Anderson 2013)	
$AlBr_3$/Al		500K/Al-5052 (Locci, 2005) 500K/Al-6061 (Locci, 2005)
$AlBr_3$/Titanium		500 K/CP-Ti (Locci, 2005)
$GaCl_3$/Superalloy		633K//Hastelloy C-22 (Anderson 2013)
$GaCl_3$/CP Ti		613K/Ti (Anderson 2013)
$SbCl_3$/Al		500K/Al-6061 (Saaski)
$SbCl_3$/Steel		476K/A-178 Steel (Saaski)
$SbBr_3$/Al		500K/Al-6061 (Locci)

Halides

A halide is a compound of the type MX, where M may be another element or organic compound, and X may be any of the Group 17 elements: fluorine, chlorine, bromine, iodine, or astatine. Starting with Saaski and Owarski (1977), a number of researchers have suggested that halides are potential heat pipe fluids. They are attractive because they are more stable at high temperatures than organic working fluids, and because their Merit number peaks in the intermediate temperature range. Information on halide properties can be found in Anderson, Rosenfeld, Angirasa, and Mi (2004) and Devarakonda and Anderson (2005).

Saaski and Owzarsky (1977) proposed an electrochemical method to predict the compatibility of halide working fluids with envelope materials. Tarau, Sarraf, Locci and Anderson (2007) found that this procedure had good agreement with the halide life tests discussed above.

Halide life tests are summarized in Table 4. Some halides are suitable for temperatures up to 673 K (400°C), and possibly at higher temperatures. Very long-term life tests show that $TiCl_4$ and $SnCl_4$ are both compatible with mild steel, and that $AlBr_3$ is compatible with superalloys. No tests to date with an aluminum envelope have been successful. This is due to the very high decomposition potential of aluminum when compared to other metals (Tarau, 2007).

TITANIUM/WATER AND MONEL/WATER LIFE TESTS

As discussed above, copper/water heat pipes are generally only used at temperatures below 150°C (425K), since the thickness of a copper envelope at higher temperatures is too large to be practical. Long term life tests have been conducted to demonstrate that water/titanium and water/Monel are compatible at temperatures up to 275°C (550K).

Anderson, Dussinger, and Sarraf (2006a) reviewed previous work and selected titanium, Monel 400, and Monel K-500 as potential heat pipe materials for high temperature water heat pipes. Titanium is attractive since titanium heat pipes have a lower mass than the equivalent Monel 400 or K-500 pipes. Second, titanium has been used with a large number of working fluids, so a titanium heat pipe could be used over a wide range of temperatures by varying the working fluid. Titanium has been used in heat pipes with the following fluids:

- Sodium (Anderson, Dussinger, and Sarraf, 2006b)
- Potassium (Lundberg, 1984, Sena and Merrigan, 1989)
- Cesium (Hartenstine, 2007)
- Dowtherm A (Heine, Groll, and Brost, 1984, Groll, 1989)

- Toluene (Heine, Groll, and Brost, 1984, Groll, 1989)
- Water (Heine, Groll, and Brost, 1984, Groll, 1989), Antoniak et al., 1991, Anderson, Dussinger, and Sarraf, 2006a)
- Ammonia (Ishizuka, Sasaki, and Miyazaki 1985)
- Nitrogen (Swanson et al., 1995)
- Cesium (Anderson, Dussinger, and Sarraf, 2006b) – in loop heat pipes.

All of these pipes have commercially pure (CP) titanium, since this is readily available in tubing. However, other titanium alloys were also examined in the current study, since they have higher strength at the heat pipe operating temperatures.

Table 5. Titanium-Water and Titanium-Monel Life Test Pipes

Wall Material	Wick	Operating Temperature	Operating Hours
Monel K 500	200x200 Monel 400 Screen 0.064 mm wire	550 & 500K	72,192 hours
CP-2 Ti	150x150CP-Ti Screen 0.069 mm wire	550 & 500K	72,192 hours
CP-2 Ti	Sintered Titanium -35+60 Mesh CP-2	550K	60,672 hours
CP-2 Ti	100 x100 CP-Ti Screen 0.05 mm wire	550K	61,064 hours
CP-2 Ti	Integral Grooves	550 K	57,170 hours
CP-2 Ti 21 S Foil Inside	100 x100 CP-Ti Screen 0.05 mm wire	550K	62,622 hours
Grade 5 Ti	100 x100 CP-Ti Screen 0.05 mm wire	550K	69,845 hours
Grade 7 Ti	100 x100 CP-Ti Screen 0.05 mm wire	550K	60,672 hours
Grade 9 Ti	100 x100 CP-Ti Screen 0.05 mm wire	550K	60,072 hours
Monel 400	120x120 Monel 400 Screen 0.05 mm wire	550K	60,168 hours
Monel K 500	120x120 Monel 400 Screen 0.05 mm wire	550K	67,536 hours
Monel 400	-100+170 Mesh Monel 400 Powder	550K	58,824 hours
Monel K 500	-100+170 Mesh Monel 400 Powder	550K	57,792 hours

Monel 400 (Monel 400, 2017) is a solid solution alloy with roughly 63% nickel and 30% copper. It is a single-phase alloy, since the copper and nickel are mutually soluble in all proportions. It can only be hardened by cold working. Monel K500 (Monel K500, 2017) is a similar nickel-copper alloy, with the addition of small amounts of aluminum and

titanium that give greater strength and hardness. The system is age-hardened by heating so that small particles of $Ni_3(Ti, Al)$ are precipitated throughout the matrix, increasing the strength of the material. This allows a recovery of the initial strength after sintering a wick in the envelope.

Water Life Tests

Anderson, Dussinger, Bonner, and Sarraf (2006) started a series of life tests with commercially pure (CP) titanium, titanium alloys, Monel 400, and Monel K-500. Table 5 shows the different life test pipes on test. The heat pipes with integral titanium grooves are intended for spacecraft thermal control. Three integrally grooved heat pipes were tested for 57,170 hours at NASA Glenn Research Center with no sign of NCG (Sanzi and Jaworske, 2011, Sanzi, 2017).

Life Test Setup

Figure 5 shows a schematic of a typical life test heat pipe set up in a heater block. The life tests are gravity aided, and cooled by natural convection. The life test pipes are instrumented with three thermocouples. One thermocouple is located just above the heater block, while the other two are located in the heat pipe condenser. During operation, the temperature difference between the evaporator and condenser are monitored to detect non-condensable gas (NCG). Any NCG is swept by the working fluid to the end of the condenser, where it forms a cold end. The temperature is normally lowered during the measurement, to facilitate detection of NCG amounts.

Figure 6 shows the titanium/water and Monel/water heat pipes set up in the heater blocks, prior to testing. One thermocouple is located just above the heater block, the other two are located in the heat pipe condenser (all of the thermocouples are under the hose clamps) in Figure 6.

During the life test, the temperatures of the evaporator and condenser are monitored for each pipe, to detect any problems. It is possible that oxygen and nitrogen can affect the outside of the titanium pipes during the test. For heat pipes in a space radiator, oxygen/nitrogen will not be a problem. To prevent this occurence during the life tests, the heat pipes are placed inside a box that is purged with argon.

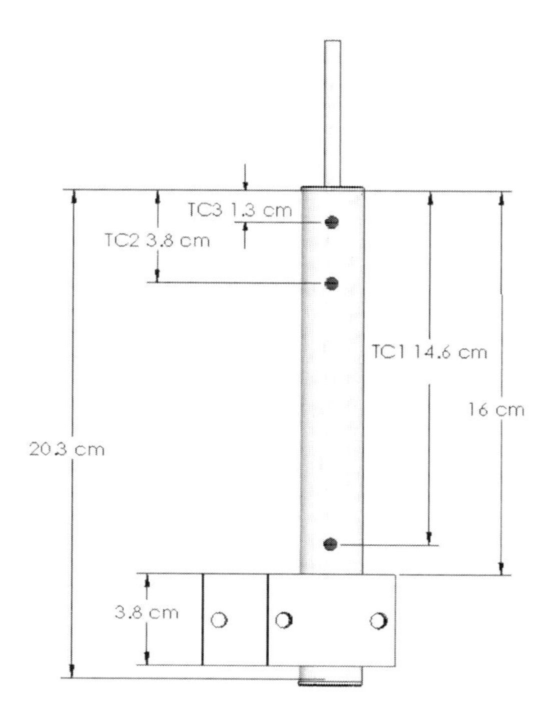

Figure 5. Typical Life Test Heat Pipe and Heater Block.

Figure 6. CP-Titanium (on left) and Monel 500 heat pipes set up in heater blocks. The fill tubes are much longer than usual, to allow for multiple purge and reseal.

Water Life Test Results

Note that the heat pipe operating temperature is dropped to 340 K (70°C), before the measurements. With this much lower temperature and vapor pressure, the NCG expands, making it easier to detect. The Monel pipes have shown almost no signs of gas generation. However, the titanium heat pipes all generated gas initially. This was believed to be a result of a passivation process that produced titanium oxide on the surface of the heat pipe.

The gas was removed from all of these pipes by heating to about 115°C and venting into the ambient. The thermocouples were monitored to verify that the NCG has been forced out of the condenser by the total pressure difference. The heat pipe fill tube is then resealed. Since they have been resealed, the pipes have all generated a small amount of gas.

HALIDE LIFE TESTS

Life Tests

A series of halide life tests were conducted at ACT. The selection criteria were discussed in Anderson et al. (2007). The fluid/envelope combinations tested and the operating conditions are shown in Table 6. The titanium pipes had a 50 x 50 mesh titanium screen wick, and the C-22 pipes had an 80 x 80 mesh C-22 wick. The other two types of pipes were bare. Note that all of the superalloy pipes had C-22 endcaps and fill tubes (due to availability).

The operating temperatures in Table 6 were set based on the vapor pressure, and the allowable stresses in each heat pipe as a function of temperature. During the life tests, the temperature of the evaporator and condenser for each heat pipe are monitored, to detect any problems. It is possible that oxygen can affect the outside of the titanium pipes during the test. To prevent this problem, the life tests are conducted inside a box that is purged with argon. During the life test, heat pipe temperatures are monitored to detect the formation of NCG.

The superalloy/$GaCl_3$ pipes all leaked at the pinchoff weld after roughly one week of operation at 360°C (633K). Note that all of the superalloy pipes used a C-22 fill tube, since that was more readily available. After roughly 11,000 hours, the Hastelloy B3 pipe with $AlBr_3$ leaked, apparently at a weld. Note that this failure is probably due to the pipe being severely overheated after the first 300 hours of testing when the heater shorted out. The maximum temperature is not known; however, it was sufficient to bubble the aluminum heater block. The titanium/$GaCl_3$, titanium/$TiBr_4$, and superalloy/$SnCl_4$ pipes all developed large amounts of non-condensable gas. After roughly 20,000 hours of operation, the pipes were taken off life test, and stored for the analyses discussed below.

Table 6. Halide Life Test Pipe Temperatures and Operating Times

Working Fluid	CP-Ti	C22	C2000	B3
$AlBr_3$	–	673K 28,704 hrs Low Gas Analyzed	673K 58,992 hrs Low Gas	673K ~11,000 hrs (Fail)[4]
$GaCl_3$	613K 19,632 hrs Analyzed	Fail	Fail	Fail
$SnCl_4$	–	553K 20,160 hrs High Gas Analyzed	553K 20,160 hrs High Gas Analyzed	553K 20,160 hrs High Gas Analyzed
$TiCl_4$	–	573K 28,560 hrs Low Gas Analyzed	573K 59,184 hrs Low Gas	573K 58,200 hrs Low Gas
$TiBr_4$	653K 20,040 hrs High Gas Analyzed	–	–	–

One superalloy/$TiCl_4$ and one superalloy/$AlBr_3$ pipe continued to run without any problem for 57,000 hours (6.7 years). The $AlBr_3$ pipes are of particular interest, since they waere running at 673 K (400°C). This is close to the temperature at which cesium starts to work.

HEAT PIPE SECTIONING AND ANALYSIS

In late 2010, several of the heat pipes were selected for destructive investigation, see Table 6. One of each pair of water life test pipes was selected, while the other one continued on life test. The $GaCl_3$ and $SnCl_4$ pipes were known to be non-compatible, since they generated large amounts of NCG. The heat pipes containing halides were neutralized with water to form a stable sediment. The sediment from the halides, and the water from the heat pipes were collected for chemical analysis. No fluid analysis was done on Pipe 4, CP-Ti/$TiBr_4$, since titanium is the only metal present in the fluid and envelope.

[4] $AlBr_3$/B3 pipe was severely overheated at 300 hours testing, melting the aluminum

To examine the cross-sections to determine the type and amount of corrosion in wicks and envelopes, the heat pipes were cut in half, pressure infiltrated with epoxy and sectioned at a location approximately one-third of the way above the bottom of the heat pipe. The sections were polished through 0.05-micrometer silica and examined using optical and scanning electron (SEM) microscopes. For more information and additional sections, see Anderson (2013).

Titanium-Water Heat Pipe Cross-Sections

Analysis of the titanium/water heat pipe cross-sections using optical microscopy revealed little if any corrosion when observed at high magnifications. Even using differential interference contrast, it was difficult to find any corrosion layer. When any evidence of corrosion was observed, the layer was typically ~1 micrometer thick. SEM imaging and EDS analysis also did not indicate any substantial corrosion layer. Further details are given in Anderson (2013).

Several heat pipes had evidence of a layer on the surface of the wires of the wick and envelope. An example is shown in Figure 7. A layer approximately 5 to 10 microns thick that is not mottled like the core can be observed in BSE images. The layer does not have a detectable difference in chemical composition. It is hypothesized that there has been a change in the grains that has reduced the contrast between grains. This is not caused by thermal exposure since the core does not exhibit the same change. Instead, it is most likely caused by the introduction of interstitial atoms of O and H into the Ti at levels too low to be detected by EDS.

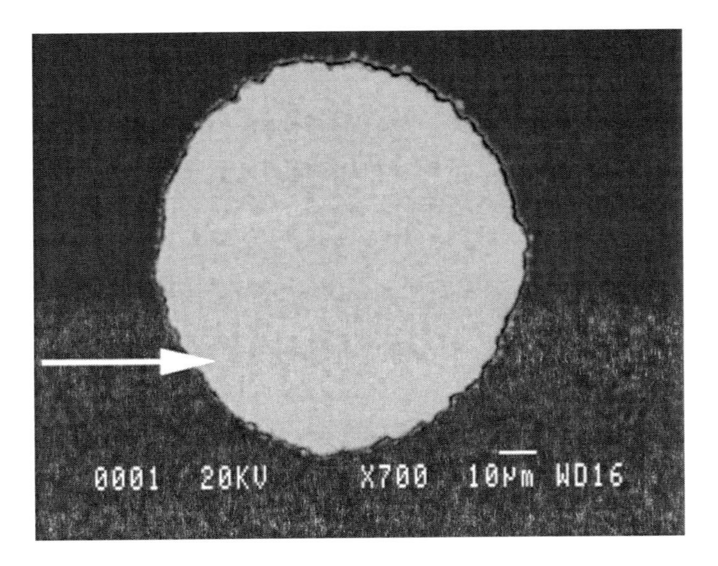

Figure 7. BSE image of CP Ti Mesh Wire Wick in Heat Pipe 124, CP-Titanium with 100x100 CP-Ti Screen. (Arrow denotes possible reaction layer).

Monel-Water Heat Pipe Cross-Sections

Several water heat pipes made with Monel K500 (Ni-30Cu-2Fe-2.7Al-0.6Ti) envelopes and Monel 400 (Ni-31Cu-2.5Fe-2Mn) wicks were tested at varying times and temperatures. While the heat pipes performed well overall, there were some surprises when the heat pipe cross-sections were examined.

Figure 8 shows an optical micrograph of the envelope and wick for one of the heat pipes that underwent the most change, Heat Pipe 136. Extensive changes, including the formation of a dark subsurface layer and bright nodules, were observed in the Monel 400 wick. While the change was the most extensive for the Monel-Water heat pipes, the morphology was typical of all Monel 400 wire mesh and powder wicks as well. Heat Pipe 134, which has a Monel 400 envelope, shows a similar dark subsurface layer. It is much thinner, in the order of 1 to 8 micrometers thick, and variable in thickness.

The Monel K500 envelope does not show an extensive change. Close examination of the envelope reveals, at most, a thin (5-10 micrometer) corrosion layer. Most likely the layer was an oxide, but it was sufficiently thin to prevent definitive identification through EDS.

Figure 9 shows a detail of one of the Monel 400 wires from Heat Pipe 107 (Monel K500-Monel 400 wick-water). EDS spot analysis was conducted on the surface nodules, the dark subsurface layer and the matrix. The results showed that the surface nodules were essentially pure Cu, the subsurface layer was Cu-depleted, and the matrix even adjacent to the dark layers retained essentially the same composition as the bulk composition of Monel 400. Additional examination of this and other heat pipes with Monel 400 screens indicated that the dark layer could be between 2 wt.% and 23 wt.% Cu compared to 31 wt.% Cu for the bulk alloy.

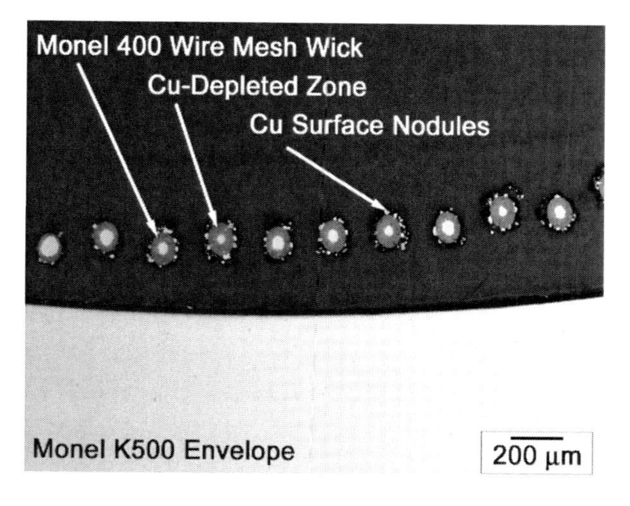

Figure 8. Optical Bright Field Micrograph of Heat Pipe 136.

Examination of the Cu-Ni phase diagram (ASM International, 1992) shows that while Cu and Ni form a solid solution, below about 627 K (354°C) there is a decomposition of the α phase to α_1+α_2 phases. This decomposition creates a Cu-rich and a Ni-rich phase. It appears that this is the driving force for the separation of the Cu and Ni that is observed. Apparently, the activity gradient was such that the Cu-rich phase preferentially moved through diffusion to the surface. The addition of several alloying elements to Monel K500 appeared to stabilize the α phase and prevent this decomposition. While no large amounts of oxygen were observed, preferential oxidation of the Ni may have also played a role in the development of the observed morphology and phases.

Since both Monel 400 and Monel K500 are extensively used in steam plant operations, the literature was searched to find similar observations. So far, no such reference has been located.

Titanium-Halide Heat Pipe Cross-Sections

The two titanium halide heat pipes examined had very different responses. The CP-TI heat pipe with a $TiBr_4$ working fluid, had minimal corrosion. There was evidence of some change at the surface as the BSE images showed the outer 10 micrometers had notably less mottling than the core. This envelope/fluid combination had a high gas generation rate, so it is not suitable.

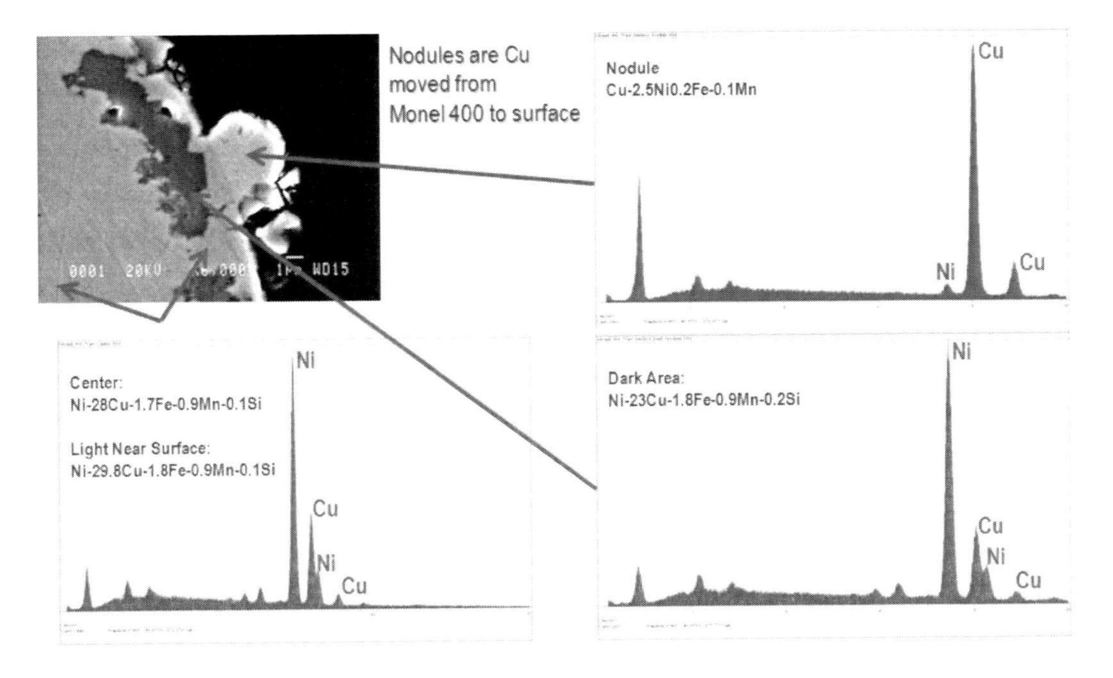

Figure 9. BSE Image of Monel 400 Wire in Heat Pipe 107 (Monel K500-Monel 400 wick-Water).

The CP-Ti heat pipe with $GaCl_3$ working fluid underwent extensive corrosion. Extensive voids and cracking were observed in the corrosion layer. EDS analysis indicated that the corrosion layer was a Ga-29.7 wt.% Ti alloy. Examination of the Ga-Ti phase diagram (National Physical Laboratory, 2012) showed that the composition is similar to the Ga_2Ti phase. Immediately adjacent to the Ga_2Ti phase on the phase diagram is Ga(l) and Ga_3Ti. Given the extensive nature of the voids, particularly on the wires, the Ga(l) and Ga_3Ti(s) phases may have been present in these voids. During neutralization of the halide, the Ga(l) and any Ga_3Ti particles within it could have been physically removed, leaving the voids behind.

Some of the corrosion layer was observed to crack and chip during polishing. The fracture surfaces were indicative of a brittle failure mode. In combination with the extensive cracking of the corrosion layer, this seems to indicate that the corrosion layer is quite brittle.

Hastelloy C-Series Superalloy-Halide Cross-Sections

As shown in Table 6, four halides were tested with the Hastelloy envelopes: $AlBr_3$, $GaCl_3$, $SnCl_4$, and $TiCl_4$. The $GaCl_3$ pipes developed leaks within one week after the testing was started. The $SnCl_4$ heat pipes had high gas generation, which was confirmed by the thick Ni-Sn-Cl corrosion layer; see Anderson (2013) for details.

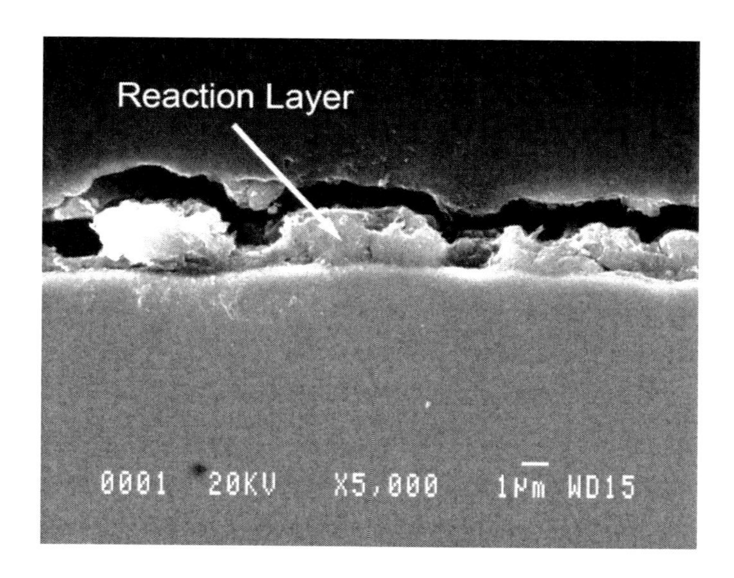

Figure 10. SE Image of C-2000 Envelope in the Hastelloy C-2000/$TiCl_4$ Heat Pipe Showing Presence of Reaction Layer.

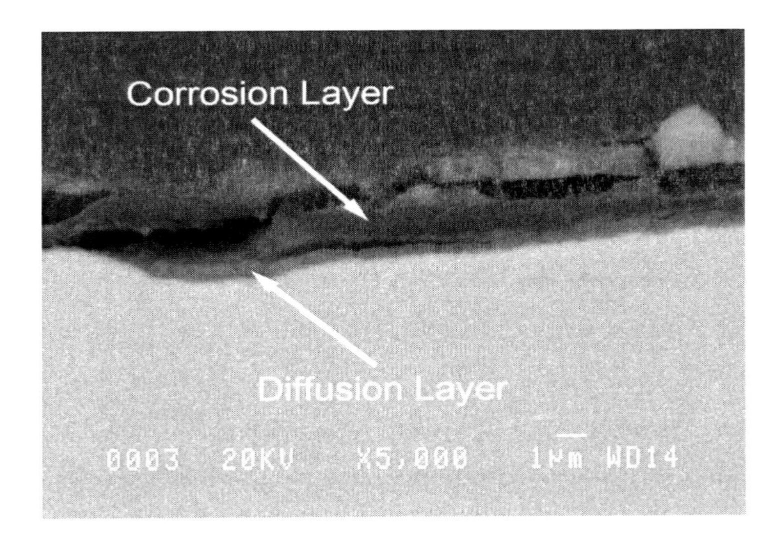

Figure 11. SE Image of the Envelope in the Hastelloy C-2000/$TiCl_4$ pipe showing the presence of a Reaction Layer.

SEM observations of the Hastelloy C-2000 with $TiCl_4$ in Figure 10 revealed a 1 to 2 micrometer thick corrosion layer on the surface. The thickness of the layer varied as the outer surface was not planar. EDS analysis showed the reaction layer consisted of Ni-33 wt.% Ti-18 wt.% Mo-18 wt.%Cr-4 wt.% Cu-2 wt.% Cl. The layer, while non-uniform does appear to be somewhat protective and adherent.

High magnification BSE images of the interface such as the one shown in Figure 11 showed the presence of a very thin (~0.5 micrometer) thick layer beneath the reaction layer that was darker than the surrounding matrix. The thinness of the layer prevented good EDS analysis, but the BSE indicates that there is a diffusion zone beneath the reaction layer that has lost the heavier elements such as Mo. This is consistent with the 18 wt.% Mo observed in the reaction layer.

The Hastelloy C-22 envelope and wick/$AlBr_3$ working fluid heat pipe exhibited a dual corrosion layer with a total thickness of 5 to 10 micrometers. The two corrosion layers were about equal in thickness, but there is variability due to a wavy interface between the two layers. EDS analysis of the two layers showed that the outer layer composition was Ni-11.5 wt.% Cr-11.9 wt.% Mo-3.6 wt.% Fe-9.4 wt.% W-0.6 wt.% Mn-1.7 wt.% Co-0.3 wt.% V-0.8 wt.% Si-9.5 wt.% Br. The inner corrosion layer composition was Ni-12.8 wt.% Cr-12.4 wt.% Mo-3.2 wt.% Fe-6.4 wt.% W-0.2 wt.% Mn-1.3 wt.% Co-0.3 wt.% V-21.9 wt.% Br. Spot EDS analysis immediately beneath the corrosion layer showed the presence of 1 wt.% Br, indicating that the Br may be diffusing into the metal substrate. Based upon these analyses, it appears that Br can react with the C-22, but it takes a considerable length of time to build up the corrosion layers, in this case 28,560 hours.

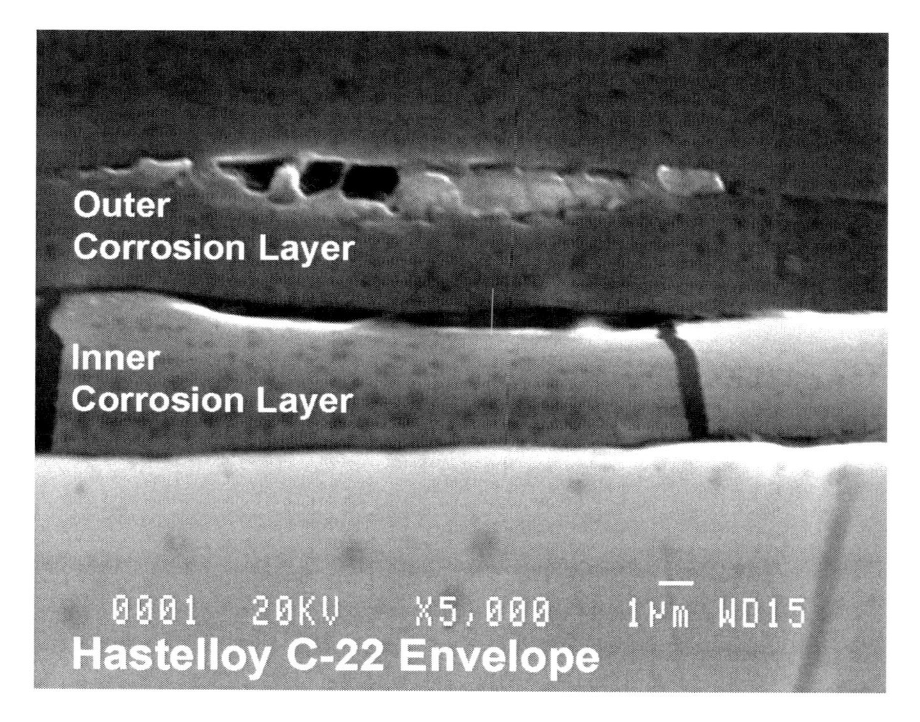

Figure 12. SE Image of Hastelloy C-22-$AlBr_3$ Heat Pipe Showing Two Corrosion Layers.

As shown in Figure 12, there is evidence of through-thickness cracking in the inner corrosion layer. This may be caused by a CTE mismatch between this layer and the substrate since there is no continuation of the crack into the outer layer and there is no evidence of any reaction on the surfaces of the cracks or development of corrosion product within the cracks. If this is the case, operating the heat pipes isothermally should result in no cracking and a relatively protective corrosion layer.

Chemical Analysis of Working Fluids

Table 7 contains the results of the chemical analysis of the working fluids. Only the elements that could be present from dissolution of the metals are listed. Since the halides are reactive, they were chemically neutralized with water prior to chemical testing to allow safe testing of the working fluids. The stable sediment was then sent out for analysis.

The chemical analyses of the heat pipes that use water as a working revealed that there was some pickup of metal from the metals, most notably Cu for the Monel heat pipes. However, the levels were in the very low ppm range and represent very minimal contamination of the water. There was no evidence of corrosion that resulted in the movement of metal from the envelopes and wicks into the working fluid. For the Monel samples, this is consistent with the diffusion of Cu to the surface rather than a dissolution/precipitation process for creating the large Cu surface nodules.

The heat pipes that used halides as a working fluid showed more contamination of the working fluids. Heat pipes 153 and 157 which appeared to form a protective corrosion layer showed some of the lowest amounts of contamination. Total contamination was on the order of 300 to 350 ppm. While indicating that some corrosion occurred, the amount was small and should be relatively insignificant.

All of the heat pipes with a high gas content had 1% or more of the envelope constituents dissolved in the halide working fluid. In comparison to Heat Pipes 153 and 157, Heat Pipes 7 and 8 which used $SnCl_4$, suffered considerably more contamination of the working fluids with Cr being the major metal present though traces of most constituents of the alloys used for the envelopes and wicks were observed. The relative amounts seemed to be consistent with the levels of attack observed with Heat Pipe 8 undergoing much more attack and reaction than Heat Pipe 7.

Table 7. Contaminants Found in Working Fluids (weight percent)

Heat Pipe	Working Fluid	Life Test Hours	Al	Co	Cr	Cu	Fe	Mn	Mo	Ni	Ti	V	W
6	$SnCl_4$	20,160	0.007		0.38		0.038	0.012	0.79	1.78			0.02
7	$SnCl_4$	20,160	0.006	0.001	0.11		0.004	0.057	0.083	0.027			
8	$SnCl_4$	20,160	0.005	0.007	0.7	0.022	0.018	0.003	0.31	0.83			0.001
9	$GaCl_3$	20,040									1.2		
10	$GaCl_3$	20,040									1.2		
153	$TiCl_4$	28,560			0.006		0.027			0.003			
157	$AlBr_3$	28,704			0.013		0.018			0.002			
100	Water	48,100									0.00013		
103	Water	48,100									0.000016		
105	Water	48,100	0.000007			0.0011	0.000031	0.00021		0.00056	0.000008		
107	Water	48,100	0.000005			0.0021	0.00002	0.0016		0.00041	0.000006		
121	Water	39,701									0.000018		
122	Water	39,701									0.000012		
123	Water	42,528	0.000011								0.000025		
124	Water	39,917									0.000037		
133	Water	34,344	0.000007			0.000021							
134	Water	35,040		0.000005		0.00064	0.000015	0.00095		0.00011			
135	Water	35,544	0.000007								0.000062	0.000005	
Di Water STD	Water		<0.000005	<0.000005		<0.000005	<0.000005	<0.000005		<0.000005	<0.000005		

Since titanium was the only metal in both the envelope and the fluid, no fluid analysis was performed for Heat Pipe 4 (CP-Ti/$TiBr_3$), which had little evidence of attack. The high level of Ti in the $GaCl_3$ for Heat Pipe 10 was consistent with the large amount of corrosion and possible Ti-containing particles in the working fluid. Recall that this pipe developed a leak in the first few hours after it was put on life test.

CONCLUSION

A survey was conducted life test results for intermediate temperature working fluids. Life tests have been conducted with 30 different intermediate temperature working fluids, and over 60 different working fluid/envelope combinations. Life tests have been run with three elemental working fluids: sulfur, sulfur-iodine mixtures, and mercury. Other fluids offer benefits over these three elemental fluids in this temperature range. Mercury is toxic, has a high density, and problems have been observed with getting the mercury to wet the heat pipe wick. Sulfur and Sulfur/Iodine have high viscosities, low thermal conductivities, and are chemically aggressive.

Life tests have been conducted with 19 different organic working fluids. As the temperature was increased, all of the organics started to decompose. Typically, they generate non-condensable gas, and often the viscosity increases. At high enough temperatures, carbon deposits can be generated. The maximum operating temperature is a function of how much NCG can be tolerated, and the heat pipe operating lifetime. Three sets of organic fluids stand out as good intermediate temperature fluids:

1. Diphenyl, diphenyl oxide, and eutectic diphenyl/diphenyl Oxide (Dowtherm A, Therminol VP, Diphyl)
2. Naphthalene
3. Toluene

A non-organic working fluid is desirable for nuclear fission space power and other applications where radioactivity can generate gas with organic working fluids. Long term life tests showed that Superalloys/$TiCl_4$ at 573 K (300°C), and Superalloys/$AlBr_3$ at 673K (400°C) are compatible. As of May 2013, the $AlBr_3$ and $TiCl_4$ tests have been running for over 59,000 hours (6.7 years).

Hastelloy C-2000 underwent little corrosion when used with TiCl4 working fluid, with the formation of only a 1-2 micrometer thick corrosion layer. Hastelloy C-22 exhibited a 5-10 micrometer thick dual corrosion layer when tested with AlBr3 working fluid. The working fluids of these two heat pipes exhibited total metal contents between 300 and 350 ppm. The results indicate that the tested envelope materials and working fluids can form viable material/working fluid combinations.

Titanium/water and Monel/water heat pipes are compatible at temperatures up to 550 K (267°C), based on ongoing life tests that have been running for up to 72,000 hours (8.2 years) as of May 2013. Analysis of titanium/water heat pipe cross-sections using optical and electron microscopy revealed little if any corrosion even when observed at high magnifications. When any evidence of corrosion was observed, the layer was typically around 1 micrometer thick. Copper depleted zones, as well as copper surface nodules formed on the Monel 400 screen wick. This was not observed on the Monel K500 envelopes. An analysis of the water working fluids showed minimal pickup of metals.

ACKNOWLEDGMENTS

The water and halide life tests were sponsored by NASA Glenn Research Center under Contracts NNC05TA36T, and NNC06CA74C. The authors would like to thank Duane Beach, Cheryl Bowman, Ivan Locci, and Jim Sanzi of NASA Glenn Research Center for helpful discussions about the fluids and materials. The authors would like to acknowledge the metallographic sample preparation and interpretation by Joy Buehler of NASA Glenn Research Center. The authors would also like to thank Laurie Anderson, Al Basiulis, Claus Busse, Don Ernst, Manfred Groll, Larry Grzyll, and Bob Reid for their generous help in locating and supplying references.

ACRONYMS

BSE	Backscatter Electron (Image)
EDS	Energy Dispersive Spectroscopy
NCG	Non-Condensable Gas
SE	Secondary Electron (Image)

REFERENCES

Anderson, WG. “Sodium-Potassium (NaK) Heat Pipe,” Heat Pipes and Capillary Pumped Loops, Ed. A Faghri, A. J. Juhasz, and T. Mahefky, *ASME HTD*, 236, pp. 47-53, 29th National Heat Transfer Conference, Atlanta, Georgia, August 1993.

Anderson, WG; Rosenfeld, JR; Angirasa, D; Mi, Y. “The Evaluation of Heat Pipe Working Fluids In The Temperature Range of 450 to 750 K,” *Proceedings*, *STAIF*, 2004, pp. 20-27, Albuquerque, NM, February 8-12, 2004.

Anderson, WG. *“Evaluation of Heat Pipes in the Temperature Range of 450 to 700 K,” STAIF*, 2005, Albuquerque, NM, February 13-17, 2005.

Anderson, WG; Dussinger, PM; Bonner III, RW; Sarraf, DB. "High Temperature Titanium-Water and Monel-Water Heat Pipes," *Proceedings of the 2006 IECEC, AIAA, San Diego, CA*, June 26-29, 2006a.

Anderson, WG; Dussinger, PM; Sarraf, DB. "High Temperature Water Heat Pipe Life Tests," *STAIF* 2006, pp. 100-107, American Institute of Physics, Melville, New York, 2006b.

Anderson, WG. "Intermediate Temperature Fluids for Heat Pipes and LHPs," *Proceedings of the 2007 IECEC, AIAA*, St. Louis, MO, June 25-27, 2007a.

Anderson, WG; Bonner III, RW; Dussinger, PM; Hartenstine, JR; Sarraf, DB; Locci, IE. "Intermediate Temperature Fluids Life Tests – Experiments" *Proceedings of the 2007 IECEC, AIAA*, St. Louis, MO, June 25-27, 2007b.

Anderson, WG; Hartenstine, JR; Sarraf, DB; Tarau, C. "Intermediate Temperature Fluids for Heat Pipes and Loop Heat Pipes," *15th International Heat Pipe Conference, Clemson, SC*, April 25-30, 2010.

Anderson, WG; Tamanna, S; Tarau, C; Hartenstine, JR; Ellis, D. "Intermediate Temperature Heat Pipe Life Tests," *43rd International Conference on Environmental Systems (ICES 2013)*, Vail, CO, July 14-18, 2013.

Antoniak, ZI; Webb, BJ; Bates, JM; Cooper, MF. "Construction And Testing of Ceramic Fabric Heat Pipe With Water Working Fluid," *Proceedings of the 8th Symposium on Space Nuclear Power Systems*, 217, American Institute of Physics, Melville, NY, pp. 125-134, Jan. 1991.

ASM International, "Cu-Ni Phase Diagram", *ASM Handbook*, Vol. 3, Alloy Phase Diagrams, Materials Park, OH, p. 2-173, 1992.

Basiulis, A; Prager, RC. "Compatibility and reliability of heat pipe materials," AIAA-1975-660, *10th AIAA Thermophysics Conference, Denver, Colo.*, May 27-29, 1975.

Basiulis, A; Fuller, M. "Operating Characteristics and Long Term Capabilities of Organic Fluid Heat Pipes," AIAA No. 71-408, *AIAA 6th Thermophysics Conference*, 1971.

Devarakonda, A; Anderson, WG. "Thermo-Physical Properties of Intermediate Temperature Heat Pipe Fluids," STAIF 2005, Albuquerque, NM, February 13-17, 2005. *NASA Report NASA/CR—2005-213582*, available from the NASA Glenn Technical Reports Server, http://gltrs.grc.nasa.gov/.

Devarakonda, A; Olminsky, JE. "An Evaluation of Halides and Other Substances as Potential Heat Pipe Fluids," *Proceedings of the 2004 IECEC, Providence*, RI, August 16-19, 2004.

Deverall, JE. "Mercury as a Heat Pipe Fluid," ASME Paper 70-HT/Spt-8, *American Society of Mechanical Engineers*, 1970.

Eastman, Y. *Personal communication*, 2007.

Ernst, DM. *Personal communication*, 2006.

Groll, M; Brost, O; Heine, D; Spendel, T. "Heat Transfer, Vapor-Liquid Flow Interaction and Materials Compatibility in Two-Phase Thermosyphons," *CEC Contractors Meeting, Heat Exchangers – Heat Recovery*, Brussels, June 10, 1982.

Groll, M; Brost, O; Roesler, S. "*Development of High Performance Closed Two-Phase Thermosyphons as Heat Transfer Components for Heat Recovery from Hot Waste Gases*," *EG-Status Seminar*, Brussels, October, 1987.

Groll, M. "Heat Pipe Research and Development in Western Europe", *Heat Recovery Systems and CHP (Combined Heat & Power)*, 9(1), pp. 19-66, 1989.

Grzyll, LR; Ramos, C; Back, DD. "Density, Viscosity, and Surface Tension of Liquid Quinoline, Naphthalene, Biphenyl, Decafluorobiphenyl, and 1,2-Diphenylbenzene from 300 to 400°C," *J. Chem. Eng. Data*, Vol. 41, pp. 446-450 1996.

Grzyll, LR; Back, DD; Ramos, C; Samad, NA. "Characterization and Testing of Novel Two-Phase Working Fluids for Spacecraft Thermal Management Systems Operating Between 300°C and 400°C," *Final Report to Phillips Laboratory*, Kirtland Air Force Base, No. PL-TR-95-1089, 1995.

Grzyll, LR; Back, DD; Ramos, C; Samad, NA. "Characterization and Testing of Novel Two-Phase Working Fluids for Spacecraft Thermal Management Systems Operating Between 300°C and 400°C," *Proceedings of the 1st Annual Spacecraft Thermal Control Symposium*, Albuquerque, NM 1994.

Grzyll, LR. "Heat Pipe Working Fluids for Thermal Control of the Sodium/Sulfur Battery," *Proceedings of the 26th Intersociety Energy Conversion Engineering Conference*, Vol. 3, pp. 390-394, American Nuclear Society, La Grange, Illinois, 1991.

Hartenstine, JR. *Personal communication*, 2007.

Heine, D; Groll, M; Brost, O. "Chemical Compatibility and Thermal Stability of Heat Pipe Working Fluids for the Temperature Range 200°C to 400°C," *8th ChiSA Congress*, Prague, September 3-7, 1984.

Ishizuka, M; Sasaki, T; Miyazaki, Y. "Development of Titanium Heat Pipes For Use In Space," *Proceedings of the Symposium on Mechanics for Space Flight*, pp. 157-165, March, 1985.

Jaworskie, D. *Personal communication*, April 5, 2007.

Kenney, DD; Feldman, KT. "Heat Pipe Life Tests at Temperatures up to 400°C," *Proceedings of the 13th Intersociety Energy Conversion Engineering Conference*, pp. 1056-1059, San Diego, CA, Aug. 20-25, 1978.

Locci, IE; Devarakonda, A; Copeland, EH; Olminsky, JK. "Analytical and Experimental Thermo-Chemical Compatibility Study of Potential Heat Pipe Materials," *Proceedings of the 2005 IECEC*, San Francisco, CA, August 15-18, 2005.

Los Alamos Scientific Laboratory, *Quarterly Status Report on the Space Electric Power R&D Program for the Period Ending April 30, 1970, Part 1*, Report No. LA-4446-MS, pp., 2-5, May, 1970.

Los Alamos Scientific Laboratory, *Quarterly Status Report on the Space Electric Power R&D Program for the Period Ending January 31, 1968, Part 1*, Report No. LA-3881-MS, pg. 4, February, 1968a.

Los Alamos Scientific Laboratory, *Quarterly Status Report on the Space Electric Power R&D Program for the Period Ending April 30, 1968, Part 1*, Report No. LA-3941-MS, pg. 2, May, 1968b.

Lundberg, LB; Merrigan, M; Prenger, FC; Dunwoody, W. "Sulphur Heat Pipes," *Energy Technology*, Los Alamos Scientific Laboratory, LA 8797-PR, October-December 1980, pp. 69-70.

Lundberg, LB. "Titanium-potassium heat pipe corrosion studies," *International Symposium on High Temperature Corrosion in Energy Systems*, Detroit, MI, Sep. 1984.

Monel 400 Technical Bulletin, Special Metals, http://www.specialmetals.com/-assets/smc/documents/alloys/monel/monel-alloy-400.pdf, accessed September 20, 2017.

Monel K-500 Technical Bulletin, Special Metals, http://www.specialmetalswiggin.co.uk-/pdfs/products/MONEL%20alloy%20K-500.pdf, accessed September 20, 2017.

National Physical Laboratory *Calculated Ga-Ti Phase Diagram*, London, UK, http://resource.npl.co.uk/mtdata/phdiagrams/gati.htm, retrieved March 12, 2012.

Polasek, F; Stulc, P. "Heat Pipe for the Temperature Range from 200 to 600°C," *Proc., Second International Heat Pipe Conference*, Bologna, Italy, 2, pg. 711, 1976.

Polasek, F. "Heat Pipe Research and Development in East European Countries," *6th International Heat Pipe Conference (1987), Heat Recovery Systems and CHP*, 9(1), pp. 3-17, 1989.

Reid, RS; Merrigan, MA; Sena, JT. "Review of Liquid Metal Heat Pipe Work at Los Alamos," *8th Symposium on Space Nuclear Power Systems*, Albuquerque, NM, January 6-10, 1991.

Saaski, EW; *Owzarski, and Two-Phase Working Fluids for the Temperature Range 50° to 350°C*, Sigma Research, Inc., Final Report, Contract NAS3-20222, *NASA Lewis Research Center*, June 1977a.

Saaski, EW; Tower, L. "Two-Phase working fluids for the temperature range 100-350°C," American Institute of Aeronautics and Astronautics, *12th Thermophysics Conference*, Albuquerque, NM, June 27-29, 1977b.

Saaski, EW; Hartl, JH. *Two-Phase Working Fluids for the Temperature Range 50 to 350°C*, Sigma Research, Inc., Phase II Final Report, Contract NAS3-21202, NASA Lewis Research Center, March, 1980.

Sena, JT; Merrigan, MA. "Niobium 1 percent Zirconium/Potassium and Titanium/Potassium Life-Test Heat Pipe Design And Testing," *Proceedings of the 7th Symposium on Space Nuclear Power Systems*, 195, American Institute of Physics, Melville, NY, Jan. 1990.

Sanzi, JL; Jaworske, DA. "*Heat Pipes and Heat Rejection Component Testing at NASA Glenn Research Center,*" *Nuclear and Emerging Technologies for Space (NETS-2011)*, Albuquerque, New Mexico, February 7–10, 2011.

Sanzi, JL. *Personal communications*, September 20, 2017.

Sarraf, DB; Bonner III, RW. "Passive Thermal Management for a Fuel Cell Reforming Process," *2006 International Energy Conversion Engineering Conference, San Diego, CA*, June 2006.

Swanson, T; Buchko, M; Brennan, P; Bello, M; Stoyanof, M. "Cryogenic Two-Phase Flight Experiment; Results Overview" proceedings of the 1995 Shuttle Small Payloads Symposium, *NASA*, pp. 111-123, Baltimore, MD, September 25-28, 1995.

Tarau, C; Sarraf, DB; Locci, IE; Anderson, WG. "Intermediate Temperature Fluids Life Tests – Theory," *Proceedings*, *STAIF*, 2007, Albuquerque, NM, February 11-15, 2007.

Timrot, DL; Serednitskaya, MA; Medveditskov, AN; Traktueva, SA. "Thermophysical Properties of a Sulfur-Iodine Binary System as a Promising Heat Transfer Medium for Heat Pipes," *Journal of Heat Recovery Systems (now Applied Thermal Engineering)*, Vol. 1(4), pp. 309-314, 1981.

Vasil'ev, LL; Volokhov, GM; Gigevich, AS; Rabetskii, MI. "Heat Pipes Based on Naphthalene," *Journal of Engineering Physics and Thermophysics*, Vol. 54, No. 6, pp. 623-626, 1988.

In: Heat Pipes: Design, Applications and Technology
Editor: Yuwen Zhang
ISBN: 978-1-53613-908-2

Chapter 8

PRESSURE CONTROLLED HEAT PIPES AND WARM-RESERVOIR VARIABLE CONDUCTANCE HEAT PIPES

Calin Tarau [*] ***and William G. Anderson***
Advanced Cooling Technologies, Inc., Lancaster, PA, US

ABSTRACT

Variable Conductance Heat Pipes (VCHPs) are used to passively control the evaporator (and associated electronics temperature) while the power and heat sink conditions vary widely. Essentially all VCHPs use an electrically heated, cold Non-Condensable Gas (NCG) reservoir, and typically control the evaporator temperature to within ±2°C. The first warm-reservoir VCHPs actually fabricated and tested are described here. In these devices, the NCG reservoir is located next to the evaporator, so the need for electrical power is eliminated, while maintaining the same ±2°C control. Eliminating power is a requirement for missions that rely on batteries during long periods of darkness, such as Lunar landers and rovers, as well as research balloons in the polar regions during the long periods of winter darkness. Pressure Controlled Heat Pipes (PCHPs) achieve tighter temperature control by changing the amount of NCG in the system, either by changing the reservoir volume, or adding/subtracting non-condensable gas. A PCHP suitable for operation in microgravity was developed using a grooved aluminum heat pipe, and controlled temperatures within ±50mK using a flexible bellows to change the reservoir volume. Even more precise temperature control is required for temperature calibration by the national laboratories. A PCHP was developed that can control the temperature within ±1mK. The setpoint can be changed by adding/removing NCG. PCHPS can also be used for high temperature power switching.

[*] Corresponding Author Email: Calin.Tarau@1-act.com.

Keywords: heat pipe, pressure controlled heat pipe, PCHP, variable conductance heat pipe, VCHP, warm-reservoir variable conductance heat pipe, temperature, calibration, isothermal, stability, variable thermal link, spacecraft thermal control, precise thermal control

INTRODUCTION TO VARIABLE CONDUCTANCE HEAT PIPES

Variable Conductance Heat Pipes (VCHPs) and Pressure Controlled Heat Pipes (PCHPs) are used when the heat pipe evaporator temperature (and associated instrumentation) must be maintained within a temperature band while the power and/or heat sink conditions vary. A standard heat pipe is filled with a two-phase working fluid, and a wick to return the condensate from the condenser to the evaporator. In a VCHP, Non-Condensable Gas (NCG) is added to the heat pipe, in addition to the working fluid. Depending on the operating conditions, the NCG can block all, some, or none of the available condenser length. When the VCHP is operating, the NCG is swept toward the condenser end of the heat pipe by the flow of the working fluid vapor. At high powers, all of the NCG is driven into the reservoir, and the condenser is fully open; see Figure 1. As the power is lowered, the vapor temperature drops slightly. Since the system is saturated, the vapor pressure drops at the same time. This lower pressure allows the NCG to increase in volume, blocking a portion of the condenser. At very low powers, the vapor temperature and pressure are further reduced, the NGC volume expands, and most of the condenser is blocked. This change in active condenser length minimizes the drop in the evaporator and associated electronics temperatures over large changes in power and evaporator sink conditions.

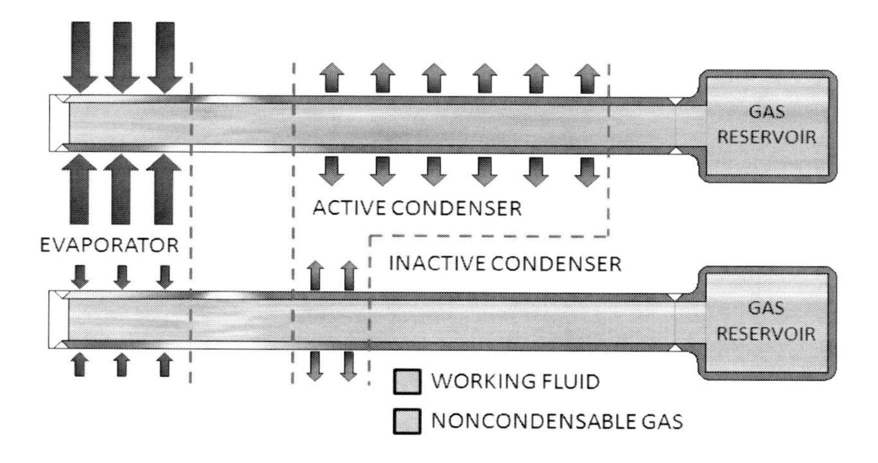

Figure 1. At high powers, the non-condensable gas in a VCHP is driven by the vapor flow into the gas reservoir. As the temperature drops slightly, the saturation pressure drops more rapidly, and the non-condensable gas starts to block the condenser.

Cold-Biased Reservoirs

For spacecraft and satellites orbiting the Earth, the power available and heat sink conditions vary widely. When the sun shines on the solar panels, electricity is produced. Since there is no atmosphere, all of the waste heat generated needs to be radiated. The amount of heat that can be rejected is dependent on the radiator orientation with respect to the sun, with minimal heat rejection when the radiator is pointed towards the sun. Spacecraft radiators are generally sized for this hot condition, with high powers to be radiated, and a warm sink.

During an eclipse, no power is generated, and any power required must be supplied by batteries. At the same time, the absence of the sun means that the radiator sees a colder sink. With lower waste heat produced, and a colder sink, the temperature of the batteries and electronics can drop, unless there is a variable thermal link between the heat sources and the radiator. VCHPs are one commonly used variable thermal link.

A typical spacecraft VCHP is shown in Figure 2. The grooved heat pipe is extruded with the attached flange for heating in the evaporator, and removing heat in the condenser. The reservoir is fabricated from stainless steel for reduced mass, and attached to the heat pipe with an aluminum/stainless transition piece. A screen wick in the reservoir allows any fluid that collects in the reservoir to be wicked back into the heat pipe.

VCHPs are generally used when the temperature of the payload that is cooled must be controlled within a few degrees Celsius. The spacecraft is designed so that the reservoir is cold-biased, meaning that its equilibrium temperature is below the required payload temperature. Electric heaters on the reservoir are then used to control the payload, typically within ±1-2°C. The heater typically requires a few Watts of power when it is on. Figure 3 shows how the VCHP in Figure 2 maintained a ±2°C temperature as the power was changed from 75 to 150W, and the sink reduced from +15°C to -65°C.

The Orbiting Astronomical Observatory satellite was the first known use of a VCHP on orbit. (Mock, Marcus, and Adelman, 1974). That VCHP maintained an electronics box within ±2K while power dissipation changed from 15 to 35 Watts. Beinert and Brennan (1971) demonstrated a VCHP with a heated reservoir and electronic feedback. The steady-state input temperature varied less than 2 K while the heat load varied from 5 Watts to 35 Watts. The transient response time was about 20 minutes.

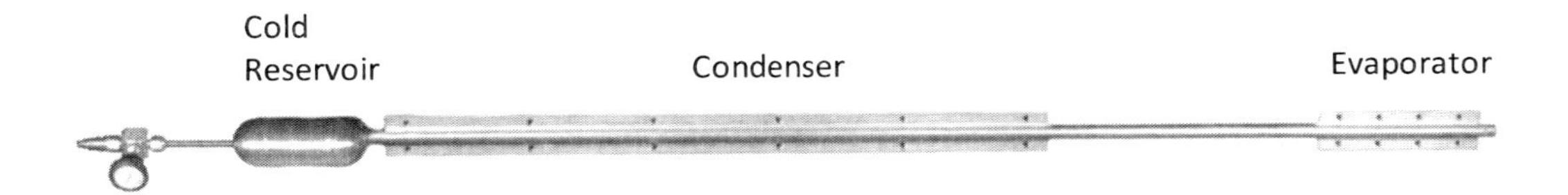

Figure 2. Spacecraft Variable Conductance Heat Pipe, with the evaporator on the right, and the condenser and cold-biased NCG reservoir on the left.

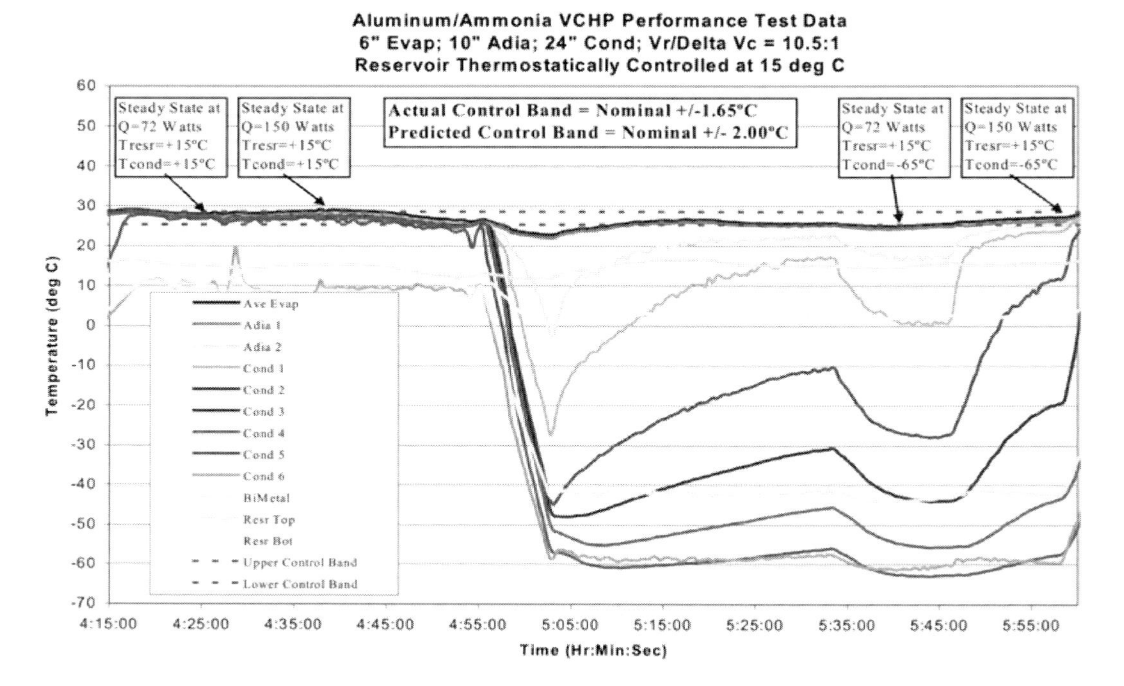

Figure 3. Performance test data for an aluminum/ammonia variable conductance heat pipe with an electrically heated reservoir. The evaporator temperature was maintained in a ±1.65°C control band, as power was varied from 72 to 150 W, and heat sink temperature from +15°C to -65°C.

In some applications with a very wide temperature band, the cold reservoir may not be heated at all. Anderson et al., (2013) developed a thermosyphon VCHP radiator for Lunar and Martian fission reactors. A single-phase pumped water loop removes waste heat from the energy convertors, and transfer the heat to the thermosyphons. During the lunar night and periods of low power production, it is imperative to prevent freezing of the water loop. The gas charge in the condenser blocks more of the condenser as the heat pipe evaporator temperature is reduced. This variable thermal link allows the heat pipe evaporators (and the water pumped loop to remain well above freezing during low power conditions. In addition, the gas blocks the sublimation of water from the heat pipe evaporator into the colder condenser, allowing the heat pipe evaporator to remain charged when the condenser is below freezing (Ellis and Anderson, 2009).

WARM-RESERVOIR VARIABLE CONDUCTANCE HEAT PIPES

As discussed above, most VCHPs have a cold-biased reservoir that is electrically heated to maintain the heat pipe evaporator (and associated electronics) at the desired temperature. Eclipse times on the solar panels are generally short, so only a small amount of battery power is required.

In some cases, it is desirable to eliminate any electrical heating of a cold reservoir. This can occur during science missions with limited power, and long periods without power. Lunar landers and rovers are one such cases, where it is estimated that providing 1 W of power over the 14-day-long Lunar night requires 5 kg of solar cells and batteries. Research balloons circling around the North or South Poles during the winter months long night is another.

Several early heat pipe handbooks briefly discuss the concept of warm (also called hot) reservoir VCHPs (Marcus, 1972, Brennan and Kroliczek, 1979). However, this VCHP configuration has never been developed any further than conceptual line drawings. To the best of authors' knowledge, the warm reservoir VCHPs discussed below are the first that have been actually fabricated and tested.

Benefits

The main benefit that a warm reservoir VCHP offers is the fact that it provides a significantly tight thermal control passively (with no electrical power). Therefore, control power can be reduced significantly or even totally eliminated. As mentioned above, control power reduction automatically translates into important battery mass savings. In addition, a hot reservoir VCHP would have a more compact configuration since the reservoir can be smaller and also, an integrated version where reservoir is wrapped around the evaporator, could save significant space and volume. A comparison between passive thermal control capability offered by a hot reservoir VCHP and a cold biased reservoir VCHP is shown both theoretically and experimentally for five different working fluids later in this Chapter.

Hot Reservoir VCHP with Active Control and Non-Integrated Configuration

As noted above, the warm reservoir VCHPs may have integrated or non-integrated reservoir-evaporator configurations. The initial warm reservoir VCHP design is shown in Figure 4 (Anderson et al., 2012). This VCHP was designed as Variable Thermal Link for Lunar Landers that must survive the long and cold Lunar nights with minimum power consumption.

The VCHP was fabricated and tested successfully in laboratory environment and some of the relevant results are presented below in Figure 5. It has a non-integrated configuration since the reservoir is not thermally connected to the evaporator's body. In this case, the VCHP operates in an active control mode since the reservoir is separately heated to follow evaporator's temperature. As seen in Figure 4, the VCHP consists of a main body (evaporator section, adiabatic section, and condenser section), a non-wicked reservoir and an NCG tube. Another fact that makes the hot reservoir VCHP design compact is the NCG tube location. The tube is internal and concentrical to the actual pipe. The reservoir, that is located near the evaporator, is connected to the tube outside of the pipe, just before the tube enters the end of the evaporator. The NCG tube then continues through the entire VCHP all the way to the end of condenser where it ends with an opening. The only gas connection between the reservoir and the evaporator is this opening.

An additional benefit for locating the NCG tube internally is the fact that, being warmer than an external version, condensation of the working fluid on the inside walls of the tube is discouraged. Also, to enhance thermal resistance during off operating mode, the aluminum pipe has a stainless-steel transition portion as adiabatic section.

The VCHP had the following specifications:

- 30.5 cm (12 inch) Condenser / ≈ 48.3 cm (19 inch) Adiabatic Section – Grooved aluminum extrusion (6063-T6 Al)
    - Bimetallic Transition – 3.17 cm (1.25 inch) 6061-T6 Al × 12.7 cm (5 inch) 304 SS × 3.17 cm (1.25 inch) 6061-T6 Al
- 22.9 cm (9 inch) Evaporator – Nickel 200 50×50 screen mesh
- NCG Tube (304 SS) – 0.32 cm (0.125 inch) outer diameter
- NCG Reservoir (304 SS) – 73.7 cm3 (4.5 inch3) internal volume
- Working Fluid (Ammonia) – 20.8 grams
- Non-Condensable Gas (Neon) – ≈0.65 grams (Neon was used instead of Argon, since the condenser temperatures could be below Argon's critical temperature)

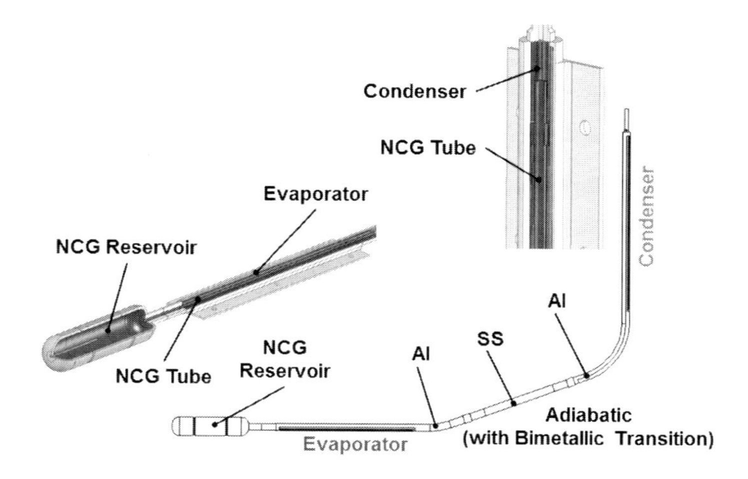

Figure 4. Schematic of the VCHP with a hybrid wick, which allows operation at different tilts for a Lunar Lander application. Placing the reservoir near the evaporator keeps the reservoir warm, minimizing the required reservoir size. Also, part of the adiabatic section is stainless steel, which minimizes heat leaks when the VCHP is inactive (low power, cold heat sink).

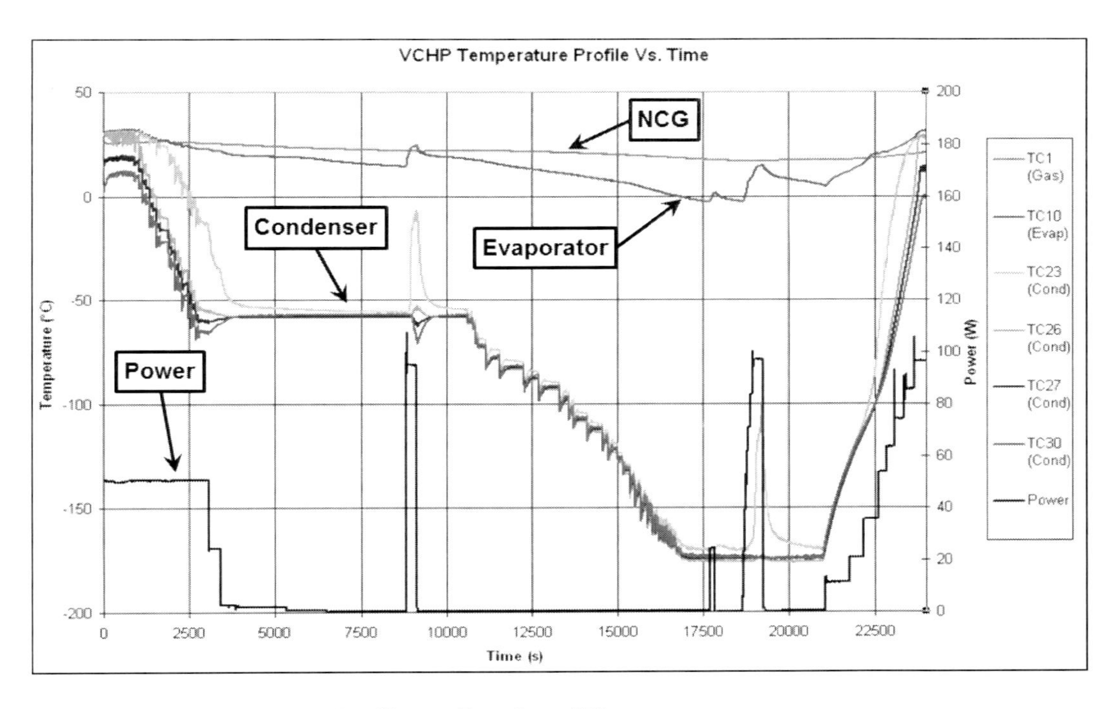

Figure 5. VCHP Temperature Profile as a Function of Time.

Table 1. VCHP Overall Conductances for Lunar Freeze/Thaw

Testing Condition	Overall Conductance (W/°C)	Power (W)
25°C Operation	4.7	95
-60°C Shutdown	0.0031	0.2
-177°C Shutdown	0.00057	0.1
9-inch Evaporator; 12-inch Condenser		

During testing, heat was supplied to the evaporator with electric cartridge heaters embedded in an aluminum block. Heat was removed from the condenser with a cold plate that was cooled with liquid nitrogen to a fixed temperature.

Figure 5 shows VCHP temperatures as a function of time during the lunar freeze/thaw test. TC1 corresponds to the gas temperature in the NCG reservoir. TC10 measures the vapor temperature of the evaporator. TC23, TC26, TC27 and TC30 detect the vapor temperature of four locations within condenser, with TC23 at the entrance of the condenser and TC30 close to the tip of the condenser. The power curve shows the electrical power input into the heater block of the evaporator. For more information on the locations of these TCs, consult Peters (2011). Initially, the pipe is operating at a nominal 25°C and 50 W. At about 6000 seconds the pipe temperature and power input are reduced to -60°C and 0.2 W, respectively. The purpose of 0.2 W of heat input was to maintain the evaporator above -10°C. At around 9000 seconds, power is temporarily increased to the full 95 W, to simulate a brief period of activity during the lunar night. After this power increase, the pipe was returned to the -60°C shutdown state. Next, the sink temperature is further reduced to -177°C (96 K, ammonia freezes at 195 K). The pipe reaches a steady-state shutdown at -177°C and 0.1 W. At approximately 17,500 seconds, the power is briefly increased to 25 W and the transient response of the frozen pipe was observed. With no indication of problems, the pipe is returned to -177°C shutdown. Power is then increased to a full 95 W for a short duration. After the full power increase, the pipe is returned to a state of shutoff until around 21000 seconds when the power is gradually increased and heat pipe startup begins. Finally, the VCHP is brought to nominal steady-state operation at 95 W and 25°C.

Table 1 lists the overall conductance (evaporator to condenser) of the VCHP during normal operation, -60°C shutdown, and -177°C shutdown. The goal of the device was to minimize conductance during shutdown, especially at very cold sink temperatures. As evidenced from Table 1, the heat pipe satisfied this goal, dropping in conductance three orders of magnitude for the -60°C shutdown state and four orders of magnitude for the -177°C shutdown states. These conductances neglect any heat in-leak from the environment. The freeze/thaw tests were of relatively short duration, and do not demonstrate conclusively that the VCHP can start up after a 14-day-long shutdown period, corresponding to the lunar night. Previous tests with titanium/water VCHPs have demonstrated startup after they were shut down with the condenser frozen for a 14-day-long period (Ellis and Anderson, 2009).

WARM-RESERVOIR VCHPS FOR HIGH ALTITUDE BALLOONS

High altitude scientific balloons provide practical and cost-effective platforms for conducting discovery science, development and testing for future space instruments. Since the balloons can reach altitudes above 36 kilometers and can stay afloat for several weeks, the payloads need a thermal management system to reject their waste heat and to maintain a stable temperature as the air (heat sink) temperature swings from as cold as -90°C to as hot as +40°C. A hot-reservoir VCHP thermal management system with an integrated reservoir was chosen. Integrating the reservoir allows the system to be designed so that the reservoir is always hotter than the evaporator, helping to keep liquid out the reservoir. (Tarau and Anderson, 2013, Weyant et al., 2016) This VCHP allows the thermal resistance to increase passively under cold operating or cold survival environment conditions, keeping the instrument section warm with minimal electric heating. This gravity-aided (thermosyphon) VCHP consists of smooth-bore, thin-wall stainless steel tubing, with methanol, toluene or pentane as working fluids. For comparison, an additional VCHP with cold reservoir was developed. Both configurations were tested with the above-mentioned working fluids and the experimental results were consistent with the modeling results.

Balloon Payload Hot Reservoir VCHP

The fabricated balloon hot reservoir VCHP is shown in below in Figure 6. The cold reservoir VCHP has similar size, however, it is not shown. Note that the NCG tube is external and also, the wicked reservoir is integrated with (wrapped around) the evaporator. Both, the non-wicked reservoir and the evaporator are made of copper. It is important to note that the hot reservoir VCHP is designed in such a way that thermal resistance between the heat source and the reservoir is slightly lower than thermal resistance between the heat source and the evaporator. As a consequence, the reservoir is always slightly warmer than the evaporator so, in addition to the NCG, only some superheated vapor can exist in the reservoir during normal operation. Since no liquid can be present, an unwicked reservoir is used.

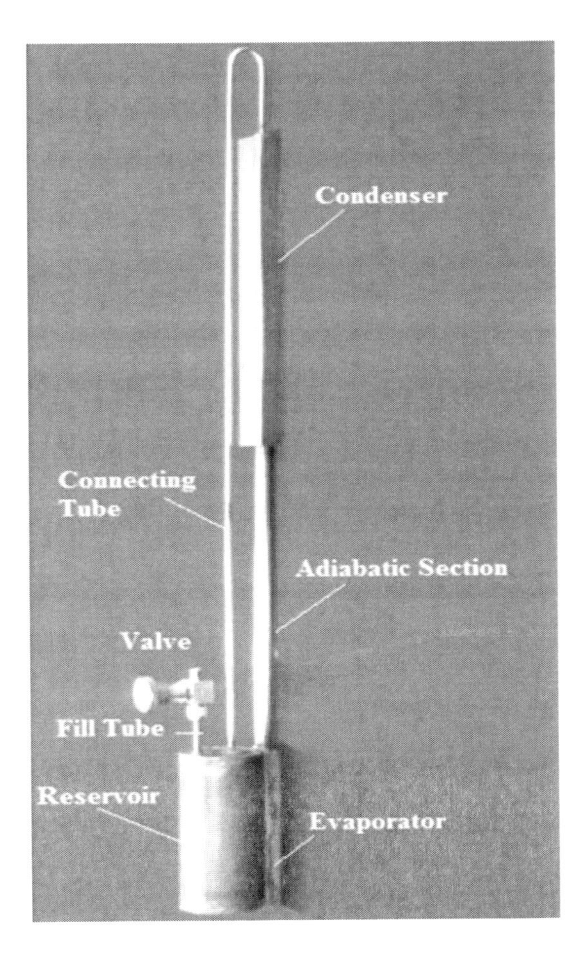

Figure 6. Hot Reservoir VCHP for High Altitude Balloon Payload Thermal Management.

MODELLING RESULTS AND PREDICTIONS

For comparison with a cold biased reservoir VCHP, modeling results for both warm-reservoir and cold-reservoir VCHPs are presented in Figure 7. Both VCHPs (including their reservoirs) have similar geometry in terms of condenser lengths, internal diameters and reservoir sizes. In addition, working conditions that include set point, power and sink temperature range are also similar. Both configurations were analyzed for five working fluids each: methanol, toluene, pentane, ammonia and propylene. Note that the first three fluids (methanol, toluene, and pentane) were also verified experimentally. As expected and shown in Figure 7, the hot reservoir configuration (Configuration 1) provides a better (tighter) temperature control than the cold reservoir configuration (Configuration 2). From the working fluid point of view, methanol is the best fluid, very closely followed by toluene. All the modeling results in terms of thermal control, for both configurations and all the five working fluids, are consistent with the fact that less volatile fluids for a given temperature range provide better thermal control.

Testing Results

Both VCHP configurations were tested in ambient conditions using liquid nitrogen to cool the condenser and cartridge heaters to heat the evaporator. For the cold reservoir VCHP (Configuration 2), the reservoir temperature was also controlled by using liquid nitrogen for cooling and cartridge heaters for heating. The reservoir temperature in Configuration 1 was not controlled since it was both thermally and physically coupled to the evaporator, and thus it closely followed the vapor temperature. As mentioned before, both VCHP configurations were tested with three working fluids: methanol, toluene and pentane. In all experimental cases, while the sink temperature was varied between -90°C and +40°C, evaporator temperature varied within a much tighter range than the required interval of -10ºC to +50ºC. The hot reservoir configuration showed the tightest temperature control. For example, the pentane based hot reservoir VCHP allowed the evaporator temperature to change only 3.7ºC from the coldest to hottest heat sinks. The largest temperature variation observed was 32.6ºC for the pentane based cold reservoir VCHP, still meeting the design requirements.

A summary of the thermal control testing results is shown in Figure 8 where the evaporator steady state temperatures are shown for both configurations and all three fluids as a function of heat sink temperature.

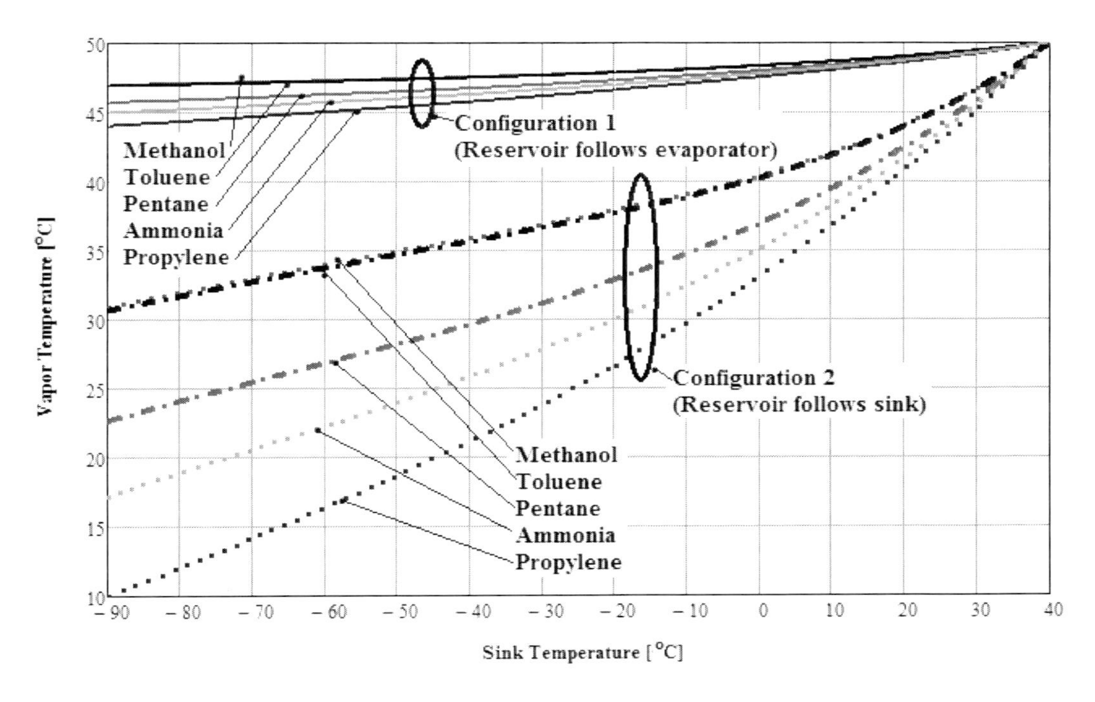

Figure 7. Evaporator temperatures variation as a function of sink temperature. Configuration 1, with a warm reservoir, has much tighter temperature control.

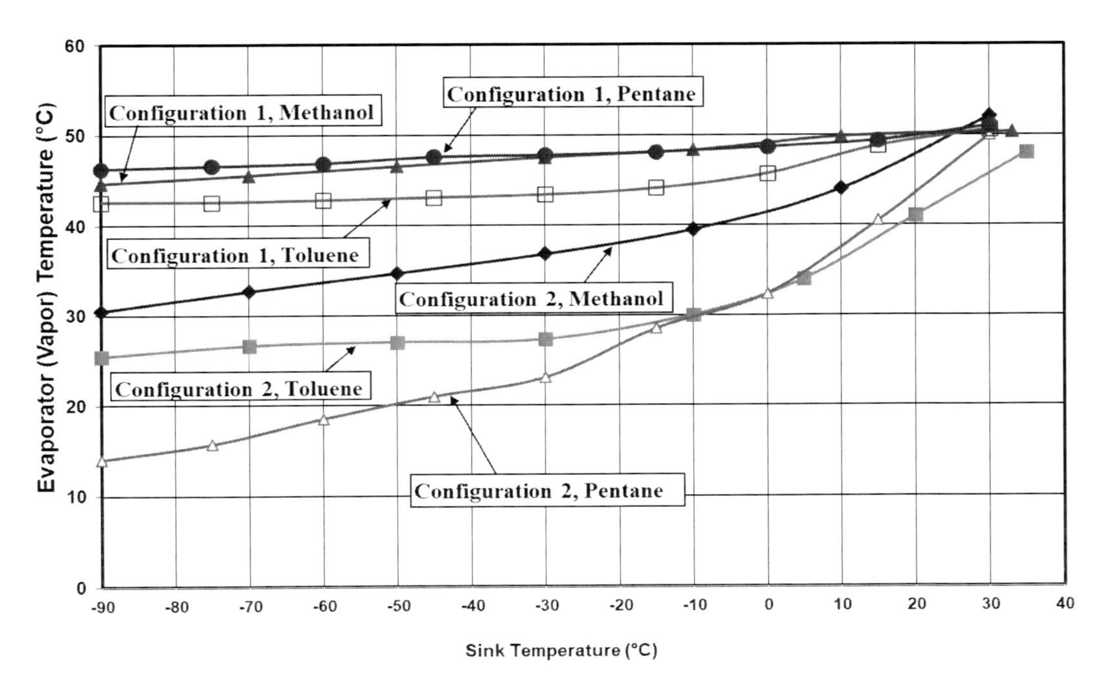

Figure 8. Steady state evaporator temperatures for both VCHP configurations and all three working fluids when sink temperature sweeps the entire required interval.

As a general conclusion, the hot reservoir VCHP (Configuration 1) shows much better temperature control than the cold reservoir VCHP (Configuration 2), both theoretically and experimentally. Moreover, both configurations with all three working fluids provide tight enough temperature control to satisfy the design requirements. The tightest temperature control was shown by pentane in hot reservoir VCHP (Configuration 1) with only a 3.7ºC (46.3 to 50ºC) temperature swing. The loosest temperature control was measured for the cold reservoir VCHP (Configuration 2) with pentane, which was 36ºC (14 to 50ºC). All other temperature swings are included in the 14 to 50ºC range, which is conservative with respect to the maximum allowed bandwidth of -5 to 50ºC.

INTRODUCTION TO PRESSURE CONTROLLED HEAT PIPES

As discussed above, VCHPs with electrically-heated, cold-biased reservoir can maintain a target temperature within ±1-2°C. The same temperature control can be demonstrated with a properly designed warm-reservoir VCHP. Pressure Controlled Heat Pipes (PCHPs) are used when even tighter temperature control, on the order of milli-Kelvins, is required for temperature calibration.

Variable Conductance Heat Pipes have a fixed amount of non-condensable gas. PCHPS achieved tighter temperature control by either changing the reservoir volume, or adding/subtracting NCG from the reservoir. Like a VCHP, the NCG is swept toward the condenser end of the heat pipe by the flow of the working fluid vapor. The NCG then blocks the working fluid from reaching a portion of the condenser, inactivating a portion of the condenser.

In a VCHP, the fraction of condenser blockage is determined by the reservoir size, the non-condensable gas charge, and the operating pressure, and cannot be adjusted once the VCHP is sealed. In contrast, the condenser blockage in a PCHP is actively controlled. In older designs, non-condensable gas is added and removed to the reservoir, allowing active control of the condenser length. In more recent designs, an actuator drives a bellows to modulate the reservoir volume. By decreasing the reservoir volume, more of the condenser is blocked. NCG may be added or removed between tests, to change the PCHP set point.

The operation of a bellows/piston type of PCHP is shown in Figure 9. Initially, the piston is withdrawn at higher powers, so that most of the condenser is open. When the heat load is reduced, the piston pushes additional gas into the condenser, helping to maintain the heat pipe at a constant temperature. While a VCHP will passively also increase the condenser blockage, PCHPs are able to react faster, and more precisely.

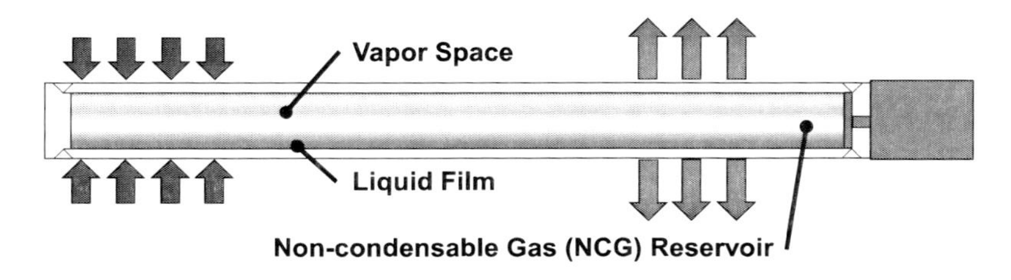

Figure 9. Pressure Controlled Heat Pipe (PCHP), where the reservoir volume is varied.

Isothermal Furnace Liners

Most PCHPs are used by Primary Standards Laboratories to calibrate reference temperature measurement equipment, which require precise temperature control, as well as a high degree of temperature uniformity inside the PCHP (Bienert, 1991, Dussinger and Tavener, 2012). These PCHPs are based on an annular heat pipe, also known as an Isothermal Furnace Liner (IFL), see Figure 10.

With a closed end, the interior can be used as a black body source to calibrate pyrometers. Thermal wells can be inserted inside the annular heat pipe to directly calibrate equipment, typically with a freeze-point cell.

During operation, the IFL is placed inside an electrically heated oven, in either a horizontal or vertical orientation. Such ovens provide non-uniform heat, both axially and radially. As shown in Figure 10, the heat vaporizes the working fluid in the wick against the outside wall. The vapor travels radially and condenses on the wick against the outside wall (as well as the wick around the thermal wells). The liquid then travels back by capillary action to the outside wick through a series of bridge wicks, and then the cycle repeats.

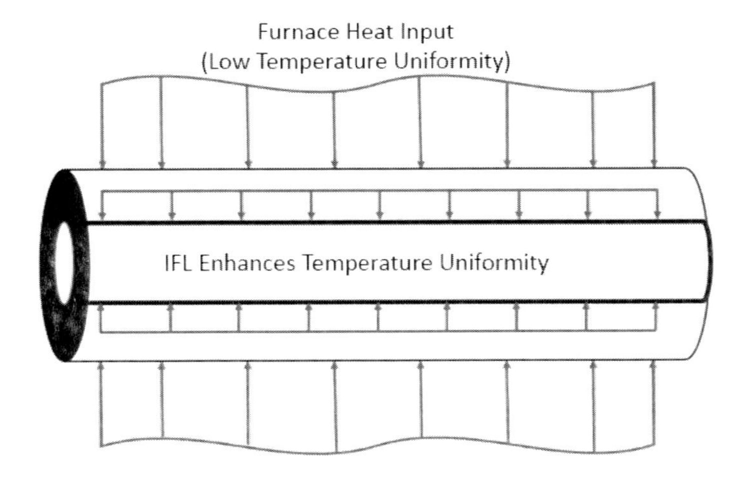

Figure 10. An Isothermal Furnace Liner is an annular heat pipe, designed so that the interior temperature is very uniform.

Most heat pipes are designed to transport large amounts of power with a minimal temperature drop. In contrast, temperature uniformity is more important in IFL applications. Most of the power in a high temperature IFL is radiated from the exterior surfaces, with just enough heat transfer to the interior cylinder to replace heat losses.

The IFL temperature is very uniform due to the very high evaporation and condensation heat transfer coefficients. Additional vapor evaporates from the outer cylinder wick where the heat flux is higher, however, the evaporation heat transfer coefficient is so high that the temperature difference is minimized. Similarly, a slightly colder patch on the inside cylinder will receive a higher heat flux until it is at the overall temperature. Bienert (1991) provides a good example of the uniformity that can be expected, which is on the order of mK.

Early PCHP Work

To the best of our knowledge, the first PCHP for temperature calibration was developed by Walter Bienert at Dynatherm (Bienert, 1991); see Figure 11. The PCHP has an annular heat pipe with a side chimney, oriented above the remainder of the heat pipe. The annular heat pipe is well insulated, and located inside the furnace. The chimney is located outside the furnace, with active cooling. The cooling means that there is a continual flow of vapor to sweeping the helium to the end of the chimney, aided by the fact that the helium is lighter than the working fluid vapor.

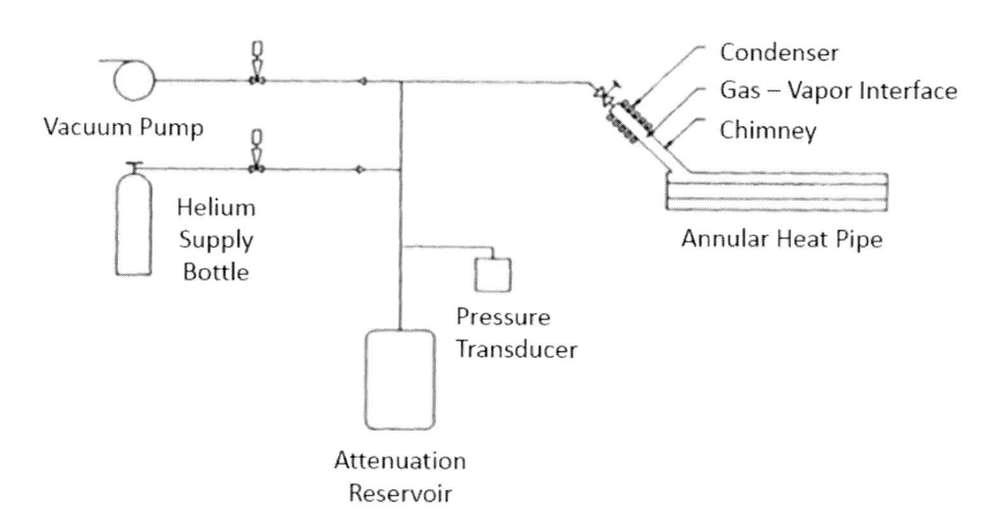

Figure 11. Early PCHPs for temperature calibration added and removed NCG to precisely control the temperature. (Bienert, 1991). The NCG gas remains in the chimney, partially due to the lower density of the helium.

The helium is supplied from a bottle, and removed with a vacuum pump. The attenuation reservoir (also known as the helium buffer volume) is much larger than the chimney, so that the pressure is minimally affected by the motion of the vapor/NCG front in the chimney. As shown in Figure 11, the PCHP is controlled using a pressure measurement device, rather than a temperature measurement device. The reason is that the heat pipe (saturation) pressure can be measured more accurately with a room temperature device, than the elevated temperature can be measured. The saturation curve is used to relate the vapor pressure and temperature.

PCHPS with Variable Reservoir Volumes

Most PCHPs for temperature calibration use the method first developed by Bienert, which adds NCG from a gas supply bottle, and removes it with a vacuum pump. These systems work well, but are not suitable for operation in space, for several reasons. First, the gas supply and pump are massive and complicated, while spacecraft systems should be simple and low mass. Second, the supply bottle limits the operating time for the PCHP, a sealed system would offer much longer operating time. As discussed above, a better design for spacecraft varies the reservoir volume (Anderson et al., 2012).

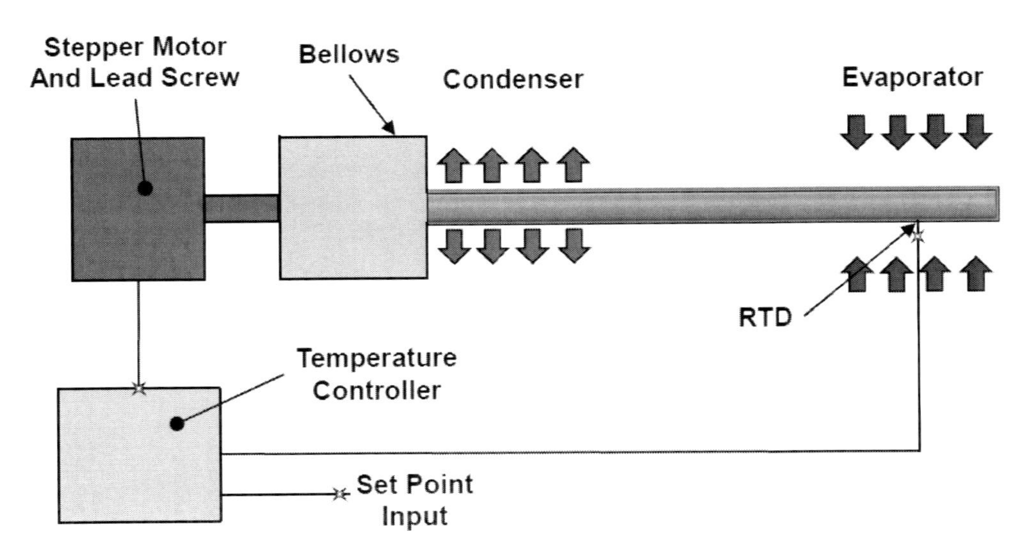

Figure 12. Schematic of Pressure Controlled Heat Pipe Showing Feedback Control of Reservoir Volume and Condenser Thermal Resistance for Precise Temperature Control.

By active control of the reservoir volume, milli-Kelvin temperature control can also be achieved, with rapid response to changing conditions. The operation of a bellows/piston type of PCHP for precise temperature control is shown in Figure 12. During operation, liquid vaporizes to cool the evaporator. The vapor travels to the condenser, then the condensate is returned by capillary forces. The non-condensable gas is pushed into the reservoir and the end of the condenser. At higher powers (or warmer sink conditions), the piston is withdrawn, so that most of the condenser is open. Less condenser area is required at lower powers, or colder sink conditions. In these cases, the piston pushes additional gas into the condenser, helping to maintain the heat pipe at a constant temperature. By using the electronics temperature to control the piston location, milli-kelvin temperature control can be reached.

PCHPS FOR PRECISE TEMPERATURE CONTROL IN MICROGRAVITY

Sarraf and co-workers developed a PCHP for precise temperature control in microgravity (Sarraf, 2008, Anderson et al., 2011, 2012). The design requirements are shown in Table 2, based on a nominal satellite application. A PCHP for operation in microgravity requires the following modifications when compared with a terrestrial PCHP:

- The heat pipe must operate in microgravity, so a grooved wick is used.
- The PCHP is required to withstand shock and vibration during launch, which is verified by shock and vibration testing prior to launch.
- The system is completely sealed with a variable reservoir volume, since the PCHP cannot be serviced after launch. This also means that the components must be designed to last for the expected life of the mission. The bellows must be designed to operate over the large number of cycles in space.
- The PCHP will operate in a vacuum environment, so must be designed to remove any heat generated in the electronics and motor during operation. Thermal vacuum testing is conducted prior to launch.
-

Table 2. Design Goals for the Microgravity PCHP

Parameter	Value
Evaporator Temperature Stability	0.001K
Input Power	50-150 Watts
Sink Temperature	-40°C to 0°C.
Life	10 years minimum, 58,400 cycles
End Use	Small Satellite on Low-Earth Orbit
Mass	Minimize
Power Consumption	Minimize
Operating Environment	Vacuum (no off-gassing)

Modifications for Operation in Microgravity

Heat Pipe Wick: To allow the PCHP to operate in microgravity, it is based on a typical spacecraft VCHP configuration that used an axially grooved aluminum extrusion. The extrusion was 1.27 cm (0.5 in) diameter x 1 meter long with a 15.24 cm (6 in) evaporator and a 30.48 cm (12 in) condenser.

Reservoir Bellows: As shown in Table 2, the PCHP design life was 10 years, or 58,400 cycles as the satellite orbits, and the PCHP compensates for the differing day-night radiator view factor. The primary limit for the bellows is fatigue. Figure 13 shows the PCHP bellows and surrounding capsule. The capsule is required to prevent squirm as the bellows are expanded and contracted. Squirming is an undesirable sideways movement of the bellows, which is prevented by the use of a guide bore. The capsule also has internal stops to prevent over extension of the bellows. During launch (and shock and vibration testing) the bellows are pushed against the stop, to minimize the bellows movement. Two of the final bellows/capsules were life tested. One bellows failed at 260,483 cycles, which is four times the design life of 58,400 cycles or 10 years on LEO. The second bellows was cycled 4.28 million times without a failure.

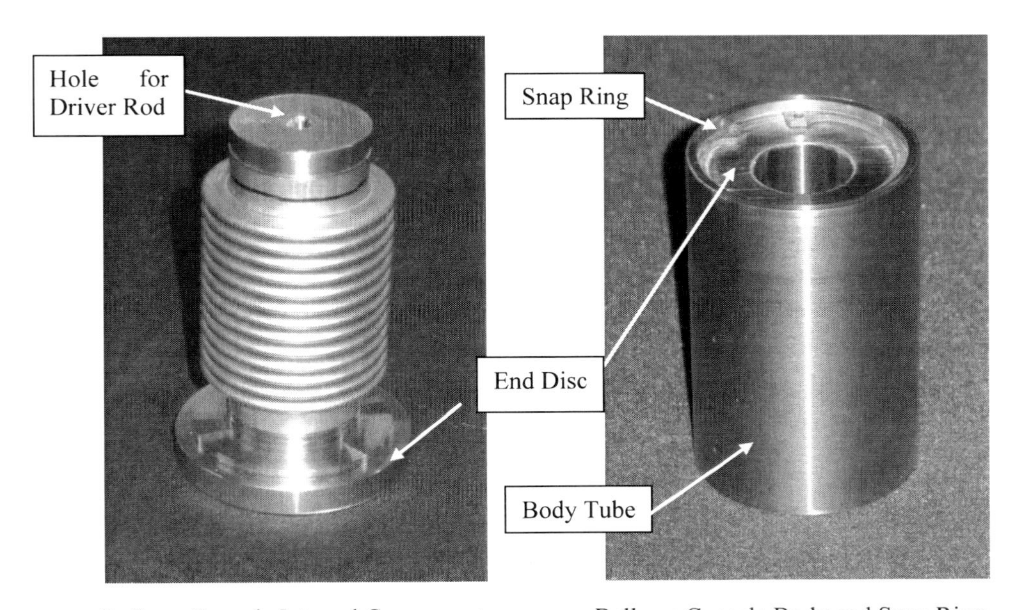

Bellows Capsule Internal Components

Bellows Capsule Body and Snap Ring

Figure 13. PCHP Bellows and Surrounding Capsule.

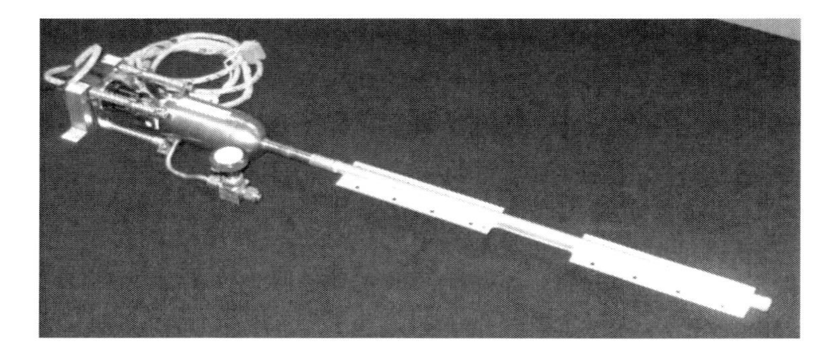

Figure 14. Modified PCHP for vibration testing, with a shorter evaporator and condenser to fit on the test stand.

Shock and Vibration: A modified PCHP was fabricated for vibration testing, see Figure 14. The bellows had the standard NCG reservoir and stepper motor/lead screw, but the evaporator and condenser were shortened to fit on the vibration test rig. A random vibration, three-axis test was conducted, using a composite of several small satellite test levels. During testing, the bellows is locked against a stop to minimize vibration. Testing was successful, with no signs of damage.

Stepper Motor and Thermal Design

The volume of the NCG reservoir bellows was controlled with a stepper motor and a lead screw. Stepper motors are a well-established technology and are relatively inexpensive. They offer the advantage of low or no power consumption during inactive or quiescent periods, which is beneficial for spacecraft operation, when electrical power is limited. The lead screw allows the system to be self-holding, requiring no power at all during periods of no motion. Since the system operates in a vacuum, and convective cooling is therefore not available, a cooling strap conducts the stepper motor heat to the heat pipe condenser.

Further details of the PCHP development and qualification testing are given in Anderson et al., (2011, 2012).

Fabrication and Testing of PCHPs Designed for Operation in Microgravity

Figure 15 shows the final aluminum/ammonia PCHP developed for precise temperature control in microgravity. The PCHP is mounted on a test fixture with electrical heaters at the evaporator and a liquid nitrogen (LN) chill block at the condenser. Heat from the stepper motor is rejected to the condenser through the cooling strap. The bellows is contained in the bellows capsule, which assures alignment and prevents overstressing due to accidental over-travel. The PCHP body is bent so that the completed device can fit in a particular vacuum system at Goddard Space Flight Center (GSFC). A heater block supplies heat to the evaporator, and a chill block is used to cool the condenser. The bellows capsule is the non-condensable gas reservoir.

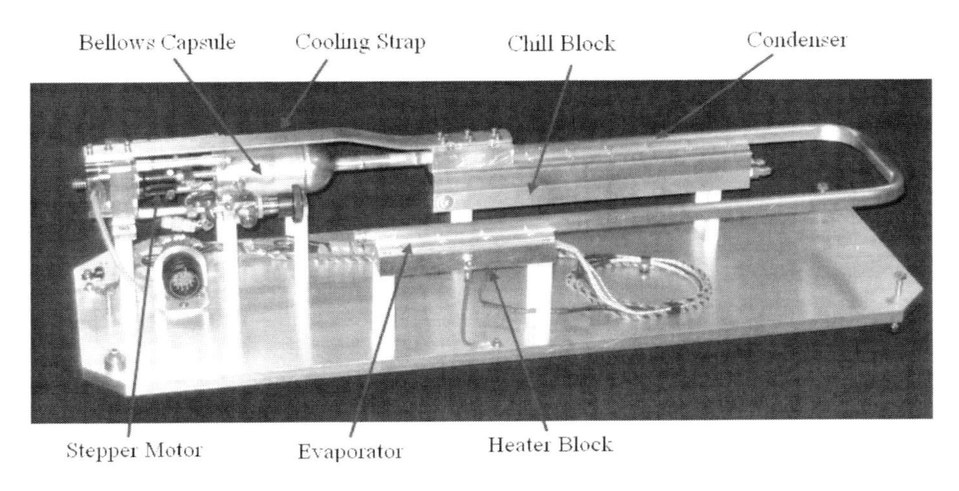

Figure 15. Aluminum/Ammonia Pressure Controlled Heat Pipe for Precise Temperature Control. The heat pipe is bent to fit into a small vacuum chamber for testing.

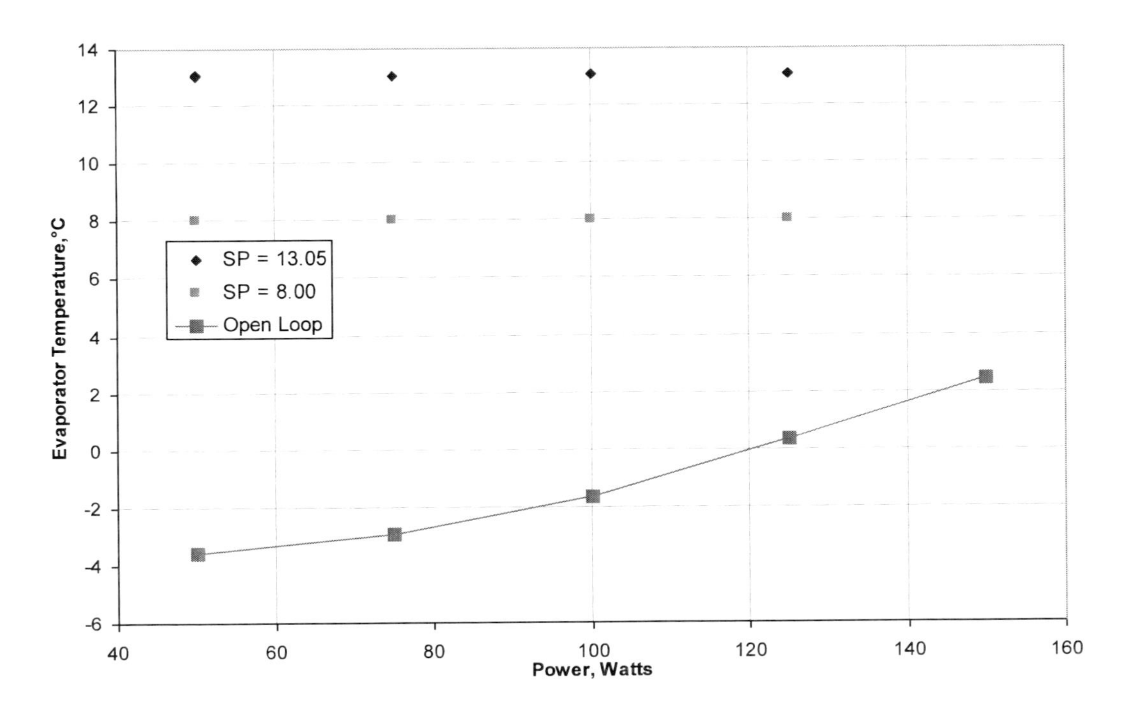

Figure 16. Comparison of Open Loop (VCHP) and Closed Loop (PCHP) Response of the PCHP in vacuum over changes of input power.

During testing, the input power ranged between 50-150 Watts and sink temperature ranged between -40°C and 0°C. Figure 16 compares the open loop and closed-loop response of an ammonia PCHP over variations of input power. This PCHP is able to maintain a steady evaporator temperature (measured by the RTD) over input powers ranging from 50 to 125 Watts. In the figures, Open Loop refers to the no control case, or behavior as a passive VCHP, and SP refers to the Set Point Temperature in closed loop mode. The temperature was controlled to ±50 mK at each setpoint. This performance is at the noise floor of the data acquisition hardware.

PCHPS WITH BOTH NCG ADDITION AND RESERVOIR VOLUME VARIATION

As discussed above, most PCHPs add and remove NCG to precisely control the temperature; see Figure 11. By varying the amount of NCG, the setpoint temperature can be changed. The PCHP for operation in microgravity discussed above modified the reservoir volume instead, allowing a smaller, closed-cycle system, and rapid response to changes in conditions. On the other hand, once sealed, the microgravity PCHP is limited to a single set point temperature.

The new PCHP calibration system discussed below uses both of these methods; see Figure 17 and Figure 18. This system was delivered to Physikalisch-Technische Bundesanstalt (PTB) in 2014 (Maxwell et al., 2016). The PCHP calibrator system for PTB was designed as an all-in-one, turn-key system with an operating temperature range from 600-1000°C and a temperature stability of ± 5mK or better. In order to meet the operating temperature specifications, a combination of sodium working fluid and Inconel 600 envelope materials were used. The PCHP geometry included an outer diameter of 4.500 in. (114mm), and 6x thermowells with an internal diameter of 0.478 in. (12.1 mm) and length of 19.68 in. (499.7mm). The thermowells are accessible through the insulation panel on top of the furnace. The PCHP also features a chimney port which connects to the bellows and helium charging sub-assembly. This chimney extends out of the furnace through the insulation panel adjacent to the thermowell entry points. One thermowell is intended to serve as a primary feedback sensor for the control system, which is a Type R thermocouple. The additional thermowells are used for the sensors to be calibrated by the comparison method. A CAD model of PCHP can be seen Figure 17. The PCHP is then installed into a commercial furnace that has been customized to accommodate the unique geometry of the PCHP (see Figure 17b).

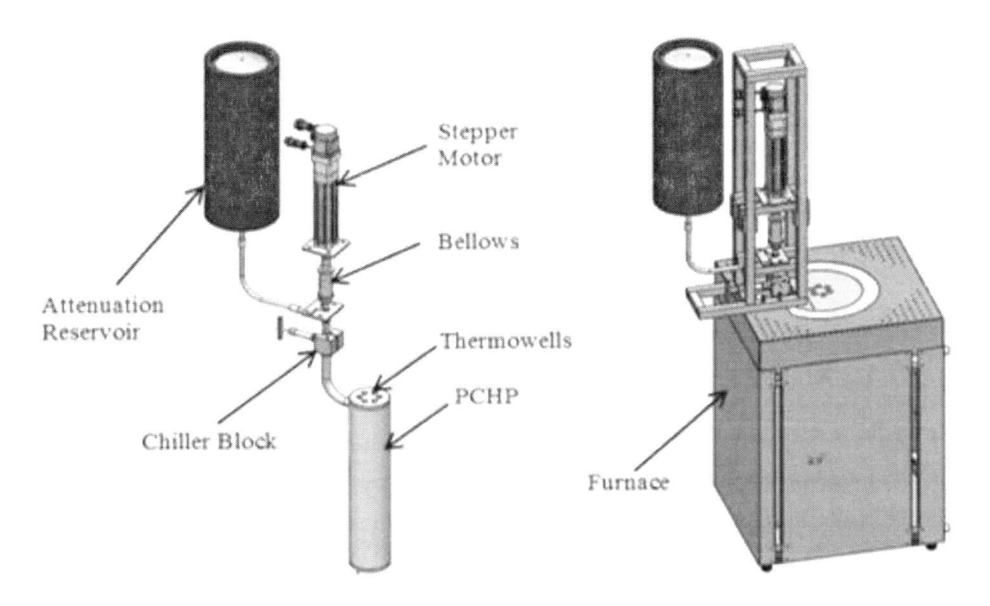

Figure 17. a) CAD representation of a Vertical Sodium PCHP b) PCHP installed into furnace. The valves to add/remove helium to change the setpoint are not shown, see Figure 18. (Maxwell et al., 2016).

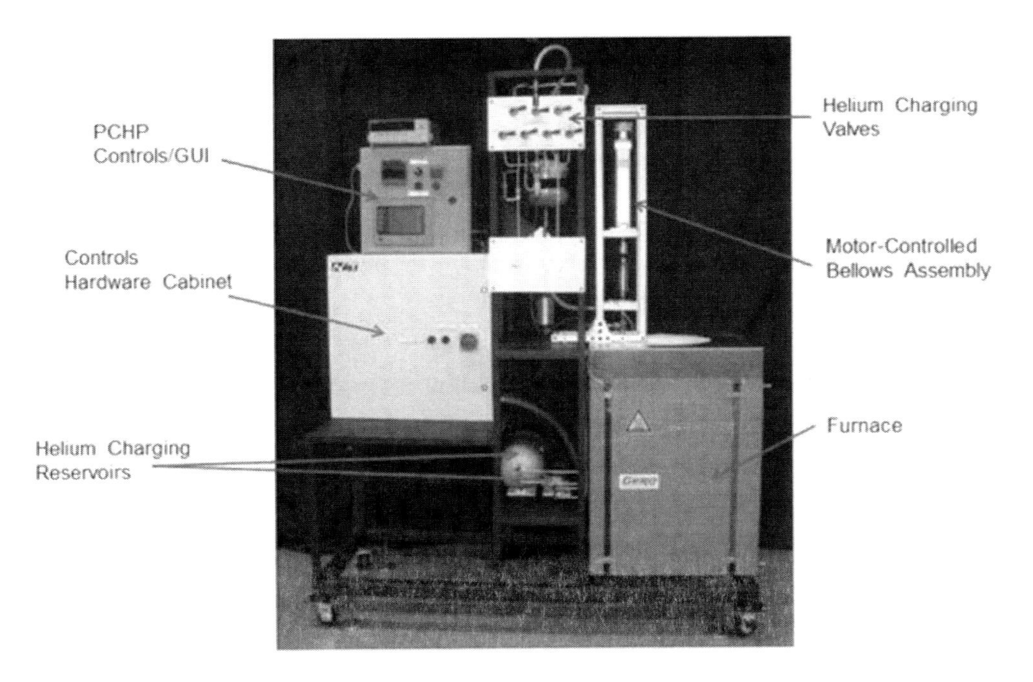

Figure 18. Sodium/Inconel 600 PCHP turn-key system with a vertical IFL. (Maxwell et al., 2016).

This helium charging sub-assembly allows the setpoint to be varied from closed-system PCHP consisted of a vertically oriented Sodium/Inconel 600 PCHP with helium as the buffer gas. This sub-assembly allows the setpoint to be varied from 600-1000°C. The PTB PCHP is also equipped with a hydro-formed stainless-steel bellows enclosed in a protective cylinder. The cylinder guides the bellows movement and prevents the bellows from squirming. The bellows movement is controlled by an electric servo motor and drive that is capable of +/-0.008 in. (0.02mm) positional accuracy. Ultra-high purity Helium (+99.999%) is used as the buffer gas to control the operating pressure, and hence, temperature of the system.

Figure 18 is a picture of the fabricated turnkey PCHP system that was delivered to PTB. The stability of the PCHP Calibrator was measured at ACT's facility prior to commissioning at PTB to verify the system met specifications. The primary feedback sensor, the Fluke Type R thermocouple, was used for the temperature measurements. The PCHP Calibrator was located in a laboratory whose air temperature and air circulation rates were relatively constant, but it was postulated that under the proper environmentally controlled conditions, such as those in a high technology metrology laboratory, the stability can be further improved beyond the results represented herein. Prior to making any measurements, the furnace was run for several hours to allow it to reach full thermal equilibrium.

As can be seen in Figure 19, the PCHP was capable of maintaining a temperature stability of 5mK at 800°C for 1 hour. It also achieved a temperature stability of slightly less than 5mK at 600°C for almost 2 hours However, since the delivery of the PCHP system to PTB, it has been reported that a stability of ± 1mK has been achieved. Furthermore, both of the tests shown in the figures above were conducted with the helium attenuation reservoir disconnected from the system, as it was found to provide no improvement in the attainable stability. It is believed that the high precision and fast response time of motor-controlled bellows is sufficient to achieve the specified stability in this application. Use of the attenuation reservoir may be more beneficial during coarse temperature control, where pressure oscillations are relatively large; in order to minimize the time required to reach steady-state.

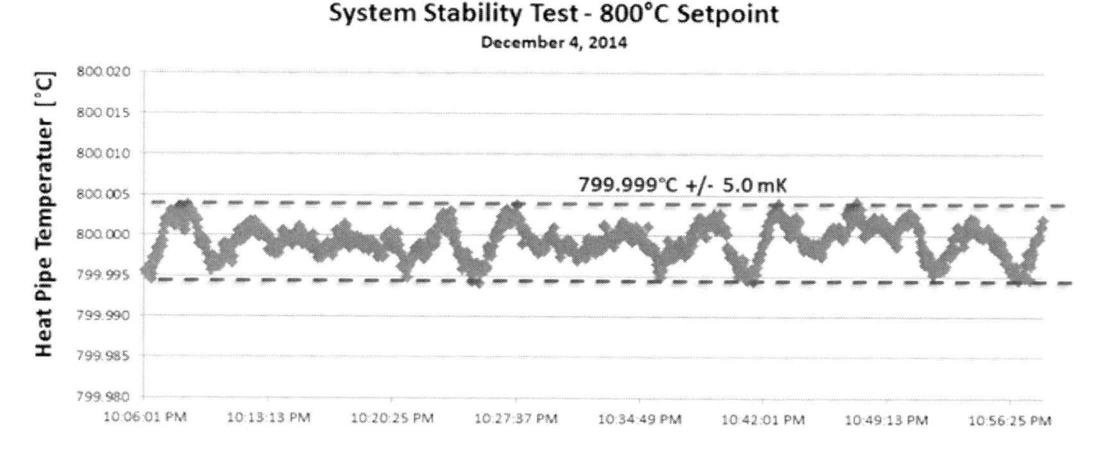

Figure 19. Temperature stability measurements of Sodium/Inconel 600 PCHP at 800°C.

PCHPS FOR HIGH TEMPERATURE POWER SWITCHING

A second application for PCHPs is their ability to switch power between different reactors at high temperatures for In-Situ Resource Utilization (ISRU) (Anderson et al., 2011, 2012). The lunar soil consists of approximately 43% oxygen that is contained within the oxides of the lunar soil. Production of oxygen is a batch process. Fresh regolith is added to a reactor, then heated up to 1050°C using roughly 4 kW of solar power. After the reactor is at 1050°C, hydrogen flows though the reactor, and reacts with the oxygen in the lunar regolith to produce water. The water is then electrolyzed to produce oxygen, and recycle the hydrogen into the process. During the oxygen production process, the reactor requires only about 1 kW of solar power to maintain the temperature. The mass of the overall system can be minimized if one solar concentrator supplies a constant rate of power to two reactors, with the power switched from one reactor to the other as fresh batches of regolith are added. In contrast to the low power PCHPs for temperature calibration, these PCHPs transfer kilowatts of power.

System Operation

The PCHP concept for using solar power to provide oxygen from lunar regolith is shown in Figure 20. The assumed base location is near the lunar South Pole, so the sunlight is always coming from near the horizon. The sunlight is focused with a solar concentrator that directs the solar energy down into the central Solar Receiver Heat Pipe (SRHP). The system components include:

1. Two Solar Receiver Heat Pipes (SRHPs), one for each regolith processing reactor.
2. Two Constant Conductance Heat Pipes (CCHPs), one for each reactor.
3. A primary condenser on each SRHP.
4. A secondary condenser on each SRHP, with bellows to vary the reservoir volumes. In other words, the SRHPs are PCHPs.

As discussed in more detail below, the solar receiver delivers a constant 4 kW of power, which is the maximum required power when one reactor is generating oxygen, and the other reactor has just started to heat up. The required power drops to a minimum of 2 kW during the cycle, so the excess 2 kW of heat must be removed. While this could be achieved by diverting some of the solar power from the receiver, this is undesirable since it would involve moving parts at high temperature, which are difficult to repair. Instead, CCHPs and PCHPs are used to transfer the power.

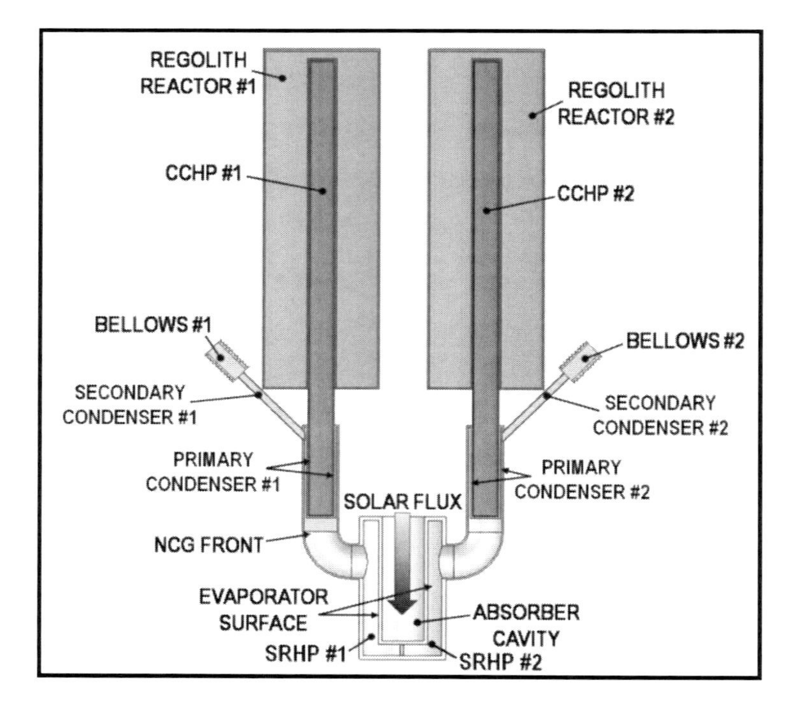

Figure 20. Schematic of the PCHPs (shown as SHRP #1 and #2) and CCHPs used to transfer heat back and forth between the two reactors.

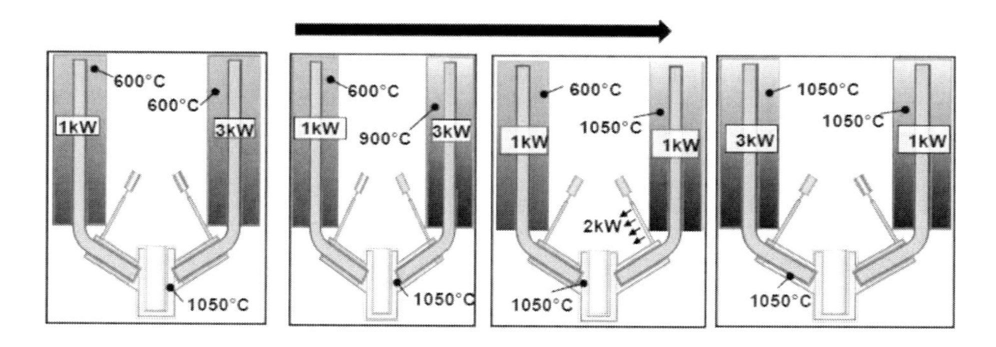

Figure 21. Schematic of the initial Lunar regolith PCHP startup. The NCG is shown in yellow.

The SRHPs are used to control the gas front location, and therefore control the power delivered to each reactor. The CCHP transfers the thermal load to the reactor from the variable conductance SRHP. CCHPs are used to provide uniform heating to the regolith. The change in the exposed length of the two condensers of each SRHP would vary as the power was transferred. The secondary condensers provide a method to reject excess heat, so that the concentrator can always deliver the full heat load to the SRHP.

A projected operational scenario for the startup and operation of a multiple reactor system is shown in Figure 21:

Step 1. The SRHPs are at 1050°C, and cold regolith (600°C) is loaded into the reactors. During the loading, the bellows are fully extended, so gas blocks the CCHP evaporators. This is done to maintain the temperature of the SRHPs, which otherwise would quickly drop to 600°C, due to the high-power capabilities of the heat pipe. Once the regolith is loaded, the gas in the Right-Hand Side (RHS) is withdrawn slightly, so that 3 kW is supplied to warm up the RHS reactor.

Step 2. As the RHS reactor warms up, the driving force between the heat pipes and the regolith diminishes. The RHS bellows is gradually withdrawn to maintain the 3-kW heating. The picture in Step 2 shows the NCG gas completely withdrawn from the RHS CCHP evaporator.

Step 3. As the RHS temperature approaches the final 1050°C temperature, the ΔT diminishes so that less than 3 kW of power can be conducted into the RHS reactor. The excess power is radiated from the secondary condenser. The picture in Step 3 shows the condition when the RHS reactor is at 1050°C: 1 kW is supplied to the reactor to make up thermal losses, and the other 2 kW is radiated. At this point, the RHS reactor is at temperature. Hydrogen then flows through the reactor to produce water, which is electrolyzed to produce oxygen.

Step 4. The 3 kW of heat is then supplied to the LHS reactor to warm up the regolith, while the RHS reactor produces oxygen. When the oxygen has been removed, fresh regolith is added, and the cycle repeats.

Regolith PCHP Design Constraints

The unique design constraints for this system are explained below.

Dual PCHPs: Dual PCHPs/CCHPs are required, rather than a single PCHP with two arms. The reason is that the sodium vapor pressure with frozen sodium is very low. This means that the NCG will be distributed throughout the system. If a single system with two arms was used, there is no control of the NCG location during start-up. We believe that it would all be driven to the side that was first started. Using two separate, hermetically sealed heat pipes guarantees that both PCHP condensers will have the required NCG charge.

Throttling Heat Below the Reactor: Our first thought was to extend the PCHPs through the reactor, with the NCG reservoir located above the reactor. However, this would make it impossible to provide uniform heating to both reactors during the transient heating after cold regolith is added to one side.

Consider the PCHPs right before fresh, cold regolith is added in step 4 of Figure 21. The RHS reactor needs 1 kW to maintain the reactor temperature during processing. 3 kW is available to heat up the LHS reactor. If the power to the LHS is not throttled, the LHS PCHP would be rapidly heated, with more than 10 kW of power supplied to the cold reactor. This is not acceptable, since the will pull the RHS down to the same temperature as the LHS. The PCHP throttles the power back to 3 kW during this process.

While it would be possible to throttle the power with a PCHP extending through the reactor, almost all of the condenser would need to be blocked initially. This will heat the regolith unevenly. In addition, it will require much larger bellows and NCG reservoir volume.

CCHP: Transferring the heat to a CCHP ensures that the throttled power is distributed uniformly along the reactor length.

Low temperature stepper motors and bellows: The bellows and stepper motors operate at much lower temperatures than the 1050°C operating temperature, since the NCG blocks condensation near them. The only heat is conducted through the heat pipe walls.

Fabrication and Testing of the Regolith Extraction System

The high-temperature double-sided system was scaled down from the full-scale regolith system to allow operation with electrical heaters, rather than a solar concentrator. The full-scale design supplies 4kW of power between the two reactors; 3kW for warm up and 1kW for processing. The scaled down design provides 1.2kW for warm up and 400W for processing (Anderson et al., 2011).

As discussed above, the high-temperature double-sided system has an annular SRHP split in two halves, each with its own side arm, CCHP, secondary condenser and bellows/stepper motor assembly; see Figure 22. The evaporator for the system is located on the exterior surface of the inner vertical pipe. Each SRHP has a separate evaporator, with the vapor space split into two halves by welding a thin Haynes 230 plate on the inside. This plate creates two separate vapor spaces for each side of the evaporator, isolating the NCG from each SRHP.

Heat input for the system was provided by a 1700 W Kanthal heater inserted into the solar receiving portion of the evaporator. In addition, supplemental heaters were installed at the base of the solar receiving portion of the evaporator to provide the remainder of the heat. The final system (before insulation was added) is shown in Figure 22.

The regolith reactor simulators used alternating layers of sandstone pebbles and steel fins to simulate the lunar regolith. Alternating layers of sandstone and carbon steel fins were placed on the inside of the regolith reactor simulators until the top of the simulator was reached.

Prior to testing, the side arms (primary condensers) were covered with 5 cm of Microtherm insulation. The vertical, annular portion of the system was insulated using the insulation package provided with the ceramic heater. Kaowool was also used as extra insulation and covered the entire system with the exception of the secondary condensers.

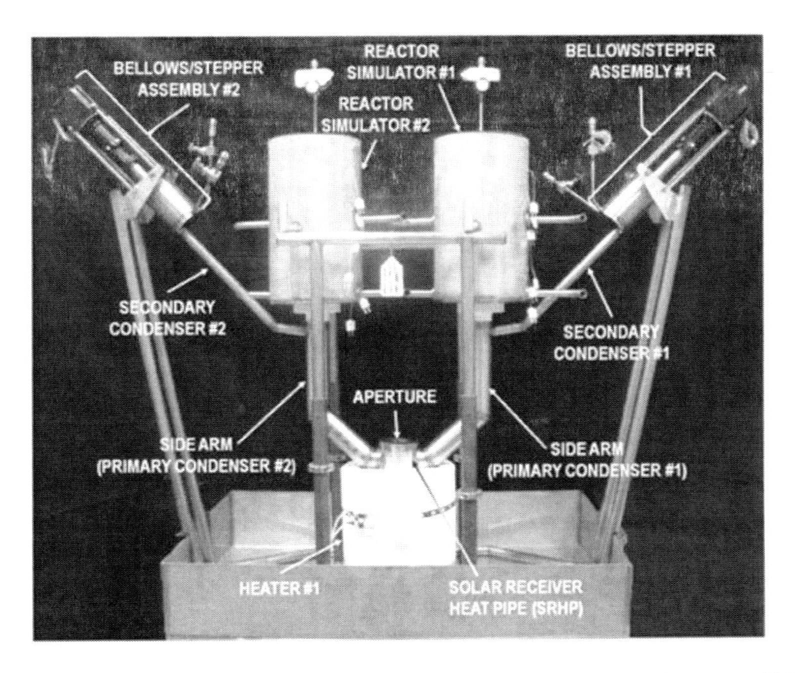

Figure 22. Lunar Regolith PCHP/CCHP Demonstration Unit, with Dual-Sided Haynes 230 PCHPs and CCHPs. The majority of the heat is supplied with an electrical heater in the aperture.

Testing of the dual sided system was conducted at lower temperatures then the nominal operating temperature of 1050°C. This decision was made to be conservative and protect the heaters from an over temperature condition. A lower operating temperature was also chosen due to the heat losses from the system. An operating temperature of 850°C was chosen. Testing of the dual sided system was performed manually: an automatic control program was not used. The vapor temperature of the SRHP evaporators was observed and the stepper motor was moved to maintain it at a constant value. In a charging state, the stepper motor was moved to pull the NCG into the bellows assembly. At the beginning of a processing state, the stepper motor was moved to push the NCG out of the bellows assembly and into the primary condenser. In the middle of the processing state, the NCG front should be moved to maintain the "regolith" temperature therefore the stepper motor was moved to either push or pull the NCG front. In a replenishment state, the stepper motor was moved to push the NCG out of the bellows assembly to blanket as much of the CCHP evaporator as possible.

The temperature set points were designated as follows:

- Regolith preheat temperature ~ 470°C
- SRHP vapor temperature ~ 850°C
- Regolith processing temperature ~ 750°C

Figure 23 shows the test results for a complete cycle for both the Left-Hand and Right-Hand Sides. Variable Conductance Heat Pipes (VCHPs) are used to passively control the evaporator (and associated electronics temperature) while the power and heat sink conditions vary widely. Essentially all VCHPs use an electrically heated, cold Non-Condensable Gas (NCG) reservoir, and typically control the evaporator temperature to within ±2°C. The first warm-reservoir VCHPs actually fabricated and tested are described here. In these devices, the NCG reservoir is located next to the evaporator, so the need for electrical power is eliminated, while maintaining the same ±2°C control. Eliminating power is a requirement for missions that rely on batteries during long periods of darkness, such as Lunar landers and rovers, as well as research balloons in the polar regions during the long periods of winter darkness. Pressure Controlled Heat Pipes (PCHPs) achieve tighter temperature control by changing the amount of NCG in the system, either by changing the reservoir volume, or adding/subtracting non-condensable gas. A PCHP suitable for operation in microgravity was developed using a grooved aluminum heat pipe, and controlled temperatures within ±50mK using a flexible bellows to change the reservoir volume. Even more precise temperature control is required for temperature calibration by the national laboratories. A PCHP was developed that can control the temperature within ±1mK. The setpoint can be changed by adding/removing NCG. PCHPS can also be used for high temperature power switching.

1. Heat up the LHS reactor (Charging)
2. Maintain the LHS reactor (Processing), and heat up the RHS Reactor
3. Cool Down the LHS reactor, to simulate adding cold regolith
4. Maintain the RHS reactor, and heat up the LHS reactor
5. Cool Down the RHS reactor, to simulate adding cold regolith
6. Maintain the LHS reactor, and heat up the RHS Reactor

As seen in the figure, the SRHP evaporator vapor temperature was maintained relatively constant for both sides through all stages of the cycle. A ΔT was experienced across the wall that separates the two SRHP vapor spaces. This temperature difference occurs as a result of the heat flux transmitted through the splitting plate from one evaporator to the other because of the unbalanced heat load through the two SRHPs. This ΔT was maintained at approximately 10°C during manual operation. The first charging stage for both the left and right-hand sides of the system took approximately 92 to 94 minutes. Charging for the second cycle took approximately 84 to 86 minutes. A ΔT of approximately 100°C was experienced for both sides between the CCHP vapor temperature and the average regolith temperature. The fluidized bed used in a true regolith reactor would greatly reduce this ΔT due to the higher effective thermal conductivity.

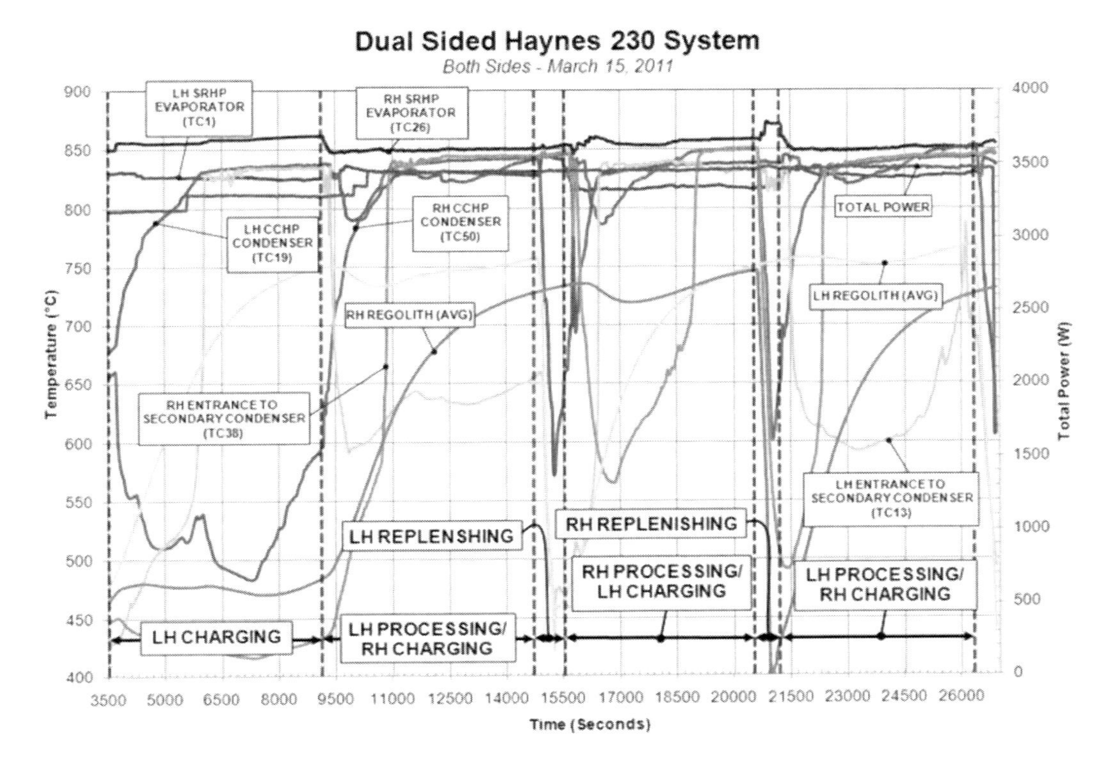

Figure 23. Operation of the Dual-Sided PCHP/CCHP system, showing a full cycle for both sides.

Soon after charging starts, a sharp increase in the secondary condenser temperature is experienced in TCs 13 and 38 for the left and right sides, respectively. This is due to the NCG front reaching the entrance of the secondary condenser and thus rejecting the excess power that cannot be conducted into the regolith. At this point in time the ΔT between the CCHP vapor temperature & SRHP vapor temperature is too small and the excess power must be rejected. This can be seen graphically as the CCHP vapor temperature at this moment has reached approximately 830°C for the left-hand side and 840°C for the right side. A drastic drop in temperature is observed in the CCHP vapor, regolith and secondary condenser during the replenishment stage of the cycle. At this moment the forced air cooling systems for the regolith reactor simulator was turned on to simulate the removal of processed regolith and addition of new regolith.

The regolith temperature was maintained near constant for all processing cycles for both the left and right-hand sides of the system. A slight dip in temperature is observed in both reactors during processing due to more power being transferred to the opposite side to aid in charging, which is the power heavy stage in the full cycle. To summarize, testing of the dual reactor system was completely successful.

CONCLUSION

Heat pipes are nearly isothermal two-phase devices, which can undergo large temperature swings at the heat sink conditions or power are varied. By adding non-condensable gas and a reservoir, the temperature swings can be greatly reduced. VCHPs normally have a cold-biased reservoir, located at the end of the condenser. The NCG gas reservoir is electrically heated using a few Watts to the desired temperature, and can typically the evaporator temperature (and associated electronics) to within ±2°C.

Eliminating this heating is required for missions that rely on batteries during long periods of darkness, such as Lunar landers and rovers, as well as research balloons in the polar regions during the long periods of winter darkness. It is estimated that providing 1 W of power over the 14-day-long Lunar night requires 5 kg of additional batteries and solar cells.

In a warm reservoir VCHP, the NCG reservoir is located next to the evaporator, eliminating the need for electrical power. The first warm-reservoir VCHPs actually fabricated and tested are discussed here. The design is more complicated than a cold-biased reservoir, since the warm reservoir must be maintained at a slightly higher temperature than the evaporator. When properly designed, they passively maintain the same ±2°C control as the standard, cold-biased reservoir.

Pressure Controlled Heat Pipes (PCHPs) are used when even tighter temperature control is required, typically for calibrating pyrometers and other reference temperature devices in National Laboratories such as NIST. They achieve this tighter temperature control by changing the amount of NCG in the system, either by changing the reservoir volume, or adding/subtracting non-condensable gas.

Until recently, PCHPs controlled the temperature by adding/subtracting the non-condensable gas, typically helium. A PCHP suitable for operation in microgravity was developed using a grooved aluminum heat pipe and ammonia working fluid. To allow for long-life without the ability to replace consumables, this PCHP was hermetically sealed. It used a flexible bellows to change the reservoir volume, and controlled temperatures within ±50mK.

Even more precise temperature control is required for temperature calibration by the national laboratories. High temperature PCHPs have been developed that can calibrate temperature devices over the 600-1000°C range. The setpoint is changed by adding/removing NCG. A flexible bellow is used to control the temperature during each calibration, controlling the temperature within ±1mK.

Dual PCHPS can also be used for to distribute power from a single source to two high-temperature, batch-process reactors. The PCHP is located below the reactors, to throttle the power when cold regolith is added to one reactor, maintaining the temperature of the other reactor. The bellows and stepper motors operate near room temperature, controlling the high-temperature, alkali-metal PCHPs.

ACKNOWLEDGMENTS

The Warm Reservoir VCHP for Lunar and Martian applications was developed on Contract No. NNX15CM03C with NASA Marshall Space Flight Center. Dr. Jeffery Farmer was the technical monitor. The Warm Reservoir VCHP for Balloon Payloads was developed for NASA Goddard Space Flight Center on Contract No. NNX11CF08P. Ms. Deborah Fairbrother was the technical monitor.

The PCHP for Precise Temperature Control in Microgravity was sponsored by NASA Goddard Space Flight Center under Contract No. NNX08CA35C. Ms. Laura Ottenstein was the Technical Monitor, and Mr. David Sarraf was the Principal Investigator. The PCHP for High Temperature Power Switching was sponsored by NASA Glenn Research Center under Contract No. NNX09CA48C. Mr. Don Jaworske was the technical monitor for this program, and Mr. John Hartenstine was the Principal Investigator.

We would like to acknowledge the Advanced Cooling Technologies, Inc. Engineers who worked on the PCHP and Warm Reservoir programs, including Richard Bonner, Darren Campo, Peter Dussinger, John Hartenstine, Taylor Maxwell, Jon Mott, Devin Pellicone, David Sarraf, and Kara Walker. Taylor Maxwell and Kara Walker put together many of the drawings.

REFERENCES

Anderson, W. G. et al., "Variable Conductance Heat Pipe Radiator for Lunar Fission Power Systems," *11th International Energy Conversion Engineering Conference* (IECEC), San Jose, CA, July 15-17, 2013.

Anderson, W. G., "Variable Conductance Heat Pipes for Variable Thermal Links," *42nd International Conference on Environmental Systems* (ICES 2012), San Diego, CA, July 15-19, 2012.

Anderson, W. G., J. R. Hartenstine, C. Tarau, D. B. Sarraf, and K. L. Walker, "Pressure Controlled Heat Pipes," *41st International Conference on Environmental Systems*, Portland, OR, July 17-21, 2011.

Anderson, W. G., J. R. Hartenstine, D. B. Sarraf, C. Tarau, and K. L. Walker, "Pressure Controlled Heat Pipe Applications," *16th International Heat Pipe Conference*, Lyon, France, May 20-24, 2012.

Anderson, W. G., J. R. Hartenstine, K. L. Walker, and J. T. Farmer, "Variable Thermal Conductance Link for Lunar Landers and Rovers," *8th International Energy Conversion Engineering Conference* (IECEC), Nashville, TN, July 25-28, 2010.

Bienert, W. B. and P. J. Brennan, "Transient Performance of Electrical Feedback Controlled Variable – Conductance Heat Pipes," ASME Paper 71-Av-27,

SAE/ASME/AIAA Life Support and Environmental Control Conference, San Francisco, California, July 12-14, 1971.

Bienert, W. B., P. J. Brennan, and J. P. Kirkpatrick, "Pressure Controlled Variable Conductance Heat Pipes," *AIAA 6th Thermophysics Conference*, Tullahoma, TN, April 26-28, 1971.

Bienert, W., "Isothermal Heat Pipes and Pressure-Controlled Furnaces," *Isotech Journal of Thermometry*, 2(1), pp. 32-52, 1991. http://www.isotechna.com/v/vspfiles-/pdf_other/journal-2.1.pdf.

Brennan and Kroliczek, "*NASA Heat Pipe Design Handbook*," prepared under Contract NAS523406, 1979. https://archive.org/details/nasa_techdoc_19810065690.

Dussinger, P. and J. Tavener, "Ultra High Temperature Isothermal Furnace Liners (IFLs) For Copper Freeze Point Cells," *9th International Temperature Symposium, Anaheim*, CA, March 2012.

Ellis, M. C. and W. G. Anderson, "Variable Conductance Heat Pipe Performance after Extended Periods of Freezing," Space, *Propulsion and Energy Sciences International Forum (SPESIF),* Huntsville, Alabama, February 2009.

Hartenstine, J. H. et al., "Pressure Controlled Heat Pipe Solar Receiver for Regolith Oxygen Production with Multiple Reactors, *9th Intersociety Energy and Conversion Engineering Conference (IECEC),* San Diego, CA, July 31 - August 3, 2011.

Marcano, P. and G. Bonnier, "Temperature Amplifier by Means of Coupled Gas-Controlled Heat Pipes, *9th International Symposium on Temperature and Thermal Measurements in Industry and Science*, Cavtat-Dubrovnik, Croatia, June 21 – 25, 2004.

Marcus B. D., "*Theory and Design of Variable Conductance Heat Pipes*," NASA Report CR-2018, April 1972. https://ntrs.nasa.gov/search.jsp?R=19720016303.

Maxwell, T., J. Mott, D. Pellicone, D. Campo, R. W. Bonner III, and H. McEvoy, "A Novel Closed System, Pressure Controlled Heat Pipe Design for High Stability Isothermal Furnace Liner Applications," *13th International Symposium on Temperature and Thermal Measurements in Industry and Science* (TEMPMEKO 2016), Zakopane, Poland, June 26 – July 1, 2016.

Mock, P. R., D. B. Marcus, and E. A. Edelman, "Communications Technology Satellite: A Variable Conductance Heat Pipe Application," in *proceedings of AIAA/ASME 1974 Thermophysics and Heat Transfer Conference*, Boston, Mass., July 15-17, 1974.

Peters, C. J., J. R. Hartenstine, Calin Tarau and W. G. Anderson, "Variable Conductance Heat Pipe for a Lunar Variable Thermal Link," *41st International Conference on Environmental Systems* (ICES 2011), Portland, OR, July 17-21, 2011.

Sarraf, D. B., S. Tamanna, and P. M. Dussinger, Pressure Controlled Heat Pipe for Precise Temperature Control, *Space Technology and Applications International Forum* (STAIF), Albuquerque, New Mexico, February 2008.

Tamba, J. and M. Arai, "Pressure-Controlled Water Heat Pipe for Precise Comparison of Platinum Resistance Thermometers," *Transactions of the Society of Instrument and Control Engineers*, Vol. 38(4), pp. 345-350, 2002.

Tarau, C. and W.G. Anderson, "Variable Conductance Thermal Management System for Balloon Payloads," *20th AIAA Lighter-Than-Air Systems Technology Conference*, Daytona Beach, FL, March, 25-28, 2013.

Weyant, J., D. Campo, and C. Tarau, "Hot Reservoir Stainless-Methanol Variable Conductance Heat Pipes for Constant Evaporator Temperature in Varying Ambient Conditions," *Joint 18th International Heat Pipe Conference and 12th International Heat Pipe Symposium*, Jeju, Korea, June 12-16, 2016.

In: Heat Pipes: Design, Applications and Technology
Editor: Yuwen Zhang
ISBN: 978-1-53613-908-2

Chapter 9

FLOW PATTERN VISUALISATION IN TWO-PHASE CLOSED THERMOSYPHON, PULSATING, LOOP HEAT PIPE AND FORCED CONVECTION

K. S. Ong*
Faculty of Engineering and Green Technology,
Universiti Tunku Abdul Rahman, Perak, Kampar

ABSTRACT

Heat pipes are efficient passive devices able to transfer large amounts of heat over long distances with small temperature differences between the heat sources and sinks. Thermal performances of heat pipes are dependent upon the type of fill liquid, fill ratio, power input, pipe inclination and pipe dimensions. The boiling and condensation processes that occur inside heat pipes are quite complex. A visual study of the internal flow patterns would be most helpful to understand the internal heat transfer phenomena. Visualization by various investigators to observe the flow patterns inside thermosyphons, pulsating, loop heat pipes and forced circulation are presented.

Keywords: heat pipe, two-phase closed themosyphon, pulsating loop heat pipe, flow pattern, visualization

* Corresponding Author Email: skong@utar.edu.my.

INTRODUCTION

Heat pipes (HPs) are efficient passive heat transfer devices able to transfer large amounts of heat over long distances with small temperature differences between the heat sources and sinks. A HP consists of a hollow metal pipe fitted with an internal mesh or sintered wick. A wickless HP is also known as a two-phase closed thermosyphon (TPCT) or simply, a thermosyphon. A cross-sectional view of a thermosyphon is shown in Figure 1. The pipe is vacuumed and filled with a small quantity of working fluid. Copper, brass, aluminium or stainless steel is usually used as pipe material. Water, acetone, methanol, and refrigerants could be utilised as fill liquid. Distilled water is the most common fill liquid used. The pipe is separated into a heating (evaporator) section and a cooling (condenser) section by an insulated section. Under normal conditions, the fluid exists as a liquid in the lower portion of the evaporator section and as a vapour in the upper section. Heat applied at the evaporator section causes the fluid to boil and vaporise, picking up latent heat of vaporisation. The vapour travels upwards along the pipe to the condenser section where it condenses and gives up its latent of condensation. The condensate then flows back to the evaporator by gravity. Heat is thus transferred from the evaporator to the condenser section by the process of evaporation and condensation of the working fluid at the respective sections of the thermosyphon. Cooling at the condenser section could be performed using a water jacket or by fins fitted external to the condenser pipe. A thermosyphon has an effective thermal conductivity many times greater than that of pure axial conduction along the pipe because of the evaporation and condensation nature of heat transfer. Its thermal performance is dependent upon the type of fill liquid, fill ratio, power input, pipe inclination and pipe dimensions. Heat transfer theory of HPs are available in Peterson [1], Reay and Kew [2], Faghri [3] and Zohuri [4].

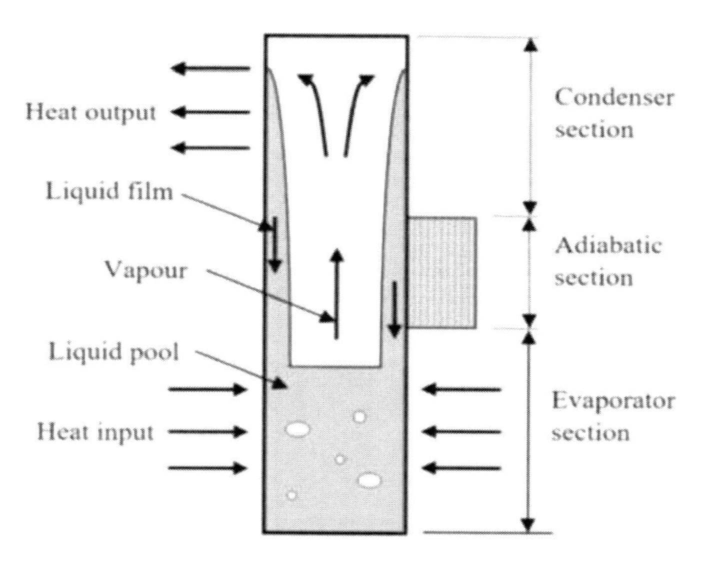

Figure 1. Cross sectional view of a thermosyphon.

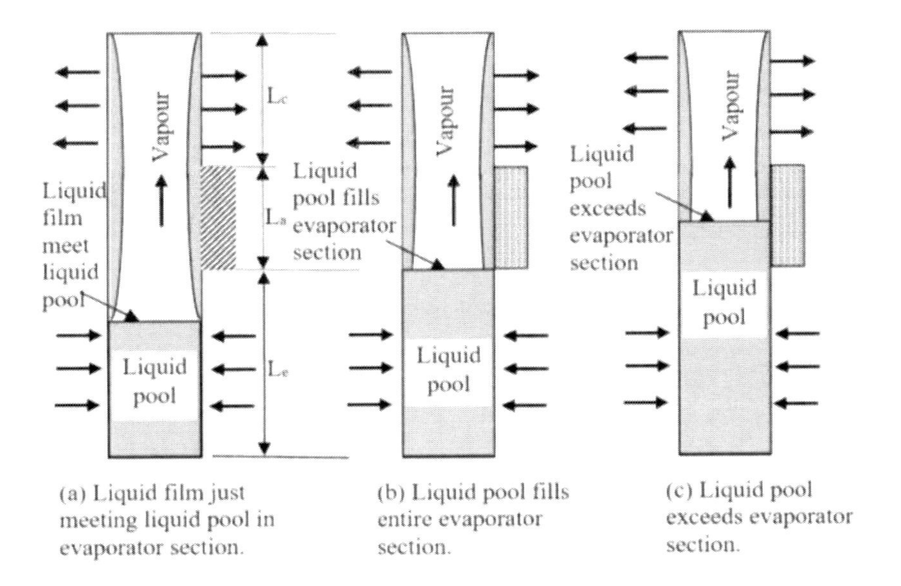

Figure 2. Fill liquid in a thermosyphon.

The fill ratio of a thermosyphon is defined as the ratio of volume of fill liquid to volume of evaporator section. Under normal operation, the falling condensate film would just meet the liquid pool in the evaporator section at the bottom of the thermosyphon as illustrated in Figure 2(a). The initial fill ratio would be just adequate to maintain this condition. It represents the minimum filling ratio for the thermosyphon to operate well at the operating condition and the heat flux is termed the critical heat flux. In the case of a thermosyphon with high fill ratio, not only does the liquid pool fill the entire evaporator section, but there is also a falling condensate film in the condenser section, Figure 2(b). During operation then, the liquid pool fills the entire evaporator section. The temperature distribution is expected to exhibit uniformity along the evaporator wall in this case. Figure 2(c) illustrates an example where there is excess initial fill liquid. Here, the level of the liquid pool exceeds the evaporator section during operation. In this case, the wall temperature along the evaporator section is expected to decrease in the axial direction.

Flow Regimes in Pool Boiling

Pool boiling occurs in a stagnant liquid pool. Flow boiling occurs when the liquid is in motion. Boiling can be classified into subcooled boiling or saturated boiling depending upon the bulk liquid temperature. Subcooled boiling occurs when the temperature of the bulk liquid is below the saturation temperature. Saturated boiling occurs when the bulk liquid is at saturated temperature. Boiling processes are complex and could be divided into different regimes, depending upon heating temperature or heat flux as illustrated in Figure

3. At low temperature and at low heat flux natural convection plays a major role in heating of the liquid. Superheated liquid near the hot surface rises towards the pool surface where it evaporates. As shown in Figure 3(a), a free convection current is formed which circulates the fluid in the pool. A small amount of nucleation occurs at the wall which contributes a minor portion to the heat transfer process. At an intermediate heat flux, bubbles are formed on the hot surface. They slide up towards the free surface. Here, a combined convection regime takes over consisting of both natural convection and nucleate boiling, Figure 3(b). In this regime, the nucleation and growth of bubbles at the wall as well as induced mixing by the sliding bubbles effectively enhance the heat transfer rate. At high wall heat flux, nucleate boiling heat transfer occurs, Figure 3(c). As the temperature increases, more bubbles are formed. Eventually, the bubbles are formed so rapidly that they cover the entire heated surface with a blanket of vapour film. The bubbles rise to the top of the pool and bursts above the pool. At a point of maximum heat flux, burn-out of the heating element could occur. A period of transition then occurs when partial nucleate boiling and unstable film boiling occurs. Stable film boiling eventually occurs at high temperature.

Condensation in a thermosyphon occurs when the condensing surface temperature is below the saturation temperature of the vapour. In vertical surfaces, condensate forms on the solid surface and flows down under gravity. Figure 4 illustrates dropwise condensation occurring when liquid droplets are formed in an ad-hoc manner and film condensation when the liquid droplets combine together to form a thin film which flows down the surface.

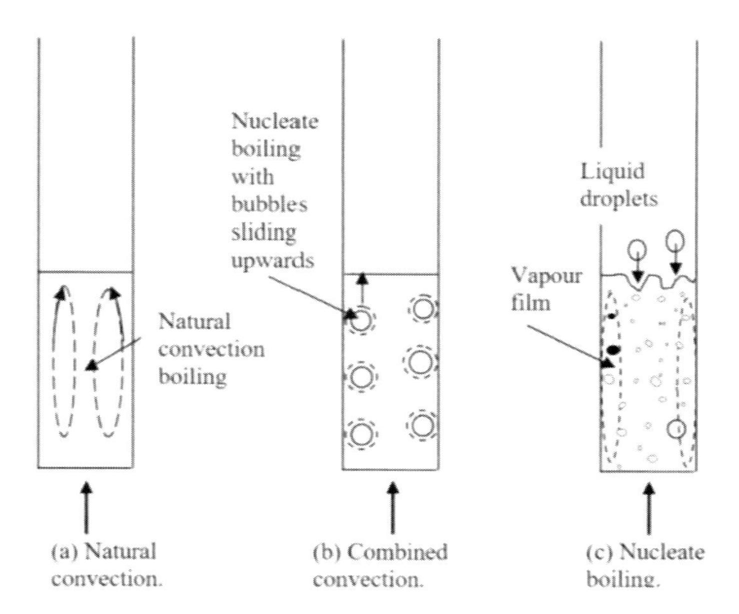

Figure 3. Flow patterns in a small heated liquid pool.

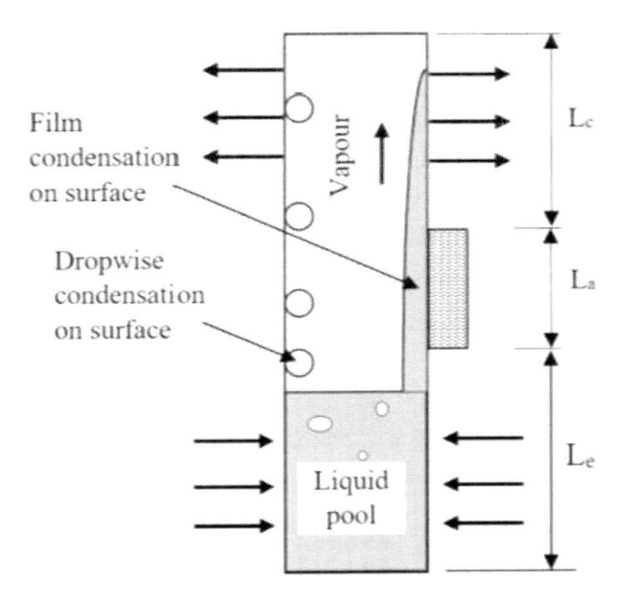

Figure 4. Condensation in a thermosyphon.

FLOW REGIMES IN 2-PHASE CLOSED THERMOSYPHON

Dry-out, burn-out or boiling limit, entrainment or flooding limit and geysering are some flow regimes observed in low- to mid-temperature range during thermosyphon operation. These phenomena would lead to non-uniform axial wall temperature distribution in the pipe, or worse still, ineffective operation. Dry-out conditions are illustrated in Figure 5. Complete dry-out of liquid in the evaporator section results when there is insufficient liquid return to the evaporator, Figure 5(a). This condition occurs especially at high heat flux and low initial fill. At larger fill ratios partial dry-out occurs between the falling liquid film and liquid pool, as shown in Figure 5(b). They occur when the radial evaporator heat flux is small and liquid return from the condenser is below that required for continuous liquid film to flow down. They could also occur in short and small diameter pipes or pipes with small evaporator length/diameter aspect ratios when there is insufficient condensation. As a result of dry-out, the evaporator wall temperature increases. Burn-out or boiling limit occurs when a stable film of vapour is formed between the liquid pool and the heated wall of the evaporator section as shown in Figure 6. The vapour film effectively insulates the wall surface from the liquid. As a result, the wall temperature rises. This phenomenon occurs especially with large fill ratio, high radial heat flux, short and small diameter pipes or small evaporator length/diameter aspect ratio. Entrainment or flooding limit in a thermosyphon results when a high heat flux at the evaporator section causes high relative velocity between upward vapour current and downward liquid flows, Figure 7. As a result, entrainment of liquid by the rising vapour prevents downward flow of liquid and the liquid film dries up and wall temperature rises. This phenomenon occurs especially with large fill ratio, high heat fluxes, long pipes, large evaporator and large evaporator length/diameter aspect ratio.

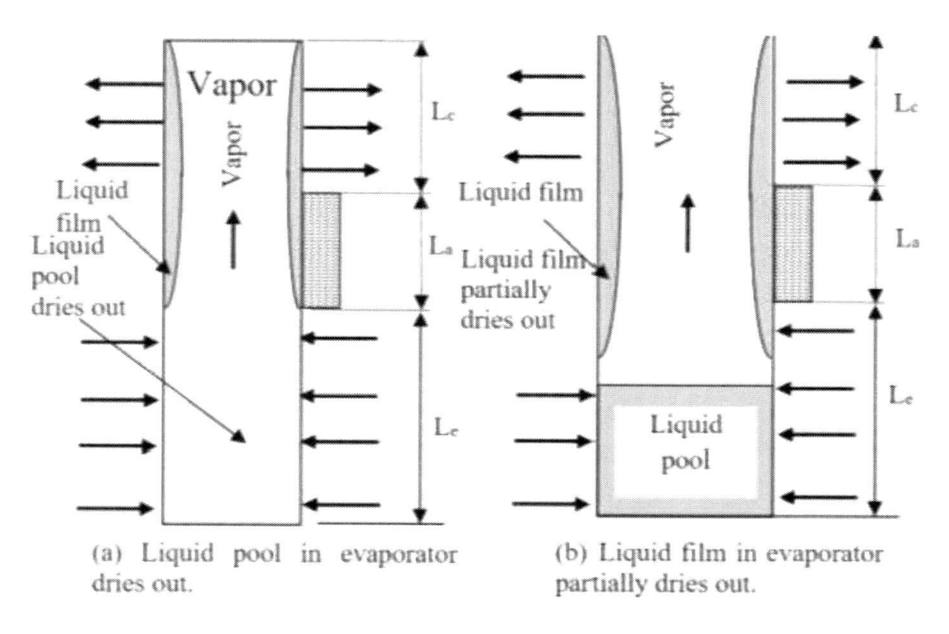

Figure 5. Dry-out in a therosyphon.

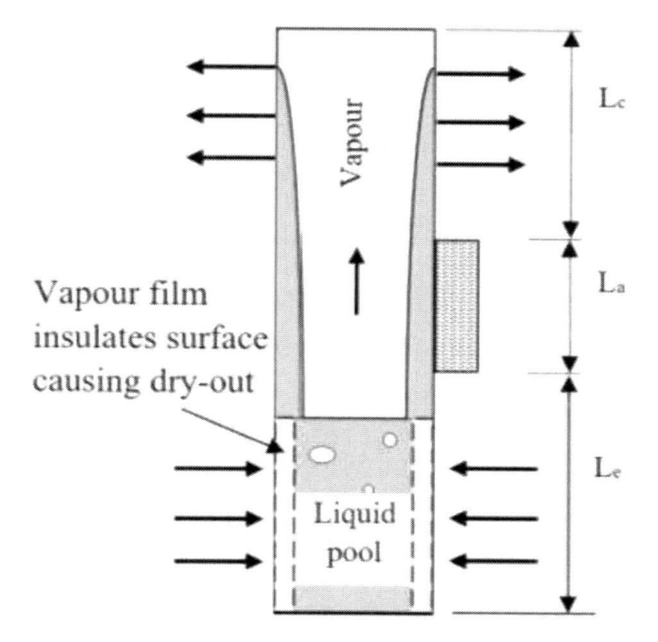

Figure 6. Burn-out or boiling limit in a thermosyphon.

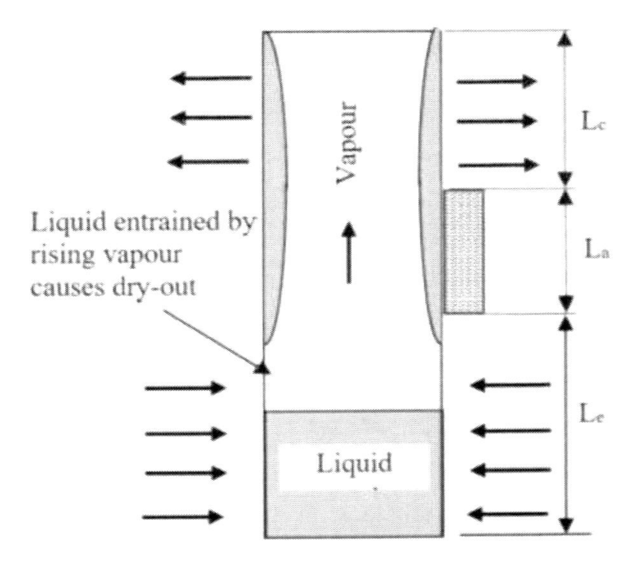

Figure 7. Entrainment or flooding limit in a thermosyphon.

Geysering, intermittent boiling or pulse boiling occurs at low heat flux. During geysering, intermittent boiling occurs when an upward superheated vapour-liquid column is suddenly propelled from the liquid pool upwards towards the top of the thermosyphon, Figure 8(a). The low heat flux is insufficient to maintain continuous boiling and evaporation of the liquid in the evaporator. As a result, the vapour bubbles tend to accumulate and coalesce until such time when a large amount of vapour is suddenly propelled to the top of the pipe where it collapses, Figure 8(b). Geysering occurs intermittently and is usually followed by a loud sound in the pipe. An oscillating flow cycle is created which results in fluctuating wall temperatures.

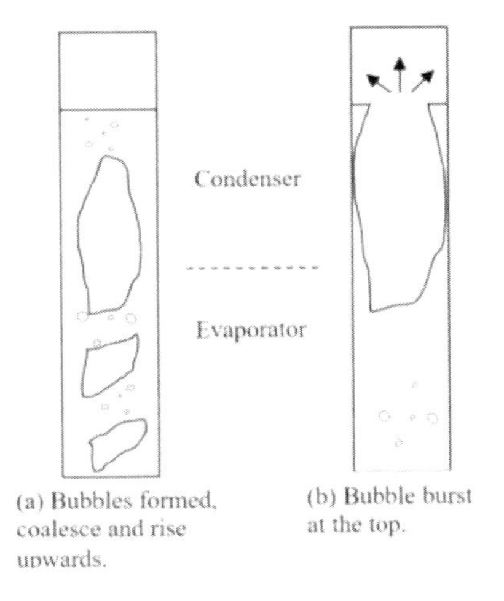

Figure 8. Geysering/intermittent/pulse boiling in thermosyphon.

El-Genk and Saber [5] compiled the boiling heat transfer correlations from various investigators for uniform-heated liquid pools of water, ethanol, methanol, Dowtherm-A, R11 and R113 in small cylindrical enclosures. A total of 731 data points covering pool diameter from 6-37 mm, heated pool height from 50-800 mm, fill ratio from 0.10-3.25 and wall heat flux from 0.7-383 kW/m^2 were correlated. They proposed correlations for natural convection, nucleate boiling and a combined convection heat transfer regime.

Imura et al. [6] performed an experimental investigation to determine the heat transfer in two 28 mm ID brass tubes with two different lengths of 643 mm and 1000 mm and with water and ethanol as fill liquids. The thermosyphons were fitted with a 300 mm long concentric tube cooling water jacket at the condenser section of each pipe. By varying the length of the heated section, various evaporator/condenser length aspect ratios were obtained. The effects of aspect ratio, fill ratio, evaporator heat input and cooling water inlet temperature on the evaporating and condensing heat transfer coefficients were determined.

El-Genk and Saber [7] developed heat transfer correlations for the liquid film region in the evaporator section of a TPCT above the fill liquid based on data obtained for ethanol, acetone, R-11 and R-113 fill fluids with wall heat fluxes from 0.99–52.62 kW/m^2, working fill ratios from 0.01-0.62, and with various dimensions of pipes and saturation temperatures. They postulated the following heat transfer regimes in the falling liquid film in the thermosyphon:

(i) Partial wetting occurs at low heat flux and low wall temperature. Here, the condensate film return is insufficient to maintain a continuous liquid film down to the liquid pool. Liquid film is depleted due to evaporation at the liquid-vapour interface before it could reach the liquid pool resulting in partial wetting of the evaporator wall. This is illustrated in Figure 5(b).

(ii) At low and intermediate heat fluxes the section of evaporator wall above the liquid pool is covered with a continuous liquid film which thickens towards the top of the pipe, Figure 9(a). The wall superheat is below that required for boiling. Heat transfer is typically laminar.

(iii) With increased heat flux, nucleate boiling at the wall results in vapour bubbles being formed on the wall which slide up along the evaporator wall as well as travelling upwards towards the vapour region, Figure 9(b). Heat transfer is by combined convection.

(iv) At high heat fluxes, nucleate boiling with dispersed liquid droplets occurs, Figure 9(c). The heat transfer here can be significantly higher than in the pool region. The growing vapour bubbles burst at the liquid-vapour interface dispersing tiny droplets of liquid into the vapour flow. This is termed geysering. Both bubble nucleation at the wall and entrainment of the tiny liquid droplets in the vapour flow results in higher heat transfer coefficient.

El-Genk and Saber [8] presented a 1-D steady state model to determine the operation envelopes of TPCTs as functions of film liquid, filling ratio, heat fluxes and pipe dimensions. Nero and Beretta [9] presented experimental results and an analytical model on the boiling mechanisms in a TPCT in order to determine the line between intermittent boiling and fully-developed boiling based on visualizing the frequency of bubble nucleation and ratio of bubble diameter to pipe diameter. They used 1000 mm long transparent pipes of 12 and 30 mm I/D with water and acetone and found that at low heat inputs, the flow pattern is characterised by unsteady operation.

Nguyen-Chi et al. [10] investigated the performance of 500 mm long TPCTs with 9 mm O/D x 7 mm I/D, 10 mm O/D x 8.5 mm I/D and 12.0 mm O/D x 10 mm I/D copper tubes filled with water and at different fill ratios. They established the existence of the dry-out and burn-out limits. These depended upon fill ratio, tube diameter, evaporator and condenser lengths as well as operating temperature. Their investigation showed that maximum heat flux due to dry-out increased with higher fill ratio and condenser length and decreased with evaporator length. The influence of operating temperature was more complex. They also found that maximum heat flux due to burn-out increased with increased operating temperature and tube diameter and decreased with evaporator length. The influence of fill ratio was rather weak. The maximum heat flux increased with fill ratio up to a maximum value. In a later investigation, Nguyen-Chi and Groll [11] investigated the entrainment or flooding limit in a TPCT in more detail. They carried out experiments with

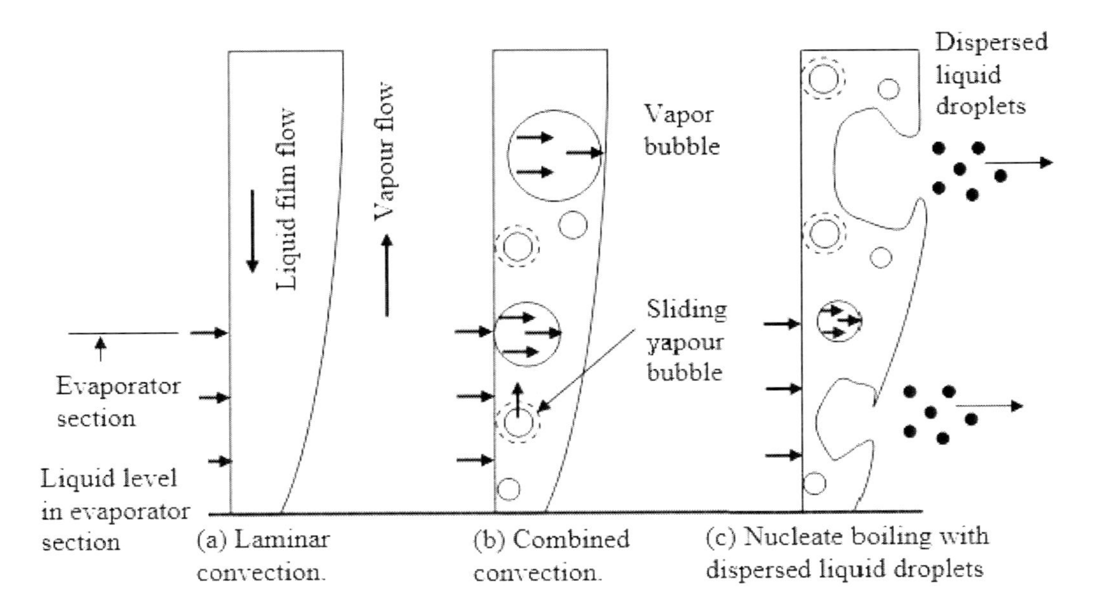

Figure 9. Heat transfer regimes along heated evaporator section of thermosyphon wall postulated by El-Genk and Saber [7].

a 2.5 m long x 20 mm O/D copper tube filled with water and derived an empirical correlation to predict the maximum performance due to flooding. They also determined the influence of liquid fill charge, inclination angle of the thermosyphon and operating temperature on the performance.

Imura et al. [12] investigated the effects of tube diameter, heated length, working fluid, liquid fill charge and the inside saturation temperature on the critical heat flux in a TCPT filled with water, ethanol and Freon 13 and proposed new correlating expressions for the condensing heat transfer coefficient. An experimental investigation was carried out by Imura et al. [13] to determine the critical heat fluxes during dry-out and burn-out limits in TPCTs. They used water, ethanol, methanol, Freon 123, Freon 113, pentane and hexane with 16 mm diameter x 0.643 m long tubes. Gross [14] showed that the heat transfer coefficient for the condensate film can generally be predicted using Nusselt's theory for filmwise condensation.

Jiao et al. [15] stated that the filling ratio of the working fluid had a predominant effect on the TPCT performance and developed a mathematical model to investigate its effect on the steady-state heat transfer performance of a vertical TPCT. They tested two different geometries of the TCPT with nitrogen as working fluid and proposed new correlations based on their experimental results.

Guo and Nutter [16] studied the effect of the axial conduction through the pipe wall on the performance of two similar R-134a filled TPCTs. One of them was fitted with a polycarbonate section that provided a thermal break in between the evaporator and the condenser section. The evaporator section was heated in a water bath at temperatures from 30-60°C and the condenser section was cooled with a water jacket at a coolant flow rate of 0.033 kg/s. Fill ratios were from 0.25-0.85. Their results showed that the axial wall temperature decreased along the thermosyphon without the thermal break. For the unit with the thermal break at the low bath temperature of 30°C the axial wall temperature increased. At bath temperatures beyond 40°C, evaporator wall temperature first decreased and then increased. They also found that the thermal break resisted axial heat conduction between the two sections and caused an increase in the overall, evaporator and condenser heat transfer coefficients of the TPCT and also the conduction effect was more dominant at low evaporator temperatures.

Jouhara and Robinson [17] experimented with water and dielectric heat transfer liquids FC-84, FC-77 and FC-3283 on a 6 m diameter x 200 mm long copper TPCT with an evaporator length of 40 mm representing an aspect ratio of 6.7 and with input power ranging from 23 – 260 W. The TPCT was over-filled in all cases (FR > 1) and also at 0.5 fill ratio for the water thermosyphon. They found that the temperature distribution was quite uniform and generally the water filled unit outperformed the others.

VISUALIZATION OF FLOW PATTERNS IN TWO-PHASE CLOSED THERMOSYPHON

In order to have a better understanding of the internal heat transfer phenomena, flow patterns using transparent glass tubes and high-speed camera recording have been performed. Flow patterns inside non-transparent thermosyphons were also visualized using various other techniques.

Roesler and Groll [18] presented experimental and analytical results on the counter-current flow (CCLF) limitation in an annular vertical pipe. Their studies included a visualization to reveal the flow patterns associated with the performance limitation process. The apparatus consisted of a 49 mm internal diameter glass tube fitted concentric outside a 39.4 mm outer diameter copper tube. Evaporator, adiabatic and condenser lengths were 450 mm, 230 mm and 690 mm, respectively. Working liquid in the copper tube was isopropylic alcohol with fill ratios of 0.2 and 0.4. The evaporator section was heated electrically while the condenser was cooled with a methanol cooling arrangement. The working or vapour temperature inside the thermosyphon was maintained from 20–70°C by controlling the coolant flow-rate. They measured the liquid film thickness based on the capacitance change of a capacitor-sensor installed flush with the condenser wall and developed an analytical model for the CCLF. By observing the flow patterns in the annular passage between the inner tube and annular section, they classified the flow into four flow regimes as:

- smooth condensate film flow
- wavy condensate film flow
- swaying condensate rivulets
- boiling liquid pool (churn or annular flow)

The regimes depended upon fill ratio and heat flux.

Liu et al. [19] carried out an experimental investigation to visualise the flow pattern of the liquid in an ethanol-filled glass TPCT by measuring the thickness of the condensate falling liquid film using an electrical capacitance tomography (ECT) technique for a range of heating rates. Their image reconstruction of the flow patterns showed that at low heating rates, annular flow predominated. As heating rate increased, liquid plugs tended to form and the flow pattern became unstable. The thickness measured by their ECT results showed thicker film than that predicted using the well-known Nusselts condensation correlation.

Visualizations of flow patterns in a transparent TPCT were conducted by Grooten et al. [20] with acetone at a filling ratio of 0.80 and heat fluxes between 14-32 kW/m^2. They used a 290 mm long 16 mm diameter glass tube with evaporator, adiabatic and condenser lengths of 100 mm, 80 mm and 110 mm, respectively. Experiments were performed with various inclination angles of 0-80 degrees from the vertical. A camcorder and a high-speed camera recorded the flow structure to show detailed transient flow patterns of the boiling mixture in the evaporator section and the condensate film in the adiabatic section. They showed that vapour plugs exist at all angles of inclination at heat fluxes below 14 kW/m^2 with a large bubble originating in the middle of the evaporator which then rose and subsequently disintegrated at the liquid vapour interface. At heat fluxes exceeding 14 kW/m^2 the flow pattern changed from plug flow to pool boiling up to 32 kW/m^2 at inclination angles of 0-80 degrees. Annular condensate film flow with a wavy structure existed at heat fluxes between 14-32 kW/m^2 and travelled down at a typical velocity of 1 m/s. These waves propagated faster with increased heat input.

Putra et al. [21] demonstrated the use of neutron radiography to capture images showing the boiling phenomena inside a TPCT. They showed the importance of a wick structure in aiding liquid return from the condenser to the evaporator sections of the pipe.

Negeshi and Sawada [22] studied the heat transfer performance of an inclined two-phase closed thermosyphon made from 330 mm long 15 mm O/D x 13 mm I/D copper pipe with water and ethanol as fill liquids. Cooling and heating were performed using concentric water pipes. They investigated the effects of fill ratio and inclination angle on the performance. A visualization was made with transparent glass windows at the ends of the thermosyphon. They determined the required amount of fill liquid depending upon the fill liquid and pipe inclination.

Geysering occurs in a thermosyphon when the fill liquid in the lower evaporator section is suddenly propelled up towards the upper condenser section at intermittent intervals. It is usually accompanied with a loud explosive noise. In addition to determining the heat transfer coefficients, Imura et al. [6] also observed the geysering in a double-wall glass thermosyphon shown in Figure 10(a). They observed that when heat flux was small, the rate of vapour generation in the evaporator section was low, Figure 10(b). The condensate film flowing in the evaporator section was thin and apt to breakdown easily. At high fill ratio, the bubbles generated in the evaporator section pushed the liquid-vapor mixture level up to the top of the condenser section, Figure 10(c). At intermediate fill ratio, film condensation occurred at both the upper and lower portions of the cooled wall. Increased heat flux resulted in an increased condensate flow rate making it more stable. They considered an optimum ratio of evaporator/condenser length to be in the range of 0.10-0.20. Imura et al. [23] investigated the geysering effect in 1.10 m long x 16 mm diameter glass and stainless steel/copper thermosyphons with water and ethanol as working fluids and at various fill ratios. They found that geysering occurred at different temperatures. Casarosa et al. (24) investigated the geyser effect by measuring the pressure and temperatures inside a water-filled 870 mm long x 46 mm OD x 42 mm ID brass tube. They also observed the physical phenomena using a glass thermosyphon and high-speed camera and showed that at low heat input, large bubbles are formed and at longer time intervals.

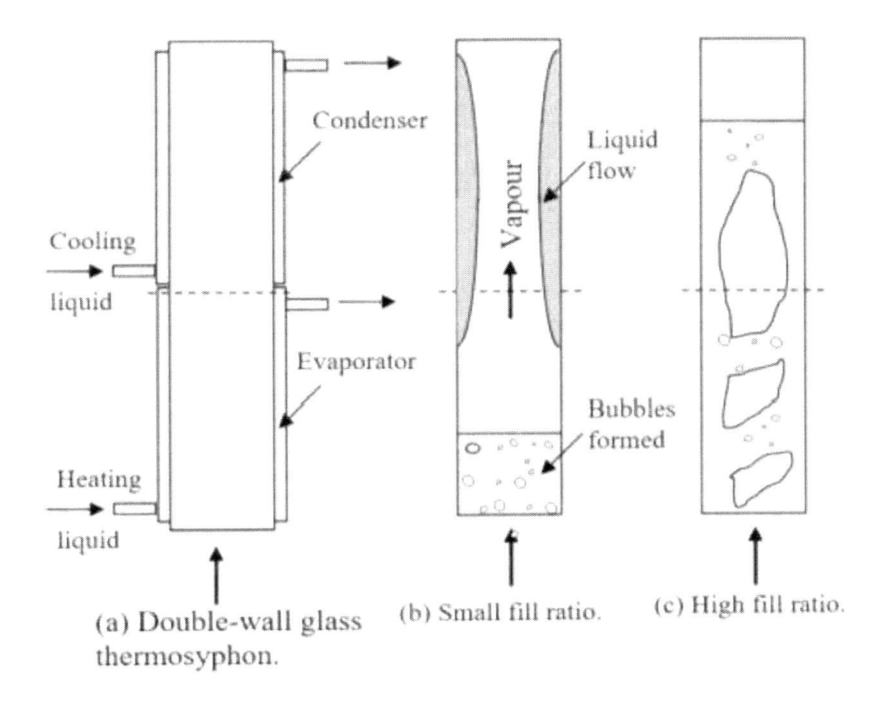

Figure 10. Flow patterns observed in glass thermosyphon by Imura [6].

Liu and Wang [25] made visual observations on the pulse boiling frequency as a criterion to determine flow patterns in 1.25 m long glass thermosyphons with diameters of 14, 20 and 30 mm. They showed that when the frequency of the pulses was less than 0.2 times per minute, the boiling process could be treated as natural convection. When the frequency exceeds 20 times perminutes, the flow pattern becomes semicontinuous froth boiling. In between these two frequencies, standard pulse boiling occurred. Kuncoro et al. [26] performed detailed visual observations on the mechanism of geysering in a 2.5 m long x 18 mm I/D glass thermosyphon with water and R113 as fill liquids. They found that temperature distribution and geysering flow patterns were strongly dependent upon geometry, physical properties of the working fluid and the heat flux. Emani et al. [27] investigated the geyser boiling phenomenon by measuring the surface temperature distribution of water-filled copper thermosyphons. They concluded that geysering depended upon fill ratio, inclination and operating temperature. Nemec and Malcho [28] observed the geysering effect in a 150 mm long x 12 mm OD x 10 mm ID straight thermosyphon filled with ethanol and placed in 90°C hot water bath. In a later experiment with 13 and 22 mm diameter water-filled glass tubes, Nemec and Malcho [29] showed that the liquid fill was pushed higher up the top of the smaller diameter tube. Dropwise condensation was observed in both tubes.

Closed Loop Pulsating or Oscillating Thermosyphon

A closed loop pulsating or oscillating heat pipe is shown in Figure 11. It consists of an array of metal pipes wound round in a continuous serpentine loop arrangement. The pipes are usually small diameter capillary tubes without wicks. The loop is vacuumed and partially filled with a working fluid just like a heat pipe. The filled liquid distributes itself naturally within the loop in the form of liquid and vapour plugs and slugs. Heat applied at the bottom end of the unit (evaporator section) is transferred to the top of the unit (condenser section) by the oscillating motion of these liquid-vapour plugs.

Khandekar et al. [30] conducted visualization experiments on an array of pulsating heat pipes consisting of 10 glass tubes each 4.2 mm O/D x 2 mm I/D x 100 mm long. The glass tubes were spaced 12 mm apart and connected at the ends by copper U-bends. Heating was provided with wrap-around electrical resistance wire and cooling was by a water bath. Fill liquids were water and ethanol. Thermocouples were attached to measure the adiabatic section temperature. The unit was mounted on a tiltable frame to vary the angle. A high-speed video was employed to identify and categorise the various types of bubble patterns formed. Generally, small bubbles are formed via nucleation as in case of the straight tube thermosyphon. These bubbles could flow upward individually in the same direction or they could coalesce to form a larger bubble. Sometimes two small bubbles flowing in the opposite direction along a common tube coalesces and flows along as a single larger bubble. Charoensawan et al. [31] studied experimentally a wide range of pulsating heat pipes with different internal diameters, number of turns, different working fluids and inclination angles. They found that a certain number of turns is required to make horizontal operation possible and that performance improves with larger diameters pipes.

Khandekar et al. [32] presented some visualization results obtained on a glass pulsating heat pipe using a high-speed video camera. They observed the frequency, distribution and size of the bubbles that were oscillating around the pipe and concluded that the heat transfer processes are complex and that available models in the literature do not truly represent the thermo-hydrodynamics of the system. They developed a semi-empirical correlation to fit their experimental data. Similar visualization studies using capillary glass tubes with high-speed photography were conducted by Xu et al. [33] with methanol and deionised water.

Loop Heat Pipes

A loop thermosyphon (LT) consists of two vertical pipes connected together by two horizontal pipes as shown in Figure 12. The LT is vacuumed and partially filled with a fill fluid. Heat is supplied to one of the verical limb of the loop while the other is cooled at the bottom.Two-phase flow occurs with vapour generation at one limb (evaporator) and liquid condensation at the other (condenser).

Agostini and Habert [34] observed the flow patterns in a 2.4 mm diameter transparent glass loop pipe (LP) partially filled with a refrigerant "Novec649". Heat was provided using a semi-transparent Indium Titanium Oxide layer deposited over a 100 mm long vertical heating section. Cooling was performed on the opposite side using a 188 mm long standard Liebig condenser. Fill ratio was kept between 44 - 78.3%. They described the flow patterns observed in the evaporator section from bubbly flow to dry-out and determined that total thermal resistance was minimum with a fill ratio of 68% at high heat flux.

Khodabandeh and Fuberg [35] studied the influence of different channel geometries on heat transfer, flow regime and instability in a partially-filled R134a LP with different heat fluxes using a high-speed camera and temperature measurements. Five 30 mm long evaporators with transparent polycarbonate cover were fabricated each with a single rectangular flow channel machined into it with widths from 5.0 – 5.3 mm and depths from 0.7 - 1.8 mm. Heat was supplied via an electrical heater and cooled with a 1 m long water-cooled jacket. They described the flow patterns in the transparent evaporator obtained under various operating conditions from bubbly to slug to churn flows.

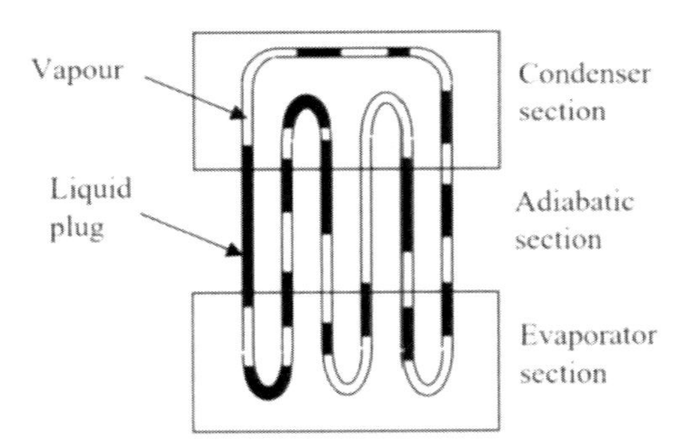

Figure 11. Cross-section view of a closed-loop pulsating/oscillating thermosyphon.

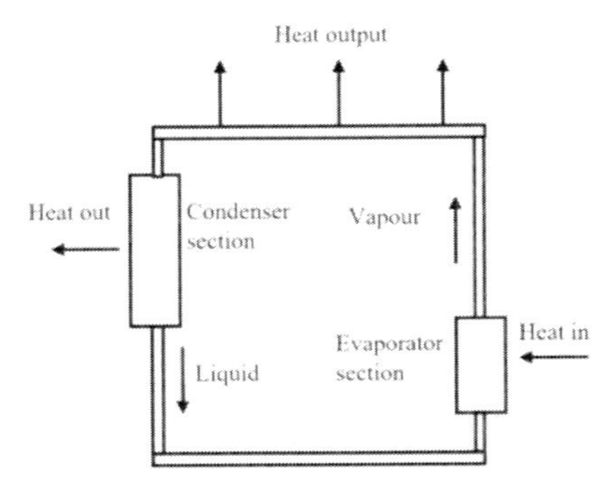

Figure 12. Cross sectional view of a loop thermosyphon.

Ruppersberg and Dobson [36] proposed a method to predict the thermosyphon flow pattern in a LP by measuring the working fluid pressure. They employed a 25 mm diameter x 2.2 m high water-filled loop pipe. Heating was provided by attaching heating elements to the finned evaporator while cooling was performed using water-cooled jackets mounted on the condenser section. Slight glasses mounted at the exit of the evaporating section and at the inlet of the condensing section allowed visual observations to be made.

Zhou and Li [37] performed a visualization study on a vertically orientated copper-water LT with a flat evaporator. The condenser section located at the top consisted of a multi-loop copper tube and cooled by forced water circulation. The evaporator located at the bottom was heated electrically up to 91°C with heat input of 550 W. They employed digital camcorders to record the flow patterns in the evaporator and condenser sections and showed that flow patterns in the condenser varied from bubbly-slug-churn-wavy-annular-mist as the heat input increased. At the evaporator section boiling, nucleate boiling and thin film liquid evaporation occurred simultaneously. Thin film evaporation became dominant at high heat flux. Partial dry-out was observed above 550 W.

Ding et al. [38] observed and measured the boiling heat transfer coefficient using a combination of visualization and measurement methods with an electrically heated 152 mm long x 12 mm O/D x 8 mm I/D copper evaporator tube and a water-cooled glass jacket. Condenser and evaporator sections were connected wing glass tubing and flow pattern observed. R134a and R22 were used as fill liquid with fill ratio > 1.0 and heating power input from 109-699 W. They observed that the flow pattern was a combination of pool boiling and flow boiling and that dry-out occurred in the evaporator section at high heat flux and low filling ratio.

2 Phase Forced Convection Flows

In addition to the previous investigations on flow patterns in two phase closed thermosyphons, closed loop pulsating or oscillating heat pipe and loop heat pipes, investigations were also carried out by numerous researchers to observe the flow patterns in two-phase forced convection flows. These studies were conducted mainly to determine the type of flow patterns in oil pipe lines in the petroleum industry. In these investigations, instead of utilising thermosyphons, air was injected into a liquid stream flowing under forced convection with various diameters transparent glass tubes and with different liquids.

Generally, the flow patterns are similar to what has been observed in the TPCT previously. These are illustrated in Figure 13:

(a) Isolated or seperated bubbly flow, Figure 13(a). Distinct and essentially spherical gas bubbles are uniformly distributed in the liquid column. Bubbles are small, less than tube diameter size and flow freely without coalescing.
(b) Confined bubbly flow, Figure 13(b). As heat flux increases, the bubbles start to coalesce together and form larger bubbles. The bubbles are distinct, distorted and non-spherical with diameters about equal to tube diameter.
(c) Bubbly flow in small diameter capillary tube, Figure 13(c). Seperated bubbles are formed and flow upwards.
(d) Slug flow, Figure 13(d), Elongated large bullet-shaped bubbles with diameter about equal to tube diameter are formed. Liquid plugs are separated by vapor slugs.
(e) Slug flow in capillary tube, Figure 13(e). Long bullet-shaped bubbles are formed.
(f) Churn flow, Figure 13(f). Vapor slugs become unstable and disrupted, leading to chaotic vapor flow through the liquid which is mainly displaced to the tube wall.
(g) Churn flow in capillary tube, Figure 13(g). Vapor slugs are formed and become elongated.
(h) Slug-annular flow, Figure 13(h). Neighbouring slugs collide leading to a wavy-annular flow pattern with deep waves that interrupt the annular flow from bottom to top.
(i) Annular flow, Figure 13(i). Vapour plugs flow continuously in central core of tube with liquid flowing as a film along the tube wall.
(j) Mist flow, Figure 13(j). Majority of the flow is entrained in the vapour core and dispersed as liquid droplets. Dry-out occurs when the liquid film dissappears completely.

Hewitt and Roberts [39] studied the two-phase flow patterns under forced convection flow using simultaneous X-ray and flash photography. Mishima and Hibiki [40], Chen et al. [41] and Huo et al. [42] presented flow patterns in small diameter vertical capillary glass tubes. Taitel et al. [43] developed models for predicting flow pattern transitions during steady gas-liquid flow in vertical tubes incorporating the effect of fluid properties and pipe sizes. Gross et al. [44] visually observed the wave frequency and its structure and formation of the fully developed falling liquid film inside the condenser section of vertical glass tubes with and without a counter-current flow of vapour inside it. Galvis and Culham [45] studied the influence of heat flux and mass flux on the flow patterns of forced convection flow of water in small single channel micro evaporators using high-speed video camera. They observed six different flow patterns and classified them as bubbly, slug, churn, annular, wavy-annular and inverted annular flow. Bubbly and slug flow tended to appear at lower heat flow and became annular and inverted at higher heat flux.

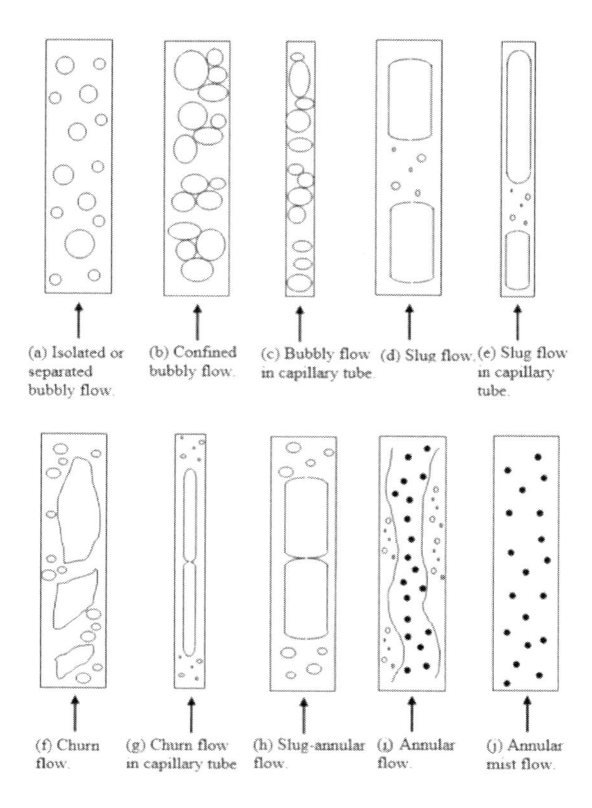

Figure 13. Flow patterns along a vertical heated pipe proposed by various investigators [36, 39, 40].

Claudi et al. [46] reported seven distinct two-phase flow patterns in a R134a filled 1.33 mm diameter quartz tube with a 235 mm heated evaporator length using a high-speed camera. The glass tube was coated with Indium tin oxide that allowed heating and visualization to be carried out simultaneously. Fluid circulation was by a magnetic gear pump. Condenser was a heat exchanger with cold water. Their photographs show similar flow patterns to those of Figure 13.

Flow patterns under forced convection for small vertical and horizontal tubes with air-water and R-134A are reported in Yang and Shieh [47]. Fukano and Kariyasaki [48] reported on the flow characteristics in 1.0 - 4.9 mm diameter glass tubes with air-water upward and downward flow in vertical and horizontal tubes. Transition flow patterns under forced convection from churn to annular for vertical tubes with upward flow of air-water liquid are reported by Mashima [49], Dasgupta et al. [50] and Posada and Waltrich [51]. Alves et al. [52] investigated the transient behaviour of churn-annular two-phase flow with high gas fraction in air-water flow in a long vertical tube. Spedding and Spence [53] observed patterns of air-water two-phase flow in a large diameter (93.5 mm) horizontal pipe. Experimental investigation of flow in a horizontal pipe was studied by Dinaryanto et al. [54] with 26 mm diameter pipe. They showed wave coalescences, wave growth mechanism and large disturbance waves in the gas-liquid slug two-phase pipe flow. Padilla et al. [55] observed flow patterns in HFO-1234yf and R134a refrigerants in horizontal bends. Investigations into the mechanism and transition from bubbly to slug flow in air-water two phase forced circulation were conducted by Zhang et al. [56] and Kong et al. [57]. A review of two-phase flow instability was reported by Boure et al. [58]. Bressani and Massa [59] observed the flow patterns in a vertical-to-horizontal bend and found that the flow around the bend behaved similar to that in a straight pipe.

VAPOUR CHAMBER

Heat pipes are normally long straight tubes with small diameters. Vapour chambers (VCs) are short and flat heat pipes with large diameters. VCs are normally attached to fin heat sinks (FHSs) as shown in Figure 14 for thermal management to cool semi-conductor chips or to reduce the effects of heat spreading. The heat source at the bottom of the VC evaporates the fill liquid which condenses on the upper surface. The condensate then drops by gravity back to the evaporator pool As a result, heat transfer occurs.

With a conventional VC-FHS device, a thermal contact resistance arises at the interface between the VC and the FHS. An improved VC-FHS is shown in Figure 15. It consists of a single or an array of hollow pipes embedded on the top of the VC to provide the condensing surface. With this design thermal contact resistance between VC and FHS is eliminated.

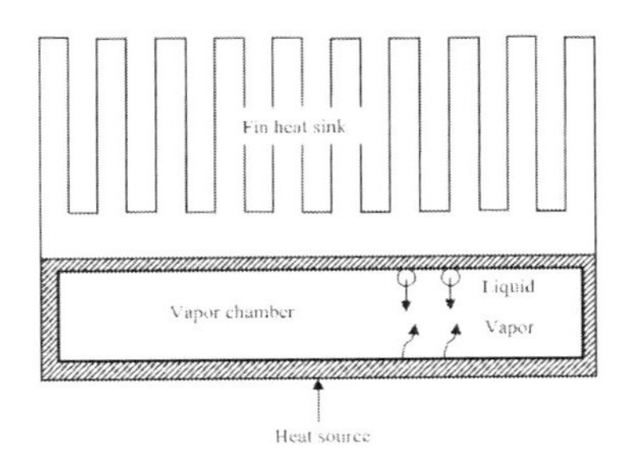

Figure 14. Vapour chamber attached to fin heat sink.

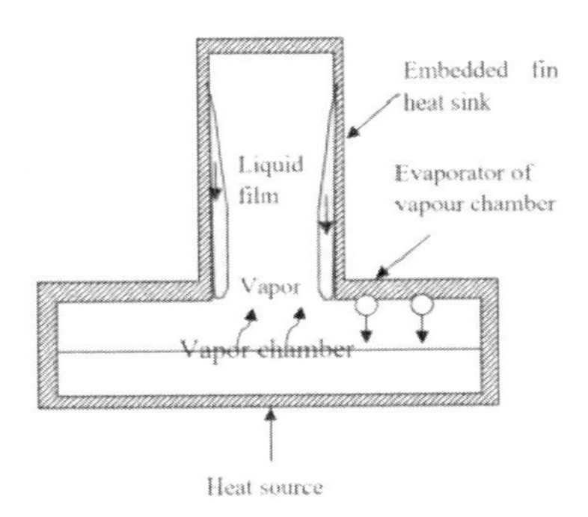

Figure 15. Vapour chamber with embedded hollow condenser tube.

Zhang et al. [60] experimentally investigated the relationship between boiling and condensation in a VC with a 26mm long x 40 mm diameter water filled stainless steel cylinder with copper cooling and heating blocks. Tests were conducted with fill depths of 10, 12, 14 and 16 mm. For their visualization observations, they used a transparent cylinder. They found that at low fill ratio, liquid condenses at the top at the same time vapour bubbles are formed on the bottom. The condensate drops to the boiling surface separately. With increasing fill, condensed liquid and vapour bubbles tended to interact. The condensing and boiling processes are strongly inter-related and have significant influences over each other. They found an optimum fill ratio existed for maximum performance. A later investigation by Zhang et al. [61] with a longer and larger VC showed similar results.

Zhang et al. [62] visually investigated the effects of a fin array machined on the boiling surface to enhance boiling. The quartz glass tube VC measured 33mm long with copper cooling and heating blocks attached. Water filled depths were at 9, 18 and 24mm. They found that boiling heat transfer was depressed while condensation heat transfer was enhanced with increase in fill liquid. Also, boiling heat transfer was significantly enhanced by the internal fin array and that condensation heat transfer coefficient was very much smaller than boiling heat transfer coefficient.

Xia et al. [63] performed a visualization study on the instabilities of phase-change heat transfer in a VC filled with water, ethanol and acetone. Effects of heat flux, fill ratio, coolant temperature and fill liquid type were investigated. Instabilities were deemed to be caused by bubble behaviour, physical properties and operating pressure. They found that natural convection, intermittent boiling and fully developed nucleate boiling were the main heat transfer modes. No intermittent boiling regime and instability were observed for ethanol and acetone. Instabilities were strongly related to heat transfer modes.

Boukhanouf et al. [64] used an IR thermal imaging camera to obtain transient and steady state temperature distributions on a 250 x 200 x 5 mm VC placed between an electrically heated heat source and an air cooled aluminium FHS condenser. The heat source surface area was about 10 times smaller than the VC. Their results showed that the thermal heat spreading resistance of the VC was about 40 times lower than a similar sized copper block.

Peng et al. [65] investigated the heat transfer performance of an aluminum VC fitted with internal fins measuring 80 x 75 x 15 mm with various heat inputs up to 300 W. Filling ratios ranged from 0.10-0.50 with water and acetone. Perforated fins were provided internally to the VC. An external FHS was fitted to the condenser and heat was dissipation by forced air convection cooling from 1.5–6.0 m/s. Their results showed that fill ratio influenced thermal performance and acetone performed better than water.

PROPOSED FUTURE STUDIES

In heating, ventilation and air conditioning (HVAC) systems, a wrap-around heat pipe heat exchanger (WHPHE) acting as a run-around coil in a conventional HVAC system increases dehumidification and reduces cooling costs. The WHPHE shown in Figure 16 provides pre-cooling and reheat during the dehumidification and cooling process. The incoming stream of ambient air which is hot and wet is first pre-cooled by the evaporator section of the WHPHE before it passes over the cooling coil of a refrigeration unit where it is cooled and dehumidified. Water vapour condenses out of the air stream and is collected by the condenser tray of the cooling coil. However, the cold dry air here is too cold for human comfort and requires to be reheated before it is supplied to the room. By passing it over the condenser section of the WHPHE it is reheated and becomes warm dry air.

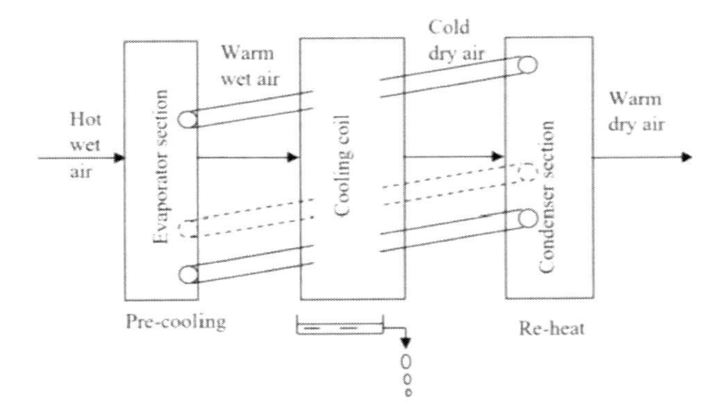

Figure 16. A wrap-around heat pipe heat exchanger.

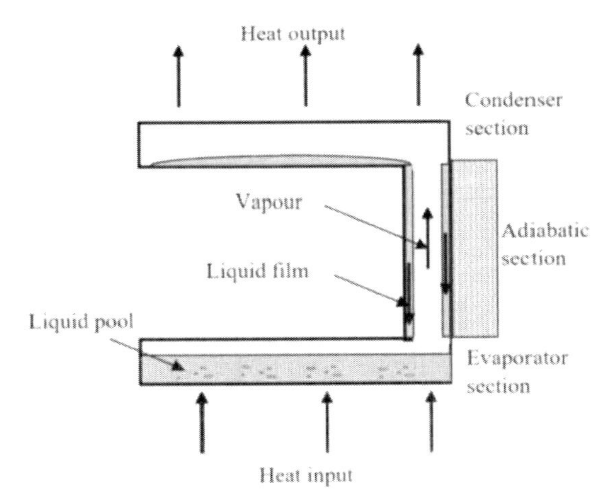

Figure 17. Cross sectional view of a C-shape thermosyphon.

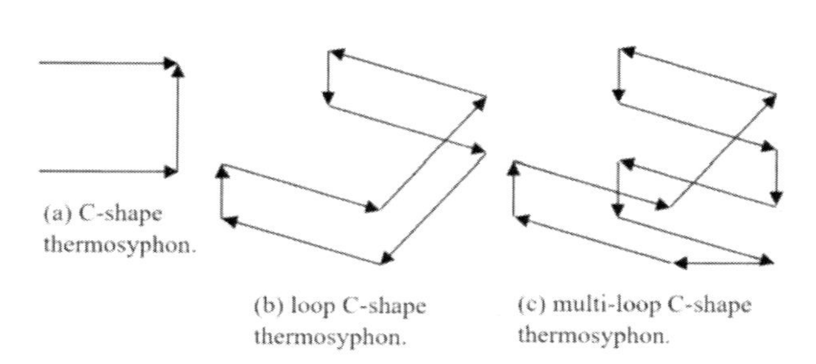

Figure 18. C-shape thermosyphons.

The conditioned air is then circulated to the room at the design comfort level condition. Each WHPHE coil consists of an array of thermosyphons which are bent into the shape of a "C" for easy insertion into the conventional existing air conditioning system. A cross-sectional view of a C-shape thermosyphon is shown in Figure 17. Evaporation and condensation of the fill liquid at the respective sections are shown. Some variations of the C-shape thermosyphon are the loop C-shape thermosyphon and the multi-loop C-shape thermosyphon shown in Figure 18. Observations of the flow patterns in these C-shape thermosyphons would enable their thermal performances to be simulated.

CONCLUSION

Visualization of the flow patterns in vertical thermosyphons, loop and pulsating heat pipes and heated tubes have been documented and recorded using high-speed cameras and various other means. The investigations enabled the flow regimes to be studied and heat transfer correlations to be determined. The flow patterns obtained show that the flow regimes are affected by the type of fill liquid, fill ratio, pipe dimensions, inclination and operating heat flux. There is no visualization investigation carried out on the C-shape thermosyphon or vapour chamber.

REFERENCES

[1] Peterson, GP. *An Introduction to Heat Pipes Modeling, Testing, and Applications*, Wiley, 1994.

[2] Reay, DA; Kew, PA. *Heat Pipes (5th Ed)*. Elsevier, 2007.

[3] Faghri. A. *Heat Pipe: Science and Technology*, Taylor and Francis, 1995.

[4] Zohuri, B. *Heat Pipe design and Technology*. CRC Press., 2011.

[5] El-Genk, MS; Saber, HH. Heat transfer correlations for small, uniformly heated liquid pools, *International Journal of Heat and Mass Transfer*, 41 (2), (1998), 261-274.

[6] Imura, H; Kusuda, H; Ogata, JI; Miyazaki, T; Sakamoto, N. Heat transfer in two-phase closed type thermosyphon, *Heat Transfer Japanese Research*, 8, (1979), 41-53.

[7] El-Genk, MS; Saber, HH. Heat transfer correlations for liquid film in the evaporator of enclosed, gravity-assisted thermosyphons, *ASME J. Heat Transfer*, 120, (1998), 477-484.

[8] El-Genk, MS; Saber, HH. Determination of operation envelopes for closed, two-phase thermosyphons, *International Journal of Heat and Mass Transfer*, 42, (1999), 889-903.

[9] Nero, A; Beretta, GP. Boiling regimes in a closed two-phase thermosyphon, *International Journal of Heat and Mass Transfer*, 33 (10), (1990), 2099-2110.

[10] Nguyen-Chi, H; Groll, M; Dang-Van, T. *Experimental investigation of closed two-phase thermosyphons*, AIAA Paper, (1979), 79-1106.

[11] Nguyen-Chi, H; Groll, M. Entrainment or flooding limit in a closed two-phase thermosyphon, *Heat Recovery Systems*, (1981), 275-286.

[12] Imura, H; Sasaguchi, K; Kozai, H; Numata, S. Critical heat flux in a closed two-phase thermosyphon, *International Journal of Heat and Mass Transfer*, 26, (1983), 1181-1188.

[13] Imura, H; Ippohshi, S; Sakamoto, M. Experimental investigation of critical heat fluxes in two-phase closed thermosyphons, *Proc. 10IHPC*, (1997), 284-289.

[14] Gross, U. Reflux condensation heat transfer inside a closed thermosyphon. *International Journal of Heat and Mass Transfer*, 35, (1992), 279-294.

[15] Jiao, B; Qiu, LM; Zhang, XB; Zhang, Y. Investigation on the effect of filling ratio on the steady-state heat transfer performance of a vertical two-phase closed thermosyphon, *Applied Thermal Engineering*, 28, (2008), 1417-1426.

[16] Guo, W; Nutter, DW. An experimental study of axial conduction through a thermosyphon pipe wall, *Applied Thermal Engineering*, 29, (2009), 3536-3541.

[17] Jouhara, H; Robinson, AJ. Experimental investigation of small diameter two-phase closed thermosyphons charged with water, FC-84, FC-77 and FC-3283, *Applied Thermal Engineering*, 30, (2010), 201-211.

[18] Roesler, S; Groll, M. Flow visualization and analytical modelling of interaction phenomena in closed two-phase flow systems, *Proc. 9IHPC*, (1992).

[19] Liu, S; Li, J; Chen, Q. Visualization of flow patterns in thermosyphon by ECT, *CP914 Multiphase Flow: The Ultimate Measurement Challenge*, (2007), 775-785.

[20] Grooten, MHM; van der Geld, CWM; Deurzen, LGM. A study of flow patterns in a thermosyphon for compact heat exchanger applications, HEAT 2008, *Fifth Int. Cong. on Transport Phenomena in Multiphase Systems*, (2008).

[21] Putra, N; Ramadhan, SD; Septiadi, WN. Sutiarso, Visualization of the boiling phenomenon inside a heat pipe using neutron radiography, *Expt Thermal and Fluid Science*, (2015), 13-27.

[22] Negishi, K; Sawada, T. Heat transfer performance of an inclined two-phase closed thermosyphon, *International Journal of Heat and Mass Transfer*, 26, (1983), 1207-1213.

[23] Imura, H; Ishii, K; Ippohshi, S. An experimental investigation of geysering in two-phase closed thermosyphons, *Proc. 11IHPC*, (1999), 166-171.

[24] Casarosa, C; Latrofa, E; Shelginski, A. The geyser effect in a two-phase thermosyphon. *International Journal of Heat and Mass Transfer*, 26, (1983), 933-941.

[25] Liu, JF; Wang, JCY. On the pulse boiling frequency in thermosyphons. *Trans. ASME*, 114, (1992).

[26] Kuncoro, H; Rao, YF; Fukuda, K. An experimental study on the mechanism of geysering in a closed two-phase thermosyphon, *Int. J. Multiphase Flow*, 21, (1995), 1243-1252.

[27] Emami, MRS; Noir, SH; Khosnoodi, M; Mosavian, MTH; Kianifar, A. Investigation of geyser boiling phenomenon in a two-phase closed thermosyphon. *Heat Transfer Engineering*, 30, (2009), 408-415.

[28] Nemec, P; Malcho, M. Visualization of working fluid flow in gravity assisted heat pipe. *EPJ Web of Conferences*, 92, 02053, (2015). Doi: 10.1051/epjconf/20159202053.

[29] Nemec, P; Malcho, M. Working fluid flow visualization in gravity heat pipe. *EPJ Web of Conferences*, 114, 02083, (2016). Doi: 10.1051/epjconf/201611402083.

[30] Khandekar, S; Groll, M; Charoensawan, P; Terdtoon, P. Pulsating heat pipes: Thermo-fluid characteristics and comparative study with single phase thermosyphon, *Proc. 12th Int Heat Transfer Conference*, 14, (2002), 459-464.

[31] Charoensawan, P; Khandekar, S; Groll, M; Terdtoon, P. Closed loop pulsating heat pipes Part A: parametric experimental investigations, *Applied Thermal Engineering*, (2003), 2009-2020.

[32] Khandekar, S; Charoensawan, P; Groll, M; Terdtoon, P. Closed loop pulsating heat pipes Part B: visualization and semi-empirical modelling, *Applied Thermal Engineering*, (2003), 2021-2033.

[33] Xu, JL; Li, YX. High speed flow visualisation of a closed loop pulsating heat pipe, *Int. J. of Heat and Mass Transfer*, 48, (2005), 3338-3351.

[34] Agostini, B; Habert, M. Measurement of the performances of a transparent closed loop two-phase thermosyphon, *Advanced Computational Methods and Experiments in Heat Transfer*, X1, (2010), 227-237, doi:10.2495/HT100201.

[35] Khodabandeh, R; Fuberg, R. Instability, heat transfer and flow regime in a two-phase flow thermosyphon loop at different diameter evaporator channel, *App. Thermal Engineering*, 30, (2010), 1107-1114.

[36] Ruppersberg, R; Dobson, T. Flow regime recognition in two-phase thermosyphon loops using pressure pulse analyses, *Proc. 10IHPS*, Taiwan, (2011), 19-23.

[37] Zhou, G; Li, J. Two-phase flow characteristics of a high performance loop heat pipe with flat evaporator under gravity. *International Journal of Heat and Mass Transfer*, 117, (2018), 1063-1074.

[38] Ding, T; Cao, HW; He, ZG; Li, Z. Visualization experiment on boiling heat transfer and flow characteristics in separated heat pipe system. *Experimental Thermal and Fluid Science*, 91, (2018), 423-431.

[39] Hewitt, GF; Roberts, DN. Studies of two-phase flow patterns by simultaneous X-ray and flash photography, Report AERA-M2159, *UKAE*, (1969).

[40] Mishima, K; Hibiki, T. Some characteristics of air-water two-phase flow in small diameter vertical tubes. *Int. J. Multiphase Flow*, 22, (1996), 703-712.

[41] Chen, L; Tian, YS; Karayiannis, TG. The effect of tube diameter on vertical two-phase flow regimes in small tubes *International Journal of Heat and Mass Transfer*, 49, (2006), 4220-4230.

[42] Huo, X; Chen, L; Tian, YS; Karayiannis, TG. Flow boiling and flow regimes in small diameter tubes, *App. Thermal Engineering*, 24, (2004), 1225-1239.

[43] Taitel, Y; Bornea, D; Dukler, AE. Modelling flow pattern transitions for steady upward gas-liquid flow in vertical tubes, *J. AIChE*, 26, (1980), 345-354.

[44] Gross, U; Storch, T; Phillip, C; Doeg, A. Wave frequency of a falling liquid film and the effect on reflux condensation in vertical tubes, *Int. J. of Multiphase Flow*, 35, (2009), 398-409.

[45] Galvis, E; Culham, R. Measurements and flow pattern visualizations of two-phase flow boiling in single microevaporators, *Int. J. of Multiphase Flow*, 42, (2012), 52-61.

[46] Claudi, MC; Palm, B; Wahib, OA. Rashid, Flow boiling visualization of R-134a in a vertical channel of small diameter, *ASME J. Heat Transfer*, 132, (2010).

[47] Yang, CY; Shieh, CC. Flow pattern of air-water and two-phase R-134a in small circular tubes, *Int. J. of Multiphase Flow*, 27, (2001), 1163-1177.

[48] Fukano, T; Kariyasaki, A. Characteristics of gas-liquid two-phase flow in a capillary tube, *Nuclear Engineering and Design*, 141, (1993), 59-68.

[49] Mashima, K. Flow regime transition criteria for upward two-phase flow in vertical tubes, *International Journal of Heat and Mass Transfer*, 27, (1984), 723-737.

[50] Dasgupta, A; Chandraker, DK; Kshirasagar, S; Reddy, BR; Rajalakshmi, R; Nayak, AK; Walker, SP; Vijayan, PkK; Hewitt, GF. Experimental investigation on dominant waves in upward air-water two-phase flow in churn and annular regime, *Expt Thermal and Fluid Science*, 81, (2017), 147-163.

[51] Posada, C; Waltrich, PJ. Effect of forced flow oscillations on churn and annular flow in a long vertical tube, *Expt Thermal and Fluid Science*, 81, (2017), 345-357.

[52] Alves, MVC; Waltrich, PJ; Gessner, TR; Falcone, G. Modeling transient churn-annular flow in a long vertical tube, *Int. J. of Multiphase Flow*, 89, (2017), 399-412.

[53] Spedding, PL; Spence, DR. Flow regimes in two-phase gas-liquid flow, *Int. J. Multiphase Flow*, 19, (1993), 245-280.

[54] Dinaryanto, O; Prayitno, YAK; Majid, AI; Hudaya, AZ; Nusirwan, YA. A. Widyaparaga, Indarto, Deendarlianto, Experimental investigation on the initiation and flow development of gas-liquid slug two-phase flow in a horizontal pipe, *Experimental Thermal and Fluid Science*, 81, (2017), 93-108.

[55] Padilla, M; Revellin, R; Bonjour, J. Two-phase flow visualization and pressure drop measurements of HFO-1234yf and R134a refrigerants in horizontal bends, *Experimental Thermal and Fluid Science*, 39, (2012), 98-111.

[56] Zhang, M; Pan, LM; Ju, P; Yang, X. M. Ishii, The mechanism of bubbly to slug flow regime transition in air-water two-phase flow: A new transition criterion. *International Journal of Heat and Mass Transfer*, 108, (2017), 1579-1590.

[57] Kong, R; Kim, S; Bajorek, S; Tien, K; Hoxie, C. Experimental investigation of horizontal air-water bubbly-to-plug and bubbly-to-slug transition flows in a 3.81 cm ID pipe. *Int. J. of Multiphase Flow*, (2017), http:dx.doi.org/10.1016/j.ijmultiphase flow.2017/04.020.

[58] Boure, JA; Bergles, AE; Tong, LS. Review of two-phase flow instability. *Nuclear Engineering and Design*, 25, (1973), 165-192.

[59] Bressani, M; Massa, RA. Two-phase slug flow through an upward vertical to horizontal transition, *Thermal and Fluid Science*, 91, (2018), 245-255.

[60] Zhang, G; Liu, Z; Wang, C. An experimental study of boiling and condensation co-existing phase change heat transfer in small confined space. *International Journal of Heat and Mass Transfer*, 64, (2013), 1082-1090.

[61] Zhang, G; Liu, Z; Wang, C. A visualization study of the influences of liquid levels on boiling and condensation co-existing phase change heat transfer phenomenon in small confined spaces. *International Journal of Heat and Mass Transfer*, 73, (2014), 415-423.

[62] Zhang, G; Liu, Y; Li, Y. Guo, Visualization study of boiling and condensation co-existing phase change heat transfer phenomenon in a small and closed space with a boiling surface of enhanced structure. *International Journal of Heat and Mass Transfer*, 79, (2014), 916-924.

[63] Xia, G; Wang, W; Cheng, L; Ma, D. Visualization study on the instabilities of phase-change heat transfer in a flat two-phase closed thermosyphon. *Applied Thermal Engineering*, 116, (2017), 392-405.

[64] Boukhanouf, R; Haddad, A; North, MT; Buffone, C. Experimental investigation of a flat plate heat pipe performance using IR thermal imaging camera. *Applied Thermal Engineering*, 26, (2006), 2148-2156.

[65] Peng, H; Li, J; Ling, X. Study on heat transfer performance of an aluminum flat plate heat pipe with fins in vapor chamber. *Energy Conversion and Management*, 74, (2013), 44-50.

INDEX

A

B

C

D

E

F

G

H

I

K

L

M

N

O

P

R

S

T